First Ahmad Dahlan International Conference on Mathematics and Mathematics Education (AD-INTERCOMME 2017)

Journal of Physics: Conference Series Volume 943

Yogyakarta, Indonesia
13 – 14 October 2017

ISBN: 978-1-5108-5847-3
ISSN: 1742-6588

Printed from e-media with permission by:

Curran Associates, Inc.
57 Morehouse Lane
Red Hook, NY 12571

Some format issues inherent in the e-media version may also appear in this print version.

Printed by Curran Associates, Inc. (2018)

For permission requests, please contact the Institute of Physics
at the address below.

Institute of Physics
Dirac House, Temple Back
Bristol BS1 6BE UK

Phone: 44 1 17 929 7481
Fax: 44 1 17 920 0979

techtracking@iop.org

Additional copies of this publication are available from:

Curran Associates, Inc.
57 Morehouse Lane
Red Hook, NY 12571 USA
Phone: 845-758-0400
Fax: 845-758-2633
Email: curran@proceedings.com
Web: www.proceedings.com

Table of contents

Volume 943

The First Ahmad Dahlan International Conference on Mathematics and Mathematics Education

13–14 October 2017, Daerah Istimewa Yogyakarta 55213, Indonesia

Accepted papers received: 18 December 2017
Published online: 5 February 2018

Preface

Papers

IOP Conf. Series: Journal of Physics: Conf. Series **943** (2017) 011001 doi:10.1088/1742-6596/943/1/011001

The First Ahmad Dahlan International Conference on Mathematics and Mathematics Education (AD-INTERCOMME)

Rully Charitas Indra Prahmana

Editor in Chief of AD-INTERCOMME's Publication, Universitas Ahmad Dahlan, Jl. Pramuka Kav. 5 Pandeyan, Umbulharjo, Yogyakarta, Indonesia.

E-mail: rully.indra@mpmat.uad.ac.id

Preface

The first Ahmad Dahlan International Conference on Mathematics and Mathematics Education (AD-INTERCOMME) is a collaborative event of Department of Mathematics and Department of Mathematics Education (Undergraduate and Postgraduate) Universitas Ahmad Dahlan. The objectives are to improve mathematics teaching and to expand mathematics contributions to the society. This international conference is a biannual conference, started in 2017. The main topics of the conference are divided into five categories namely Analysis, Statistics, Algebra, Applied Mathematics, and Mathematics Education.

The keynote presentations are provided especially to show the contribution of Mathematician and Mathematics Educator in the world of mathematics and mathematics education towards research and knowledge sharing where our conference theme for this year is 'Current Issues in Mathematics and Mathematics Education'. We have four keynote speakers coming from Goethe-University Frankfurt, Germany, Prof. Dr. Mathias Ludwig and Dr. Xenia-Rosemarie Reit, South East Asian Ministers of Education Organization (SEAMEO) QITEP in Mathematics, Assoc. Prof. Allan Leslie White, and Universitas Ahmad Dahlan, Indonesia, Assoc. Prof. Dr. Suparman. We also have two keynote speakers in workshop session coming from Universitas Gadjah Mada, Dr. Indah Emilia Wijayanti, and Universitas Ahmad Dahlan, Dr. Rully Charitas Indra Prahmana.

AD-INTERCOMME 2017 was an overwhelming success, attracting the delegates, speakers and sponsors from many countries and provided great intellectual and social interaction for the participants. Without their support, the conference would not have been successfully organized. I trust that all the participants found their involvement in the Conference both valuable and rewarding. Our wish is that all participants would enjoy this conference, contribute effectively toward it and take back with you knowledge, experiences, contacts and happy memories of this conference and especially with this beautiful kingdom of Yogyakarta.

Dr. Rully Charitas Indra Prahmana

IOP Conf. Series: Journal of Physics: Conf. Series **943** (2017) 011002 doi:10.1088/1742-6596/943/1/011002

The Committees of The First Ahmad Dahlan International Conference on Mathematics and Mathematics Education (AD-INTERCOMME)

Rully Charitas Indra Prahmana

Editor in Chief of AD-INTERCOMME's Publication, Universitas Ahmad Dahlan, Jl. Pramuka Kav. 5 Pandeyan, Umbulharjo, Yogyakarta, Indonesia.

E-mail: rully.indra@mpmat.uad.ac.id

Organizing Committee

Muhammad Sayuti	Chairman
Puguh Wahyu Prasetyo	Co-Chair
Uswatun Khasanah	Treasurer
Afit Istiandaru Fariz Setyawan	Secretary

Advisory Committee

Kasiyarno	Rector
Trikinasih Handayani	Dean of Faculty of Teacher Training and Education
Suparman	Vice Dean of Faculty of Teacher Training and Education
Aris Thobirin	Dean of Faculty of Mathematics and Natural Sciences

Steering Committee

Abdul Taram	Head of Mathematics Education Department
Yudi Ari Adi	Head of Mathematics Department
Sunaryo	Lecturer of Mathematics Education Department
Sugiyarto	Lecturer of Mathematics Department

IOP Conf. Series: Journal of Physics: Conf. Series **943** (2017) 011002 doi:10.1088/1742-6596/943/1/011002

The committees of the First Ahmad Dahlan International Conference on Mathematics and Mathematics Education (AD-INTERCOMME) would like to express gratitude to all advisory editorial board and scientific and reviewer Committee for the volunteering support and contribution in the editing and reviewing process.

Advisory Editorial Board

Scientific and Reviewer Committee

IOP Conf. Series: Journal of Physics: Conf. Series **943** (2017) 011002 doi:10.1088/1742-6596/943/1/011002

Esther Yook Kin Loong	Deakin University
Dony Permana	Universitas Negeri Padang
Al Jupri	Universitas Pendidikan Indonesia
Tri Sri Noor Asih	Universitas Negeri Semarang
Danang Teguh Qoyyimi	Universitas Gadjah Mada
Zaleha Ismail	Universiti Teknologi Malaysia
Yeni Susanti	Universitas Gadjah Mada
Rina Oktaviyanthi	Universitas Serang Raya
Fatia Fatimah	Universitas Terbuka
Dini Afsyah Destina	Universitas Negeri Jakarta
Adi Nur Cahyono	Universitas Negeri Semarang
Muhammad Wakhid Musthofa	UIN Sunan Kalijaga
Ahmad Wachidul Kohar	Universitas Negeri Surabaya

IOP Conf. Series: Journal of Physics: Conf. Series **943** (2017) 011003 doi:10.1088/1742-6596/943/1/011003

The Photographs of The First Ahmad Dahlan International Conference on Mathematics and Mathematics Education (AD-INTERCOMME)

Rully Charitas Indra Prahmana

Editor in Chief of AD-INTERCOMME's Publication, Universitas Ahmad Dahlan, Jl. Pramuka Kav. 5 Pandeyan, Umbulharjo, Yogyakarta, Indonesia.

E-mail: rully.indra@mpmat.uad.ac.id

Keynote Speakers

Mathias Ludwig	Goethe-University Frankfurt
Xenia-Rosemarie Reit	Goethe-University Frankfurt
Allan Leslie White	SEAMEO QITEP in Mathematics
Suparman	Universitas Ahmad Dahlan

Figure 1. Mathias Ludwig and Xenia-Rosemarie Reit from Goethe-University Frankfurt, Germany, delivering their keynote talk

IOP Conf. Series: Journal of Physics: Conf. Series **943** (2017) 011003 doi:10.1088/1742-6596/943/1/011003

Figure 2. Allan Leslie White from SEAMEO QITEP in Mathematics and Suparman from Universitas Ahmad Dahlan delivering their keynote talk

Figure 3. One of the participants delivering her talk in parallel session

IOP Conf. Series: Journal of Physics: Conf. Series **943** (2017) 011004 doi:10.1088/1742-6596/943/1/011004

Peer review statement

All papers published in this volume of *Journal of Physics: Conference Series* have been peer reviewed through processes administered by the proceedings Editors. Reviews were conducted by expert referees to the professional and scientific standards expected of a proceedings journal published by IOP Publishing.

IOP Conf. Series: Journal of Physics: Conf. Series **943** (2017) 012001 doi:10.1088/1742-6596/943/1/012001

The relationships within the mathematical content of teachers' lesson sequences

M Shahrill[1], R C I Prahmana[2] and R Roslan[1]

[1] Sultan Hassanal Bolkiah Institute of Education, Universiti Brunei Darussalam, Bandar Seri Begawan, Brunei Darussalam
[2] Universitas Ahmad Dahlan, Jl. Pramuka Kav. 5, Pandeyan, Yogyakarta, Indonesia

Email: masitah.shahrill@ubd.edu.bn

Abstract. This study explored how mathematics content is carried through by means of the problems presented during lessons. Following the definitions and the coding criteria from the TIMSS 1999 Video Study, a total of 163 mathematics problems were identified in the video-recorded lesson sequences of four Bruneian mathematics teachers teaching at the Year 8 level. These problems were classified according to the four basic kinds of relationships: mathematically related, thematically related, repetition and unrelated. Drawing on the mathematical content of the teachers' lesson sequences, the findings revealed variations among the mathematical problems coded as repetition and thematically related, between the four Brunei classes. The aggregated results obtained from the four classes highlighted several points of discussion, such as the relatively higher proportion of repetition problems (52%) from one teacher in particular; the percentage similarities of thematically related problems for all four classes (ranging from 26% to 33%); and the incredibly varied results for mathematically related problems across the four Brunei classes.

1. Introduction

A teacher can attest that what happens when they teach is not just about conveying new or even recalling previously taught information to students. A lot of planning will occur prior to stepping foot in the classroom, for instance, how best to teach a specific topic, which relevant resources to use as reference, what examples to show, the amount of exercises to give and so on [1-6]. The same can be said about teaching mathematics. Much of the content in a single mathematics lesson involves solving mathematical questions or problems assigned to students by the teacher.

From a researcher's perspective, how we view the content of a mathematics lesson may help us understand the factors influencing the clarity and the coherence of the lesson. And one such influence is on how the problems are related [5-11]. A thematic approach to teaching can also provide coherence to the classroom occupants [12], and there may be obstacles in implementing mathematics courses based on a thematic approach [13]. It was suggested that if a thematic approach is to be practical, then the thematic application must serve the purpose by assimilating it fully across the syllabus [14].

The general aims of the 1998-2000 Third International Mathematics and Science Video Study (henceforward mentioned as the TIMSS 1999 Video Study) was to examine and provide teaching practice descriptions concerning Year 8 mathematics classrooms. The seven countries that participated in the TIMSS 1999 Video Study were Japan, Hong Kong SAR, the United States, Switzerland,

Australia, the Netherlands and the Czech Republic. In the TIMSS 1999 Video Study, random samples of Year 8 video-recorded mathematics lessons were collected from these seven countries. The six countries Japan, Hong Kong SAR, Switzerland, Australia, the Netherlands and the Czech Republic were considered as 'high-achieving' countries because of their relatively high achievement performance in comparison to the United States on the TIMSS 1995 Year 8 mathematics assessment [7, 15].

One of the features investigated in the TIMSS 1999 Video Study was on the mathematical content of lessons. Furthermore within this certain feature, the mathematics problem analysis specialist group coded [8] the mathematical relationship among all the problems that were presented during the lessons. In the TIMSS 1999 video study, the term 'problem' was used to indicate any mathematical questions assigned by the teacher to the students. According to Jacobs and her colleagues, "The main requirement of a problem was that a mathematical operation must be necessary in order to arrive at the intended answer, and the problem needed to require some degree of thought by an eight-grade student" [8]. Subsequently, each problem, with the exception of the first problem, was categorised using strictly one of the four basic types of relationships: mathematically related, thematically related, repetition and unrelated. The descriptions for the four kinds of relationships are provided in Figure 1.

• *Mathematically related:* The problem was related to a preceding problem in the lesson in a mathematically significant way. This included using the solution to a previous problem for solving this problem, extending a previous problem by requiring additional operations, highlighting some operations of a previous problem by considering a simpler example, or elaborating a previous problem by solving a similar problem in a different way.

• *Thematically related:* The problem was related to a preceding problem only by virtue of it being a problem of a similar topic or a problem treated under a larger cover story or real-life scenario introduced by the teacher or the curriculum materials. If the problem was mathematically related as well, it was coded only as mathematically related.

• *Repetition:* The problem was the same, or mostly the same, as a preceding problem in the lesson. It required essentially the same operations to solve although the numerical or algebraic expression might be different.

• *Unrelated:* The problem was none of the above. That is, the problem required a completely different set of operations to solve than previous problems and was not related thematically to any of the previous problems in the lesson.

Figure 1. Classifications of the four basic kinds of relationships (taken from [7]).

Furthermore, Hiebert et al. [5] reported that problems that are mathematically related links the content of the mathematics lesson through diverse mathematical connections, while problems classified as repetition involves students to think less since they were able to solve the preceding problem. In contrast, any coded unrelated problems imply lessons being split into mathematically unrelated fragments.

2. Method

This study explores in what way mathematics content is carried through by means of the mathematics problems presented during lessons. Although the design of the present study followed the style of the TIMSS 1999 Video Study [7, 8], for example in relation to the codings and its subsequent analyses, several aspects of the research design from the Learner's Perspective Study [16] was also incorporated in the data collection approach, such as video-recording each competent teacher [17] for a relatively longer period.

In relation to the data collection approach, several continuous series of mathematics lessons, ranging from four to six consecutive lessons or lesson sequences, by four teachers teaching at the Year 8 level in two of the secondary schools in Brunei Darussalam were video-recorded [11, 18]. In the first school, teachers Andy and Henri each had 5 lessons that were video-recorded consecutively, and from the second school, Natalia with 4 and Yasmin had 6 lessons. Presented in Table 1 are the characteristics of the classes and the lessons belonging to the four teachers. Among them, only Henri was considered more experienced, at the time of study, in terms of mathematics teaching experience and qualifications. For Andy in particular, he was teaching the Year 8 express class, which meant that his students completed three years of syllabus within two years. Furthermore, at the time of recording, all his lessons were on reviewing or revising the entire topic that had been covered.

Table 1. Characteristics pertaining to the four mathematics teachers.

Class/Lesson characteristics	Andy	Henri	Natalia	Yasmin
Teaching experience (at time of study)	Two years	Six years	Two years	Two years
Qualification background	Bachelor*	Bachelor & Postgrad Cert	Bachelor*	Bachelor*
Classes responsible & Students' ability	High-achieving (Express Class)	Average	High-achieving	High-achieving
Nature of lesson content	5 AR	2 HR & HN + 3 N	2 HR & HN + 2 AR	3 HR & HN + 2 N + 1 AR
Delivery pace	High	Relaxed	High	High

NOTE: * denotes the field of the degree is specifically in Education. The nature of the lesson content with AR stands for 'All Review' lessons, HR & HN stands for 'Half Review and Half New' lessons, and N stands for 'New' lesson.

In total, twenty video-recorded lessons were collected for further analysis. Relevant supplementary documents were also obtained, such as photocopied lesson plans, notes and exercise worksheets given to their students and the mathematics syllabus. These collected materials were necessary in providing as reference guides in analysing the data. The topic areas covered by the teachers categorised in Table 2 were based on the five major and sub-category topics as listed by Hiebert et al. [7] in Figure 2.

- ***Number:*** Whole numbers, fractions, decimals, ratio, proportion, percent, and integers;
- ***Geometry:*** Measurement (perimeter and area), two-dimensional geometry (polygons, angles, lines, transformations, and constructions), and three-dimensional geometry;
- ***Statistics:*** Probability, statistics, and graphical representation of data;
- ***Algebra:*** Operations with linear expressions, linear equations, inequalities and graphs of linear functions, and quadratic and higher degree equations; and
- ***Trigonometry:*** Trigonometric identities, equations with trigonometric expressions.

Figure 2. The five major and sub-category topics (taken from Hiebert et al. [7, p 68]).

Table 2. The mathematics topic areas covered by the teachers.

Topic Area	Andy	Henri	Natalia	Yasmin
Number	✓	✓	✓	✓
Geometry	✓	✕	✓	✓
Algebra	✓	✕	✓	✓
Statistics	✓	✓	✓	✓
Trigonometry	✕	✕	✕	✕

IOP Conf. Series: Journal of Physics: Conf. Series **943** (2017) 012001 doi:10.1088/1742-6596/943/1/012001

This paper focuses mainly in addressing the relationships of the mathematics problems presented by the four teachers in their lessons. A problem is defined as a statement that requires one to apply a mathematical operation to solve. Subsequently, using the definitions and the coding criteria and schemes provided in the technical report by the mathematics problem analysis specialist coding group [8] and the main report [7] of the TIMSS 1999 Video Study, initially there were 183 mathematical problems coded and identified from the present study's video-recorded lesson sequences data and the collected relevant supplementary documents. However, in order to proceed in coding for relationship, Hiebert et al. [7] indicated, "The first problem in each lesson was not coded for a relationship because the coding scheme defined relationship in terms of problems that preceded a given problem, and no problem preceded the first problem". Since the first problem in each lesson was not taken into account, the total number of mathematical problems to be analysed from the coded data was then reduced to 163. Each of the 163 problems was subsequently classified according to the four basic kinds of relationships. The classifications were done following the original definitions shown earlier in Figure 1. A reliability check was also conducted to confirm the agreement in the coding categories. The examples on how problems were coded as Repetition and Thematically Related are provided in Table 3.

Table 3. Examples of the two-dimensional geometry problems coded as Repetition and Thematically Related (from Andy's fourth lesson).

1(a) In each of the following figures, O is the centre of the circle. Find the unknown length in each case. (All lengths are in cm)

(i)

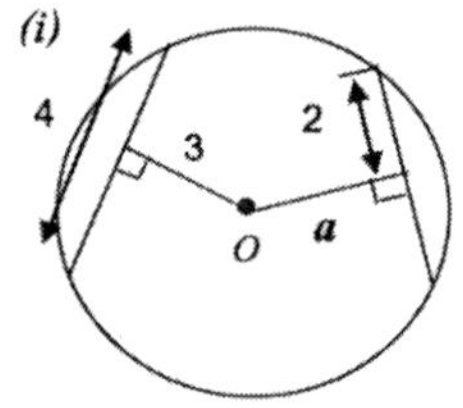

1a (i) Finding the unknown length '***a***' is the first problem in Andy's fourth lesson; hence it was not coded for a relationship.

(ii)

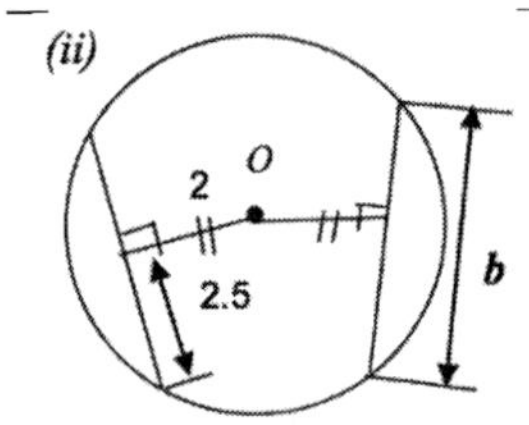

1a (ii) Finding the unknown length '***b***' is the second problem and it was coded as Repetition because it was essentially the same as the first problem.

1(b) Find the unknown angle marked in each of the following figures where O is the centre of the circle.

(i)

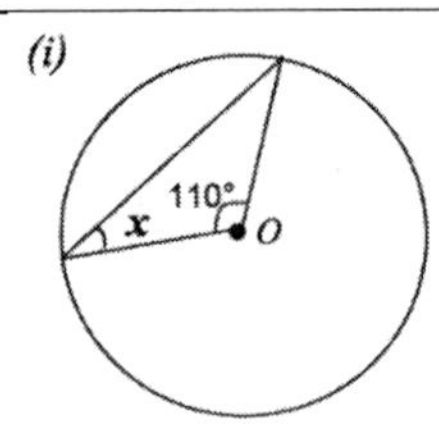

1b (i) Finding the unknown angle '***x***' is the third problem in Andy's fourth lesson and it was coded as Thematically Related because this third problem was related to the first and second problems above by means of it being of a similar topic (two-dimensional geometry problem) and the operation to solve this problem was different to preceding problems.

(ii)

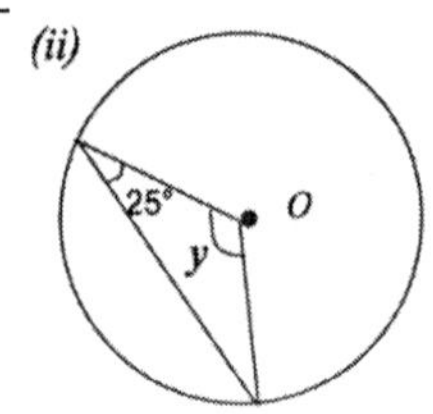

1b (ii) Finding the unknown angle '***y***' is the fourth problem and it was coded as Repetition because this fourth problem was essentially the same as third problem.

Displayed in Table 3 above are examples of four problems (under the topic two-dimensional geometry), taken from Andy's fourth video-recorded lesson, coded as Repetition and Thematically Related. These examples were obtained from the 'Revision Paper 8' exercise worksheet depicted in the textbook by Tay et al. [19]. The examples shown in Table 3 are typical of the problems coded as Repetition or Thematically Related in the coded data, especially within the topics statistics and two-dimensional geometry.

3. Result and discussion

3.1 How mathematics is related within the lesson sequences

From the 163 mathematics problems within the twenty video-recorded lessons there were a total of 32 problems coded as 'Mathematically Related', while 'Thematically Related' and 'Repetition' each had 54, and 'Unrelated' was 23. Table 4 shows the breakdown of the mean (or average) percentage of problems for each lesson in relation to the four kinds of relationship for the four teachers studied in Brunei. From Table 4, apart from the thematically related results, significant differences can be seen between the results of individual teachers. As a reminder, the difference between Mathematically Related and Thematically Related was that a problem was coded as Mathematically Related because it was related to a preceding problem in a mathematically significant way (for example, using the solution of a previous problem to solve the next problem), whereas for a Thematically Related problem to be coded, it required a completely different procedure from the previous problem.

Table 4. Mean percentage of mathematics problems (except the first problem) for each lesson related to previous problems, for each teacher.

Relationship	Classroom (*N–163*)			
	Andy (*n =69*)	Henri (*n=19*)	Natalia (*n=31*)	Yasmin (*n=44*)
Mathematically Related	16	28	5	35
Thematically Related	32	33	32	26
Repetition	52	39	35	31
Unrelated	0	0	28	8

NOTE: '*N*' stands for the overall number of mathematical problems (except the first problem in each video-recorded lesson) in all four classrooms and, '*n*' is the overall number of mathematical problems (except the first problem in each video-recorded lesson) in each classroom. Due to rounding, the percentages may not amount to 100%. For each teacher, the average or mean percentage was calculated as the sum of the percentage within each lesson, divided by the number of lessons.

From Table 4, there are four identified points that will be discussed. Firstly, the relative proportion of Repetition problems for Andy was higher (52%) compared to the other three teachers because most of the mathematical problems that he distributed to his students in all five of his review lessons were related to a single topic (either 'two dimensional geometry', 'statistics' or 'solutions and graphs of linear equalities and inequalities'). The operations to solve these problems were essentially the same (or mostly the same) as the previous problem (within the particular topic area). It seems reasonable that a revision lesson would have a higher concentration of repetition problems, so Andy's high repetition proportion should not be unexpected. The difference between Andy's lessons and the other three teachers' lessons was that their lessons were a mixture of reviewing and introducing or practicing new content (refer also to the 'Nature of lesson content' in Table 1). However, consistent with earlier findings where the lessons were short involving fast-paced utterances or exchanges [11, 20], the students in the classrooms of these three teachers were given less opportunity to practice procedures for solving the mathematical problems [20, 21].

Secondly, the percentages of Thematically Related problems were very similar for all four classrooms. Further investigation revealed that 34 out of the 54 problems coded as Thematically Related belonged to the mathematics topics of two-dimensional geometry (*n = 15*, but for Andy only) and statistics ($n = 18$, for all four teachers). The problems within these two topics were required to be solved using different types of procedures, for example, finding the mean, mode and median of a set of numbers (for statistics), or finding the image of an object under reflection, rotation or vector translation (for two-dimensional geometry). Additionally, referencing the coding criteria given in Table 3 above as examples, each of the 34 Thematically Related problems was coded based on the criteria of having an apparent thematic connection where "the problem was related to a preceding problem only by virtue of it being a problem of a similar topic" [7] or, "the problem required operations that were much different than the first. However the mathematics topic was similar" [8]. It should be noted that Repetition and Thematically Related are mutually exclusive codes, where for each problem, with the exception of the first problem, was categorised using strictly one of the four basic types of relationships. Based on this analysis, each teacher appeared to have a certain number of Thematically Related problems in lessons covering these two topics. Nonetheless, since the lessons were short and fast-paced, along with the students getting less mathematical work time in a week, this would indicate that the teachers had to expose their students to many different types of procedures with which to solve the mathematical problems.

The third point is that the results for Mathematically Related problems were incredibly varied across the four Brunei classrooms. Accordingly, 27 out of the 32 Mathematically Related problems were coded as 'Extension', a Mathematically Related code category that indicated "the problem required many of the same operations as the preceding problem plus some important additional operations" [8]. Andy and Yasmin typically gave two or three problems of this Mathematically Related-Extension code category type within their single topic lessons, whereas for Henri, a majority of the problems were confined within the topic 'integers'. In contrast, there was only one Mathematically Related problem coded for Natalia.

For the final point, the result attained from Natalia's video-recorded lesson was unusual because her lessons contained a much larger proportion of Unrelated tasks (28%) in comparison to the other three teachers. The reason was, unlike Andy's review lessons where each review lesson had a specific topic that was covered, Natalia's third and fourth review lessons consisted of several review problems from a variety of Year 8 topics. Thus, the revision problems were mathematically unrelated or entirely different operations had to be used in the process of solving the previous problems.

3.2 Mathematics problems related to a single topic

Similar to Hiebert et al. [7], another kind of investigation into the mathematical relatedness within lessons was carried out, but this time it was in relation to the mathematical topics and the shifts among topics. The number mathematics lessons for each of the four teachers containing problems associated to a one particular major topic area are presented in Table 5.

Table 5. The number of lessons containing problems associated to a one specific topic.

Major Topic Area	Andy	Henri	Natalia	Yasmin
Number	0	1	0	0
Geometry	2	0	0	0
Algebra	1	0	0	0
Statistics	1	4	2	5
Trigonometry	0	0	0	0
Total (N=16)	4	5	2	5
Total of video-recorded lessons (N=20)	5	5	4	6

The results reveal that a large percentage of the lessons (16 out of 20 lessons or 80%) included problems associated to a one specific topic. Thus indicating that at least 20% of the overall video-recorded lessons contained problems associated to two topics or more. Although there were fewer shifts among the topics made by these teachers, what is of concern here is whether the topic shifts within a lesson may interrupt (or break) the flow of the mathematics content being taught. Hiebert et al. [7] stated "...parts of lessons can become unrelated when teachers switch from one topic to another...Of course, a teacher might also connect different topics mathematically or thematically". Comparing the results obtained here with the number of unrelated problems for each lesson for all four teachers combined, the result is negligible. Thus, taking these results into consideration, the topic shifts within a lesson did not necessarily lead to content fragmentation mainly because a majority of the mathematics problems per lesson were related.

4. Conclusion

Drawing on the mathematical content of the teachers' lesson sequences, the findings revealed variations among the mathematical problems coded as repetition and thematically related, between the four Brunei classes. The aggregated results obtained from the four classes highlighted several points of discussion, such as the relatively higher proportion of repetition problems (52%) from Andy's lessons; the percentage similarities of thematically related problems for all four classes (ranging from 26% to 33%); and the incredibly varied results for mathematically related problems across the four Brunei classes. The additional question prompted from the above findings is, what are the advantages for students, in their mathematics lessons, if they are given more thematically related problems in comparison to mathematically related problems or repetition problems, and how does this affect or even contribute to the coherence and the clarity of the lessons in Brunei? Possibly, an advantage for having more thematically related problems in the lessons will be, if the mathematics content of the lesson is contained in a single topic, then students will be able to work on the different types of related-single-topic-problems given within that lesson, such as the difference between the examples given in Table 4. Nevertheless, to make full use of the mathematics content so as to contribute to the clarity and coherence of a lesson, perhaps, teachers should incorporate tasks that are mathematically related as well. By doing so, students will be able to see how those problems are related to one another in a more elaborate mathematically significant way [22, 23].

There were observations of a teacher's practice that would have been misleading if only one of the lessons were video-recorded and analysed [11, 16-18, 24, 25]. For example, the results relating to the four kinds of relationships presented in Table 4 for each teacher would have been vastly different if they had not been averaged out over the five lessons. In the case of Andy, if only the results of his first video-recorded lesson were analysed, it would have shown 33%, 0%, 67% and 0% for the types Mathematically Related, Thematically Related, Repetition and Unrelated respectively, as opposed to 16%, 32%, 52% and 0% where each kind of relationship had been averaged out over his five lessons. However, as his lesson sequence progressed, variations were observed in the proportions of the mathematics problems coded as Mathematically Related, Thematically Related and Repetition. A much more representative picture of Andy's practice was obtained by documenting a sequences of lessons instead of only one lesson. As shown in this present study, video-recording a teacher's lesson sequence provided a better indication of the diversity needed to characterise a teacher's practice [11, 18, 26-28].

References

[1] Goos M, Stillman G and Vale C 2007 *Teaching Secondary School Mathematics: Research and Practice for the 21st Century* (New South Wales: Allen & Unwin)
[2] Krantz S G (2015) *How to Teach Mathematics* (Rhode Island: American Mathematical Society)
[3] Johnston-Wilder S, Johnston-Wilder P, Pimm D and Lee C 2011 *Learning to Teach Mathematics in the Secondary School: A Companion to School Experience* (New York: Routledge)
[4] George M 2014 Mathematics teaching as a narrative art *Mathematics Teacher* **108** 266
[5] Stacey K, Almuna F, Caraballo R M, Chesné J F, Garfunkel S, Gooya Z, Kaur B, Lindenskov

L, Lupiáñez J L, Park K M, Perl H, Rafiepour A, Rico L, Salles F, and Zulkardi 2015 PISA's Influence on Thought and Action in Mathematics Education *Assessing Mathematics Literacy* ed K Stacey and R Turner (Switzerland: Springer)

[6] Putri R I I and Zulkardi 2017 Fraction in shot-put: A learning trajectory AIP *Conference Proceedings 1868 Research and Education for Developing Scientific Attitude in Sciences and Mathematics* ed C. Kusumawardani, A M Abadi, S Suyanto, and I Wilujeng (New York: AIP Publishing)

[7] Hiebert J, Gallimore R, Garnier H, Givvin K B, Hollingsworth H, Jacobs J, Chui A, Wearne D, Smith M, Kersting N, Manaster A, Tseng E, Etterbeek W, Manaster C, Gonzales P and Stigler J 2003 *Teaching Mathematics in Seven Countries: Results from the TIMSS 1999 Video Study* (Washington: National Center for Education Statistics)

[8] Jacobs J, Garnier H, Gallimore R, Hollingsworth H, Givvin K B, Rust K, Kawanaka T, Smith M, Wearne D, Manaster A, Etterbeek W, Hiebert J and Stigler J 2003 *Third International Mathematics and Science Study 1999 Video Study Technical Report: Volume 1: Mathematics* (Washington: National Center for Education Statistics)

[9] Jacobs J K, Hiebert J, Givvin K B, Hollingsworth H, Garnier H and Wearne D 2006 Does eighth-grade mathematics teaching in the United States align with the NCTM Standards? Results from the TIMSS 1995 and 1999 video studies *Journal for Research in Mathematics Education* **37** 5

[10] Hollingsworth H 2004 From Research to practice: Using the TIMSS video study findings to enhance Australian mathematics teaching *Keynote Paper from the Mathematical Association of Victoria Annual Conference Melbourne Australia*

[11] Shahrill M 2009 From the general to the particular: Connecting international classroom research to four classrooms in Brunei Darussalam *Doctoral Dissertation* (Melbourne: University of Melbourne)

[12] Lipson M Y, Valencia S W, Wixson K K and Peters C W 1993 Integration and thematic teaching: Integration to improve teaching and learning *Language Arts* **70** 252

[13] Handal B and Bobis J 2004 Teaching mathematics thematically: Teacher's perspectives *Mathematics Education Research Journal* **16** 3

[14] Tessier L and Tessier J 2015 Theme-based courses foster student learning and promote comfort with learning new material *Journal for Learning through the Arts: A Research Journal on Arts Integration in Schools and Communities* **11** 1

[15] Hollingsworth H, Lokan J, and McCrae B 2003 *Teaching mathematics in Australia: Results from the TIMSS 1999 Video Study* (Melbourne: ACER)

[16] Clarke D J 2006 The LPS Research Design *Mathematics Classrooms in Twelve Countries: The Insider's Perspective* ed D J Clarke, C Keitel and Y Shimizu (Rotterdam: Sense Publications)

[17] Blömeke S, Kaiser G, and Clarke D 2015 Preface for the special issue on "video-based research on teacher expertise *International Journal of Science and Mathematics Education* **13** 257

[18] Shahrill M 2017 Examining the methodological implications of international classroom research *International Journal of Interdisciplinary Educational Studies* **12** 27

[19] Tay C H, Riddington M and Grier M 2004 *New Mathematics Counts for Secondary 3 Normal (Academic)* (Singapore: Marshall Cavendish International)

[20] Shahrill M and Clarke D J 2014 Brunei teachers' perspectives on questioning: Investigating the opportunities to 'talk' in mathematics lessons *International Education Studies* **7** 1

[21] Salam N H A and Shahrill M 2014 Examining classroom interactions in secondary mathematics classrooms in Brunei Darussalam *Asian Social Science* **10** 92

[22] Shahrill M, Mahalle S, Matzin R, Hamid M H S, and Mundia L 2013 A comparison of learning styles and study strategies used by low and high math achieving Brunei secondary school students: Implications for teaching *International Education Studies* **6** 39

[23] Othman R, Shahrill M, Mundia L, Tan A and Huda M 2016 Investigating the relationship of

student's ability and learning preferences: Evidence from year 7 mathematics' students *The New Educational Review* **44** 125

[24] Prahmana R C I and Kusumah Y S 2016 The hypothetical learning trajectory on research in mathematics education using research-based learning *Pedagogika* **123** 42

[25] Prahmana R C I and Kusumah Y S 2017 Didactic trajectory of research in mathematics education using research-based learning *Journal of Physics: Conference Series* **893** 012001

[26] Clarke D J, Emanuelsson J, Jablonka E, and Mok I A C 2006 The Learner's Perspective Study and International Comparisons of Classroom Practice *Making Connections: Comparing Mathematics Classrooms Around the World* ed D J Clarke, J Emanuelsson, E Jablonka and I A C Mok (Rotterdam: Sense Publications)

[27] Clarke D J, Keitel C, and Shimizu Y 2006 *Mathematics Classrooms in Twelve Countries: The Insider's Perspective* (Rotterdam: Sense Publications)

[28] Clarke D J, Mesiti C, O'Keefe C, Xu L H, Jablonka E, Mok I A C, and Shimizu Y 2007 Addressing the challenge of legitimate international comparisons of classroom practice *International Journal of Educational Research* **46** 280

IOP Conf. Series: Journal of Physics: Conf. Series **943** (2017) 012002 doi:10.1088/1742-6596/943/1/012002

Exploring local plausible reasoning: the case of inequality tasks

I Rofiki, T Nusantara, Subanji and T D Chandra

State University of Malang, Jl. Semarang 5 Malang 65145, Indonesia

E-mail: rofiki.math@gmail.com

Abstract. This study explores local plausible reasoning in solving inequality tasks. The study was conducted to 78 students of mathematics education major from a university in Surabaya, Indonesia. Data were collected through written tasks and interviews. The data were analysed by a constant comparative method. The results of the study were the characteristics of local plausible reasoning shown by these behaviours: (1) students applied plausible reasoning in the local part of task solving or (2) students involved a conceptual understanding or a relational understanding in a few part of entirely argumentations. Educators can overcome the students' behaviours by designing a meaningful learning strategy which develops students' plausible reasoning in the whole of task solving.

1. Introduction

Existing studies about plausible reasoning (PR) have been widely documented in the literature [1-5]. Those studies relate to task solving. The result showed that a majority of students did not involve PR. In fact, PR is an essential competency to be applied in conjecturing, problem solving, and proving.

Polya [6] used the term PR to differentiate a more reasonable guess from a less reasonable guess. Meanwhile, Lithner [7] analysed the qualities of mathematical reasoning into 2 categories, namely PR and reasoning based on established experience (EE). Lithner [8, 9] added 2 new categories, namely local plausible reasoning (LPR) and global plausible reasoning (GPR). Thus, there are 4 types of mathematical reasoning, namely EE, LPR, GPR, and PR. The qualities of mathematical reasoning are not focused on a right or wrong answer, but they focus on the process of argumentations during task solving. Relating to Lithner's idea, the authors define EE as reasoning by giving argumentation based on ideas and procedures constructed from student's previous experience without deep understanding. This understanding refers to conceptual understanding [10] or relational understanding [11]. Meanwhile, PR is defined as reasoning by giving argumentation based on intrinsic mathematical properties. LPR is defined as PR applied locally in the whole of task solving, whereas GPR is defined as PR applied globally in the whole of task solving.

The intrinsic property is relevant to task solving or is accepted by mathematical society as a correct. It is vital in particular contexts and situations [8, 9]. The intrinsic property is opposite of a surface property. The surface property has no or little relevance to task solving. It is not the principal part of task solving because it does not consider the understanding of mathematical ideas. As an example, given that a fraction task to determine a greater fraction between $\frac{7}{16}$ and $\frac{3}{5}$. The quotient of the fraction is an intrinsic property, but the numerical values (7, 16, 3, and 5) are a surface property which is insufficient to be considered in the task solving. An argumentation is based on a surface property such as "$\frac{7}{16}$ is greater than $\frac{3}{5}$ because 7 and 16 are greater than 3 and 5", whereas an argumentation is based on an intrinsic property such as "since $\frac{7}{16}$ is equal to 0.4375 and $\frac{3}{5}$ is equal to 0.6, $\frac{3}{5}$ is greater than $\frac{7}{16}$." Other examples of argumentations based on intrinsic mathematical properties such as (1) "because in the fraction $\frac{7}{16}$ the value of the numerator is less than a half of the denominator, so the value of $\frac{7}{16}$ is less than $\frac{1}{2}$. Whereas in the fraction $\frac{3}{5}$ The value of the numerator is greater than a half of the denominator, so the value of $\frac{3}{5}$ is greater than $\frac{1}{2}$. Hence, $\frac{3}{5} > \frac{7}{16}$."

In the fraction task, student may get a true answer, but his/her argumentation based on established experience without deep understanding. For instance, "according to my teacher's explanation, the task can be solved by cross multiplication in two fractions $\frac{7}{16}$ and $\frac{3}{5}$ so that it is obtained $7 \times 5 = 35$ and $16 \times 3 = 48$. Because 35 is less than 48, so $\frac{7}{16}$ is less than $\frac{3}{5}$." Student only applies the rule and cannot give a reason conceptually. In this case, the student's reasoning is EE.

The results of several earlier studies on the qualities of mathematical reasoning indicated that many students used EE or LPR rather than GPR or PR [7-9]. Most students did not consider the intrinsic mathematical property in mathematical task solving. Those studies only identified the qualities of mathematical reasoning, but they did not examine the characteristics of LPR. Therefore, the aim of this study is to reveal LPR in the inequality task solving.

The inequality concept underlies the understanding of the concept in almost all mathematical fields [12]. Students who do not understand it clearly may lead to a great difficulty when dealing with concepts such as function, monotonicity and concavity. If students understand the inequality concept well, the tasks become easier to be solved. On the contrary, it is difficult for students to make sense the inequality concept.

The finding presented in this study will expand educators' knowledge about students' reasoning in the inequality task solving. This study is significant because it provides a contribution associated with the characteristics of LPR. The finding can be used by educators as a guide in designing the learning strategies or learning tools to improve students' LPR.

2. Literature review

Reasoning has become an essential topic to be promoted to students [13]. Students need to apply their reasoning in order to gain success in mathematics learning and mathematical task solving. By applying reasoning, mathematics can be meaningfully understood. Yet, the fact is that many students do rote learning. Students learn by memorising without knowing the meaning and the reason why a rule or a procedure works. Ultimately, their learning process becomes meaningless.

Reasoning can be seen as thought process, the product of thought process, or both [14]. In this study, the reasoning refers to the thinking process. The reasoning is the line of thinking used by someone to produce statements and reach a conclusion in task solving. To solve a mathematical task, a student can solve a set of subtasks which have the different character. The mathematical task can be categorised into two types, i.e. an exercise (a routine task) or a problem (a non-routine task). A mathematical task is classified as a problem if a student understands the mathematical task and desires to solve it (challenging to solve the task), but the complete solution scheme is not available immediately on the student's mind. While if the student has a complete procedure solution (a complete solution scheme) immediately, then the mathematical task is considered as a routine task. The routine task frequently can be solved by the direct application of one or more algorithms. The term algorithm means that a systematic procedure to solve a mathematical task.

Lithner [8, 9] defines PR in mathematical task solving if the argumentation: (i) is based on intrinsic mathematical properties of the component involved in the reasoning, and (ii) is intended for guiding toward what probably is the truth, without necessarily complete or correct. The component relates to mathematical objects (e.g. facts, concepts, definitions, operations, principles, or procedures) and heuristic. Reasoning in mathematical task solving is called as EE if the argumentation: (i) is based on the ideas and procedures established by the one's previous experience of the learning environment, and (ii) is intended for guiding toward what probably is the truth, without necessarily complete or correct [7-9]. Condition (ii) in the definition of EE is same as in the definition of PR because the goal of the reasoning is the same. The fundamental difference between the definition of PR and EE is on the argumentation as described by the condition (i). Generally the argumentation in EE is the transfer of property from one situation of familiar task solving to another situation that has some similarity. EE is superficial reasoning, while PR is mathematically well-founded reasoning [9].

PR and EE are the extension of analytical thought process and pseudo-analytical thought process introduced by Vinner [15]. The analytical thought process occurs when students faces a structure of a complicated problem and their scheme does not reach it, so they will solve the problem into simpler parts that can be reached out. The difference between the analytical thought process and PR is the degree of certainty in reasoning. The degree of certainty in PR is higher than the analytical thought process. Whereas, the difference between pseudo-analytical thought process and EE is on the degree of analyticity. The pseudo-analytical thought process is not the analytical thought process, but EE has little bit analytical thought content. Students who apply the pseudo-analytical thought process may produce a wrong or right solution.

Other types of mathematical reasoning are LPR and GPR. The difference between LPR and GPR is the range of PR, whereas the similarity is the existence of PR. GPR relates to applying PR globally from the entirely solution, while LPR relates to applying PR locally from the entirely solution. If a mathematical task is impossible to be solved by EE, LPR, or GPR, then PR in the entirely solution needs to be applied.

The authors further make the position of Lithner's reasoning characterisation based on the lens of PR in the whole of task solving. It includes EE, LPR, GPR, and PR. These four types of mathematical reasoning are a continuum. It can be illustrated in a straight line as presented in Figure 1. The two extremities of reasoning are EE and PR. EE is opposite of PR. PR represents the peak of reasoning with a rich network of argumentations based on intrinsic mathematical properties, whereas EE represents the lowest reasoning with argumentations based on the ideas and procedures from previous experience without a deep understanding.

Figure 1. Continuum of mathematical reasoning

3. Method

This exploratory study with a qualitative approach was undertaken to second-year students of a private university located in Surabaya, Indonesia. The participants were 78 students, consisting 63 females and 15 males. The participants' age ranged from 18 to 20 years. All the participants were students of mathematics education major who attended multivariate calculus course. Data were obtained by giving an instrument of task sheet which consists 3 items of inequality tasks as shown in Figure 2. The participants solved the inequality tasks independently. Some notes were taken by the authors during the session of task solving.

> Find the set of all real numbers x which satisfies the following inequality tasks.
> Task 1. $3x^2 \leq 0$ Task 2. $x^2 < 2x$ Task 3. $2 - 8x < 4(1 - 2x)$

Figure 2. The instrument of task sheet

An interview was carried out for collecting data of students' argumentation. The authors recorded the students' conversation and behaviour during the interview. Further, the authors transcribed the recording of the interview. The authors analysed the data using a constant comparative method. It is called the constant comparative method because in this study the data analysis compares the data with the other data constantly, and then it compares the category with the other categories regularly [16, 17]. The method was done to get the characteristics of LPR.

4. Results and discussion

Distribution of students' reasoning in solving inequality tasks is shown in Table 1. In all tasks, many students applied LPR. The highest proportion of LPR was given in Task 3 (53% of the students), whereas the lowest proportion of LPR was given in Task 1 (38% of the students) and 44% of students applied LPR in Task 2. It is noteworthy that a few students applied PR in the whole solution of each task (17%, 10%, 6%).

Table 1. Distribution of students' reasoning in each task

Inequality task	Frequency (percentage) of students' reasoning				Total
	EE	LPR	GPR	PR	
Task 1	18 (23%)	30 (38%)	17 (22%)	13 (17%)	78 (100%)
Task 2	25 (32%)	34 (44%)	11 (14%)	8 (10%)	78 (100%)
Task 3	12 (15%)	41 (53%)	20 (26%)	5 (6%)	78 (100%)

The construction of LPR is next analysed based on students' responses in each task separately. Following the authors report and discuss the analysis of each task. In Task 1, many students considered intrinsic mathematical properties in the local part of task solving. For instance, S1 factorised $3x^2 \leq 0$ to $3x.x \leq 0$. Subsequently, S1 used her previous knowledge about the solution procedure of the equation to the solution procedure of inequality so that it is obtained $(3x \leq 0$ or $x \leq 0) \Leftrightarrow (x \leq 0$ or $x \leq 0)$. S1 determined the solution set of Task 1, i.e. $\{x|\, x \leq 0, x \in \mathbb{R}\}$. S1's written answer to Task 1 is shown in Figure 3.

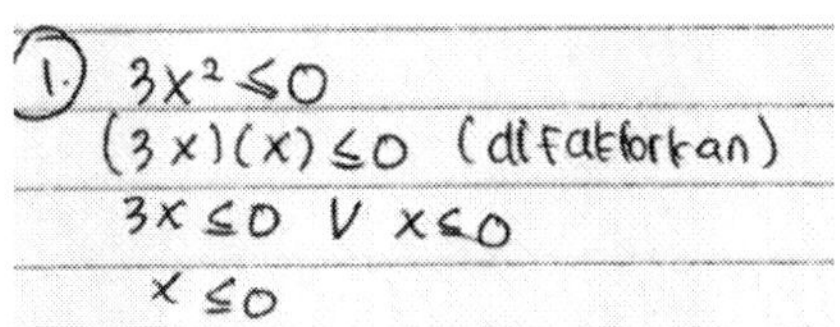

Figure 3. S1's written answer to Task 1

S1 applied PR locally on the part of the whole solution, i.e. involving the concept of factorisation. S1's argumentation related to applying the solution procedure based on her previous experience. If it is viewed from Vinner's terminology [15], S1 only identified the superficial similarity between the solution procedure of $3x.x = 0$ and the solution procedure of $3x.x \leq 0$. S1 did not understand that the solution procedure in the domain of the equation could not be applied in the domain of the inequality. This could result in the improper solution procedure. According to Tall's terminology [18-20], S1 did not realise that the met-before of the solution procedure in the domain of the equation does not work in the domain of the inequality. In other words, a problematic met-before occurred when S1 used met-before outside the domain of validity. Here is the excerpt of S1's argumentations in the interview process.

R: *Your solution set of Task 1 is* $\{x|\, x \leq 0, x \in \mathbb{R}\}$. *How can it be?*

S1: *I factorised* $3x^2 \leq 0$ *so that I got* $3x.x \leq 0$. *At this moment, I faced a problem because I did not know how to solve it. After I recalled my previous experience, I knew how to proceed. I considered the way of equation solving to be applied in the inequality solving. I got* $3x \leq 0$ *or* $x \leq 0$. *For* $3x \leq 0$ *will be* $x \leq 0$. *Because the last part of the solution is* $x \leq 0$ *or* $x \leq 0$, *so the solution set is* $x \leq 0$.

R: *How did you guarantee that your way worked? Justify it.*

S1: *I often used this way to solve inequality. This way is same as a way to solve inequality.*

R: *How do you convince that your answer is correct?*

S1: *I was sure that my answer was correct because I used my previous knowledge about the way of equation solving.*

Note. R = Researcher

In Task 2, an example of student's LPR is represented by S2 behaviour. S2 added both sides of $x^2 < x$ by $-2x$ so that it is obtained $x^2 - 2x < 0$. S2 factorised $x^2 - 2x < 0$ to $x(x-2) < 0$. S2 solved $x(x-2) < 0$ by following the solution procedure from his teacher without relational understanding. S2 determined the values for which each factor is zero, i.e. $x = 0$ or $x = 2$. S2 made three intervals on a real number line, i.e. $(-\infty, 0)$, $(0,2)$, or $(2, \infty)$. S6 tested the values of x at each interval. S2 substituted each of the values $x = -1$, $x = 1$, and $x = 3$ to $f(x) = x(x-2)$ so that it is obtained $f(-1) = 3$ (positive), $f(1) = -1$ (negative), and $f(3) = 3$ (positive). Since $x(x-2) < 0$ has a negative value, S2 decided $\{x|\ 0 < x < 2, x \in \mathbb{R}\}$ as the solution set of Task 2. S2's written answer to Task 2 is shown in Figure 4.

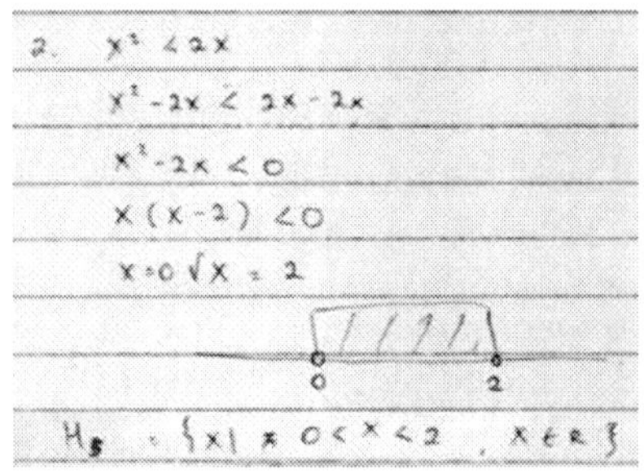

Figure 4. S2's written answer to Task 2

S2 applied PR locally on the part of the whole solution, i.e. involving the rule of adding both sides of the inequality by an expression and the concept of factorisation. S2 understood the rule and the concept. For the next solution steps, S2 applied the solution procedure based on his previous experience from his teacher or textbook. S2 only followed the solution procedure of inequality without knowing the reason why the procedure worked. In this task, S2 got the correct answer, but he was unable to show a right reasoning. According to Vinner's terminology [15], his thought process is categorised as pseudo-analytical. S2's argumentations which support the response are shown in the following excerpt.

R: *How did you get the solution set of Task 2? Explain it.*

S2: *I did it because I have memorised the solution procedure of the inequality. I have learned it from my teacher. I change the symbol "*$>$*" by "*$=$*" for finding the values of* x. *It can be determined the values for which each factor is zero.*

R: *Why can you change the symbol "*$>$*" by "*$=$*"? Justify it.*

S2: *I got the solution procedure of inequality from my teacher. My teacher said that to solve inequality task, we change the symbol of inequality by the symbol of the equation. I also studied the solution procedure of inequality in the textbook.*

R: *Any other reasons? Convince me that your procedure works.*

S2: *I do not know, sir. I only apply the procedure from my learning experience.*

R: *How do you convince that all values in* $0 < x < 2$ *satisfy Task 2?*

S2: *I was very sure that my answer is correct because my teacher said that to find the solution set of the inequality we can do test value in the interval and take one value. If one value of x satisfies the inequality, we decide all values in the interval as the solution. Because for testing the value of x=1 in the* $0 < x < 2$ *satisfied the Task 2, so all values in* $0 < x < 2$ *also satisfy Task 2.*

In Task 3, an example of student's LPR is represented by S3 behaviour. S3 simplified $2 - 8x < 4(1 - 2x)$ to $2 - 8x < 4 - 8x$ by applying the distributive property of multiplication over addition. Afterwards, S3 grouped the expression on the left-hand side of the inequality, while the quantity was grouped on the right-hand side of the inequality. By applying the rule "moving side mean changing sign", S3 simplified the solution process so that it is obtained $0 < 2$ as the last part of the solution. This rule is often done by S3 on the solution of inequalities. This indicates that S3 only focuses on imitating the rule that has been taught by her teacher without the effort of relational understanding. S3 faced a problematic situation. S3 thought for a few minutes and solved task again. S3 was unable to continue the last step of the solution. S3 revealed that there is no solution for Task 3 because the variable x is missing. S3's written answer to Task 3 is shown in Figure 5.

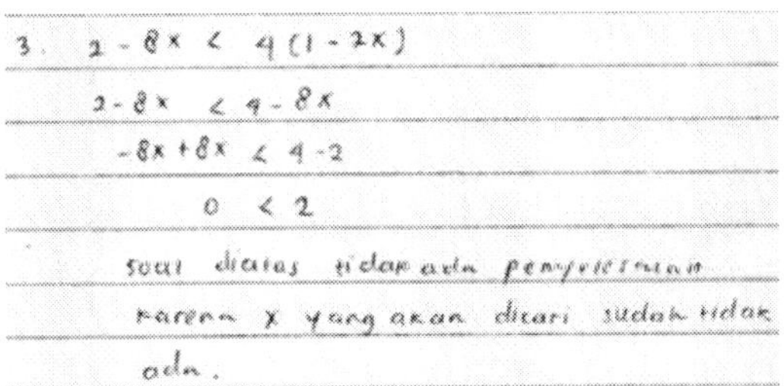

Translation:
The task has no solution because there is no variable x.

Figure 5. S3's written answer to Task 3

S3 could not find the solution set of Task 3 because her conceptual understanding was poor. S3 did not conduct a controlled association in her reasoning process so that she failed to evaluate her reasoning. This reasoning association mechanism which involves the uncontrolled reaction and association without understanding is called as the uncontrolled association [15]. Actually, all real numbers x satisfies $2 - 8x < 4(1 - 2x)$. In other words, the solution set of Task 3 is a real number set. Here is the excerpt of S3's argumentations in the interview process.

R: *Could you explain your solution to Task 3?*
S3: *I distributed the right-hand side of* $2 - 8x < 4(1 - 2x)$ *so that it was obtained* $2 - 8x < 4 - 8x$. *I grouped the variable and the number to the left-hand side and right-hand side, respectively. I used the rule moving side mean changing the sign. I operated and got the result* $0 < 2$.
R: *How did the rule moving side mean changing the sign work? Give reason.*
S3: *I often solved inequalities tasks. The rule is always like that. I got it from my teacher in the past.*
R: *Could you convince that your answer is correct?*
S3: (The student is quiet for a few minutes and try to solve Task 3 again) *I was not sure, sir. I got stuck. I could not solve the task because there is no variable* x *in the last solution. I was confused to determine the solution set. The task was strange.*

The constant comparative method was used to analyse the data of students' LPR so that it was obtained the same characteristics of LPR. The characteristics were that (1) students applied plausible reasoning in the local part of task solving or (2) students' argumentations were mainly founded on their previous experience by considering a few conceptual understanding or relational understanding.

In general, the structure of LPR in task solving can be illustrated in Figure 6. The orange and purple circles represent argumentations, and the lines joining the argumentations represent the relationship or connections between argumentations. The purple circle is an argumentation based on intrinsic mathematical property, whereas the orange circle is an argumentation based on ideas and procedures from previous experience without deep understanding.

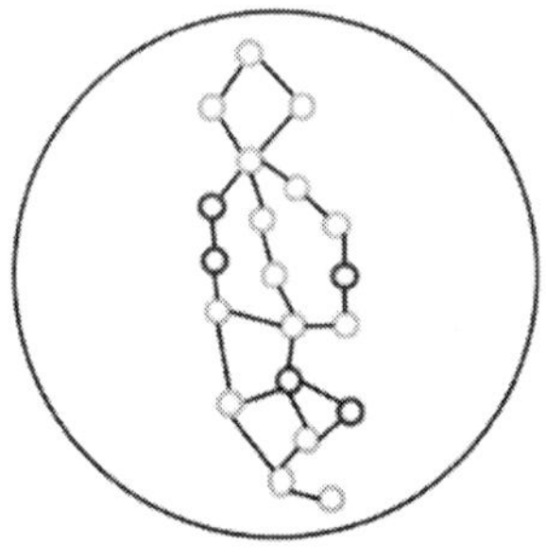

Figure 6. Structure of local plausible reasoning in task solving

These results extend previous results [1-9] to a new situation, namely exploring the characteristics of local plausible reasoning rather than investigating students' difficulty or misconception. Besides, the present study showed that the highest percentage of students' local plausible reasoning was given in the inequality task which has a real number set as the solution set. Such task was called by the students as a strange task or an undefined task. Students experienced an extreme difficulty. Some students stated that (1) the value of x cannot be determined, (2) the task cannot be solved, or (3) the task has no solution set. The finding is accordance with the results of existing studies [2, 3] which reported that an inequality producing a real number set for the solution is very problematic for students. Another finding of the present study is that some students had an intuitive belief that equations and inequalities can be solved by the same procedures. The finding is consistent with earlier studies [4, 5] which reported that the effect of the intuitive belief solving inequalities and equations are the same was extensively evident in many students' solution.

For educational implications, a meaningful learning strategy seems to be an effective way to teach and learn the inequality concept. By the meaningful learning strategy, students can understand the new knowledge well by relating to their existing knowledge. Students' reasoning process is developed through the process of meaningful learning [21]. In addition, a process-oriented instruction can be implemented in mathematics classrooms to promote students' mental models. It is an instructional model in which students are taught thinking strategies to construct, modify and utilise their knowledge or mental models [22].

In this study, some students used their gesture to clarify their reasoning. For further research, it is crucial to examine the contribution of gestures toward students' reasoning. This is supported by the fact that gestures can be utilised as a tool for reasoning [23-25]. The influence of gender is not investigated in this study. Therefore, comparisons of students' local plausible reasoning ability based on gender differences are possible in a future study.

5. Conclusion

In summary, the present study showed that the proportion of local plausible reasoning was dominant. Most students focused on mimicking the rules or procedures without understanding the inequality per se. Moreover, students did not realise the mistake in applying the solution procedure of the equation to the solution procedure of inequality. Therefore, in the instructional process educators should train and develop students' plausible reasoning optimally. Students need to be familiarised for explaining the process of task solving, justifying the steps of task solving (i.e. giving reasons why the students think or answer like that, why students use this strategy, or why the rule/procedure works), and convincing the truth of the obtained results. In addition, educators also can provide scaffolding for students who apply local plausible reasoning in order to have plausible reasoning in the whole of solution task.

References

[1] Almog N and Ilany B S 2012 *Educational Studies in Mathematics* **81** 347
[2] Fujii T 2003 *Proceedings of the 27th Conference of the International Group for the Psychology of Mathematics Education* vol 1 eds N. A. Pateman et al (Honolulu, HI: University of Hawai'i) pp 49–65
[3] Tsamir P and Almog N 1999 *Proceedings of the 23rd Conference of the International Group for the Psychology of Mathematics Education* (Haifa, Israel: PME) p 328
[4] Tsamir P and Almog N 2001 *International Journal of Mathematical Education in Science and Technology* **32** 513
[5] Tsamir P and Bazzini L 2004 *International Journal of Mathematical Education in Science and Technology* **35** 793
[6] Polya G 1954 *Mathematics and Plausible Reasoning: Vol. I. Induction and Analogy in Mathematics* (Princeton, New Jersey: Princeton University Press)
[7] Lithner J 2000 *Educational Studies in Mathematics* **41** 165
[8] Lithner J 2003 *Educational Studies in Mathematics* **52** 29
[9] Lithner J 2004 *Journal of Mathematical Behavior* **23** 405
[10] Hiebert J and Carpenter T P 1992 *Handbook of Research on Mathematics Teaching and Learning* ed D. A. Grouws (New York: Macmillan Publishing Company) pp 65–97
[11] Skemp R R 1976 *Mathematics Teaching* **77** 20
[12] Rofiki I, Nusantara T, Subanji and Chandra T D 2017 *IOSR Journal of Research & Method in Education* **7** 101
[13] Rofiki I, Subanji, Nusantara T and Chandra T D 2015 *Proceeding of International Conference on Educational Research and Development* (Surabaya: State University of Surabaya) pp 358-365
[14] Lithner J 2008 *Educational Studies in Mathematics* **67** 255
[15] Vinner S 1997 *Educational Studies in Mathematics* **34** 97
[16] Creswell J W 2012 *Educational Research: Planning, Conducting, and Evaluating Quantitative and Qualitative Research* (Boston: Pearson Education, Inc)
[17] Kolb M S 2012 *Journal of Emerging Trends in Educational Research and Policy Studies (JETERAPS)* **3** 83
[18] Tall D O 2008 *Mathematics Education Research Journal* **20** 5
[19] Lima R N de and Tall D O 2008 *Educational Studies in Mathematics* **67** 3
[20] Tall D O, Lima R N de and Healy L 2014 *Journal of Mathematical Behavior* **34** 1
[21] Ausubel D P 2000 The Acquisition and Retention of Knowledge: A Cognitive View (Dordrecht: Springer-Science+Business Media, B.V.)
[22] Vermunt J D 1995 *European Journal of Psychology of Education* **10** 325
[23] Soto-Johnson H and Troup J 2014 *Journal of Mathematical Behavior* **36** 109
[24] Williams-Pierce C, Pier E L, Walkington C, Boncoddo R, Clinton V, Alibali M W and Nathan M J 2017 *Journal for Research in Mathematics Education* **48** 248
[25] Nurhasanah F, Kusumah Y S and Sabandar J 2017 Concept of triangle: Examples of mathematical abstraction in two different contexts *International Journal on Emerging Mathematics Education* **1** 53

IOP Conf. Series: Journal of Physics: Conf. Series **943** (2017) 012003 doi:10.1088/1742-6596/943/1/012003

A study of local domination number of $Sn \trianglerighteq H$ graph

Dafik[1,2], **Ika Hesti Agustin**[1,3], **Dwi Agustin Retno Wardani** [1,4], **Elsa Yuli Kurniawati** [1,3]

[1] CGANT University of Jember, Indonesia
[2] Mathematics Edu. Depart. University of Jember, Indonesia
[3] Mathematics Depart. University of Jember, Indonesia
[4] Mathematics Edu. Depart. IKIP PGRI Jember, Indonesia

E-mail: d.dafik@unej.ac.id, ikahesti.fmipa@unej.ac.id, 2i.agustin@gmail.com

Abstract. All graphs in this paper are undirected, connected and simple graph. Let $G = (V, E)$ be a graph of order $|V|$ and size $|E|$. We define a set D as a dominating set if for every vertex $u \in V - D$ is adjacent to some vertex $v \in D$. The domination number $\gamma(G)$ is the minimum cardinality of dominating set. By a locating dominating set of graph $G = (V, E)$, we define for every two vertices $u, v \in V(G) - D$, $N(v) \cap D \neq \emptyset$. Locating dominating set is a special case of dominating set with an extra constrain above. The minimum cardinality of a locating dominating set is locating dominating number $\gamma_L(G)$. The value of locating dominating number is $\gamma_L(G) \subseteq V(G)$. This paper studies locating dominating set of edge comb product of graphs, denoted by GH. The graph $G \trianglerighteq H$ is a graph obtained by taking one copy of G and $|E(G)|$ copies of H and grafting the i-th copy of H at the edge e to the i-th edge of G, where G is star graph S_n and H is any special graph.

1. Introduction

Let $G = (V(G), E(G))$ be a simple graph with vertex set $V(G)$ and edge set $E(G)$. A subset D of $V(G)$ is called a vertex dominating set of G if every vertex not in D is adjacent to some vertices in D. A graph $G = (V, E)$ is called a locating dominating set if for every two vertices $u, v \in V(G) - D$, $N(v) \cap D \neq \emptyset$. Slater [5], [6] defined the locating-dominating number $\gamma_L(G)$ of a graph G is the minimum cardinality of a locating-dominating set of G. The value of locating dominating number is $\gamma_L(G) \subseteq V(G)$. In this paper, we will initiate to analyze locating dominating set of edge comb product of two graphs, denoted by $G \trianglerighteq H$, where G is path graph and H is any special graph.

Saputro *et. al* [7] firstly introduction a comb product of graph. Let G and H be two connected graphs. Let o be a vertex of H. The comb product between G and H, denoted by $G \triangleright H$, is a graph obtained by taking one copy of G and $|V(G)|$ copies of H and grafting the i-th copy of H at the vertex o to the i-th vertex of G. By the definition of comb product, we can say that $V(G \triangleright H) = \{(a, v)|a \in V(G), v \in V(H)\}$ and $(a, v)(b, w) \in E(G \triangleright H)$ whenever $a = b$ and $vw \in E(G)$ and $v = w = o$.

A natural extension of comb product of graph is an edge comb product of graph. Let G and H be two connected graphs. Let e be an edge of H. The edge comb product between G and H, denoted by $G \trianglerighteq H$, is a graph obtained by taking one copy of G and $|E(G)|$ copies of H and grafting the i-th copy of H at the edge e to the i-th edge of G. By the definition of comb

IOP Conf. Series: Journal of Physics: Conf. Series **943** (2017) 012003 doi:10.1088/1742-6596/943/1/012003

product. We can say that $p = |V(G \trianglerighteq H)| = q_1(p_2 - 2) + p_1$ and $q = |E(G \trianglerighteq H)| = q_1 q_2$, see [8] for detail.

Let G be path P_n with vertex set $V(P_n) = \{x_i; 1 \leq i \leq n\}$, and edge set $E(G) = x_i x_{i+1}; 1 \leq i \leq n-1$ so $|V(P_n)| = n$, $|E(P_n)| = n-1$ and H is helm graph H_m with vertex set and edge set are $V(H_m) = \{A\} \cup \{x_i; 1 \leq i \leq m\} \cup \{y_i; 1 \leq i \leq m\}$, $E(H_m) = \{Ax_i; 1 \leq i \leq m\} \cup \{x_i x_{i+1}; 1 \leq im - 1\} \bigcup \{x_m x_1\} \cup \{x_i y_i; 1 \leq i \leq m\}$. Thus $|V(H_m)| = 2m + 1$, $|E(H_m)| = 3m$. Furthermore, the graph $P_n \trianglerighteq H_m$ has a vertex set $V(P_n \trianglerighteq H_m) = \{A_i; 1 \leq i \leq n-1\} \cup \{x_{i,j}; 1 \leq i \leq n; 1 \leq j \leq m-1\} \cup \{y_{i,j}; 1 \leq i \leq n-1; 1 \leq j \leq m\}$ and edge set $E(P_n \trianglerighteq H_m) = \{x_{i,j} x_{i+1,j+1}; 1 \leq i \leq n-1; 1 \leq j \leq m-2\} \cup \{A_i x_{i,j}; 1 \leq i \leq n-1; 1 \leq j \leq m-1\} \cup \{x_{i,j} y_{i,j}; 1 \leq i \leq n; 1 \leq j \leq m-1\}$. The order and size of $(P_n \trianglerighteq H_m), n \geq 3, m \geq 3$ are $|V(P_n \trianglerighteq H_m)| = 2nm - m - 2$ and $|E(P_n \trianglerighteq K_m)| = 3mn - 3m$. We can say that $p = V|G \trianglerighteq H| = q_1(p_2 - 2) + p_1$ and $q = |E(G \trianglerighteq H)| = q_1 q_2$. See Figure 1 as an example of edge comb product of graphs.

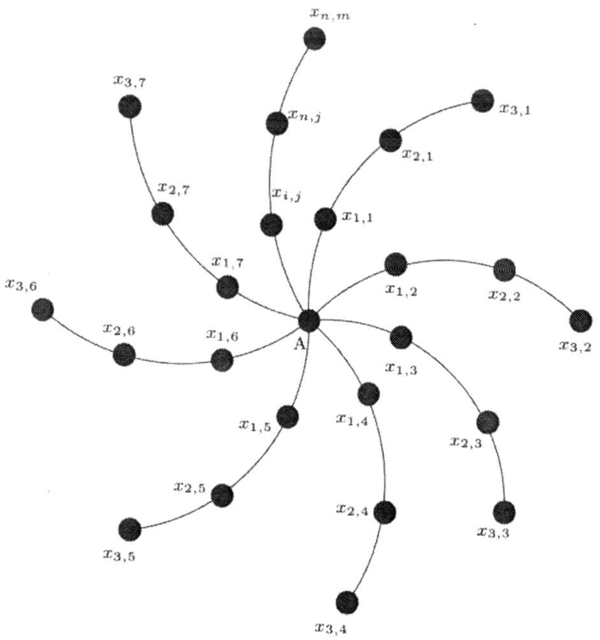

Figure 1. Locating dominating set of edge comb product $S_8 \trianglerighteq P_4$

2. Main Results

In this section, we determine the exact values of locating dominating number of some edge comb product of graphs, namely $S_n \trianglerighteq H$. In this paper, H are complete graph K_m, star graph S_m, triangular book Bt_m and path graph P_n.

Theorem 2.1. *Let $x_i x_{i+1}$ be an edge of K_m, as well as be a grafting edge of K_m. The locating domination number of comb product graph $S_n \trianglerighteq K_m$ is $\gamma_L\ (S_n \trianglerighteq K_m) = nm - 2n$.*

Proof. An edge comb product graph $S_n \trianglerighteq K_m$, $n \geq 3$ and $m \geq 4$, is a connected graph with vertex set $V(S_n \trianglerighteq K_m) = \{A\} \bigcup \{x_{i,j}; 1 \leq i \leq n; 1 \leq j \leq m-1\}$ and edge set $E(S_n \trianglerighteq K_m) = \{Ax_{i,j}; 1 \leq i \leq n; 1 \leq j \leq m-1\} \bigcup \{x_{i,j} x_{i,j}; 1 \leq i \leq n; 1 \leq j \leq m-1\}$. The order and size of $(S_n \trianglerighteq K_m), n \geq 3, m \geq 4$ are $|V(S_n \trianglerighteq K_m)| = nm - n + 1$ and $|E(S_n \trianglerighteq K_m)| = (n)(\dfrac{m(m-1)}{2})$.

First we analysis the lower bound. We claim the lower bound is $\gamma_L(S_n \trianglerighteq K_m) \geq nm - 2n$. To show that is the lower bound, we prove by contradiction. Assume $\gamma_L(P_n \trianglerighteq K_m) < nm - 2n$, we choose locating dominating set of $(S_n \trianglerighteq K_m)$ namely $D = \{x_{i,j}; 2 \leq i \leq n-1; 1 \leq j \leq m-1\} \cup \{x_{1,j}; 1 \leq j \leq m-2\}$ with cardinality of D is $|D| = nm - 2n - 1$. The vertex set without locating dominating set of $(S_n \trianglerighteq K_m)$ is $V - D = \{A\} \cup \{x_{i_j}; 2 \leq i \leq n; j = m-1\} \cup \{x_{1,j}; 1 \leq j \leq m-2 \leq j \leq m-1\}$. The intersection between vertex set $\forall v \in (V - D)$ and D are as follows.

$$\begin{aligned} N(A) \cap D &= \{x_{i,j}; 1 \leq i \leq n; 1 \leq j \leq m-2\} \\ N(x_{i,j}) \cap D &= \{x_{i,j}; 1 \leq i \leq n; 1 \leq j \leq m-2\} \\ N(x_{1,j}) \cap D &= \{x_{1,j}; 1 \leq j \leq m-2\} \end{aligned}$$

For the vertices $x_{1,m-1}, x_{1,m-2} \in (V - D)$, it can be seen that $N(x_{1,m-1}) \cap D = N(x_{1,m-2}) \cap D$, and D does not comply the properties of locating dominating set. So, we can conclude that $\gamma_L(S_n \trianglerighteq K_m) < nm - 2n$ is a contradiction. Hence, the location domination number of $(S_n \trianglerighteq K_m)$ is $\gamma_L(S_n \trianglerighteq K_m) \geq nm - 2n$.

Furthermore, we show that $\gamma_L(S_n \trianglerighteq K_m) \leq nm - 2n$, by choosing locating dominating set of $(S_n \trianglerighteq K_m)$ is $D = \{x_{i,j}; 1 \leq i \leq n; 1 \leq j \leq m-2\}$ with $|D| = nm - 2n$. The vertex set without locating dominating set of $(S_n \trianglerighteq K_m)$ is $V - D = \{A\} \cup \{x_{i,j}; 1 \leq i \leq n; j = m-2\}$. Intersection between vertex set $\forall v \in (V - D)$ and D are as follows.

$$\begin{aligned} N(A) \cap D &= \{x_{i,j}; 1 \leq i \leq n; 1 \leq j \leq m-3\} \\ N(x_{i,j}) \cap D &= x_{i,j}; 1 \leq i \leq n; 1 \leq j \leq m-3\} \end{aligned}$$

For the vertices $x_A, x_{i,j} \in (V - D)$ can be seen that $N(x_A) \cap D \neq \emptyset$ and $N(x_{i,j}) \cap D \neq \emptyset$, D is locating dominating set of $(S_n \trianglerighteq K_m)$ because D satisfied the definition 1. In the otherhand $N(x_A) \cap D \neq N(x_{i,j}) \cap D$, it conclude that D satisfied to be locating dominating set, so D satisfied the definition 2. Hence the location domination number of $(S_n \trianglerighteq K_m)$ is $\gamma_L(S_n \trianglerighteq K_m) \leq nm - 2n$. Because $\gamma_L(S_n \trianglerighteq K_m) \geq nm - 2n$ and $\gamma_L(S_n \trianglerighteq K_m) \leq nm - 2n$, then we can say that $\gamma_L(S_n \trianglerighteq K_m) = nm - 2n$. □

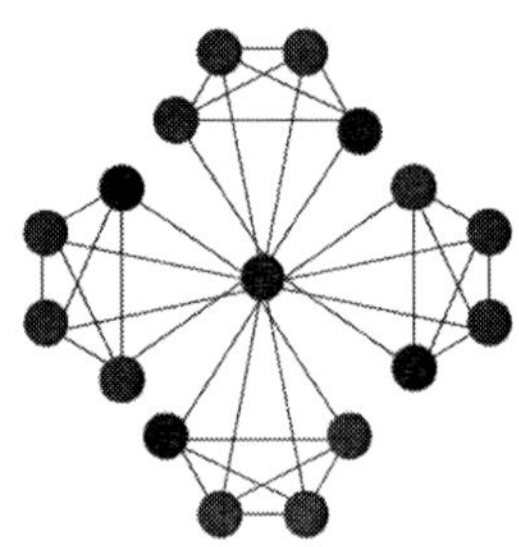

Figure 2. Edge Comb Product $S_4 \trianglerighteq K_5$

Theorem 2.2. *Given that a comb product graph $S_n \trianglerighteq S_m$. Suppose Ax_i is a grafting edge of S_m. Then the locating domination number of γ_L $(S_n \trianglerighteq S_m) = nm - n$.*

Proof. An edge comb product graph $S_n \trianglerighteq S_m$ $n \geq 3$ and $m \geq 3$, is a connected graph with vertex set $V(S_n \trianglerighteq S_m) = \{A\} \cup \{x_i; 1 \leq i \leq n\} \cup \{x_{i,j}; 1 \leq i \leq n; 1 \leq j \leq m-1\}$ and edge set $E(S_n \trianglerighteq S_m) = \{Ax_i; 1 \leq i \leq n\} \cup \{x_i x_{i,j}; 1 \leq i \leq n; 1 \leq j \leq m-1\}$. The order and size of $(S_n \trianglerighteq S_m), n \geq 3, m \geq 3$ are $|V(S_n \trianglerighteq S_m)| = nm + 1$ and $|E(S_n \trianglerighteq S_m)| = nm$.

First we analysis the lower bound. We claim the lower bound is γ_L $(S_n \trianglerighteq S_m) \leq nm - n$. To show that is the lower bound, we prove by contradiction. Assume $\gamma_L(S_n \trianglerighteq S_m) < nm - n$, we choose locating dominating set of $(S_n \trianglerighteq K_m)$ namely $D = \{x_{i,j}; 2 \leq i \leq n-1; 1 \leq j \leq m-1\} \cup \{x_{1,j}; 1 \leq j \leq m-2\}$ with cardinality of D is $|D| = nm-2n-1$. The vertex set without locating dominating set of $S_n \trianglerighteq S_m$ namely $D = \{x_i; 1 \leq i \leq n-1\} \cup \{x_{i,j}; 1 \leq i \leq n, 1 \leq j \leq m-2\}$ with cardinality of D is $|D| = nm - n - 1$. The vertex set without locating dominating set of $S_n \trianglerighteq S_m$ is $V - D = \{A\} \cup \{x_n\} \cup \{x_{i,j}; 1 \leq i \leq n; j = m-2\}$ Intersection between vertex set $\forall v \in (V - D)$ and D are as follows.

$$\begin{aligned} N(A) \cap D &= \{x_i; 1 \leq i \leq n-1\} \\ N(x_{i,j}) \cap D &= \{x_i; 1 \leq i \leq n-1\} \\ N(x_n) \cap D &= \emptyset \end{aligned}$$

For the vertices $x_n \in (V - D)$, it can be seen that $N(x_n) \cap D = \emptyset$, and D not satisfied properties of locating dominating set. So, we can conclude that $\gamma_L(S_n \trianglerighteq S_m) < nm - n$ is a contradiction. Hence, the location domination number of $(S_m \trianglerighteq S_m)$ is $\gamma_L(S_n \trianglerighteq S_m) \geq nm - n$.

Furthermore, we show that $\gamma_L(S_n \trianglerighteq S_m) \leq nm - n$, by choosing locating dominating set of $(S_m \trianglerighteq S_m)$ is $D = \{x_i; 1 \leq i \leq n\} \cup \{x_{i,j}; 1 \leq i \leq n; 1 \leq j \leq m-2\}$ with $|D| = nm - n$. The vertex set without locating dominating set of $(S_m \trianglerighteq S_m)$ is $V - D = \{A\} \cup \{x_{i,m-1}; 1 \leq i \leq n\}$. Intersection between vertex set $\forall v \in (V - D)$ and D are as follows.

$$\begin{aligned} N(A) \cap D &= \{x_i; 1 \leq i \leq n\} \\ N(x_{i,m-1}) \cap D &= \{x_i; 1 \leq i \leq n\} \end{aligned}$$

For the vertices $A, x_{i,m-1} \in (V - D)$ can be seen that $N(A) \cap D \neq \emptyset$ and $N(x_{i,m-1}) \cap D \neq \emptyset$, D is locating dominating set of $S_n \trianglerighteq S_m$ because D satisfied the definition 1. In the otherhand $N(A) \cap D \neq N(x_{i,m-1}) \cap D$ it conclude that D satisfied to be locating dominating set, so D satisfied the definition 2. Hence the location domination number of $S_n \trianglerighteq S_m$ is $\gamma_L(S_n \trianglerighteq S_m) \leq nm - n$. Because $\gamma_L(S_n \trianglerighteq S_m) \geq nm - n$ and $\gamma_L(S_n \trianglerighteq S_m) \leq nm - n$, then we can say that $\gamma_L(S_n \trianglerighteq S_m) = nm - n$. □

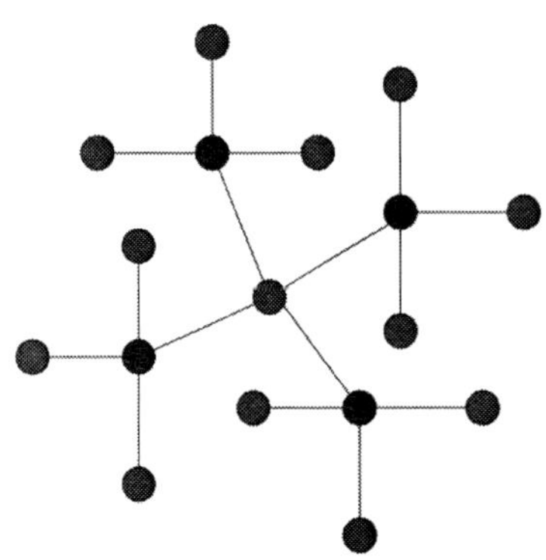

Figure 3. Edge Comb Product $S_4 \trianglerighteq S_4$

Theorem 2.3. *Given that a comb product graph $S_n \trianglerighteq Bt_m$ and AB is a graft edge of Bt_m. Then locating domination number of γ_L $(S_n \trianglerighteq Bt_m) = nm$.*

Proof. An edge comb product graph $S_n \trianglerighteq Bt_m$, $n \geq 3$ and $m \geq 3$, is a connected graph with vertex set $V(S_n \trianglerighteq Bt_m) = \{A\} \cup \{x_i; 1 \leq i \leq n\} \cup \{x_{i,j}; 1 \leq i \leq n; 1 \leq j \leq m\}$, and edge set $E(S_n \trianglerighteq Bt_m) = \{Ax_i; 1 \leq i \leq n\} \cup \{Ax_{i,j}; 1 \leq i \leq n; 1 \leq j \leq m\} \cup \{x_i x_{i,j}; 1 \leq i \leq n; 1 \leq j \leq m\}$. The order and size of $(S_n \trianglerighteq Bt_m), n \geq 3, m \geq 3$ are $|V(S_n \trianglerighteq Bt_m)| = n + nm + 1$ and $|E(S_n \trianglerighteq P_m)| = n + 2nm$.

First we analysis the lower bound. We claim the lower bound is $\gamma_L(S_n \trianglerighteq Bt_m) \geq nm$. To show that is the lower bound, we prove by contradiction. Assume $\gamma_L(S_n \trianglerighteq Bt_m) < nm$, we choose locating dominating set of $S_n \trianglerighteq Bt_m$ namely $D = \{x_{i,j}; 2 \leq i \leq n-1; 1 \leq j \leq m\} \cup \{x_{1,j}; 2 \leq m\}$ with cardinality of D is $|D| = nm - 1$. The vertex set without locating dominating set of $S_n \trianglerighteq Bt_m$ is $V - D = \{A\} \cup \{x_{1,1}\}$ Intersection between vertex set $\forall v \in (V - D)$ and D are as follows.

$$\begin{aligned} N(A) \cap D &= \{x_{i,j}; 1 \leq i \leq n; 1 \leq j \leq m\} \\ N(x_{1,1}) \cap D &= \emptyset \end{aligned}$$

For the vertices $x_{1,1} \in (V - D)$, it can be seen that $N(x_{1,1}) \cap D = \emptyset$, and D does not comply the properties of locating dominating set. So, we can conclude that $\gamma_L(S_n \trianglerighteq Bt_m) < nm$ is a contradiction. Hence, the location domination number of $(S_n \trianglerighteq Bt_m)$ is $\gamma_L(S_n \trianglerighteq Bt_m) \geq nm$.

Furthermore, we show that $\gamma_L(S_n \trianglerighteq Bt_m) \leq nm$, by choosing locating dominating set of $(S_n \trianglerighteq Bt_m)$ is $D = \{x_{i,j}; 1 \leq i \leq n; 1 \leq j \leq m\}$ with $|D| = nm$. The vertex set without locating dominating set of $(S_n \trianglerighteq Bt_m)$ is $V - D = \{A\} \cup \{x_i; 1 \leq i \leq n\}$. Intersection between vertex set $\forall v \in (V - D)$ and D are as follows.

$$\begin{aligned} N(A) \cap D &= \{x_{i,j}; 1 \leq i \leq n; 1 \leq j \leq m\} \\ N(x_i) \cap D &= \{x_{i,j}; 1 \leq i \leq n; 1 \leq j \leq m\} \end{aligned}$$

For the vertices $A, x_{(i)} \in (V - D)$ can be seen that $N(A) \cap D \neq \emptyset$ and $N(x_i) \cap D \neq \emptyset$, D is locating dominating set of $S_n \trianglerighteq Bt_m$ because D satisfied the definition 1. In the otherhand $N(A) \cap D \neq N(x_i) \cap D$ it conclude that D satisfied to be locating dominating set, so D satisfied the definition 2. Hence the location domination number of $S_n \trianglerighteq Bt_m$ is $\gamma_L(S_n \trianglerighteq Bt_m) \leq nm$. Because $\gamma_L(S_n \trianglerighteq Bt_m) \geq nm$ and $\gamma_L(S_n \trianglerighteq Bt_m) \leq nm$, then we can say that $\gamma_L(S_n \trianglerighteq Bt_m) = nm$. □

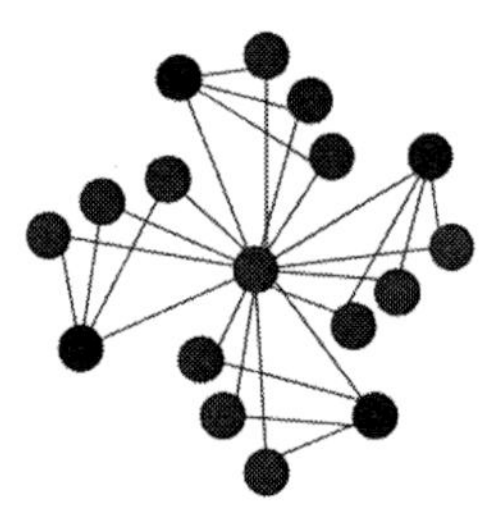

Figure 4. Edge Comb Product $S_4 \trianglerighteq Bt_3$

Theorem 2.4. *Given that a comb product graph $S_n \trianglerighteq P_m$ and x_1x_2 is a grafting edge of P_m. Then locating domination number of γ_L $(S_n \trianglerighteq P_m) = n(\lfloor \frac{2m}{5} \rfloor)$.*

Proof. An edge comb product graph $S_n \trianglerighteq P_m$, $n \geq 3$ and $m \geq 5$, is a connected graph with vertex set $V(S_n \trianglerighteq P_m) = \{A\} \cup \{x_{i,j}; 1 \leq i \leq n; 1 \leq j \leq m-1\}$, and edge set $E(S_n \trianglerighteq P_m) = \{Ax_{i,j}; 1 \leq i \leq n; j = 1\} \cup \{x_{i,j}x_{i,j+1}; 1 \leq i \leq n; 1 \leq j \leq m-2\}$. The order and size of $(S_n \trianglerighteq P_m), n \geq 3, m \geq 5$ are $|V(S_n \trianglerighteq P_m)| = nm - n + 1$ and $|E(S_n \trianglerighteq P_m)| = nm - n$.

First we analysis the lower bound. We claim the lower bound is $\gamma_L(S_n \trianglerighteq P_m) \geq n(\lfloor \frac{2m}{5} \rfloor)$. To show that is the lower bound, we prove by contradiction. Assume $\gamma_L(S_n \trianglerighteq P_m) < n(\lfloor \frac{2m}{5} \rfloor)$, we choose locating dominating set of $S_n \trianglerighteq P_m$ namely $D = \{x_{i,j}; 1 \leq i \leq n-1; j \equiv 2 + 0$ mod

$2\} \cup \{x_n, j; j \geq 3; j \equiv 2+0 \bmod 2\}$ with cardinality of D is $|D| = n(\lfloor \frac{2m}{5} \rfloor) - 1$. The vertex set without locating dominating set of $S_n \trianglerighteq P_m$ is $V - D = \{A\} \cup \{x_{i,j}; 1 \leq i \leq n; j \equiv 1 \bmod 2\} \cup \{x_{n,j}; 1 \leq j \leq 2\}$ Intersection between vertex set $\forall v \in (V - D)$ and D are as follows.

$$\begin{aligned} N(A) \cap D &= \{x_{i,1}; 1 \leq i \leq n-1\} \\ N(x_{i,j} \cap D &= x_{i,j}; 1 \leq i \leq n; j \equiv 2+0 \bmod 2 \\ N(x_{n,j}) \cap D &= \emptyset \end{aligned}$$

For the vertices $x_{n,j} \in (V - D)$, it can be seen that $N(x_{n,j}) \cap D = \emptyset$, and D does not comply the properties of locating dominating set. So, we can conclude that $\gamma_L(S_n \trianglerighteq P_m) < n(\lfloor \frac{2m}{5} \rfloor)$ is a contradiction. Hence, the locating dominating number of $(S_n \trianglerighteq P_m)$ is $\gamma_L(S_n \trianglerighteq P_m) \geq n(\lfloor \frac{2m}{5} \rfloor)$.

Furthermore, we show that $\gamma_L(S_n \trianglerighteq P_m) \leq n(\lfloor \frac{2m}{5} \rfloor)$, by choosing locating dominating set of $(S_n \trianglerighteq P_m)$ is $D = \{x_{i,j}; 1 \leq i \leq n; j \equiv 2+0 \bmod 2\}$ with $|D| = n(\lfloor \frac{2m}{5} \rfloor)$. The vertex set without locating dominating set of $(S_n \trianglerighteq P_m)$ is $V - D = \{A\} \cup \{x_{i,j}; 1 \leq i \leq n; j \equiv 1 \bmod 2\}$. Intersection between vertex set $\forall v \in (V - D)$ and D are as follows.

$$\begin{aligned} N(A) \cap D &= \{x_{i,1}; 1 \leq i \leq n\} \\ N(x_{i,j} \cap D &= \{x_{i,j}; 1 \leq i \leq n; j \equiv 2+0 \bmod 2\} \end{aligned}$$

For the vertices $x_A, x_{i,j} \in (V - D)$ can be seen that $N(x_A) \cap D \neq \emptyset$ and $N(x_{i,j}) \cap D \neq \emptyset$, D is locating dominating set of $S_n \trianglerighteq P_m$ because D satisfied the definition 1. In the otherhand $N(x_A) \cap D \neq N(x_{i,j}) \cap D$ satisfied to be locating dominating set, so D satisfied the definition 2. Hence the location domination number of $S_n \trianglerighteq P_m$ is $\gamma_L(S_n \trianglerighteq P_m) \leq n(\lfloor \frac{2m}{5} \rfloor)$. Because $\gamma_L(S_n \trianglerighteq P_m) \geq n(\lfloor \frac{2m}{5} \rfloor)$ and $\gamma_L(S_n \trianglerighteq P_m) \leq n(\lfloor \frac{2m}{5} \rfloor)$, then we can say that $\gamma_L(S_n \trianglerighteq P_m) = n(\lfloor \frac{2m}{5} \rfloor)$. □

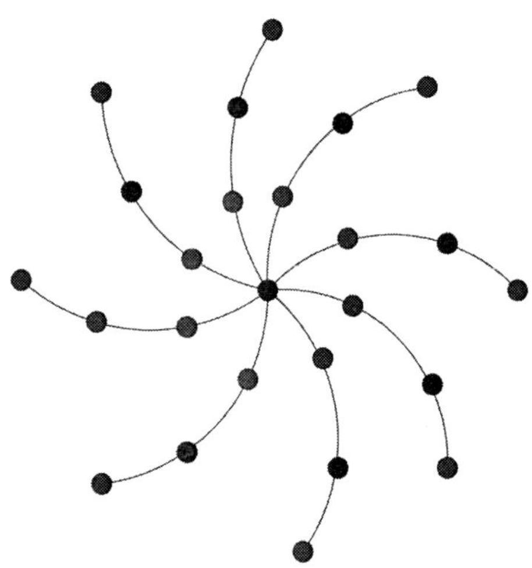

Figure 5. Edge Comb Product $S_8 \trianglerighteq P_4$

IOP Conf. Series: Journal of Physics: Conf. Series **943** (2017) 012003 doi:10.1088/1742-6596/943/1/012003

3. Concluding Remarks

In this paper, we have obtained the exact values of locating dominating number of some edge comb product graph, namely $S_n \trianglerighteq H$. In this paper, we studied H as complete graph K_m, star graph S_m, triangular book Bt_m and path graph P_n. We have found he exact values of their locating dominating number. However for H is any graph we have found any result yet. Therefore we proposed the following open problem.

Open Problem 3.1. *Determine the sharp lower bound or upper bound of locating dominating number of $S_n \trianglerighteq H$ for H is any graph.*

3.1. Acknowledgments

We gratefully acknowledge the support from CGANT University of Jember of year 2017.

References

[1] A. Finbow, B.L. Hartnell 1988 On locating dominating sets and well-covered graphs, Congr. Numer. **65** 191-200.
[2] C.J. Colbourn, P.J. Slater, L.K. Stewart 1987 Locating-dominating sets in series-parallel networks, Congr. Numer. **56** 135-162.
[3] Chartrand G and Lesniak L 2000 *Graphs and digraphs 3rd* ed (London: Chapman and Hall).
[4] Foucaud, F., Henning, M. A., Lawenstein, C. and Sasse, T 2016 Locating dominating sets in twin-free graphs, Discrete Applied Mathematics. **200** 52-58.
[5] P.J. Slater 1987 Dominating and location in acyclic graphs, Networks **17** 55-64.
[6] P.J. Slater 1988 Dominating and reference sets in graphs, J. Math. Phys. Sci. **22** 455.
[7] Saputro, S. W. 2013. The Metric Dimension of Comb Product Graphs. Graph Theory Conference in Honor of Egawas 60th Birthday. 1-2.
[8] Wardani, D. A. R., Dafik, A. C. Prihandoko, and A. I. Kristiana 2016 On The Total r-Dynamic Coloring of Graph: A New Graph Coloring Study. International Conference on Mathematics: Education, Theory and Aplication. ICMETA. Surakarta.

IOP Conf. Series: Journal of Physics: Conf. Series **943** (2017) 012004 doi:10.1088/1742-6596/943/1/012004

Visualizer's representation in functions

F Setyawan, R C I Prahmana, A istiandaru, and A hendroanto

Mathematics Education Department, Universitas Ahmad Dahlan, Jalan Prof. Dr. Soepomo, S.H. Warungbroto, Yogyakarta, 55164 Indonesia

E-mail: fariz.setyawan@pmat.uad.ac.id

Abstract. Students' understanding of a mathematics objects can be seen by their perspective of a mathematics topic. In other words, the study of a specific case can enrich one's conceptualization of the general. This study is a descriptive qualitative research. The visualizer's paperwork is described based on their preference in representing a function. Visualizer's paperwork and interviews provide opportunities to get additional description about her mathematical conceptions and how her understanding about functions by using representation. Visualizer tends to connect their imagination of a picture and represents it based on her ideas. Visualizer defines functions by its perceptual unit that each elements of the domain connected with exactly one element of the codomain. In modelling graph representation, object imager uses all information in the diagram and represent it in symbol notation. In interpreting functions, object imager prefers diagram representation rather than word situation and symbol representation.

1. Introduction

Some people are prefer at processing words and some people are prefer at processing pictures. Most studies have investigated individual preferences based on visual dan verbal information since 1970s partly due to its difficulties in conceptualizing and measuring the underlying both visualizer and verbalizer cognitive style [1–3]. However, considerable cognitive and neuroscience research conducted by Levine [4] gives a new perspective that imagery is unitary. Blajenkova [5] stated that there are two types of imagery in visualizer cognitive styles. Both of them are object and spatial imagers that encode and process visual information in different ways. A perspective of an object is related to the representations of the appearances of an objects in terms of their form, size, shape, colour and brightness. Besides, spatial imagery is related to the abstract representations of the spatial relations amongst objects, parts of objects, locations of objects in space, movements of the objects and the object parts and other complex spatial transformations [6]. This study visualizer subject is seen as the object imagery. The representation of the subject is seen by her understanding about a topic in mathematics.

Visualizer's understanding of a mathematics objects can be seen by their perspective of a mathematics topic. Considering visualizers's thinking about a mathematical topic, allows one better understand the broader domain of visualizers's mathematical thinking and its influence on teaching and learning process. Previous studies conducted by Setyawan [7] investigate that students' representation of a mathematical topic contributed significantly to their conceptual understanding of a linear equation in one variable (LEOV). In other words, the representation about a mathematical topic can be used to construct the understanding of the visualizer subject. Representation of the teacher will affect the students' understanding about a concept and will be a representation of the students has learned. The student's representation can be visual, verbal, and mathematical. This process will continue to be modified in the mind of the student [8].

Most studies has focused on how representations benefit students' understanding about a topic in mathematics [8, 10, 11]. Functions, as one of topics in mathematics, can be seen in picture, words and symbol representations. Within this literature, there is some aspects of the functions concept are most crucial for deep understanding. This present study identify how visualizer subject defines functions as the relation between domain and codomain, gives examples and non-examples of functions, interprets functions by using three kinds of representation such as diagram, situation description, and symbols representation.

This study was designed to get greater depth of information concerning visualizer's understanding about functions. It is describing visualizers's perspective, especially object imager, in defining functions using their own beliefs, giving examples and non-examples of functions, and interpreting functions represent by diagram, situation description, and symbols.

2. Method

This case study, a part of a larger qualitative study on investigating cognitive style's representation, aims to validate visualizer's understanding related to functions when dealing with diagram, situation description and symbol representations. The visualizer's paperwork is described based on her definition about functions, examples which is written and preference in representing a function. The visualizer subject is a 3rd-year college student of Universitas Ahmad Dahlan.

By using a diagram, situation description, and symbols representation, the subject explains her understanding and perspective about functions. The information about conceptions of visualizer's representation in functions was obtained during the first 3 weeks of study, using a written instrument and two interviews. Visualizer's paper work and interviews provide opportunities to get additional insight concerning her mathematical conceptions and perspective influenced those conceptions.

Table 1. Visualizer's tasks on representing functions

No	Visualizers tasks	Representation
1.	Define functions by using own word	Word representation
2.	Give example and non-example of functions	Symbol representation
3.	Choose appropriate representation to make best use in defining functions	Subject's preference in representation

3. Result and discussion

Table 2 summarize visualizer's paperwork in representing the functions. For example, visualizer's perspective about function is seen by using diagram representation. This study found that visualizer's word representation about functions is described based on the visualization of diagram representation. Most of the examples and non-examples is written using diagram.

Table 2. Visualizer representation in Functions

No	Visualizers tasks	Representation
1.	Define functions by using own word	Connecting the elements of domain to exactly one elements of codomain
2.	Give example and non-example of functions	Using diagram representation rather than both word and symbol representation
3.	Choose appropriate representation to make best use in defining functions	Subject's preference in diagram representation

For the task 1, visualizer defines functions by connecting the elements of the domain with exactly one element of the codomain. In the written instrument, visualizer tends to connect their imagination of a picture when she was in senior high school and represents it based on her ideas (Figure 1). For example, she was explained that all of the elements of the domain should have relations with the codomain. In addition, all the elements of the domain should have only one relation with the codomain. This explanation is confirmed by the interviews that she remembers the definition of functions based on her memory in Senior High School while her teacher explains the definition by using a diagram.

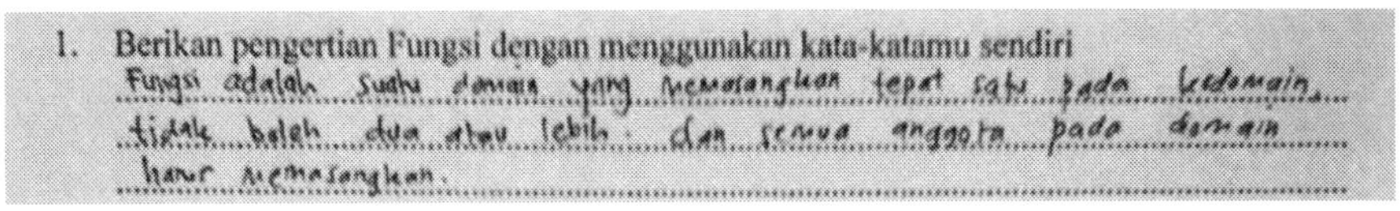

Figure 1. Visualizer's word representation about functions

For the task 2, visualizer gives examples of functions by using diagram and explains it by using her definition about the functions. It is interesting while she explains that the examples related to the linear equation or linear functions. Kozhevnikov [5] stated that the perspective of an object imager, visualizer subject, refers to representations of the appearances of an objects in terms of their precise form, size, shape, colour and brightness. By confirming her written work, the researcher found that she connecting its definition with linear equation because it has equals sign (=). It also found that the subject uses the diagram to represent the examples of functions and non-functions (Figure 2). Visualizer also mentions that one example of functions is linear and quadratics function from the interview.

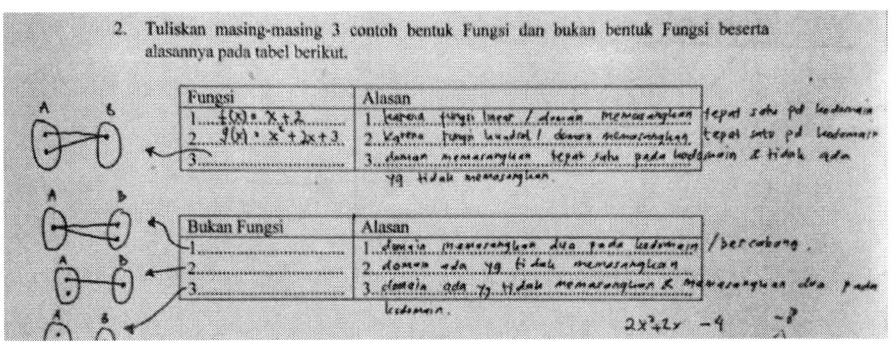

Figure 2. Visualizer's word representation about functions

For task 3, visualizer prefers diagram representation rather than word situation and symbol representation (Figure 3). She thought that diagram representation was better in interpreting the definition of function concept than others. Her preference is relevant with the results of Kozhevnikov [6] whom found that object imagers encoded and processed the images holistically. They processed the images as a single perceptual unit. In comparison with the previous study [11, 12], diagram representation helps visualizer to understand the concept of function. Students' representation is the result of its decision from a phenomenon which was experienced. Students' representation of a concept is influenced by two factors such as internal factors and external factors. Internal factors is the level of student development that form mental model which is displayed through their representation of a concept [13]. Besides, external factors is adopted from the environment either directly or indirectly, qualitatively or quantitatively [13].

IOP Conf. Series: Journal of Physics: Conf. Series **943** (2017) 012004 doi:10.1088/1742-6596/943/1/012004

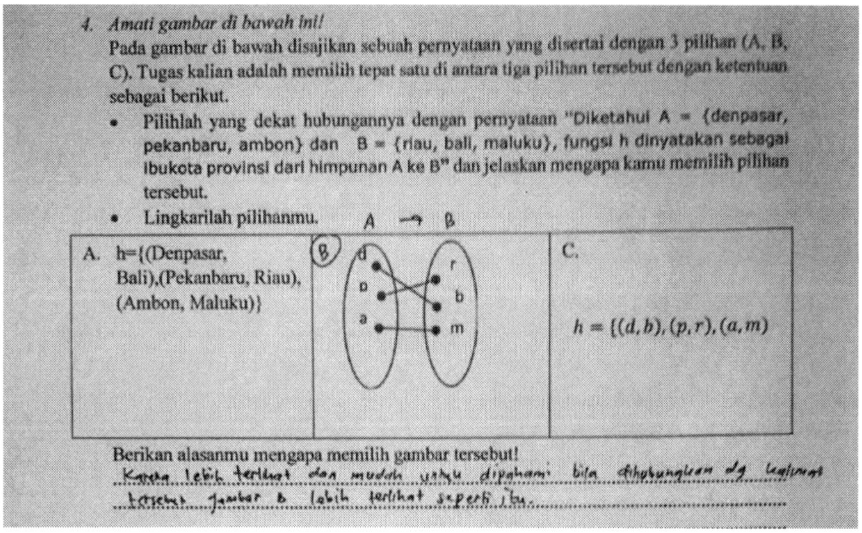

Figure 3. Visualizer's preference in representation

Since representation of the teacher will also greatly affect the students' understanding and will be a representation of the concept he or she has learned. The visualizer understanding about function is described based on their explanation of functions by using multiple representation. The visualizer's representation can be visual, verbal, and mathematical. This process will continue to be modified in the mind of her [8]. In this study, diagram representation helps visualizer better than the other representation to construct the concept of function. Visualizer explained the similarity between word, diagram and symbol representation (Figure 4). She defines the domain and the codomain of the functions in both word and symbol representation. She connects the concept of function in word and symbol representation with the diagram representation.

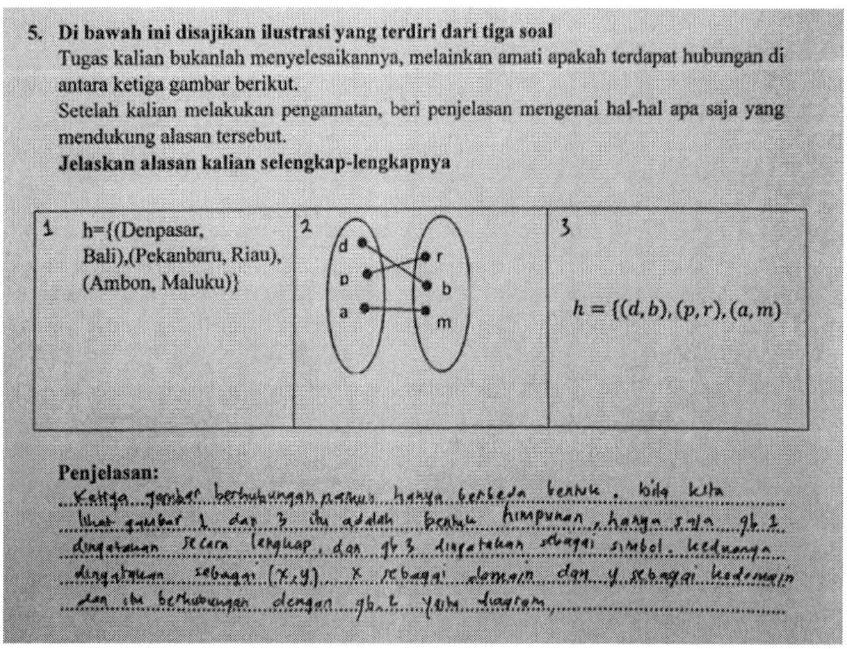

Figure 4. Visualizer's understanding of function using representation

Based on the data, the diagram representation helps visualizer to construct the understanding about functions. Several researchers agree that in mathematics education the term representation refers to the construction, abstraction, and demonstration of mathematical knowledge [9-11, 14-15]. Panasuk [10] stated that the students' ability to create, recognize, interpret, make connection, and translate among representations are powerful communication tools for mathematical thinking. It is relevant with the data observed that representation helps the students to construct the understanding of a topic in mathematics [16]. Visualizer's mathematical thinking in representing functions is related to an image as memory that she learned [15]. In this study, visualizer's preference to describe the concept of functions is by using diagram representation. For object imager, the representation is refer to the literal appearances of a concept.

Since visualizer prefer using diagram representation to both word and symbol representation, the further study can describe and make comparison between object imagers and spatial imagers. The existence of two distinct visualizer learning styles is classified as object and spatial imagery. Both of them is seen as a subsystems that encode and process visual information in different ways [5].

4. Conclusion

For the conclusions, visualizer tends to connect their imagination of a picture and represents it based on her ideas. Visualizer defines functions by its perceptual unit that each element of the domain connected with exactly one element of the codomain. In interpreting functions, visualizer prefers diagram representation rather than word situation and symbol representation. Based on the data, diagram representation helps visualizer to construct the understanding about functions. Visualizer defines the functions from domain to codomain by connecting the elements of domain to exactly one elements of codomain. Visualizer's preference is using diagram representation rather than both word and symbol representation. Furthermore, it is potential to compare the representation between two imagers, object and spatial imagers, above to confirm the findings and also the cognitive style understands among the other relevant studies.

References

[1] Hollenberg C K 1970 Functions of Visual Imagery in the Learning and Concept Formation of Children *Child Dev.* **41** 1003

[2] Paivio A 1972 *Imagery and Verbal Processes* vol 176, ed R and Holt and Winston (New York: AAAS)

[3] Richardson A 1977 Verbalizer-visualizer: A cognitive style dimension *J. Ment. Imag.* **1** 109

[4] Levine D N, Warach J and Farah M 1985 Two visual systems in mental imagery: dissociation of what and where in imagery disorders due to bilateral posterior cerebral lesions. *Neurology* **35** 1010

[5] Blazehenkova O and Kozhevnikov M 2008 The New Object-Spatial-Verbal Cognitive Style Model: Theory and Measurement *Appl. Cogn. Psychol.* **23** 638

[6] Kozhevnikov M, Kosslyn S and Shephard J 2005 Spatial versus object visualizers : A new characterization of visual cognitive style *Memory & Cognition* **33** 710

[7] Setyawan F 2017 Profil Representasi Siswa SMP terhadap Materi PLSV Ditinjau dari Gaya Belajar KOLB *Journal of Medives* **1** 82

[8] Anam R S, Widodo A and Sopandi W 2017 Representation of Elementary School Teachers on Concept of Heat Transfer *Journal of Physics: Conf. Series* p 895

[9] Stylianou D A 2010 Teachers' conceptions of representation in middle school mathematics *J. Math. Teach. Educ.* **13** 325

[10] Panasuk R M 2010 Three Phase Ranking Framework for Assessing Conceptual Understanding in Algebra Using Multiple Representations *Education* **131** 235

[11] Panasuk R M 2011 Taxonomy for assessing conceptual understanding in Algebra using multiple representations. *Coll. Stud. J.* **45** 219

[12] Schnotz W and Bannert M 2003 Construction and interference in learning from multiple

representation *Learn. Instr.* **13** 141
[13] Justi R S and Gilbert J K 2002 Modelling, teachers' views on the nature of modelling, and implications for the education of modellers *Int. J. Sci. Educ.* **24** 369
[14] Setyawan F 2015 Conceptual Understanding Profile of LEOV Junior High School Students Based on Kolb's Learning Style *International Conference on Mathematics, Science, and Education (ICSME)* (Semarang: Universitas Negeri Semarang) pp 61–3
[15] Lamichhane B R and Belbase S 2017 Images of Mathematics Held by Undergraduate Students *International Journal on Emerging Mathematics Education* **1** 147
[16] Tanujaya B, Prahmana R C I, and Mumu J 2017 Mathematics instruction, problems, challenges, and opportunities: A case study in Manokwari regency, Indonesia *World Transactions on Engineering and Technology Education* **15** 287

IOP Conf. Series: Journal of Physics: Conf. Series **943** (2017) 012005 doi:10.1088/1742-6596/943/1/012005

Mathematics creative thinking levels based on interpersonal intelligence

R H Kuncorowati, Mardiyana and D R S Saputro

Universitas Sebelas Maret, Jl. Ir. Sutami No.36 A, Surakarta 57126, Indonesia

E-mail: ririnherpina@gmail.com

Abstract. Creative thinking ability was one of student's ability to determine various alternative solutions toward mathematics problem. One of indicators related to creative thinking ability was interpersonal intelligence. Student's interpersonal intelligence would influence to student's creativity. This research aimed to analyze creative thinking ability level of junior high school students in Karanganyar using descriptive method. Data was collected by test, questionnaire, interview, and documentation. The result showed that students with high interpersonal intelligence achieved third and fourth level in creative thinking ability. Students with moderate interpersonal intelligence achieved second level in creative thinking ability and students with low interpersonal intelligence achieved first and zero level in creative thinking ability. Hence, students with high, moderate, and low interpersonal intelligence could solve mathematics problem based on their mathematics creative thinking ability.

1. Introduction

Mathematics is one of important learning objectives since it has been used importantly in sciences and technologies development. Mathematics subjects are given to all learners to equip them in logical, analytical, systematic, critical and creative thinking, and the ability to cooperate [1]. Based on those aspects, it is known that the potential learners that need to be developed and improved is High Oerder Thinking Skills (HOTS). HOTS is indispensable to learners, because the real life problems are complex, unstructured, complicated, new, and require more thinking skills than just applying what has been learned. Learners should be creative in solving problems to be faced in society. Everyone is assumed to be creative, but the level of creative thinking is different [2]. Those facts could be seen from contextual phenomenon by looking at people who create technology and people who use technology. Learning to know our creativity ability is one of the most significant aspects of our life, for everything we do, is affected by our thinking abilities [3]. Currently, not a few students who view mathematics as a subject that is boring and terrify. Many students find difficulty to understand mathematics. It affects students' low achievement. The absorption capacity of junior high school students in Karanganyar District in mathematics subjects for geometry and measurement materials has the lowest absorption of 42.01% at the district level, 40.26% at the provincial level, and 47.19% at the national level [4].

In the learning of geometry,needs student's creativity to solve the problem. Creativity is classified by multiple levels as if it is related to intelligence levels as well. Since creativity is one of creative thinking process, so that creative thinking also has levels [5]. The fourth level in creative thinking is (very creative). In this level, students are able to solve problem with multiple answers and solutions due to various states fluently in a flexible way [6]. The third level in creative thinking is (creative). Students are able to create new answer with different solution (flexible) though they are not fluent in creating different various answers (not flexible). Besides, students are able to create new problem fluently (fluent) though single problem could lead to new divergent problem. Students who are capable in pursuing this level are called creative. The second level in creative thinking is (enough creative). Students are able to create one different answer or problem from common habit though students do this without flexibility or fluent way, or students are able to show various different solutions though the answer is not new [6]. The first level in creative thinking is (less creative). Students are not able to create different answer of problem (new), though one condition is fulfilled, regarding to create different solutions (flexible) or various answers or problems (fluent) [6]. The zero level in creative thinking is (not creative). Students are not able to create alternative answer or solution or different problem fluently (fluent) and flexible [6].

Student's creative thinking ability can be enhanced by giving various models of questions and problems which have a lot of solutions. Open-ended approach could be used effectively when teachers give exercises after course. Open-ended approach is started by giving open problem with multiple correct answers to students [7]. In open-ended problem solving, students are allowed to cooperate with their friends so that various alternative solutions will arise. The ability to think creatively can be obtained through interaction with the other people [8]. Creative students have many opportunities and readiness to accept and absorb ideas from various sources [9]. In this context, social environment will definitely influence student's role to deal with a bunch of scientific and mathematics problems and create creative ideas. Regarding to instructional learning process, social environment consists of teachers or school mates with higher intelligence or competencies [10]. In the other hand, Piaget's theory as one of stress cognitive theories is focused on student's apperception to create ideas and reduce social environment's influence as well [11]. One of related intelligence in social environment is interpersonal intelligence. Interpersonal intelligence is referring to individual's ability to communicate cooperatively in group, whether it is verbal or non-verbal. Besides, people who have effective interpersonal intelligence tend to be sensitive with feelings and emotions from other people around them [12].

Students with high interpersonal intelligence would be easy to socialize with friends, so that they can solve any problem by collecting multiple ideas. When students are faced with open-ended questions are questions with several solution, students can build their creative ideas by discussing with other friends so that the role of interpersonal intelligence is needed. Creative thinking abilities of students led to the student's skills in solving problems with various solution. These skills must be trained with exercises that have some alternative solution. Perrya Anna state that granting practice aim to develop creativity through learning and practicing creative thinking strategies and through transforming information and experiences from training into new knowledge and skills [13]. The results of the research conducted by Leikin merely to measure creativity possessed by the students but did not provide conclusions about the level of creative thinking that is owned by the students [14]. Therefore, at this researchers will analyze of mathematics creative thinking levels of students based on student interpersonal intelligence.

2. Method

Qualitative descriptive method is used in this research to explain the influence of student's interpersonal intelligence toward their mathematics creative thinking ability. This research is held during second term of 2016/2017 in SMP Negeri 3 Karanganyar. Purposive random sampling technique is used to choose subjects based on student's interpersonal intelligence category in VII C class with 32 students. Based on the result, three subjects are obtained to represent each category of interpersonal intelligence which is classified as high, moderate, and low. Subjects are given by test to determine creative thinking ability level of each student. Then, interview is held to determine student's creative thinking ability. Subject 1 is symbolized (S1), subject 2 is (S2), and subject 3 is (S3). Table 1 is method in determining the levels of student's interpersonal intelligence.

Table 1. Measuring Scale of Student Interpersonal Intelligence Category

No.	Interval	Category
1	$x_i > \bar{x} + 0.5\ s$	High
2	$\bar{x} - 0.5\ s \leq x_i \leq \bar{x} + 0.5\ s$	Moderate
3	$x_i < \bar{x} - 0.5\ s$	Low

Labels:

x_i : student's interpersonal intelligence score in each respondent
$\bar{x}$: mean score of student's interpersonal intelligence in samples
s : standard deviation in samples [15]

3. Result and discussion

Interpersonal intelligence category is derived from total score of measurement scale of interpersonal intelligence. Based on the result, it has been obtained that the score $x_i > 98.47$ means high interpersonal intelligence, $91.21 \leq x_i \leq 98.47$ means moderate interpersonal intelligence, and $x_i < 91.21$ means low interpersonal intelligence. Essay test with open-ended type is used in this research to measure student's mathematics creative thinking ability levels. Question given has multiple alternative solutions. Interview is held to confirm student's written answer, so that researcher could get in depth information. As Sowder and Wheeler based interviews around solutions to estimation problems from hypothetical students in order to elicit explanations [15].

3.1 Analysis of subject with high interpersonal intelligence(S1)

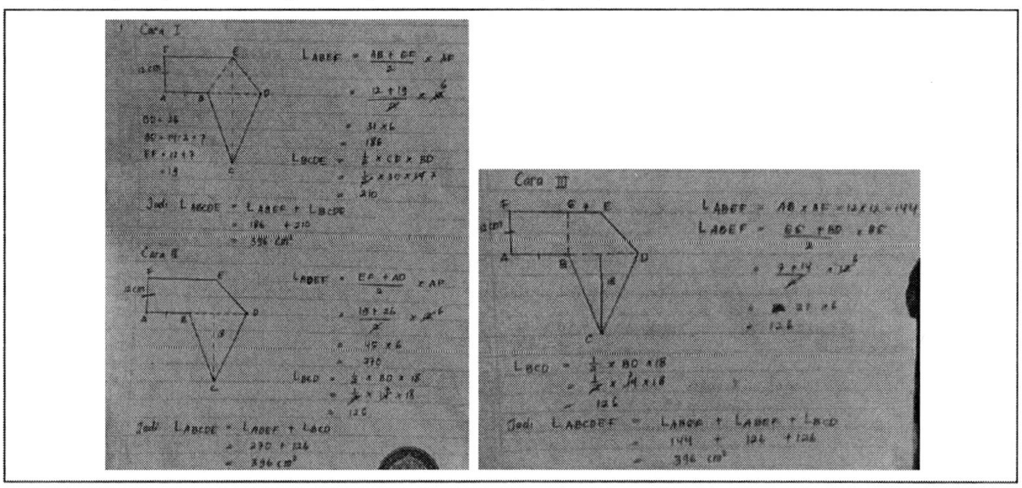

Figure 1. S1 answer for number 1

Figure 1 shows that S1 has finished question number 1 in three different ways (flexible). S1 made the auxiliary line,so the shape looks like a combination of plane. S1 was able to show 3 different ways of completion (flexible). Based on the description above, it means that S1 able to understand the meaning of the problem, then plan some alternative solutions and solve the problem by combining some concepts that have been obtained (fluent).

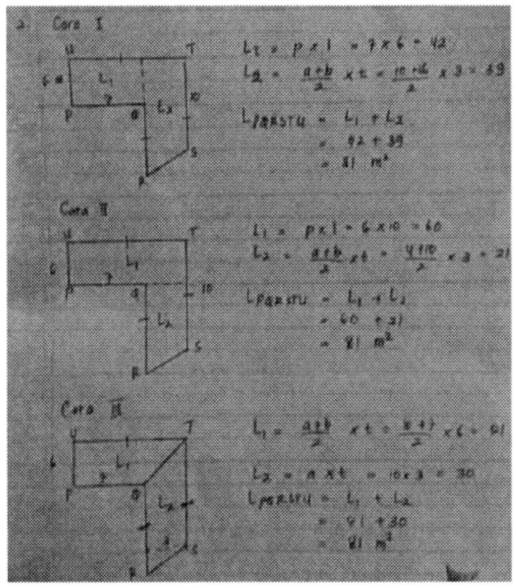

Figure 2. S1 answer for number 2

Figure 2 shows that S1 also completedin 3 different answer alternatives(flexible). As well as in question number 1, S1 created auxiliary lines to divide the combined plane into 2 planes so it can calculate the area of each of its planes. S1 wrote in a row the area ofplane to be searched (fluent).

Based on the interview, S1 was able to explain what was written smoothly and mastered the area of plane material. S1 said that he likes to try to think of ways that are different from what has been taught by the teacher. From the description above, it can be said that S1 classified as having creative thinking. In line with Laycock's view, creative thinking is the ability to analyze a given problem in many ways, to observe patterns, to see similarities and differences, to generate ideas and to decide on appropriate methods for coping with foreign mathematical situations [16].

From the results of student work and the interview results can be concluded that S1 is at the level of creative thinking 3. Creative thinking always seems to show flexibility [5]. It is also in accordance with the opinion of Tatag, the student was able to show a new answer in a different solution (flexible), although he is not fluent or make new answers in different ways (inflexible) [6].

3.2 Analysis of subject with moderate interpersonal intelligence (S2)

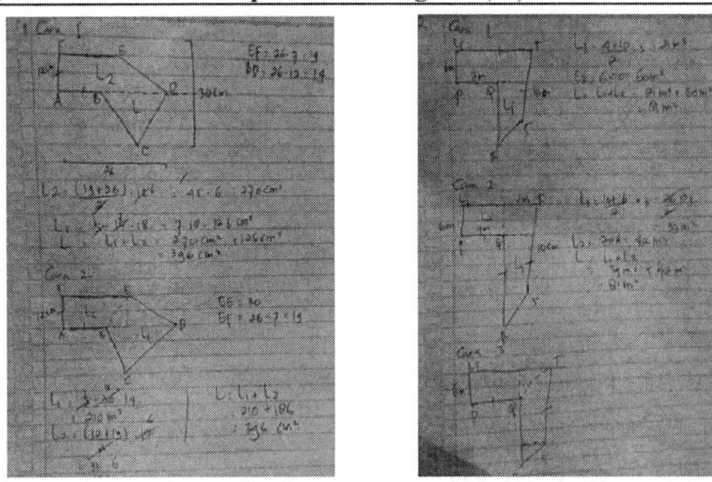

Figure 3. S2 answers for number 1 and 2

Figure 3 shows that S2 was able to solve the problem in 2 different ways (flexible) but S2 did not write the formula used to calculate each of these builds. In this work requires creativity that is how students can divide the combined plane into a few planes, so it can be calculated the area of each planeeasily, and how the students can think of other ways that can be used to solve the problem. On the processing of number 2, S2 is shown to illustrate the third way to calculate the area of combined plane but not proceed to count it (not fluent).

From the results of interviews, S2 had already been understand and memorized the formula of plane, so hedid not write the formula. S2 was able to understand the purpose of the problem, and found 2 different ways of settling. S2 tried to make an auxiliary line to calculate in the third way but still found difficulty incounting of each plane. So it can be said that in this case S2 is not fluent. In line with Tatag's opinion, the students are able to make one answer or different problem from common practice, though not flexibly or fluently, or able to show different ways of fluently different resolution even if the answer is not new [6] S2 is at the level of creative thinking 2.

3.3 Analysis of subject with low interpersonal intelligence (S3)

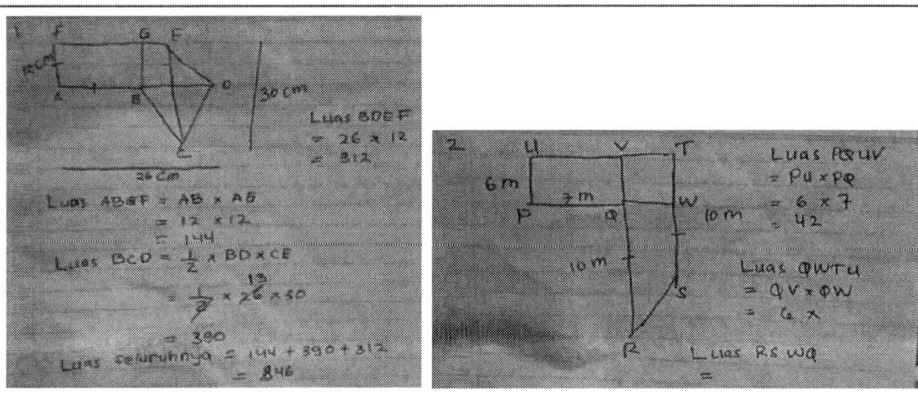

Figure 4. S3 answers for number 1 and 2

Figure 4 shows that on the work of number 1, S3 has done the work by 1 way of settlement. S3 step is to divide the combined plane into 5 parts. But when it came into calculation, S3 only wrote 3 areas of planes. S3 did the correct calculation in the first area of ABGF. When S3 calculated the second area of BCD triangle, S3 wrote incorrect number of triangle base and height. When S3 calculated the third area of BDEG trapezoid, S3 had a mistake in the calculation and S3 didn't write the formula of trapezoid as well. S3 only wrote the multiplication of sides. After the interview was held, S3 stated that S3 has experienced difficulty in determining the height of BCD triangle so that S3 considered that the height was 30 cm. In the third calculation of planes area, S3 has experienced difficulty to determine the formula used to get the correct answer. Based on that result, S3 was able to divide the combined planes into parts but S3 was not able to determine each area. S3 didn't create different alternative solutions. So, S3 had experienced various difficulties with only one solution.

In calculate number 2, S3 divided the combined planes into 3 parts. In the first calculation of PQUV rectangular, S3 wrote the formula and calculated the problem correctly. In QWTV calculation, S3 was only able to write the length of QV and not able to write the length of QW. So, S3 was not able to continue the solution to get the right answer. S3 didn't calculate the third plane which was RSWQ trapezoid. From the result of interview, S3 stated that S3 has experienced difficulty in determining the length of QW. S3 only got it from side length in the question without any effort to try finding out more since S3 was confused to solve the problem. S3 didn't calculate RSWQ area since S3 didn't know the length of plane side. S3 couldn't write the formula of planes area to find the answer. S3 didn't write different alternative solutions.

From written answer and interview result of S3, it can be concluded that S3 is classified to fit into not creative level (zero level of creative thinking ability). This result is suitable with Tatag's statement, student who is fitted into zero level of creative thinking ability (not creative), is not able to create new alternative answer through different solution or state different problem in a fluent or flexible way [6].

4. Conclussion

Based on the results of this research, on the level of students' mathematical creative thinking based on the interpersonal intelligence of grade VII students, it can be concluded that students with high interpersonal intelligence are at the level of creative 3 (creative) thinking. Students with interpersonal intelligence are at the level of creative thinking 2 (creative enough), and students with low interpersonal intelligence are at the creative thinking level 0 (not creative). The ability to think creatively is not only acquired through individuals, but creatively can be obtained through social relationships [8] which focus on interpersonal intelligence. Creativity will come with the knowledge received through the socialization process with the surrounding environment [17, 18] so that students with low interpersonal intelligence tend not to be creative. The results of this study are expected to provide information to the school, especially teachers about the level of creative thinking that students have different, so that teachers should be able to design learning that can train students' creative thinking skills to achieve learning objectives.

5. Acknowledgements

The authors thank to Universitas Sebelas Maret and local government in Karanganyar regency due to gave permission for this study.

References

[1] BSNP *2006 Standar isi untuk satuan pendidikan dasar dan menengah: Standar kompetensi dan kompetensi dasar SMP/MT (*Jakarta: BSNP)
[2] Solso and Robert L 1995 *Cognitive Psychology* (Needham Heights, MA: Allyn & Bacon)
[3] Idris N and Nor N M 2010 Mathematical creativity: usage of technology *Procedia Social and Behavioral Sciences* **2** 1963
[4] BSNP 2016 *Laporan Hasil Ujian Nasional Tahun Pelajaran* 2015/2016
[5] Haylock and Derek 1997 Recognising mathematical creativity in school children *ZDM Electronic Edition* **29** 1615
[6] Siswono T Y E 2006 Konstruksi Tentang Berfikir Kreatif Siswa Dalam Matematika *Seminar Konferensi Nasional Matematika XIII dan Konggres Himpunan Matematika Indonesia* (Semarang: UNESA)
[7] Inprasitha M 2006 Open-Ended approach and teacher education *Tsukuba Journal of Educational Study in Mathematics* **25** 170
[8] Magno C 2008 Explaining the Creative Mind *The International Journal of Research and Review* **3** 15
[9] Murray R and Moore S 2006 The *Handbook of Academic Writing A Fresh Approach* (New York : Open university Press)
[10] Joni S and Masrayati 2016 Students' creative thinking process stages: Implementationof realistic mathematics education *Thinking Skills and Creativity* **24** 118
[11] Schunk D H 2012 *Learning theories An Education perspective* (New York: The University of North California at Greensboro)
[12] Macnamara B N and Rupani N S 2017 The relationship between intelligence and mindset *Procedia - Social and Behavioral Sciences* **64** 52

IOP Conf. Series: Journal of Physics: Conf. Series **943** (2017) 012005 doi:10.1088/1742-6596/943/1/012005

[13] Anna P and Elena K 2017 *Efficacy of teaching creative thinking skills: A comparison of multiple creativity* (USA: Colorado State University)
[14] Waynberg A L and Leikin R 2007 Multiple Solution Tasks As a Magnifying Glass for Observation of Mathematical Creativity *Proc of the 31st Conference of the International Group for the Psychology of Mathematics Education* (Israel: University of Haifa)
[15] Budiyono 2015 *Pengantar Penilaian Hasil Belajar* (Surakarta: UNS Press)
[16] Laycock M 1970 Creative mathematics at Nueva *The Arithmetic Teacher* **17** 325
[17] Moses B and Yaron D 2000 Using Portfolio to Enhance Creative Thinking *The Journal of Technology Studies Summer-Fall 2000* **26** 2
[18] Maharani H R, Sukestiyarno, and Waluya B 2017 Creative thinking process based on Wallas model in solving mathematics problem *International Journal on Emerging Mathematics Education* **1** 177

Improved structural pricing model for the fair market price of Sukuk Ijarah in Indonesia

D Rosadi[1] and Muslim[2]

[1]Department of Mathematics, Universitas Gadjah Mada, Indonesia
[2]Departmenf of Mathematics Universitas Jambi, Indonesia

E-mail: dedirosadi@ugm.ac.id

Abstract. Shariah financial products are currently developing in Indonesia financial market. One of the most important products is called as Sukuk which is commonly referred to as "sharia compliant" bonds. The type of Sukuk that have been widely traded in Indonesia until now are Sukuk Ijarah and Sukuk Mudharabah. In [1], we discuss various models for the price of the fixed-non-callable Sukuk Ijarah and provide the empirical studies using data from Indonesia Bonds market. We found that the structural model considered in [1] cannot model the market price empirically well. In this paper, we consider the improved model and show that it performs well for modelling the fair market price of Sukuk Ijarah.

1. Introduction

Shariah financial products are currently developing in sharia financial market, including in Indonesia bond market. One of the most important products is called as Sukuk which is commonly referred to as "sharia compliant" bonds. The types of Sukuk that have been widely traded in Indonesia until now are Sukuk Ijarah and Sukuk Mudharabah. Sukuk *Ijarah* is an Islamic bond using *Ijarah* contract. *Ijarah* is "essentially" a rental or lease contract to use the benefits of a good or service with a reward. This means that the lessor would entitle the other party to use the objects, but the tenant has the obligation to provide compensation in accordance with the agreement. In *Ijarah* contract basically there is a transfer of temporary benefits, but without any transfer of ownership. Ijarah object used as the underlying in the issuance of sukuk is fixed assets (e.g. ships, electrical networks, land, machinery, buildings, circuits, vehicles, business spaces, engine power plants,) or services (contract/ purchase agreement, transport services). The differences between Sukuk Ijarah and the conventional bonds are described in the Table 1.

In [1], we discuss various models for the price of the fixed-non-callable Sukuk Ijarah and provide the empirical studies using data from Indonesia Bonds market. We found that the structural model considered in [1] cannot model the market price empirically well. Here we propose the new method to improve the accuracy of the model.

The rest of this paper is organized as follows. In Section 2 of this paper, we consider an approach for modeling the price of the fixed - non callable Sukuk Ijarah using a structural method as described in [2]. Here we also present our proposal to improve the accuracy of [2] model. In Section 3 we provide empirical study using data from Indonesia market. In Section 4 we conclude our presentation.

Table 1. Comparison between Sukuk Ijarah and the Conventional Bonds

	Sukuk Ijarah	**Conventional Bonds**
Contract	Ijarah/lease	No
Type of transaction	Certainty Contract	-
Nature of instrument	Ownership certificate of an asset	Acknowledgement of Debt Instruments
Coupons/Income	Remuneration/Fee	Interest/Usury
Returns	Predetermined	Float/fixed
Underlying asset object agreement	Important	No need

2. Modeling the price of Sukuk Ijarah using structural model

2.1. Structural model

The structural method for modelling the price of Sukuk Ijarah was described in [2]. From [2], we can see the cash flow of Sukuk Ijarah as follows

1. At time zero, the sukuk holders give the proceeds (Qo) of Sukuk Ijarah to the issuing "special purpose vehicle" (SPV), an independent board of management between issuer and sukuk holder.
2. During the contract the sukuk holders receive periodic distribution amounts (L) from the SPV which are equal to the leasing payment received from the borrower.
3. Finally, at the end of the contract, the SPV sell back the asset to borrower and borrower pays the repurchase price of the asset (Pt) to the SPV. The SPV distributes the dissolution distribution amount to sukuk holders.

From the cash flows above, we can see that the Sukuk Ijarah cash flows are similar to conventional bond cash flows. The periodic distribution amounts in Sukuk Ijarah replace the coupon payments in conventional bonds, and the repurchase price or the dissolution distribution amount substitutes for the principal.

It is also known in the literature (see e.g. [2]) that the sale and lease transaction in Sukuk Ijarah is not real, and the core of the Sukuk Ijarah structure is a conventional bond. In other words, Sukuk Ijarah is an expensive version of the conventional bond. It is said to be an expensive version because the transaction cost in this instrument is considered significant. The Sukuk Ijarah transaction costs are: the fees of the SPV or the investment bank, lawyers' and Shari'ah Boards' fees, maintenance expenses and insurance expenses in the leasing transaction.

The transaction involved in Sukuk Ijarah can be separated as the following parts:

Part 1. Periodic distribution amounts

1. From time 1 to time T (maturity) the SPV makes regular payments or periodic distribution amounts (L) to sukuk holders which are equal to the lease payment received by the SPV from the borrower.
2. Present value of the lease payments or the periodic distribution amounts are:

$$PV = \frac{L}{(1+r)^1} + \frac{L}{(1+r)^2} + \frac{L}{(1+r)^3} + \ldots + \frac{L}{(1+r)^T} \tag{1}$$

The discount rate used in this transaction cost is the cost of capital in a non-interest economy as discussed by [3]. Therefore, the discount rate is:

$$r = \rho = \frac{Y}{V}(1 - d + dq) \tag{2}$$

where

Y = Expected value of firm's accounting earnings
V = Market value of firm
b= expected value from retention rate of the company

s= expected value of rate stock financing
d=b+s
C= cost of replacement (current asset, saving, security of the company, cost of property, ground and tools)
q=V/C

Part 2. Repurchase price

1. If the asset follows geometric Brownian motion (gBm), the percentage in asset price is independent and identically distributed. The asset price changes independently from the movement of past price. Therefore the monthly expected asset price at time t is: $P_t = Po.e^{\mu t}$, where Po is the asset price at time 0.
   The repurchase price held at maturity is: $P_T = Po.e^{\mu T}$
2. In some agreements, the repurchase price is predetermined, so the repurchase price is simply "PT", which is equal to the price agreed at the beginning of the contract transaction

We obtain the Present Value of repurchase price as
a. Asset repurchase with certainty:
If the asset will be repurchased with certainty at maturity, the present value of the repurchase price is:

- Geometric Brownian motion (gBm) $PV = \dfrac{Po.e^{\mu T}}{(1+r)^T}$ (3)

- Predetermined price $PV = \dfrac{P_T}{(1+r)^T}$ (4)

b. Asset repurchase with call option given to the borrower is not discussed here; we may use the Black Scholes formula

Part 3. Transaction cost

1. Fee of the SPV or investment bank at the beginning of the contract (time 0). The SPV or investment bank's fee is based on the annual asset price given by the borrower (Po).
   The SPV fee is:
   $$Fee = Po.a \tag{5}$$
   a is the proportion of the annual asset price.
2. Fee of the Legal and Shari'ah Board at the beginning of the contract (time 0)
   This is the dollar amount fee charged by the Legal and Shari'ah Board for this transaction, expressed by "S".
3. Operating expense, such as maintenance and insurance fee from time 1 to maturity (T):
   $$PV = \frac{O_t}{(1+r)^1} + \frac{O_t}{(1+r)^2} + \frac{O_t}{(1+r)^3} + \ldots + \frac{O_t}{(1+r)^T} \tag{6}$$

The price of Sukuk Ijarah is the net present value from all cash flow of this instrument. Combining all of the transaction parts, after simplification in the equation (1) - (6), we obtain

1. Repurchase transaction with certainty and with predetermined price:
   $$Qo = \frac{L+Ot}{r} + Po.a + S + \left(Pt - \frac{L+Ot}{r} \right) + (1+r)^{-T} \tag{7}$$
2. Repurchase transaction with certainty and the asset price follows geometric Brownian motion (gBm):

$$Qo = \frac{L+Ot}{r} + Po.a + S + \left(Po.e^{\mu T} - \frac{L+Ot}{r} \right) + (1+r)^{-T} \tag{8}$$

2.2. Improved Structural model

Based on empirical study in [1], we found that when the structural model (7) and (8) are applied to the real data, they tend to be extremely over price. For instance, in reality we found that the real price is slightly higher than 100 percent, but the model found that the predictive price is somewhere like 300 percent. One of the possible problems that causes this inaccuracy is the quality of the financial reporting of the company, which makes the estimate of the discount rate values calculated by [3] method above, see equation (2), will be too far from their possibly reasonable values. In this paper, we propose to replace the value of discount rate based on [3] into the value of expected yields kd. The values of kd depend on the various factors, such as the premium to maturity, the default premium, inflation rate, liquidity rate, etc. For simplification, we estimate the value of kd as the average of yields of bond (corporate) with the same rating with the Sukuk Ijarah that we consider. The future values of kd can be assumed to be fixed or it can be predicted using monte carlo simulation.

3. Empirical study

For empirical example, we consider using the Sukuk Ijarah SIKISAT04B with transaction date June 14, 2010, the summary of input parameters is given in Table 2. For this Sukuk Ijarah, unfortunately the structural model performs poorly; we cannot obtain the price of Sukuk Ijarah with reasonable values. When we assumed that the values of r=kd=16% and using the model with the predetermined values, we obtain the value of Q_0 is Rp. 258.241.284.575,66.

Table 2. Summary of the input parameters for SIKISAT04B

Parameters	Value	Parameters	Value
Po	Rp172,000,000,000.00	V	Rp543,393,000,000.00
Pt	Rp172,000,000,000.00	C	Rp51,568,856,000,000.00
L	Rp20,210,000,000.00	Y	Rp1,648,069,500,000.00
a	5.00%	B	2.72%
S	Rp8,600,000,000.00	S	275.72%
Ot	Rp15,303,029,258.08	Q	0.010537232
T	6.48	σly	0
D	10.00%	d	278.44%
σ^2	8.57%	m	24.8%

4. Conclusions

In this paper we consider the structural models for modeling the price of Sukuk Ijarah. We found that the standard model in equation (7) and (8) can not perform well in practice, but the new improved model gives more reasonable values of the fair market price for Sukuk Ijarah.

5. Acknowledgements

We acknowledge the research support from Hibah Stranas DIKTI 2017

References

[1] Rosadi D, Wahyuni H, and Redjeki S 2016 Modelling the Fair Market Price of Sukuk Ijarah in Indonesia, Presented in *Asian Mathematical Conference* 2016, Bali, July, 25-29
[2] Rohmatunnisa D 2008 Design of Ijarah Sukuk *Thesis* (Nottingham: University of Nottingham)
[3] Mirakhor A 1996 *Islamic Economic Studies* **4** 35

IOP Conf. Series: Journal of Physics: Conf. Series **943** (2017) 012007 doi:10.1088/1742-6596/943/1/012007

Students' thinking level based on intrapersonal intelligence

Rahadian Sholikhati, Mardiyana and Dewi Retno Sari Saputro

Universitas Sebelas Maret, Jl. Ir. Sutami No.36 A, Surakarta, Indonesia

E-mail: diant.113@gmail.com

Abstract. This research aims to determine the students' thinking level based on bloom taxonomy guidance and reviewed from students' Intrapersonal Intelligence. Taxonomy bloom is a taxonomy that classifies the students' thinking level into six, ie the remembering, understanding, applying, analyzing, creating, and evaluating levels. Students' Intrapersonal Intelligence is the intelligence associated with awareness and knowledge of oneself. The type of this research is descriptive research with qualitative approach. The research subject were taken by one student in each Intrapersonal Intelligence category (high, moderate, and low) which then given the problem solving test and the result was triangulated by interview. From this research, it is found that high Intrapersonal Intelligence students can achieve analyzing thinking level, subject with moderate Intrapersonal Intelligence being able to reach the level of applying thinking, and subject with low Intrapersonal Intelligence able to reach understanding level.

1. Introduction

Mathematics is the main source of other knowledge since it has been used importantly in sciences and technologies development [1, 2]. Mathematics is a required learning objective and attached in every class purposively, starting from elementary school, high school, and college, especially in engineering field due to its importance [3, 4]. Mathematics is genuinely related in problem solving [5]. Mathematical knowledge plays an important role in supporting a large number of engineering courses and subsequently, it is vital for engineering students to embrace a strong mathematical problem solving abilities that can keep their motivation for reasonable progress of their engineering programs [6]. According to NCTM, mathematics learning is focus on problem solving [7, 8]. The current emphasis is on problem solving as a means to learn mathematical content and processes [7, 8]. However, mathematics teachers strongly agree if problem solving ability in mathematics is important as well. In teaching and learning process of mathematics, students who are able to solve each problem given to them, they are considered to be able to understand the underlying material.

Materials in mathematics are always related one to another so that students have to master perquisite concepts before jumping into the next material. For instance, the more complex the problem is given, the more perquisite concepts are needed to solve the problem [5]. This problem solving process is correlated to student's cognitive domain. Metacognition in problem solving helps to recognize the presence of a problem that needs to be solved, to discern what exactly the problem is, and to understand how to reach the goal (solution) [9]. So, student's ability to solve any problem in learning process is really important. For the successful solution of any complex problem-solving task, a variety of metacognitive processes is necessary; regulatory activities of planning, monitoring, testing, revising, and evaluating throughout problem solving, especially in making the mental representation and selecting and assessing the effectiveness of the strategies employed [9]. Due to different characters in every mathematics problem, student's responses are varied as well. Bloom stated that real recitation is actually the lowest level in thinking ability. There were higher levels which have to be mastered in order to achieve student's competencies. Process antedates the Bloom's taxonomy, wherein the level of difficulty increases gradually from a mere 'recall' to an application of mind in 'evaluation' [10]. The taxonomy developed by Benjamin S has been revised by a group of educationists [11]. Classification of learning objectives within education that educators set for students use bloom taxonomy. To verify a student's cognitive level educator design cognitive domain within this taxonomy [12]. So, Bloom's Taxonomy is a hierarchy structure to identify skills, starting from the lowest up to the highest. To be able to achieve the higher level, students have to pass the requirements in the lower level as well. Bloom's taxonomy of educational objectives happens to be one of the most often used models while designing the different training, learning and examination methods [13]. Blooms' Taxonomy cognitive thinking levels define into six domains from simplest to complex, are remembering, understanding, applying, analyzing, creating, and evaluating levels [14]. Those six levels from Bloom's Taxonomy are described in brief explanation here: 1) Remembering, the lowest level is used for recitation ability in the latest materials which have been studied before. 2) Understanding, in this second level, understanding materials means to be comprehension ability. 3) Applying, in this third level, application directs to implementation ability toward information received by students in new contextual way in real life. 4) Analyzing, this fourth level means to be elaboration ability in order to sort learning objectives out into each aspects. 5) Creating, this level is appeared to be creation ability. 6) Evaluating, the highest level in Bloom's Taxonomy means to be evaluation ability, making students able to give some sort of values and achievements whether it is good or bad in learning objectives based on indicators given [14]. Nevertheless, one of methods in describing and determining the quality of student's thinking level in solving mathematics problem is using Bloom's Taxonomy.

Problem solving ability of students is also influenced by other factors, such as intrapersonal intelligence. Intrapersonal intelligence is an ability to understand self and identity. Intrapersonal intelligence is the ability to understand oneself as well as one's strengths, weaknesses, moods, desires, and intentions [15]. This includes such skills as understanding how you are similar to or different from others, reminding yourself to do something, knowing about yourself as a language learner, and knowing how to handle your feelings [15]. Someone's ability in understanding themselves would influence their own acts to find out the solution of a problem. The person with Intrapersonal intelligence causes one to distinguish their different feelings [16]. Intrapersonal factor significantly predicted personal goals too [17]. Therefore, it can be seen clearly through how someone tries to solve any problem by looking into their intrapersonal intelligence.

Build Flat Forms is one of materials which are delivered to students in VIII class. This learning objective is one of geometry material. Various questions based on problem solving ability could be created to determine student's intrapersonal intelligence precisely so that it can be seen how far student's thinking level is.

Hence, it can be concluded that the research aims to determine how far student's thinking level to solve multiple problems in mathematics due to Bloom's Taxonomy based on student's intrapersonal intelligence in Build Flat Forms learning objective.

2. Method

This study used grounded theory, which is a suitable to analyze large quantities of unstructured or semi-structured data is qualitative [18, 19, 20]. To be able to determine how far student's thinking level, this research is regarding to Bloom's Taxonomy and based on student's intrapersonal intelligence. Subjects are taken from the result of student's intrapersonal intelligence questionnaire. Questionnaire consists of 40 questions. Based on the result, subjects are classified by high intrapersonal intelligence, moderate intrapersonal intelligence, and low intrapersonal intelligence which represent each level of intrapersonal intelligence. Those three levels of subjects then are given by 2 questions of problem based test. The results are analyzed by Bloom's Taxonomy. Triangulation method is used to examine validity of analysis result by using interview to each subject as well [21].

3. Result and discussion

Student's intrapersonal intelligence questionnaire was used as instrument of determining subjects. The result from questionnaire was used to classify the subjects into three levels of intrapersonal intelligence namely high intrapersonal intelligence (HII), moderate intrapersonal intelligence (MII), and low intrapersonal intelligence (LII). Three students were chosen as representative of each level to do problem based test and give each clarification. The analysis of result showed that HII subject was able to achieve four levels in Bloom's Taxonomy namely remembering, understanding, applying, and analyzing. MII subject was able to achieve three levels namely remembering, understanding, and applying. Then, LII subject was able to achieve two levels namely remembering and understanding.

3.1. High intrapersonal intelligence (HII)

High intrapersonal intelligence subject (HII) was able to achieve four levels in Bloom's Taxonomy which are remembering, understanding, applying, and analyzing. The details can be seen in Figure 1 and Figure 2.

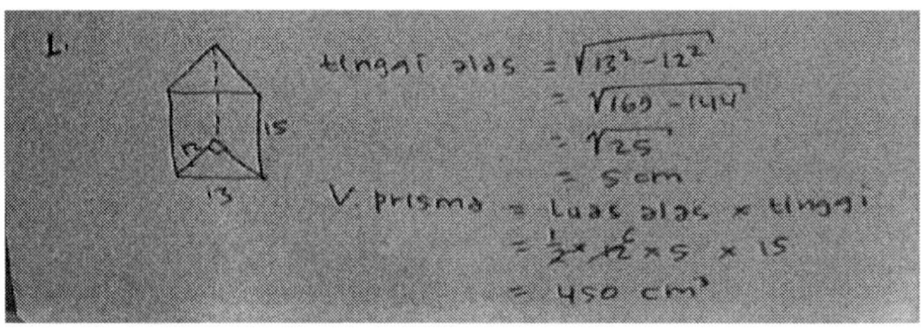

Figure 1. HII subject's answer for number 1

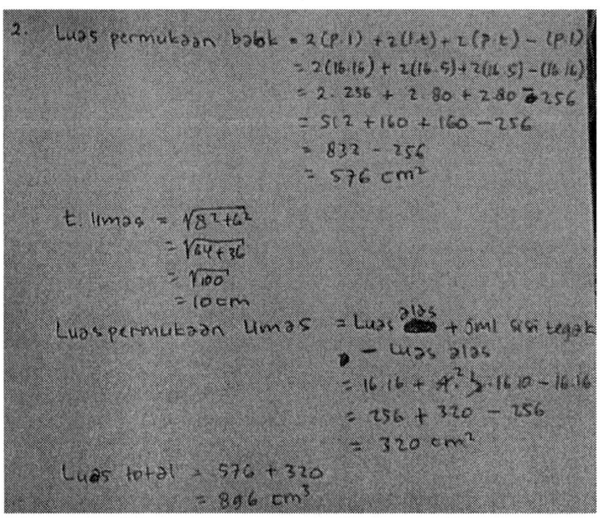

Figure 2. HII subject's answer for number 2

Figure 1 showed that HII subject was able to describe question scheme and put the size precisely. In this step, HII subject has already achieved the second level, understanding the question. HII subject then was looking for additional information to support the solution using interpretation of the question as well. The student was finding out the triangle's height of prism's base. In this step, HII subject has already achieved analyzing level of information which has not been given by the question to find out the volume of prism. Then, student was applying the formula to determine the volume of prism as a solution of the question. In this step, student was coming back to the lowest level of, remembering the formula and applying it to the question.

Figure 2 showed that HII subject has already achieved in understanding the question, which is finding out the area of surface of stacked cuboids and pyramid. The first step done by HII subject was writing down the formula to determine cuboids area of surface. In this step, student has been remembering and has been applying the formula to solve the problem. Student was writing the area of surface formula of cuboids. Then student subtracted it with the top area of surface in cuboids in order to get it off the solution. The same step was used by HII subject to calculate pyramid's area of surface. HII subject subtracted with pyramid's area of base in order to get it off the solution. In this step, HII subject has already achieved analysis level. Subject knew that pyramid's base and cuboids top were not included to stacked build's area of surface. HII subject could also analyze additional information which has to be found before applying the formula by finding out the height of triangle as upright side of pyramid.

Based from interview result, HII subject could explain in detail about what he did. From the first question, HII subject explained that he had not solved the prism's volume directly yet, but he had to find out the base triangle's height first using Pythagoras theorem. Then, he could find out the prism's volume. From the second question, HII subject could explain in detail due to surface area of stacked cuboids and pyramid, so that the area of top cuboids and base pyramid would not be calculated as well. Hence, HII subject substracted the total surface area with the area of top cuboids and base pyramid in his answer sheet.

Based on analysis result of student's answer and interview, it can be obtained that students with high intrapersonal intelligence (HII) could achieve up to fourth level which is analyzing level based on Bloom's Taxonomy, where student could explain an object learning into its definitive parts in this level [12, 14].

3.2. Moderate intrapersonal intelligence (MII)

Subject with moderate intrapersonal intelligence (MII) was able to achieve three levels in Bloom's Taxonomy which are rremembering, understanding, and applying. The details can be seen in Figure 3 and Figure 4.

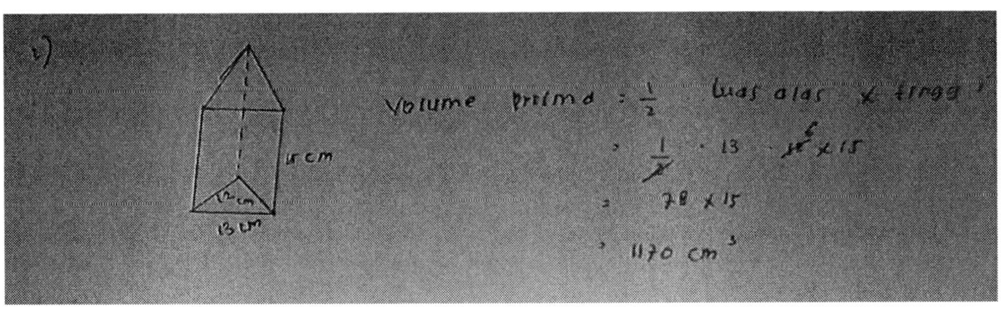

Figure 3. MII subject's answer for number 1

IOP Conf. Series: Journal of Physics: Conf. Series **943** (2017) 012007 doi:10.1088/1742-6596/943/1/012007

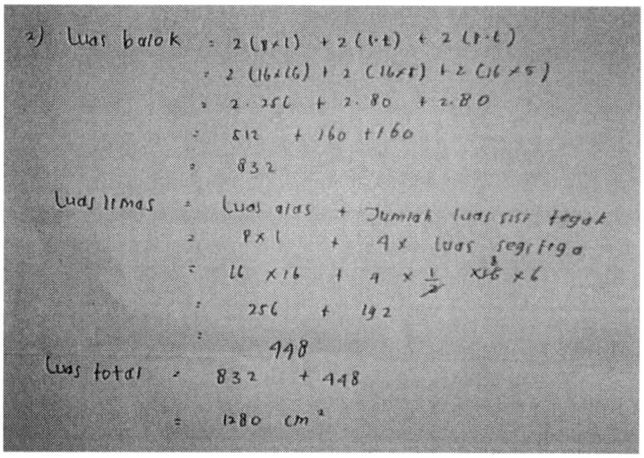

Figure 4. MII subject's answer for number 2

Figure 3 showed that MII subject was able to describe question scheme and put the size precisely. In this step, MII subject has already achieved the second level, understanding the question. MII subject then wrote down the formula of prism's volume directly without looking for additional information. In this step, student was coming back to the lowest level of, rremembering the formula and applying it to the question. The written formula was right, but MII subject didn't analyze the additional information, so that MII subject assumed that the hypotenuse of base triangle as the height of triangle as well.

Figure 4 showed that MII subject has already achieved in understanding the question, which is finding out the area of surface of stacked cuboids and pyramid. The first step done by MII subject was writing down the formula to determine cuboids area of surface. In this step, student has been remembering and has been applying the formula to solve the problem. The same step was done by MII subject in determining pyramid's area of surface, but MII subject didn't analyze that cuboids top and pyramid base area were not be calculated as well since the builds were stacked.

Based from interview result in the first question, MII subject could explain the direction of the problem in a good way, but MII subject coud not analyze the information given and additional informations from the question first to solve the problem. So, MII subject was using the formula to calculate pyramid's volume directly right after reading the question by using triangle hypotenuse as triangle height. From the second question, MII subject could explain all of the forms incleded in the question. MII subject tried to solve the problem using outlines one by one, starting from the surface area of cuboids and pyramid. But, MII subject did not analyze that the cuboids and pyramid were stacked so those two actually didn't need to be calculated. MII subject added their surface area directly inorder to solve the problem.

Based on analysis result of student's answer and interview, it can be obtained that students with moderate intrapersonal intelligence (MII) could achieve up to third level which is applying level based on Bloom's Taxonomy, where student could use his concept in passing new situation and condition in this level [12, 14].

3.3. Low intrapersonal intelligence (LII)

Subject with low intrapersonal intelligence (LII) was able to achieve two levels in Bloom's Taxonomy which are reciting and understanding. The details can be seen in Figure 5 and Figure 6 as well.

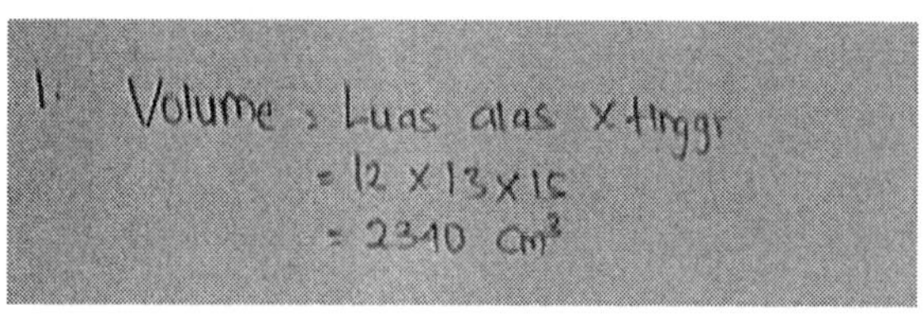

Figure 5. LII subject's answer for number 1

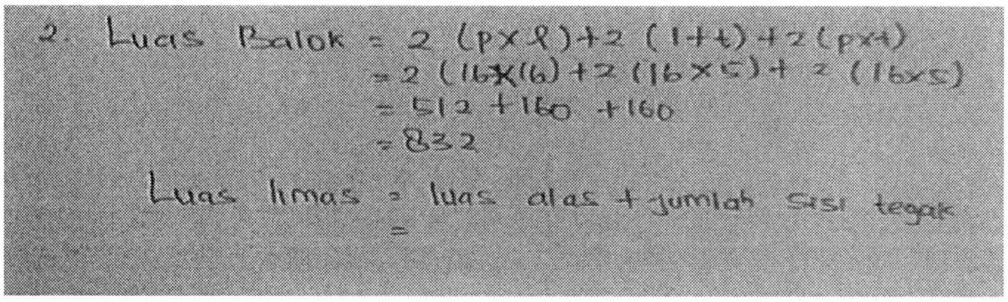

Figure 6. LII subject's answer for number 2

Figure 5 showed that LII subject was not able to describe question scheme. LII subject wrote down the formula of prism's volume directly. In this step, LII subject has already achieved the second level, understanding the question and remembering the formula of prism's volume. The written formula was right, but LII subject didn't analyze the additional information, so that LII subject assumed that the hypotenuse of base triangle as the height of triangle as well.

Figure 6 showed that LII subject has already achieved in understanding the question, which is finding out the area of surface of stacked cuboids and pyramid. The first step done by LII subject was writing down the formula to determine cuboids area of surface. In this step, subject rremembers the formula to solve the problem. LII subject was able to write the right formula, but LII subject didn't analyze that cuboids top was not be calculated as well since the builds were stacked. LII subject wrote down the formula to find pyramid's area of surface, but LII subject didn't continue the steps to find the solution.

Based from interview, LII subject seemed still to be in doubt due to his answer. From the first question, LII subject could write the formula of prism's volume correctly (base multiplied to height of the forms), but LII subject did not understand that difference in prism's volume could change its formula as well. In his worksheet, LII subject multiplied all of the informations given directly without further analysis of prism's shape and additional informations from the question. From the second question, LII subject could mention formula of surface area of cuboids and pyramid, but he could not find out the correct answer of pyramid's surface area. So, LII subject could not solve the problem.

Based on analysis result of student's answer and interview, it can be obtained that students with low intrapersonal intelligence (LII) could achieve up to second level, understanding level based on Bloom's Taxonomy, where student could understand the direction of the problem in this level [4, 5, 12, 14].

4. Conclusion

Based on the aim of research and data analysis toward student's thinking level in solving mathematics problem in Build Flat Forms learning objective using Bloom's Taxonomy, it can be concluded that students with high intrapersonal intelligence could achieve fourth level of Bloom's Taxonomy which is analyzing level. Then, students with moderate intrapersonal intelligence could achieve applying level and students with low intrapersonal intelligence could achieve understanding level.

Since each student has their own thinking level, it can be concluded that student's ability in problem solving is different as well. Teachers are expected to determine student's intrapersonal intelligence due to its importance as one of essential factors of student's achievement in learning mathematics. Furthermore, if teachers are well prepared in designing the whole teaching and learning process and determining each student's intrapersonal intelligence, the achievement in learning mathematics will definitely be optimal.

5. Acknowledgements

The authors thank to Universitas Sebelas Maret and local government in Sukoharjo regency for giving permission for this study

6. References

[1] Young J R 2017 Technology integration in mathematics education: Examining the quality of meta-analytic research *International Journal on Emerging Mathematics Education* **1** 71
[2] Fathurrohman M, Porter A L, and Worthy A L 2017 Teachers' real and perceived of ICT supported-situation for mathematics teaching and learning *International Journal on Emerging Mathematics Education* **1** 11
[3] Sundayana R, Herman T, Dahlan J A and Prahmana R C I 2017 Using ASSURE learning design to develop students' mathematical communication ability *World Transactions on Engineering and Technology Education* **15** 245
[4] Tanujaya B, Prahmana R C I, and Mumu J 2017 Mathematics instruction, problems, challenges, and opportunities: A case study in Manokwari regency, Indonesia *World Transactions on Engineering and Technology Education* **15** 287
[5] Widodo S A, Purnami A S, and Prahmana R C I 2017 Team accelerated instruction, initials and problem-solves ability in junior high school *International Journal on Emerging Mathematics Education* **1** 193
[6] Asshaari I, Othman H, Bahaludin H and Ismail N A 2012 Appraisal on bloom's separation in final examination question of engineering mathematics courses using rasch measurement model *Procedia social and behavioral sciences* **6** 172
[7] Bandura A and Adams N E 1997 Analysis of self efficacy theory of behavioral change *Cognitive theory and research* **1** 287
[8] Pajares F and Miller M D 1994 Role of self-efficacy and self concept beliefs in mathematical problem solving: a path analysis *Journal education of psych.* **86** 193
[9] Kuzle A 2013 Patterns of metacognitive behavior during mathematics problem-solving in a dynamic geometry environment *International electronic journal mathematics education* **8** 20
[10] Bhalerao S 2017 Bloom's taxonomy reiterates pramana *Journal ayurveda and integrative medicine* **8** 56
[11] Bumen N T 2007 Effects of the original versus revised bloom's taxonomy on lesson planning skills: a turkish study among pre-service teachers *Review of education* **53** 439
[12] Omar N, Haris S S, Hassan R, Arshad H, Rahmat M and Zainal N F A 2012 Automated analysis of exam questions according to bloom's taxonomy *Procedia social and behavioral science* **59** 297
[13] Singh D, Tripathi P K and Patwardhan K 2016 What do ayurveda postgraduate entrance examinations actually assess? - results of a five-year period question-paper analysis based on bloom's taxonomy *Journal ayurveda and integrative medicine* **7** 167
[14] Krathwohl D R 2002 A Revision of bloom's taxonomy: An Overview *Theory into practice* **41** 212
[15] Vongkrahchang S 2016 Effects of personal intelligence reading instruction on personal intelligence profiles of thai university students *Kasetsart journal social sciences* **37** 7
[16] Behjat F 2012 Interpersonal and intrapersonal intelligences: Do they really work in foreign-language learning? *Procedia social and behavioral sciences* **32** 351
[17] Rampullo A, Licciardello O and Castiglione C 2015 Intrapersonal factors effects on professional orientation and environmental representations *Procedia social and behavioral Sciences* **205** 422
[18] Lawrence J and Tar U 2014 The use of grounded theory technique as a practical tool for qualitative data collection and analysis *Journal travel & tourism marketing* **31** 417
[19] Prahmana R C I and Kusumah Y S 2016 The hypothetical learning trajectory on research in mathematics education using research-based learning *Pedagogika* **123** 42-54
[20] Parakash E S, Narayan K A and Sethuraman 2010 Student perceptions regarding the usefulness of explicit discussion of structure of the observed learning outcome taxonomy *Adv. Phsyiol. Educ.* **34** 145
[21] Prahmana R C I, Kusumah Y S, and Darhim 2017 Didactic trajectory of research in mathematics education using research-based learning *J. Phys.: Conf. Ser.* **893** 012001

IOP Conf. Series: Journal of Physics: Conf. Series **943** (2017) 012008 doi:10.1088/1742-6596/943/1/012008

Implementation of problem-based learning in geometry lessons

S N S H Ahamad[1], H-C Li[1], M Shahrill[1] and R C I Prahmana[2]

[1] Sultan Hassanal Bolkiah Institute of Education, Universiti Brunei Darussalam, Bandar Seri Begawan, Brunei Darussalam
[2] Universitas Ahmad Dahlan, Jl. Pramuka Kav. 5, Pandeyan, Yogyakarta, Indonesia

E-mail: masitah.shahrill@ubd.edu.bn

Abstract. The aim of this study is twofold. Firstly, it aims to examine the effects of the Problem-Based Learning (PBL) approach on students' performance in the learning of geometry. Secondly, it seeks to gain insights from the students regarding the implementation of PBL in geometry lessons. The participants were 22 students from one Year 10 class in a co-educational secondary school in Brunei Darussalam. A mixed method design was employed with data collected from the pre-, post- and retention tests, and interviews. The findings from this study revealed positive influences on students' performance in learning geometry as gain and retention of knowledge was observed. Meanwhile, mixed responses from the interviews implied that in terms of 1) learning attitudes, students favoured the idea of independent learning but some critiqued that the process of PBL might be time-consuming; 2) learning difficulties, some students struggled in assimilating information leading to poor decision-making; and 3) knowledge and skills, some students believed to have nurtured some skills such as communication and research skills.

1. Introduction

The development of Problem-Based Learning (PBL) started at McMaster University in Canada in the 1960s to facilitate medical students in their learning [1]. PBL is generally described as a student-centred learning environment where students construct their own knowledge and work collaboratively to solve problems, which drive the learning. The teacher facilitates learning with him or her acknowledging the students' efforts, ideas and prior knowledge [2-6]. In PBL, there are expectations for students to be engaged in collaborative activities to solve problems [7]. PBL has been widely adapted in many research studies worldwide and invited discussions among educators, psychologists and researchers. According to Barrows and Tamblyn [1], the fundamental aspect of PBL is based on social constructivism. Since its implementation, several PBL models for student-centred, problem-based, collaborative learning have been developed widely. Considering much empirical evidence showing the advantages of PBL over traditional teacher-centred teaching, there has been adaptation phenomenon of PBL across many levels and subject areas.

1.1. The benefits of PBL

PBL as a student-centred learning strategy has been implemented and developed in many different educational levels [8]. One of which relates to the effectiveness of PBL approach in learning, specifically in the increase of information or knowledge retention [9]. In comparison to traditional teaching methods, PBL allows students to work collaboratively and construct their knowledge through social interaction [10-12]. Group composition is one of the essential features in PBL that may affect learning outcomes and motivation [13]. Small group structure in PBL lesson helps distribute the cognitive tasks among group members [14]. It was also suggested that small group discussions in PBL lessons improves problem solving and higher order thinking, and promote shared knowledge construction [15]. According to Hmelo-Silver [16], critical reflection can provide students with feedback that can be utilised to make improvements during the collaboration process. It was also noted that if the students struggle reflecting on their own learning hence a teacher would be required to facilitate the students in the reflection process.

Previous studies have highlighted the benefits of PBL in developing the students' mathematical knowledge [17-20]. In a recent report by Botty et al. [19], they found improvements on students' performances in learning mathematics based on their pre- and post-test results. This shows that PBL implementation could positively impact the students' progress in the areas relating to real-life and syllabus oriented knowledge. PBL could also promote the learners' gain in the content of the subject and in solving real-life problems [3]. Furthermore, a study on PBL approach in developing Taiwanese students' understanding of fractions also observed improvements in students' learning performance, which implies that the PBL intervention may be effective in developing students' understanding of fractions [21]. In learning using PBL, students were more engaged in learning and had positive attitude towards their learning [22]. With these PBL strengths, it has been used in multiple domains that include mathematics education [23]. Schettino [4] conducted a study based on her strong belief in PBL curriculum and observed higher learning enthusiasm as students were not restricted to create their own solutions to a problem, and also through collaborative works, they get to exchange ideas and feedback which further improves the quality of learning. Furthermore, the use of PBL could be more effective with lower ability students [24].

1.2. The learning of geometry

In geometry lesson, the teachers tend to explain to students the properties associated with geometrical shapes and the properties, and subsequently, requiring the students to attempt the given exercises to indicate whether they have understood the topic or simply responding from memorising the facts [25]. There were few made attempts in encouraging the students to explain their reasoning and to make connections logically. Developing the students' thinking and geometrical senses is important and they will require this by possessing a solid understanding of the facts in geometry. This problem can be tackled by varying teaching approaches such as PBL that can support students in making the necessary connections using the various depictions of geometrical ideas and the interdisciplinary concepts in other area of mathematics [4, 26]. This is to ensure that students gain a full understanding of the concepts and the processes within and not by just learning the rules. This will enable students to retain knowledge and skills and provide confidence when solving new mathematical problems.

2. Methodology

This present study utilises an action research framework involving mixed method approaches of data gathering. An action research approach, using prominent stages such as problem identification, action planning, implementation, evaluation, and reflection, is well suited to the need of the study as it enables a systematic collection of data that provides the basis for social change [27] to seek improvement on students' learning. Additionally, all the data collected were triangulated [28] to provide a better idea on the effects of PBL on students' performance in learning geometry and their perceptions towards the PBL approach.

The aim of this study is twofold; firstly, it aims to examine the effects of PBL approach on students' performance in the learning of geometry, and secondly, it seeks to gain insights from the students regarding the implementation of PBL in geometry lessons. This study was conducted in a co-educational secondary school situated in Bandar Seri Begawan, the capital city of Brunei Darussalam.

The sample consisted of 22 students from a Year 10 class. According to Mason [29], a sample size of 20 to 30 is feasible in studies involving action research. The sampled students were collected by means of random sampling. All the relevant permissions from the school, parents and students were appropriately obtained before the start of the study.

2.1. Design of intervention

The topic of the intervention lessons (PBL lessons) was on symmetrical properties of circles, which was included in the Year 10 geometry syllabus in Brunei. The PBL lessons were conducted over a period of two weeks, and met three times a week. The first author, who implemented all the PBL lessons during the course of the intervention, also designed the PBL lessons. The PBL model by Lee and Bae [30] was adapted to design the PBL lessons (see Figure 1).

Lesson 1: Introduction, Understand the Problem, and Searching for Information (on-going)

→ Students are introduced to understand and analyse the given problem.

→ Students make inquires and perform searches individually or collaboratively to gain understanding of the problem.

→ Facilitator monitors students' progress and make sure that they are on the right track.

Lesson 2: Construct and Gather Solution

→ Students gather all the necessary information, and discuss to produce a draft of their possible solutions to solve the given problem.

Lesson 3: Presentation and Reflection

→ Students prepare for a 5-7 minutes presentation to share their findings in front of the whole class.

→ Facilitator checks that students have completed their work and asks students additional questions upon closure of PBL session.

Figure 1. PBL lesson intervention design.

How to Fix a Broken Piece Ceramic Plate?

During a geography field trip, Sarah found a broken piece of an antique ceramic plate and she strongly believes that the plate is in a circular shape. In order to make a replica of the entire plate for her geography project, Sarah knows that she needs to find the radius of the broken plate but she has no idea how to find it. Can you and your team help Sarah identify the step-by-step ways on how to find the radius of the broken plate?

Your task is to come up with convincing strategies (step-by-step ways) on how to find the radius of the broken plate.

As a reminder, you are not required to find the exact value (length of the radius) for this task. You are expected to do a presentation to discuss your findings with justifications.

Figure 2. The PBL problem.

The PBL intervention allowed students to work in groups of 4 or 5 students in solving the presented PBL problem [14]. There were 5 groups of students in this study and each group had to nominate a leader, a writer and a presenter to ensure that the tasks were evenly distributed among the group members. A PBL Activity Booklet was given to each group. This booklet consisted of a cover page, the PBL model [30], a designed problem, a PBL Facts List and blank A4 papers to record the findings. The PBL problem (see Figure 2) given during the lessons presented students with real-life situation and had typically not been covered in class. Specifically, the PBL problem aims to allow students to solve it in several manners. As recommended by Li [31], the content was carefully constructed as problem. Furthermore, the PBL problem was designed based on three important aspects of a good PBL problem, which are ill structured, contextualised and related to real-life situations [32].

The students were encouraged to perform active interactions within their group members where students explained their reasoning when solving the problem. Wandering facilitation model was performed as a facilitation strategy according to Hmelo-Silver [16] to successfully facilitate multiple groups. Memo pads were also given to each group so that students could record all their queries to be addressed later during facilitation process. Once the students understood the context of the problem, they were asked to identify the statement of the problem such as, what information is given, and what information is needed. These were then recorded in the adapted PBL Facts List. The searching for information phase took a full 30-minute lesson in which students searched for relevant resources in finding the solutions. They were also to access and assess multiple resources such as textbook and notebook. This was carried out as an on-going process where students could perform this task at home after school. In the next lesson, the students worked collaboratively to analyse all the gained information and produce a draft on their approach and methods in solving the problem. At this stage, the students also evaluated their group members' findings and decided the best possible approach and methods to solve the given problem. The presentation was conducted when all group members had completed their proposal of findings. A presenter was appointed from each group to present their group's proposal for about 5 to 7 minutes. A reflection sheet was also given to each student to evaluate his or her peer's overall presentation. A whole-class discussion was conducted after each presentation to allow interactions in giving constructive feedback followed by a wrapping up session [19]. The entire PBL lessons were video-recorded for analysis and to examine the process of PBL implementation in the classroom.

2.2. Data collection instruments and analysis

2.2.1. Pre-test, post-test and retention test. The purpose of the pre- and post-tests were to examine whether there were any differences observed in the students' academic achievement after the conduct of the PBL lessons. The post-test was given to the students about a week after the pre-test was administered. Following a gap of about more than four weeks, the same students were again given the retention test questions. The main aim of the retention test was to examine how much knowledge from the intervention lessons the students have retained. There was 7 items in all the three test instruments. The test items consisted of questions that examined the students' prior-knowledge on basic geometrical terms of a circle, their geometrical representation skills and procedural skills in finding unknown angles and sides in a circle. Scores were allocated for each question that weighted 11 marks in total, and the time given was 25 minutes to complete the test.

To validate the instruments, two experienced mathematics teachers with more than eight years of teaching experiences were invited to review the test questions and face validation was appropriately done. A pilot study was also conducted which involved a different set of students at the same school. This was done to approximate the length of time required to answering the tests and the appropriateness of the questions with the students' ability. The reliability of the test items was measured using Cronbach Alpha and it yielded a correlation coefficient of 0.628, which suggested acceptable reliability [33]. After the pilot study, the test questions were revised and improved in order to fine-tune the test instrument.

The number of pre-tests collected was 21 out of 22 students; hence one data was missing due to an absentee. However, the number of post-tests collected was 22 with no missing data. In order to do comparison, the post-test from that particular student who was absent during the pre-test had to be discarded. Hence the number of each pre-tests and post-tests was left with 21. As for the retention test, there were only 15 students who took the test. To examine knowledge retained, scores from the retention test were compared with the scores from the post-test. Therefore, the post-tests from 15 students who were present during the retention test were only used for comparison and analysis. For each correct answer, 1 mark was awarded and for each wrong answer or incomplete answer, it was awarded 0 mark. However, there was a case where the student got the correct answer but the working was incorrect. This was given 0 mark because the student portrayed a misconception. The difference scores between the pre- and post-tests, and the differences between the post- and retention tests were calculated using the SPSS statistical software.

2.2.2. Semi-structured interviews. Generally, the questions in semi-structured interviews were aimed to gain students' perspectives that analysed their understanding, learning attitudes, and learning difficulties [34]. There were three individual interviews and three group interviews with 5, 5 and 2 students in each group. Each interview lasted for about less than 10 minutes. All the interviews were conducted with confidentiality and appropriate interview etiquettes. In reporting the results, pseudonyms were used to replace the students' names.

All the interviews were audio-recorded and transcribed for analysis. The analysis of the video recorded lessons was conducted using the seven steps process [35] to support and gain a sense of the students' perceptions on PBL experience obtained during the interviews. The video was viewed and some prominent excerpts of interactions during the lessons were transcribed. The students' sampled work from the PBL facts list were also examined to understand the extent of group efforts and participation and how well they organised their work in a problem-solving task.

3. Results

3.1. The effects of the PBL approach on students' performance in the learning of geometry

Assumptions of independence and normality were examined using SPSS statistical software [36] as illustrated in Table 1. It shows that the data was not normal ($p<0.05$) hence the assumptions of normality was violated. Thus, non-parametric tests were utilised using the Wilcoxon Signed Rank Test [37] to examine whether there were statistically significant differences in the mean scores between the pre- and post-tests, and between the post- and retention tests.

Table 1. Test of normality for the pre- and post-test scores.

	Shapiro-Wilk		
	Statistic	df	Sig (p)
Difference	.719	21	.000

Following the comparisons shown in Table 2, there was a slight increase in the average mean scores of the students in the post-test. Similarly, an increase was observed during the retention test.

Table 2. Mean scores for the pre-, post- and retention tests.

	Pre-test (N=21)	Post-test (N=21)	Retention test (N=15)
Average mean scores	8.81	9.19	10.6

Further analysis using Wilcoxon Signed Ranks Test was done to examine the significant difference of mean scores between the pre- and post-tests since assumption of normality was violated. Table 3 shows that after the intervention, four students performed worse while five students showed improvements, and 12 students did not show any changes in academic performance. The analysis did not confirm any significant differences between the pre- and post-tests.

Table 3. Analysis using Wilcoxon Signed Ranks test (N=21).

	Negative ranks	Positive ranks	Ties	Sig. (p)
Post-test – Pre-test	4	5	12	0.298

Results from the Wilcoxon test (see Table 4) shows that during the retention test, none of the students have performed badly. While seven students performed better and eight maintained their scores. Moreover, the test revealed a statistically significant difference in the mean scores between the post-test and retention test ($z = 2.43$, $p = 0.015$), with a medium effect size ($r = .44$). According to Pallant [36], for Wilcoxon test, the most appropriate method to calculate the effect size is by dividing the critical-z value by the square root of N, such that N is the number of samples over the two time points $r = \frac{z}{\sqrt{N}}$.

Table 4. Analysis using Wilcoxon Signed Ranks test (N=15)

	Negative Ranks	Positive Ranks	Ties	Sig. (p)
Retention test – Post-test	0	7	8	0.015

Among the seven questions in the test instrument, Questions 1 and 2(a) examined the students' ability to identify the line of symmetry and the key terms such as diameter, radius, chord and segment (major and minor), Questions 2(b) and 5(a) were concerned about the students' ability to recognise the geometric properties of circles, and Questions 3, 4 and 5(b) were associated with the students' geometrical procedural skills in calculating unknown sides and angles. The percentages of correct responses in each question were compared as illustrated in Table 5.

Table 5. Comparisons of percentages of correct responses in the pre- and post-tests.

Areas	Questions	Percentage of correct responses (%)	
		Pre-test	Post-test
Prior knowledge	1 and 2(a)	77.4	78.6
Geometrical representations	2(b) and 5(a)	85.7	95.3
Procedural skills	3, 4 and 5(b)	59.5	67.5

It is worth noting that all questions observed an increase in correct percentage response: 95.3% of the students correctly responded to Questions 2(b) and 5(a) in the post-test, a 9.6% increase when compared with the results in the pre-test. This indicated that the students had improved in their geometrical representation ability. Examining the students' answers for Question 2(b), during the pre-test, majority of the students did not provide any answers and for those who provided answers, the responses were incorrect. However, there were more students with correct answers in the post-test for Question 2(b). The increase in the mean percentage of Question 2(b) appeared to be highly associated with students' ability to recognise the geometrical representation particularly on the first symmetrical property of a circle which is equal chords are equidistant from the centre. Figure 3(a) shows a sample of a student's work on Question 2(b) who incorrectly solved this question in the pre-test. Initially, she thought that the length of CN was equal to the length of AB, which is 8 cm. However, in the post-test she successfully realised that the length of CN was actually halves the length of AB as shown on the correct labeling in Figure 3(b).

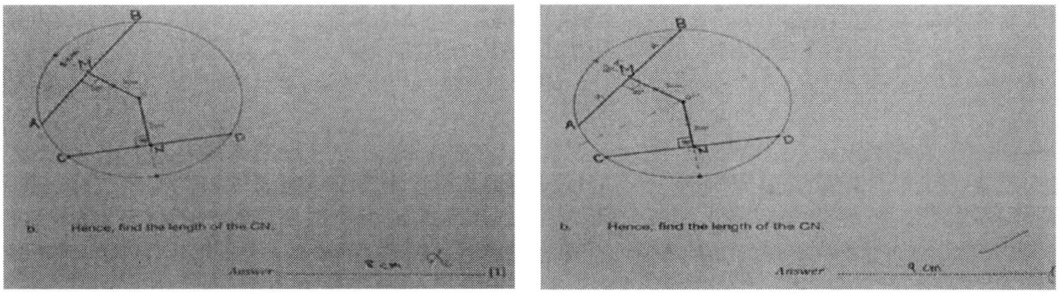

(a) (b)

Figure 3. Samples of a student's work for Question 2(b).

In this present study, the intervention had a positive impact on student's performance in terms of acquisition of knowledge and knowledge retention in learning geometry. A significant improvement in students' achievement was seen during the retention test, which implies that PBL intervention could have a positive influence on students' performance particularly in developing long-term knowledge retention in learning geometry. This finding supports a review reported by Norman and Schmidt [38] as they noted that an in-depth process of PBL information seems to promote greater retention knowledge if done over a substantial period. Similarly, Robbs and Meredith [9] associated effectiveness of PBL with greater retention of information.

The students collaboratively worked in small groups of four or five, thus allowing for a fair distribution of task and participation during the activity. The students shared facts and thoughts through discussions and responses to questions. Based on their conversations during the discussion, it was seen that sometimes they repeatedly discuss similar ideas. The following excerpts exemplify this example, where the students discussed about diameter and the shape of circle repetitively. This subsequent recall strengthens understanding of the concept, making it easier to remember.

[Start of conversation]

Fay: *That's a sector... How do you find a diameter from a sector?*
Des: *What's the point of finding the missing piece?*
Aldo: *It does not even look like a circle [laughs]*
Des: *A quar...a quarter of a circle.*
Fay: *It's a plate! Not a quarter of circle.*
Des: *Oh! A quarter of a plate.*

[Middle of conversation]

Aldo: *What do we need to know?*
Kayla: *The diameter is... from one end to one end right?*
Des: *Yes.*
Aldo: *Diameter... yeah.*
Kayla: *So for me I think we need to find the diameter.*
Aldo: *Why do keep saying...*

3.2. Students' insights on the use of PBL in geometry lessons

During the interviews, students shared their personal insights on PBL experience in terms of learning attitudes, learning difficulties, and knowledge and skills. Data from video-recordings of classroom observations and the teacher's notes were also used to support the findings obtained from the interviews with the students.

3.2.1. Learning attitudes. The students were asked if they preferred the PBL method or the traditional method in general in order to seek their preferences of the two learning styles. From the students' interviews, there were mixed responses observed. Some students favoured PBL generally because they believed that PBL would provide them with some valuable skills and also greater level of knowledge acquisition. For instance, as shown in the following excerpt, Starla considered that the PBL method was supposedly a relevant learning approach that she should practice because it would gave her the opportunity to take control of her own learning. She felt that it would be necessary gain such skills that will help her in future learning.

Well...maybe for this time, I think the...method we do just now is supposed to be the method we should do because its more...us doing the work, so we know what to do, because in the other time not everyone will tell us what to do, we need to be independent on finding our own information. So starting from now I guess... it is better so that we know how we will do it for the other time... so yeah. (Starla)

Similarly, Samsun claimed that he learned better in PBL mainly because he gained better understanding of the content knowledge as he searched the information by himself. He also described that in the traditional learning, he mostly listened to the teacher's explanation and sometimes he merely understood the knowledge if he was not fully attentive to the teacher. The following excerpt illustrates his perception of PBL.

Umm...I think this method (PBL) is better because... like from the traditional method you just listen and then sometimes you are like... when you don't give a 100% attention... you just like... not really understand. But when you do it yourself, you understand more because you are the one that searching for the information... so its not spoon-fed or anything. (Samsun)

Based on the findings, the students felt that PBL allowed them to be more independent in their learning. For Starla, PBL method would give her the opportunity to be more independent in her learning, and she also believed that the skills to search for information effectively would foster a greater learning competency in her future learning experiences. Relatively, Samsun felt more attentive in PBL because it has given him a deeper understanding as he searched and learnt the information by himself. These findings support the result by Huang [39] that reported 70.6% of the students had positive perceptions by learning independently in PBL.

In contrast, some students preferred the traditional approach to PBL because of two aspects, direct instructions and shorter time duration. For example, Wani and Gaby claimed that it was much easier for them to understand the concept (refer to the following excerpts).

Wani: *Traditional... it's umm... a more easier to understand and you can solve the problems... like different situations... [one concept]*

Gaby: *We understand more in traditional... yeah.*

Laila: *Well... the... the old one, because this one may take too long.*

This may imply that students gained better understanding if the teacher directly provided the answers or explanations. Having said that, this study did not obtain further explanation from the students as evidence to support the assumption. This may neither ensure a much deeper understanding of the knowledge content nor the development of the students' thought. It is vital for the students to acquire in-depth understanding in the subject content so as to master the necessary skills to apply knowledge to other situations in real-life context. Here, it is the responsibility of the teacher who should try to provide authentic and active learning environment that enhance deeper understanding of content knowledge and consequently develop other valuable lifelong learning skills [40].

In terms of implementation, as shown above, Laila argued that too much time might be needed for PBL as compared to the traditional approach. As stated by Mansor et al. [41], large amount of time spent in PBL may be observed particularly during teacher supervision and giving advices to groups. In this present study however, based on the teacher's notes and classroom observations, wandering facilitation strategy [16] was performed, and this enabled the teacher to facilitate the students effectively. Hence, it may be inferred that there might be other reasons that contributed to the student's concerns in terms of time allocation in PBL. Perhaps this warrants further investigation in the future.

3.2.2. Learning difficulties. From the interviews, it seemed that some students encountered difficulties during the PBL process. Some of the issues given by students from the interviews were, difficulty in processing information and struggling to generate ideas at the beginning of the activity. For instance, Jenny and Laila who were both in the same group claimed that there was a lot of information as each group members shared different ideas. So they struggled to process all the information that was acquired during the discussions in order to select the relevant and accurate information. Jenny also claimed that this created some disagreements among the group members. The following excerpts illustrate these examples:

There is one problem... during the activity... in the group work... there is five of us in the group... and each of us have different ideas of how to explain... the activity at that time. There is so much information... so we don't know which one is correct... so we always arguing about it. (Jenny)

And it was a bit hard during the discussion... because everybody has their own... ideas... so it's hard to choose which one. (Laila)

Referring to the students' conversations during the group discussions prompts such as "why" or "can you explain further" were barely observed during the conversations. The students were mostly extracting their prior knowledge and ideas without providing any profound reasons. The following excerpts exemplify this example.

Nancy: *If you want to find it... We need to find the "thing" and then we times it by four. [Silent pause]*
Fez: *That's nice!*
Kayla: *For me for the next question... what do we need to know... that ... [pointing to the diagram]*

This may imply that most students showed poor process skills such as reasoning skills, critical thinking skills, and argument skills that could be one of the causes of the students' difficulties. Moreover, referring to the previous excerpts, Fez purely accepted Nancy's suggestion based on its logic nature without a deep notion of the idea. This shows that the students lack the ability to make judgment (evaluation), which could also be the reason why the students argued during the discussion. Nevertheless, argument skills can be used for learning when we treat reasoning as a process of argumentation [42]. Moreover, a study by Nussbaum [43] reported that argumentation might promote the retention of course content, which supported the quantitative finding in this present study. Importantly, critical thinking and argument skills are the significant elements in decision-making [44]. Thus, poor thinking skills and argument skills may lead to bad decision-making. Therefore, it is necessary for the teacher to put adequate emphasis on the students in developing and practicing relevant process skills in order to maximise the students' learning. Then again, Zabit [45] argued that it is rather challenging to develop the students' such skills especially when the teachers may not have developed the skills fully themselves. Consequently, it is also important that teachers should also constantly try to develop these important skills to be able to give full support in the student's learning.

The second learning difficulty was that some students struggled to generate ideas at the beginning of the activity. As shown in the excerpt below, Samsun noted that he did not have any ideas on how to start solving the problem. From the teacher' notes, at first, the student was seen to be timid.

Samsun: *Well... umm... at the start of the... the... problem, we were clueless on how we are going to start. Other than that... umm...we have no problem during the discussion... a lot of ideas but it was okay.*

It is also worth noting that the students had no previous experience with PBL. So, it seems that the student needed more guidance and greater facilitation from the teacher even though the full implementation of the PBL lessons by the teacher was clearly conducted. Initially, the teacher let them discuss among the group to understand the problem first before she assisted the students. This could be the cause for the difficulty. Moreover, the teacher also noted that initially, none of the students were observed using their notes or textbooks to search for information. Thus, this could be another cause for the difficulty. The teacher managed to assist the students by asking open-ended questions such as "what do you understand about the problem?" and she also used probing questions to encourage students' active participation in discussion. Furthermore, she encouraged her students in searching for answers with the use of textbooks and notes. In doing so, the teacher observed gradual group collaboration. The classroom environment steadily changed as students started talking about the given problem with each other among their respective groups. The teacher also noted that the students were seen making some progress as they managed to record some of their thoughts on the columns provided in the PBL Fact List. Appropriate inquiry techniques are crucial in an inquiry-based classroom especially at the early stages [46]. Expertise in facilitation skills is essential in attaining effective PBL learning. Hence, proper training workshops may be required to develop a clear understanding of effective facilitation strategies such as role-playing, in order to facilitate a better PBL experience.

3.2.3. Knowledge and skills. Evidence from the presentations suggests the students' ability in explaining the procedures to solve the problem correctly using the appropriate strategies. However, it was less likely they gave explicit justifications and reasoning to their procedures. The following excerpts reveal these examples.

To find the centre of the circle... we used the two chords to find the perpendicular bisector... the perpendicular bisectors must pass through the centre which gives us the centre of the circle... and the radius must be the same. (Group 3)

First of all just draw a long chord and then find... like you did in maths... find the arcs like this... and then... and then if you check it will be 90 degrees because it is a perpendicular bisector... and draw the lines and that will make a triangle. After that draw another arc on both lines and just joint them up. And all the perpendicular bisectors will touch the centre... they all intersect at a point which indicates the centre of the circle. (Group 4)

It can be seen that most students were able to explain the steps to find the radius of the circle. For instance, they were able to note that they first needed to find the centre of the circle in order to find the radius. This was followed by an explanation on the ways to find the centre of the circle that eventually derived the radius of a circle. In addition, Group 4 also portrayed some understanding on the concept of perpendicular bisector as they justified that the angle between the chord and perpendicular bisector equals 90 degrees. Hence, it was observed that the students gained procedural skills. It was also observed that fewer students validated their strategy, as they did not explain which symmetrical properties were being applied to find the centre of a circle. For instance, Group 5 did not give any reasons for using two chords to find the centre, as shown in the following excerpt.

To find the radius we need to find the centre of a circle. To find the centre of a circle, we need to find the perpendicular bisector of chord. We have to use two chords... so we just choose anywhere from the broken plate which is like half of them... and then after we draw the chord, we use a compass to find the arc. The arc is to find the perpendicular bisector of chords. So this line here... from the centre to this line is the radius. (Group 5)

In this case, the students were expected to mention that both property 1 and property 2, which are equal chords, are equidistance from the centre and the perpendicular bisector of a chord passes through the centre respectively, could be applied to find the centre of a circle, but they failed to do so. This may imply that the students did not seem to notice that some of the procedures are the symmetrical properties of a circle. The teacher also observed that initially most groups only constructed a single chord with its perpendicular bisector. At this point, they got stuck and clearly had no idea how to proceed to find the centre of a circle. The teacher encouraged the students to search for more information. However, the students still did not manage to relate both property 1 and property 2 to apply into the problem. Thus, the teacher had to prompt the students by questioning them "what do you think will happen if you use more than one chords?" in a way to assist the students to perform trial and error and make comparisons. The students were then seen to construct the perpendicular bisector of another chord and they eventually noticed that the intersection point of two perpendicular bisectors was indeed the centre of a circle. Apparently, further questionings from the teacher were not facilitated. Thus this could be a possible for the insufficient understanding of the concepts. However, no further evidence was observed associating the students' lack of conceptual understanding.

Conceptual understanding is equally as essential as procedural understanding to allow students to develop proficiency in mathematics. This includes strong understanding of mathematical concepts, operations and relations [47]. According to the National Research Council [48], students who own conceptual understanding in mathematics are less likely to make critical errors particularly during problem solving. Teachers must give the students opportunities to communicate their conceptual understanding [47]. Therefore, it is necessary for the teacher to fully acquire the student' thoughts. This may involve using several strategic questionings or promptings [49] to avoid giving students direct instructions to proceed. It is also equally important to guide students to reason critically and also to assess their knowledge.

In relation to skills, from the interviews with the students, they all stated that they have nurtured some learning skills after the PBL intervention. The most mentioned learning skill was communication skills. Some students also mentioned other skills, for instance leadership skills, critical-thinking skills and information skills, which they have developed throughout the PBL intervention. For example, Laila mentioned that she has improved her research skills and communication skills. However, she also emphasised that during collaborative discussion, sometimes their conversation was off topic and the group tend to lose focus. The following excerpts illustrates this example:

Well... searching... for information and for communication skills... well... the problem about our group is they talked a lot not about the thing... not about the question and they get distracted... too distracted. (Laila)

This may imply that students have developed some communication skills through the PBL experience. In terms of communication skills, good problems in PBL foster the development of communication skills when the students presented their findings to their peers. Based on the teacher's notes, during the presentation stage, the students were able to deliver various findings confidently in front of the audiences, listen to their peers effectively and most importantly they were able to gain audiences' attention by making eye contact. The students were very fluent in their command of the English language during the delivery of their presentations. These reflect the students' good communication competency. However, the only limitation in doing presentation is that in general, the students who volunteered to be the presenter are among the extrovert students. Hence, communication competency may not apply to the quiet students. Moreover, during the presentation, the students were observed using only the whiteboard to present their findings. The teacher should have encouraged her students to perform their presentations using technological tools found on Internet or even other presentation software [50].

Analysing the case for Laila, the intervention from the teacher may be necessary. Hence, teachers should guide the students to be better communicators that will help them to be more productive and effective in communicating [51]. Furthermore, teachers should also consider giving students feedback on their communication skills during the presentations in addition to giving feedback on the content. Apart from communication skills, some students also believed to have developed research skills as they search for the information on their own [52, 53]. During the PBL lessons, the students used different resources in searching for information. Out of five groups, only Group 4 utilised the use of Internet to search for information and others, such as Group 2 only used the textbook and notebook. It was evident that Group 4 utilising the use of Internet were able to gain broader information. They also proposed several solutions that were related to real-life applications. This is in contrast to Group 2, which displayed solutions that were mostly subject-based. Thus, this may imply that the use of Internet [54, 55] may enhance the integration of knowledge.

4. Conclusion

This study reported evidence on how students' learning and their insights on PBL were positively influenced with regards to the implementation of PBL in the context of learning geometry. The results from Wilcoxon test show a significant increase in students' mean scores during the retention test. From the interviews, the students noted some challenges during learning process, which consisted of difficulties in decision-making and in generating ideas at the start of problem solving. The mixed responses from the interviews implied that in terms of learning attitudes, students favoured the idea of learning on their own but some critiqued that the process of PBL might be time-consuming. In terms of learning difficulties, some students struggled in assimilating information, which leads to poor decision-making. Nevertheless, the PBL was well perceived by students mainly due to the experience of independent learning and the development of communication and research skills, which was observed during the learning process leading up to the presentations of the group work.

Some limitations of this study were observed in the intervention lessons. Firstly, there was a lack of proper introduction at the beginning of the study. Due to time constraints and lack of technical support, the students were instructed to watch two educational videos on PBL at home prior to first lesson. These two videos provide students an introduction to PBL. However, only few students watched the videos, as little response from students was observed during questionings in the first lesson. Therefore regardless of any limiting factors, a proper introduction in class should be implemented to ensure better understanding of PBL practice in classroom [56]. Secondly, the post-test results did not show significant improvements directly after the intervention lessons because the test was administered to the students only about a week after they were given the pre-test, and due to time constraints, only one cycle of the action research was conducted. In relation to time factors, it is recommended that for future research should implement PBL in a longer time period. This will allow teachers and students to familiarise and adapt with the PBL environment effectively. Besides, more support is necessary to help teacher to handle any challenges and limitations to improve teachers' knowledge and ability to implement a successful PBL lesson [6]. The implementation of PBL in the context of teaching and learning of geometry is still relatively sparse in the Brunei secondary schools. There is limited amount of resources on PBL problems especially in geometry topics and designing one was rather difficult for the teachers. This present study only offered one PBL problem for the students to work on during the activity, which may reflect inadequate gain of content knowledge. Hence, further support from other on-going research studies is needed in guiding mathematics teachers to design relevant PBL problems particularly in geometry topics.

References

[1] Barrows H S and Tamblyn R S 1980 *Problem-Based Learning: An Approach to Medical Education*. (New York: Springer)

[2] Savery J R 2006 Overview of problem-based learning: Definitions and distinctions *The Interdisciplinary Journal of Problem-Based Learning* **1** 9–20

[3] Burris S and Garton B L 2007 Effect of instructional strategy on critical thinking and content knowledge: Using Problem-Based Learning in the Secondary Classroom *Journal of Agricultural Education* **48** 106–16

[4] Schettino C 2012 Teaching geometry through problem-based learning *Mathematics Teacher* **105** 346–51

[5] Khalid M and Rahman H 2014 Problem-based learning to enhance students' understanding and motivation for learning mechanical science *IIUM Journal of Educational Studies* **2** 56–66

[6] Caesar M I M, Jawawi R, Matzin R, Shahrill M, Jaidin J H and Mundia L 2016 The benefits of adopting a problem-based learning approach on students' learning developments in secondary geography lessons *International Education Studies* **9** 51–65

[7] Wang H and Posey L 2011 An inquiry-based linear algebra class *US-China Education Review* **B4** 489–94

[8] Torp L and Sage S 2002 *Problems as Possibilities: Problem-Based Learning for K-16 Education* (2nd Ed) (Alexandria: Association of Supervision and Curriculum Development)

[9] Robbs J and Meredith S 1994 The problem-based learning curriculum at Southern Illinois University School of Medicine [online]

[10] Polanco R, Calderón P and Delgado F 2004 Effects of a problem-based learning program on engineering students' academic achievements in a Mexican university *Innovations in Education and Teaching International* **41** 145–55

[11] Goodnough K 2006 Enhancing pedagogical content knowledge through self-study: An exploration of problem-based learning *Teaching in Higher Education* **11** 301–18

[12] Sungur S, Tekkaya C and Geban Ö 2006 Improving achievement through problem-based learning *Journal of Biological Education* **40** 155–60

[13] Schmidt H G abd Moust J H 2000 Factors affecting small-group tutorial learning: A review of research *Problem-Based Learning: A Research Perspective on Learning Interactions* eds D Evensen and C E Hmelo (New Jersey: Erlbaum) pp 19–52

[14] Heller P and Hollabaugh M 1992 Teaching problem solving through cooperative grouping Part 2: Designing problems and structuring groups *American Journal of Physics* **60** 637–44

[15] Vye N J, Goldman S R, Voss J F, Hmelo C and Williams S 1997 Complex mathematical problem solving by individuals and dyads *Cognition and Instruction* **15** 435–84

[16] Hmelo-Silver C E 2000 Knowledge recycling: Crisscrossing the landscape of educational psychology in a problem-based learning course for preservice teachers *Journal on Excellence in College Teaching* **11** 41–56

[17] Abdullah N I, Tarmizi R A and Abu R 2010 The effects of problem based learning on mathematics performance and affective attributes in learning statistics at form four secondary level *Procedia-Social and Behavioral Sciences* **8** 370–76
[18] Botty H M R H and Shahrill M 2015 Narrating a teacher's use of structured problem-based learning in a mathematics lesson *Asian Journal of Social Sciences & Humanities* **4** 156–64
[19] Botty H M R H, Shahrill M, Jaidin J H, Li H-C and Chong M S F 2016 The implementation of problem-based learning (PBL) in a year 9 mathematics classroom: A study in Brunei Darussalam *International Research in Education* **4** 34–47
[20] Han S H, Shahrill M, Tan A, Tengah K A, Jaidin J H and Jawawi R 2016 Administering problem-based learning (PBL) approach in the teaching of college-level mathematics *Turkish Online Journal of Educational Technology* November Special Issue 197–07
[21] Li H-C 2011 The development of Taiwanese students' understanding of fractions: A problem-based learning approach *Proceedings of the British Society for Research into Learning Mathematics* **31** 25–30
[22] Walker J T and Lofton S P 2003 Effect of a problem based learning curriculum on students' perceptions of self directed learning *Issues in Educational Research* **13** 71–00
[23] Teoh B T, Preechaporn W and Leong C K 2010 Problem-based learning the 4 core areas (PBL4C) in search of excellence in mathematics education *Proceedings of the 5th East Asia Regional Conference on Mathematics Education Tokyo Japan*
[24] Ridlon C L 2009 Learning mathematics via a problem-centered approach: A two-year study *Mathematical Thinking and Learning* **11** 188–25
[25] Jones K 2002 Issues in the teaching and learning of geometry *Aspects of Teaching Secondary Mathematics: Perspectives on Practice* ed L Haggarty (London: Routledge) pp 121–39
[26] Schettino C 2003 Transition to a problem-solving curriculum *Mathematics Teacher* **96** 534–37
[27] Bogdan R C and Biklen S K 1992 *Qualitative Research: An Introduction to Theory and Methods* (Needham Height: Allyn & Bacon)
[28] Creswell J W 2003 *Research Design: Qualitative, Quantitative, and Mixed Methods Approaches* (Los Angeles: SAGE Publications)
[29] Mason M 2010 Sample size and saturation in PhD studies using qualitative interviews *Forum Qualitative Sozialforschung/Forum: Qualitative Social Research* **11** Art 8
[30] Lee H and Bae S 2007 Issues in implementing a structured problem-based learning strategy in a volcano unit: A case study *International Journal of Science Mathematics Education* **6** 655–76
[31] Li H-C 2012 Implementing problem-based learning in a Taiwanese elementary classroom: A case study of challenges and strategies *Research in Mathematics Education* **14** 89–90
[32] Goodnough K and Hung W 2008 Designing effective problems: Evaluation of 3C3R 9-step design process *Interdisciplinary Journal of Problem-based Learning* **2** 61–90
[33] Loewenthal K M 2001 *An Introduction to Psychological Tests and Scales* (Hove: Psychology Press)
[34] van den Hurk M M, Wolfhagen H A P, Dolmans D H J M and van der Vleuten C P M 1998 The relation between time spent on individual study and academic achievement in a problem-based curriculum *Advances in Health Sciences Education* **3** 43–49
[35] Rubin H J and Rubin I S 2011 *Qualitative Interviewing: The Art of Hearing Data* (Thousand Oaks: Sage)
[36] Pallant J F 2007 *SPSS Survival Manual: A Step-by-Step Guide to Data Analysis with SPSS* (New York: McGraw Hill)
[37] Leech N L, Barrett K C and Morgan G A 2012 *IBM SPSS for Intermediate Statistics: Use and Interpretation* (New York: Routledge)
[38] Norman G R and Schmidt H G 1992 The psychological basis of problem-based learning: A review of the evidence *Academic Medicine* **67** 557-65
[39] Huang R 2005 Chinese international students' perceptions of the problem-based learning

experience *Journal of Hospitality, Leisure, Sport and Tourism Education* **4** 36–43
[40] Hmelo-Silver C E 2004 Problem-based learning: What and how do students learn? *Educational Psychology Review* **16** 235–66
[41] Mansor A N, Abdullah N O, Wahab J A, Rasul M S, Nor M Y M, Nor N M and Raof R A 2015 Managing problem-based learning: Challenges and solutions for educational practice *Asian Social Science* **11** 259–68
[42] Kuhn D 1992 Thinking as argument *Harvard Educational Review* **62** 155–79
[43] Nussbaum E M 2008 Collaborative discourse, argumentation, and learning: Preface and literature review *Contemporary Educational Psychology* **33** 345–59
[44] Klaczynski P A 2004 A dual-process model of adolescent development: Implications for decision making, reasoning, and identity *Advances in Child Development and Behavior* **32** 73–23
[45] Zabit M N M 2010 Problem-based learning on students' critical thinking skills in teaching Business education in Malaysia: A literature review *American Journal of Business Education* **3** 19–32
[46] Shariff A 2009 Reflections on a first time experience of problem-based learning *New Directions in the Teaching of Physical Sciences* **5** 48–54
[47] National Council of Teachers of Mathematics 2000 *Principles and Standards for School Mathematics* (Reston: NCTM)
[48] National Research Council 2001 Adding it up: Helping children learn mathematics *Mathematics Learning Study Committee, Center for Education, Division of Behavioral and Social Sciences and Education* eds J Kilpatrick, J Swafford and B Findell (Washington: National Academy Press)
[49] Shahrill M 2013 Review of effective teacher questioning in mathematics classrooms *International Journal of Humanities and Social Science* **3** 224–31
[50] Ismail Z, Matzin R, Jawawi R, Shahrill M, Jaidin J H, Mundia L and Mahadi M A 2017 The effectiveness of using an online presentation platform in the teaching and learning of history *Silpakorn University Journal of Social Sciences, Humanities, and Arts* **17** 75–96
[51] Sundayana R, Herman T, Dahlan J A and Prahmana R C I 2017 Using ASSURE learning design to develop students' mathematical communication ability *World Transactions on Engineering and Technology Education* **15** 245
[52] Prahmana R C I, Kusumah Y S and Darhim 2017 Didactic trajectory of research in mathematics education using research-based learning *J. Phys.: Conf. Ser.* **893** 012001
[53] Prahmana R C I and Kusumah Y S 2016 The hypothetical learning trajectory on research in mathematics education using research-based learning *Pedagogika* **123** 42
[54] Ronis D L 2007 *Problem-Based Learning for Math & Science: Integrating Inquiry and the Internet* (Thousand Oaks: Corwin Press)
[55] Young J R 2017 Technology integration in mathematics education: Examining the quality of meta-analytic research *International Journal on Emerging Mathematics Education* **1** 71
[56] Dion L 1996 But I Teach a Large Class… *A Newsletter of the Center for Teaching Effectiveness* (Newark: University of Delaware)

Geographically weighted regression model on poverty indicator

I Slamet[1], N F T A Nugroho[2] and Muslich[2]

[1]Department of Statistics, Universitas Sebelas Maret, Jl. Ir. Sutami no. 36-A, Kentingan, Surakarta, 57126, Indonesia
[2]Department of Mathematics, Universitas Sebelas Maret, Jl. Ir. Sutami no. 36-A, Kentingan, Surakarta, 57126, Indonesia

E-mail: isnandarslamet@staff.uns.ac.id

Abstract. In this research, we applied geographically weighted regression (GWR) for analyzing the poverty in Central Java. We consider Gaussian Kernel as weighted function. The GWR uses the diagonal matrix resulted from calculating kernel Gaussian function as a weighted function in the regression model. The kernel weights is used to handle spatial effects on the data so that a model can be obtained for each location. The purpose of this paper is to model of poverty percentage data in Central Java province using GWR with Gaussian kernel weighted function and to determine the influencing factors in each regency/city in Central Java province. Based on the research, we obtained geographically weighted regression model with Gaussian kernel weighted function on poverty percentage data in Central Java province. We found that percentage of population working as farmers, population growth rate, percentage of households with regular sanitation, and BPJS beneficiaries are the variables that affect the percentage of poverty in Central Java province. In this research, we found the determination coefficient R2 are 68.64%. There are two categories of district which are influenced by different of significance factors.

1. Introduction

Poverty is a problem that has not been resolved completely in Indonesia. The condition of the poor can be known based on the lack of income ability to meet the standard of living (Nugroho[5]). Poverty can be identified by low income ability to meet basic needs in the form of food, health, housing or housing, and education. Low income ability is also defined as low purchasing power or the ability to consume.

Central Java province is a province with high poverty level in Indonesia. Currently, the number of poor people in Central Java province amounts to 4.5 million people, or 13.32 percent in 2016. The percentage is higher than the national poverty rate of 11.13 percent. With the increase of the population that happens every year is possible the number of poor people is increasing. So it is necessary to determine an analysis that shows what factors affect poverty in Central Java Province.

In determining a region classified as poor or not, it is necessary to apply appropriate model analysis to each region. In this case a geographically weighted regression model can be used to analyze the data. The geographically weighted regression model is a model used to analyze spatial data resulting in estimation of local model parameters where each location is different from other locations

(Fotheringham *et al.*[4]). The geographically weighted regression model is part of spatial analysis by weighting based on the position or distance of one observation location with another observation location.

In previous research, Yusuf showed that geographically weighted regression models are more effectively used than global regression models. Sinaga in 2013 examined the geographically weighted regression for poverty data in North Sumatra Province with the weighted function of the kernel Gaussian. In this research will be shown in the form of geographically weighted regression model with Gaussian kernel weighted function on poverty percentage data in Central Java Province

2. Theoritical Basis

2.1. *Multiple Linear Regression*

Multiple linear regression model is a regression model used to examine the relationship of two or more independent variables to the dependent variable. The method that can be used to estimate the parameters of multiple linear regression model is the least squares method (OLS). A common form of multiple linear regression model with p-1 independent variables is presented in the equation below

$$Y_i = \beta_0 + \beta_1 X_{i,1} + \beta_2 X_{i,2} + \cdots + \beta_{p-1} X_{i,p-1} + \varepsilon_i \tag{1}$$

In matrix form the equation can be written as

$$\boldsymbol{Y} = \boldsymbol{X}\widehat{\boldsymbol{\beta}} + \boldsymbol{\varepsilon} \tag{2}$$

where $\boldsymbol{Y}$ is the dependent variable vector $(nx1)$, $\boldsymbol{X}$ is the independent variable matrix (nxp), $\widehat{\boldsymbol{\beta}}$ is vector parameter $(px1)$, $\boldsymbol{\varepsilon}$ is an error vector $(px1)$ mean zero and variance $\sigma^2 \boldsymbol{I}$, $\boldsymbol{I}$ is the identity matrix.

The parameter estimate of $\widehat{\boldsymbol{\beta}}$ in can be expressed by

$$\widehat{\boldsymbol{\beta}} = (\boldsymbol{X}^T\boldsymbol{X})^{-1}\boldsymbol{X}^T\boldsymbol{Y} \tag{3}$$

The testing statistic $\boldsymbol{F}$ of regression model is

$$\boldsymbol{F} = \frac{SSR}{SSE} \tag{4}$$

With $SSR = \frac{\Sigma(\hat{y}-\bar{y})^2}{p}$ and $SSE = \frac{\Sigma(\hat{y}-\bar{y})^2}{n-p}$. H_0 is rejected if $F > F_{(\alpha;p-1,n-p)}$.

Partial test of regression parameter is used to know the effect of regression parameter significantly in model. The testing statistic t of regression model is

$$t = \frac{b_k}{s(b_k)} \tag{5}$$

H_0 is rejected if $|t| > |t_{(\frac{\alpha}{2};n-p)}$.

2.2. *Spatial Heterogeneity*

Spatial heterogeneity suggests a diversity of regions. According to Anselin [1], testing of spatial heterogeneity may use the Breusch-Pagan test (*BP* test) below

$$BP = \frac{1}{2}[h^T Z(Z^T Z)^{-1} Z^T h] \sim \chi_k^2 \tag{6}$$

Spatial heterogeneity occurs when the value of $BP > \chi_k^2$.

2.3. *Geographically Weighted Regression.*

Geographically weighted regression (GWR) is an expansion of the linear regression model with the assumption of spatial heterogeneity in the error. In geographical weighted regression the dependent variable depends on the location of the region. The weighted regression model is geographically defined as

$$y_i = \beta_0(u_i, v_i) + \sum_{k=1}^{p-1} \beta_k(u_i, v_i) x_{i,k} + \varepsilon_i, i = 1,2, \ldots, n, \tag{7}$$

where y_i is the value of the observation of dependent variable at the location i, $x_{i,k}$ is the observation value of independent variable at the location i, $\beta_0(u_i, v_i), \beta_1(u_i, v_i), \dots, \beta_k(u_i, v_i)$ are the parameters of the weighted regression model, (u_i, v_i) denotes the coordinate point (longitude, latitude) destination location i, and ε_i is error on the trend location \$i\$ (Fotheringham *et al.* 4[]). The GWR model parameter estimation is done using the weighted least square (WLS) method by assigning different weight to each location. With a weighted location in \$i\$ against another location j is $w_{ij}(u_i, v_i)$ then the sum of squared errors that have been weighted based on

$$\sum_{j=1}^{n} w_{ij}\varepsilon_j^2(u_i, v_i)\, y_i = \beta_0(u_i, v_i) + \sum_{k=1}^{p-1}\beta_k(u_i, v_i)x_{i,k} + \varepsilon_i, i = 1,2, \dots, n, \quad (8)$$

so that obtained GWR model parameter estimation for each location is

$$\hat{\boldsymbol{\beta}} = (\boldsymbol{X}^T\boldsymbol{W}(u_i, v_i)\boldsymbol{X})^{-1}\boldsymbol{X}^T\boldsymbol{W}(u_i, v_i)\boldsymbol{Y} \quad (9)$$

where $\boldsymbol{W}(\boldsymbol{u}_i, \boldsymbol{v}_i) = diag[w_{i1}, w_{i2}, \dots, w_{in}]$, $\boldsymbol{X}$ is independent variable matrix, and $\boldsymbol{Y}$ is the dependent variable matrix.

The spatial weights are used to estimate the parameters of the GWR model. The following Gaussian kernel function will be used

$$w_{ij} = exp[-\frac{1}{2}\left(\frac{d_{ij}}{b}\right)^2] \quad (10)$$

where $d_{ij} = \sqrt{(x_i - x_j)^2 + (y_i - y_j)^2}$. d_{ij} is the euclidean distance between locations i and j while b is bandwidth. Bandwidth is the circle radius where the point within it is considered still influential in the formation of model parameters. The optimum bandwidth used is the one that produces the minimum cross validation (CV) value, by the formula

$$CV = \sum_{i=1}^{n}\sum_{j=1}^{n}[y_i - \hat{y}_{j\neq i}(b)]^2 \quad (11)$$

where $\hat{y}_{j\neq i}(b)$ is the estimated value of the dependent variable with on-site observations i omitted from the prediction process (Fotheringham[4]).

2.4. Mapping Models.

Model mapping is done by determining the factors that affect the dependent variable. To determine the factors that influence significantly on the dependent variable used partial test parameters. Partial test is used to know the significance of parameter $\beta(u_i, v_i)$ against the dependent variable partially on the GWR model with test statistic

$$t = \frac{\hat{\beta}_k(u_i,v_i)}{SE(\hat{\beta}_k(u_i,v_i))} \quad (12)$$

Where $\hat{\beta}_k(u_i, v_i)$ is an estimated parameter of the GWR model, and $SE(\hat{\beta}_k(u_i, v_i))$ is a standard GWR model error obtained from $\sqrt{Var(\hat{\beta}_k(u_i, v_i))}$. H_0 is rejected if $|t| > |t_{(\frac{\alpha}{2};n-p)}$. Partial test of parameters is used to determine the factor map affecting poverty in Central Java Province based on district or city location.

3. Research Methodology

3.1. Data Source

The data used in this study is the percentage of poverty and factors affecting poverty in Central Java Province, for the year 2015 taken from Central Bureau of Statistics (BPS [2]). The data consist of 12 independent variables: labor force participation rate (X_1), percentage of population working as farmers

(X_2), and HDI (X_3) as influencing factors terms of employment. Furthermore, population growth rate (X_4) as an influential factor in terms of population.

The percentage of the population who does not have an elementary school certificate $(\boldsymbol{X_5})$, the percentage of the population graduated from primary school $(\boldsymbol{X_6})$, the percentage of the population graduated from junior high school $(\boldsymbol{X_7})$ percentage of the population graduated from senior high school $(\boldsymbol{X_8})$as a factor affecting poverty in education. In terms of food needed factors affecting poverty ie minimum wage work $(\boldsymbol{X_0})$, per capita expenditure $(\boldsymbol{X_{10}})$. The percentage of households with regular sanitation $(\boldsymbol{X_{11}})$ and BPJS beneficiaries $(\boldsymbol{X_{12}})$ as influential factors in terms of health.

3.2. *Analysis Method*

The research is done by first determining global regression model with stepwise regression method and estimating regression coefficient parameter with ordinary least squares method (OLS). The next step is to test the classical assumption of regression analysis. In addition to the classical assumption test, the next test is to identify the spatial effects on the data by using spatial heterogeneity tests. If the assumptions of spatial heterogeneity are met then we can use geographically weighted regression to overcome the heterogeneity problems contained in the model. Furthermore, the geographical weighted regression coefficient parameter is estimated using the under weighted least square method with the kernel weighting function Gaussian. For deriving the model we used R programing.

4. Analysis and Discuccion

4.1. *Ordinary Least Square*

By using the least square method, we obtained linear regression model for poverty percentage data in Central Java in 2015 which is

$$Y = 3.736 + 0.084X_2 - 11.622X_4 + 0.120X_{11} + 0.090X_{12} \tag{13}$$

From the model it can be seen that from 12 independent variables there are only four independent variables that influence significantly. There are percentage of population working as farmer (X_2), population growth rate (X_4) percentage of households with regular sanitation (X_{11}), and BPJS beneficiaries (X_{12}).

The normality and multicolinearity assumptions in this model has been fulfilled. From the test of spatial heterogeneity, we obtained BP value of 11.438 which results greater than the value of $\chi^2_{(0.1;4)} = 7.779$ which means there is spatial heterogeneity. For that, we use geographically weighted regression to determine the model.

4.2. *Geographically Weighted Regression.*

The geographically weighted regression model is formed based on four independent variables that have been obtained from the global regression model i.e. (X_2), (X_4), (X_{11}), (X_{12}). In determining the estimation value of GWR model, parameters depends on the weighting of each location. For that, it needs to be determined first bandwidth optimum obtained from the minimum CV.

The bandwidth value with the gaussian kernel weighted function is 2.7939 with minimum CV 285.3083. Then the bandwidth value is used to determine the weighting matrix of each GWR. Based on the equation $\boldsymbol{W}(\boldsymbol{u_i}, \boldsymbol{v_i}) = diag[w_{i1}, w_{i2}, \ldots, w_{in}]$ we have a weighted value for each location. The weighted value is then used to determine the estimated parameters of the GWR model according to the equation (9).

From geographically weighted regression model it can be arranged classification of area in Central Java Province based on factors that affect poverty as in Figure 1.

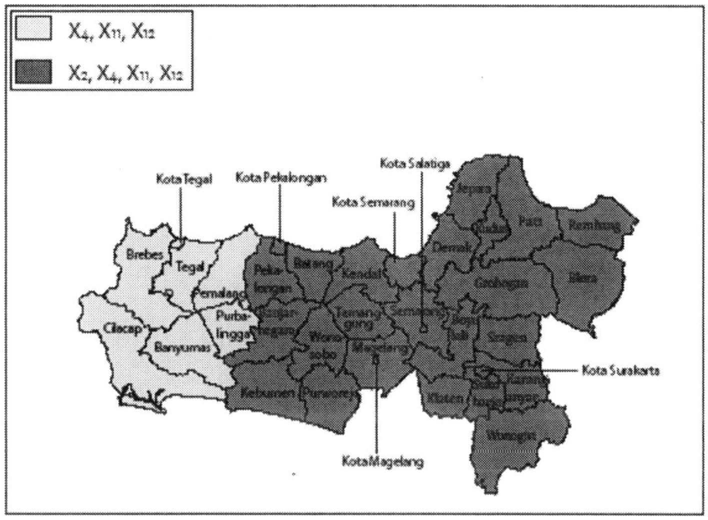

Figure 1. The division of regions is based on factors affecting the poverty of districts in the province of Central Java.

Figure 1 shows region mapping based on geographically weighted regression model with kernel weighted function Gaussian. With the color difference shows the independent variables that affect the percentage of poverty in the area.

References

[1] Anselin L 1998 *Spatial Econometrics : Methods and Models* (Dordrecht : Kluwer Academic Publishers)
[2] Badan Pusat Statistik 2016 *Provinsi jawa tengah dalam angka tahun 2016* (Jakarta)
[3] Chaudry A, and Wimer C 2016 Poverty is Not Just an Indicator: The Relationship between Income, Poverty, and Child Well-Being *Academic Pediatrics* **16** S23–29.
[4] Fotheringham S, Brunsdon C, and Charlton M 2002 *Geographically Weighted Regession - The Analysis of Spatially Varying Relationships* (Chichester : Wiley)
[5] Nugroho H 1995 *Kemiskinan, Ketimpangan, dan Kesenjangan* (Yogyakarta: Aditya Media)

IOP Conf. Series: Journal of Physics: Conf. Series **943** (2017) 012010 doi:10.1088/1742-6596/943/1/012010

Utilizing geogebra in financial mathematics problems: didactic experiment in vocational college

Saiful Ghozi[1] and Suci Yuniarti[2]

[1]Balikpapan State Polytechnic, Jl.SoekarnoHatta KM.8, Balikpapan, Indonesia
[2]Balikpapan University, Jl. Pupuk Raya, Gn. Bahagia, Balikpapan, Indonesia

Email: saiful.ghozi@poltekba.ac.id

Abstract. GeoGebra application offers users to solve real problems in geometry, statistics, and algebra fields. This studydeterminesthe effect of utilizing Geogebra on students understanding skill in the field of financial mathematics. This didactic experiment study used pre-test-post-test control group design. Population of this study were vocational college students in Banking and Finance Program of Balikpapan State Polytechnic. Two classes in the first semester were chosen using cluster random sampling technique, one class as experiment group and one class as control group. Data were analysed used independent sample t-test. The result of data analysis showed that students understanding skill with learning by utilizing GeoGeobra is better than students understanding skill with conventional learning. This result supported that utilizing GeoGebra in learning can assist the students to enhance their ability and depth understanding on mathematics subject.

1. Introduction

The development of information and communication technology (ICT) affects many sectors all over the world, including higher education [1, 2]. The ICT in higher education are being used for developing course material; delivering and sharing content; communication between learners, teachers and the outside world; creation and delivery of presentation and lectures; academic research; administrative support; student enrolment etc. [2]. ICT can be an effective tool in supporting teaching and learning when it is used appropriately [3, 4].

Utilizing ICT especially computer application in college students will enhance understanding skill and assist a depth discussion in the higher education classroom. When working with multimedia applications, students are engaged in activities which develop higher order thinking skills such as problem solving, reflecting, recognizing relationship, analyzing and creatively synthesizing information, as they deal with complex and realistic information [5]. These activities will facilitate students to understand the problems in particular subject, including mathematics.

Students must learn mathematics with understanding, actively building new knowledge from experience and prior knowledge [6]. Understanding has, in effect, become a means and a goal of mathematics education [7]. Furthermore, it seems reasonable to characterize mathematical understanding as a matter of having a world of dynamic mental models that are consistent with the conceptual systems of mathematics. The dynamic nature of mathematical understanding and the corresponding needs for multiple representations serve as a theoretical foundation for the integration of technological tools such as GeoGebra [8].

GeoGebra is a community-supported open-source mathematics learning environment that integrates multiple dynamic representations, various domains of mathematics, and a rich variety of computational utilities for modeling [9] and simulation [10]. It is dynamic mathematics software for all levels education that brings together geometry, algebra, spreadsheets, graphing, statistics and calculus in one easy-to-use package. GeoGebra includes a number of features-most notably, sliders and CAS capabilities-that aren't available in typical spreadsheet programs such as Excel or Calc [11]. This study focuses on effect of utilizing GeoGebra on students understanding skill in the field of financial mathematics.

2. Difficulties in financial mathematics

Financial mathematics is one of compulsory subjects in Banking and Finance Program, Polytechnic State of Balikpapan, East Borneo, Indonesia. The students attend this course in the first semester. Several students who attend this course were still had difficulties to understand the materials. One of the reasons is some students do not have enough background on mathematics in their high school. In addition, most problems in this field are real problems so students must make a mathematical model from the problems before they solve it.

The short definition of mathematical model and modeling is "translation between mathematics and reality". The reality according to "rest of the world" outside mathematics including nature, society, everyday life and other scientific discipline [12]. In this paper, GeoGebra will be used to visualize mathematical model based on the problems about Compound Value, Present Value, and Future Value. The descriptions of the materials as follows:

2.1. Compound Value, Present Value and Future Value

Let $\boldsymbol{P}$(Principal) refers to principal of loan which is charged by compound interest,$\boldsymbol{S}$ is the Compound Value, $\boldsymbol{n}$is the number of compounding periods, and i is the monthly interest in its decimal form. Then the formula of Compound Value is:

$$S = P(1+i)^n \tag{1}$$

Let $\boldsymbol{PV}$(Present Value) mean total amount of loan, $\boldsymbol{A}$ is the instalment amount, $\boldsymbol{n}$ is the number of the instalment, and i is the monthly interest in its decimal form. The $\boldsymbol{PV}$equation is:

$$PV = \frac{(1-(1+i)^{-n})}{i}A \tag{2}$$

And let $\boldsymbol{FV}$(Future Value) means total savings, and $\boldsymbol{A}$is amount of each annuity payments (saving), then the formula is:

$$FV = \frac{((1+i)^n-1)}{i}A \tag{3}$$

The examples of problems in the particular materials above as follows:

2.2. Problem 1

The following is a table of Credit for Businesses installment (*Kredit Usaha Rakyat*, or *KUR*) issued officially by *Bank Negara Indonesia* (BNI). The information listed states that the interest rate is 0.412% flat per month. How much is the real interest rate of the annuity considering that the system used by bank is the annuity interest rate?

IOP Conf. Series: Journal of Physics: Conf. Series **943** (2017) 012010 doi:10.1088/1742-6596/943/1/012010

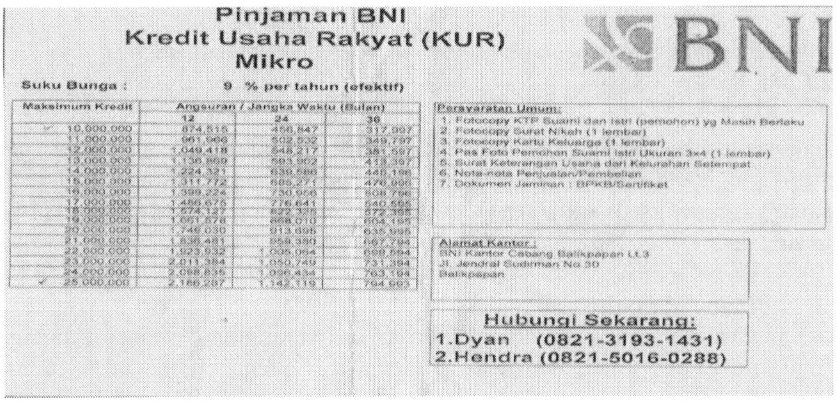

Pinjaman BNI
Kredit Usaha Rakyat (KUR)
Mikro

Suku Bunga : 9 % per tahun (efektif)

Maksimum Kredit	Angsuran / Jangka Waktu (Bulan)		
	12	24	36
✓ 10,000,000	874,515	456,847	317,997
11,000,000	961,966	502,532	349,797
12,000,000	1,049,418	548,217	381,597
13,000,000	1,136,869	593,902	413,397
14,000,000	1,224,321	639,586	445,196
15,000,000	1,311,772	685,271	476,996
16,000,000	1,399,224	730,956	508,796
17,000,000	1,486,675	776,641	540,595
18,000,000	1,574,127	822,325	572,395
19,000,000	1,661,578	868,010	604,195
20,000,000	1,749,030	913,695	635,995
21,000,000	1,836,481	959,380	667,794
22,000,000	1,923,932	1,005,064	699,594
23,000,000	2,011,384	1,050,749	731,394
24,000,000	2,098,835	1,096,434	763,194
✓ 25,000,000	2,186,287	1,142,119	794,993

Persyaratan Umum:
1. Fotocopy KTP Suami dan istri (pemohon) yg Masih Berlaku
2. Fotocopy Surat Nikah (1 lembar)
3. Fotocopy Kartu Keluarga (1 lembar)
4. Pas Foto Pemohon Suami istri Ukuran 3x4 (1 lembar)
5. Surat Keterangan Usaha dari Kelurahan Setempat
6. Nota-nota Penjualan/Pembelian
7. Dokumen Jaminan : BPKB/Sertifikat

Alamat Kantor :
BNI Kantor Cabang Balikpapan Lt.3
Jl. Jendral Sudirman No.30
Balikpapan

Hubungi Sekarang:
1.Dyan (0821-3193-1431)
2.Hendra (0821-5016-0288)

Figure 1.Table of credit for businesses installment (*Kredit Usaha Rakyat*, or *KUR*) issued officially by BNI.

To determinethe interest rate annuity from the instalment table manually needs a quite long iteracy. From equation (2) above, it is obtained that the polynomial equation of degree$\boldsymbol{n+1}$:

$$\boldsymbol{PV(1+i)^n i - A(1+i)^n + A = 0} \tag{4}$$

By replacing equation (2) with the number of the instalment$\boldsymbol{n}$ = 12, $\boldsymbol{PV}$= 10,000,000 and $\boldsymbol{A}$= 874,515 from the table (see Figure 1), it will form the polynomial equation of degree 13:

$$\boldsymbol{10{,}000{,}000\,(1+i)^{12} i - 874{,}515(1+i)^{12} + 874{,}515 = 0}$$

The features of CAS on GeoGebracan be utilized to solve the equation above. *Computer Algebra System (CAS)* is one of features in GeoGebra which is possible to perform mathematical calculation in a manner similar to manual one. The presentation of CAS feature to solve the problem above is as follows.

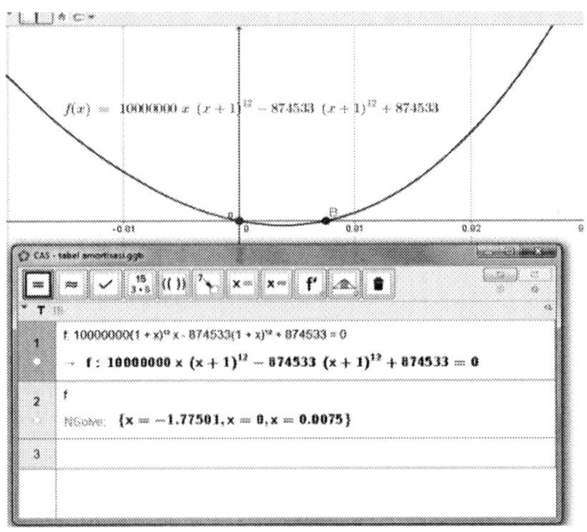

Figure 2. Determining value of *i* through CAS feature in GeoGebra

It is seen clearly from Figure 2 that a positive solution of polynomial equation $\mathbf{10,000,000(1+i)^{12}i - 874,515(1+i)^{12} + 874,515 = 0}$ is i = 0.0075. So it is concluded that the effective interest rate in a year = 0.0075 x 12 x 100% = 9%.

2.3. *Problem 2*

Determine the loan balance on tenor 15 years between loan Rp500,000.00 which is charged by compound interest 8% p.a, and fixed savings Rp4,000.00 permonth with interest rate annuity 8% p.a, and when the total of loan will be paid off ?

The Compound Value ($\boldsymbol{S}$) on the Problem 2 can be obtained by replacing equation (4) with parameters $i = \frac{8\%}{12}$ p.a, $\boldsymbol{P}$ = 500,000 and $\boldsymbol{n}$ = 15x12=180, such that:

$$\boldsymbol{S = 500,000(1 + \frac{0.08}{12})^{180}} = 1{,}653{,}460.7$$

And the Future Value ($\boldsymbol{FV}$) can be obtained by replacing equation (4) with $= \frac{8\%}{12}$, p.a, $\boldsymbol{A}$ = 4,000 and $\boldsymbol{n}$ = 15×12=180, such that:

$$\boldsymbol{FV} = \frac{\left(\left(1+\frac{0.08}{12}\right)^{180} - 1\right)}{\frac{0.08}{12}}\boldsymbol{4,000} = 1{,}384{,}152.9$$

A loan balance is the amount left to pay on the loan. Every loan that is taken out will have a loan balance up until the loan is completely paid off [10]. The formula of loan balance can be generated as follows. Let a means to the value of loan balance, then

$$\boldsymbol{a = S - FV} \tag{5}$$

Such that, $\boldsymbol{a}$=1,653,460.7 – 1,384,152.9 = 269,307.9. So the first question in Problem 2 is solved.

The next question is to determine when the total of loan will be paid off. The debt settlement is achieved when the value of loan balance equal to zero ($\boldsymbol{a = 0}$). It can be written as follows.

$$\boldsymbol{S - FV = 0}$$

Then:

$$\boldsymbol{S = FV} \tag{6}$$

Such that:

$$\boldsymbol{P(1+i)^n} = \frac{((1+i)^n - 1)}{i}\boldsymbol{A} \tag{7}$$

After replacing equation (7) with parameter $\boldsymbol{P}$= 500,000, $\boldsymbol{A}$= 4,000, and $\boldsymbol{i}$= 0.08, we get

$$\boldsymbol{500,000\left(1+\frac{0.8}{12}\right)^n} = \frac{\left(\left(1+\frac{0.8}{12}\right)^n - 1\right)}{\frac{0.8}{12}}\boldsymbol{4,000}$$

We can solve equation above numerically, so that the value of $\boldsymbol{n}$ = 269.66 months = 22.4717 years ≈ **22.5**years. Furthermore, we can compute the Compound Value ($\boldsymbol{S}$) which is equal to Future Value ($\boldsymbol{FV}$), such that

$$\boldsymbol{S = 500,000\left(1+\frac{0.08}{12}\right)^{269.66} \approx 3000,000}$$

Several steps to solve Problem 2 above can be visualized in GeoGebra and can be analyzed geometrically as follows.

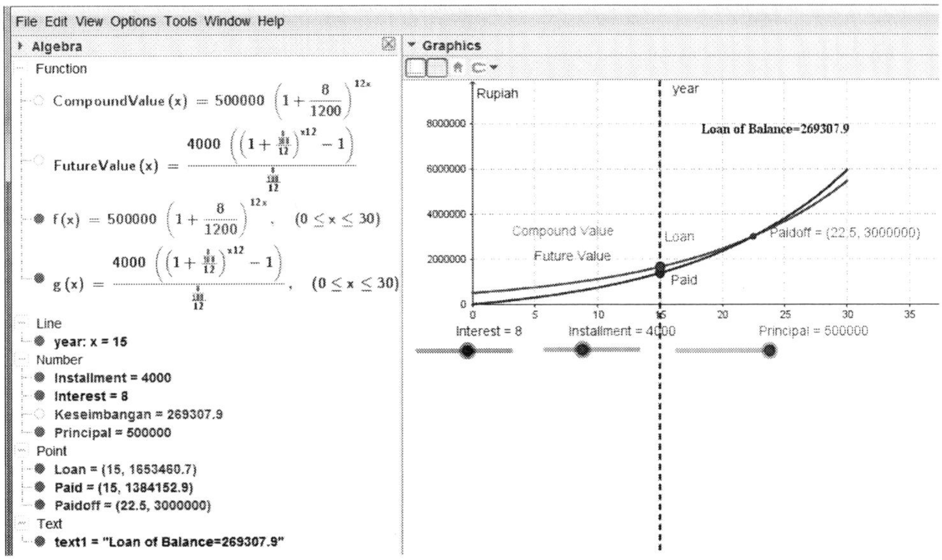

Figure3.Applet of balance between compound value and future value [14]

Based on the Figure 3 above, the value of interest rate, instalment, and principal of loan are movable using slider in GeoGebra. The value of loan balance will be obtained automatically by moving parameter value $\boldsymbol{i}$, $\boldsymbol{A}$, $\boldsymbol{P}$and $\boldsymbol{n}$. According toProblem2 above, it is seen that the value of loan balance obtained is 269,307.9and the intersection between the Compound Value and the Future Value is achieved when$n = 22.5$. It means that the compound loan will be paid off on the 23thyear.In other words, it produces the same value as previous calculation without using GeoGebra. However, GeoGebra can show the dynamic movements of the interest rate, annuity instalment, principal of loan so it offers wider explorations to see the influence of characteristics of these three parameters on the value of loan balance.

3. Method

This experiment study used pre-test-post-test control group design. Population of this study were vocational college students in Banking and Finance Program of Balikpapan State Polytechnic. Two classes in the first semester were chosen as sample of the study using cluster random sampling. Class 1KP1 was chosen as control group with conventional learning and Class 1KP2 was chosen as experiment group with learning by utilizing GeoGebra. A pre-test was conducted to measure the initial ability of two groups before the treatment. Then, a post-test was conducted to measure students understanding skill after the treatment.

Independent sample t-test was used to analyse the data from the post-test. The fulfilment of the assumptions is required before running the analysis. The assumptions were normality and homogeneity test. The normality of the post-test score was assessed by Kolmogorov Smirnov test and Levene test were used to assess the homogeneity of variance from the post-test score.

4. Result and discussion

4.1. Result

Descriptive statistics were conducted to describe the mean, standard deviation and standard error mean from the post-test scores. The result is provided on Table 1.

Table 1. Result of descriptive statistics

Performance	Group	N	Mean	Std. Deviation	Std. Error Mean
Financial Mathematics Post-test	Control	33	7.2061	2.05502	.35773
	Experiment	32	8.1750	1.67120	.29543

According to Table 1, mean value of control and experiment groups are 7.2061 and 8.1750 respectively. It means that the mean value of experiment group is higher than the mean of control group.

The t-test is used to discover whether there are statistically significant differences between the means of two groups. In this study, the independent t-test was used to determine the differences between control and experiment groups. The experiment group used GeoGebra while the control group did not. Data of post-test score was normally distributed and had equal variances based on the result of Kolmogorov Smirnov test and Levene test. The result of independent sample t-test showed that the value of t_0 is -2.082 and the value of p is 0.041. It means that there is significant difference between the means of two groups, control and experiment groups.

4.2. Discussion

The study was conducted in Banking and Finance Program of Balikpapan State Polytechnic at the first semester. Before the treatment, both groups of students were given a pre-test to measure their initial abilities. The result of pre-test showed that both groups had same initial abilities. During the treatment, the experiment group discussed the real problems about financial mathematics using GeoGebra. Students were fully engaged in the learning process as they were interested with the tools. Apagu and Wakili's study [15] revealed that using ICT in technical college makes the teaching and learning become interesting.

GeoGebra software also can be used as a powerful tool to visualize relationship between compound value, present value, and future value. During the discussion, students made a mathematical model from the real problems given using GeoGebra. The dynamic movements of the interest rate, annuity instalment, and compound value of the main loan can enhance wider explorationson the influence of characteristic parameters. It means that GeoGebra helped students in modelling the real problem and in broadening their explorations and visualization skills through the process of constructing visual images to analyse the problem, taking into account their informal and visual conceptions [16].

In the last meeting, they were given a post-test to measure their understanding skills in the course.The result of independent sample t-test showed that there is difference on students understanding skill between group with learning by utilizing GeoGeobra and group with conventional learning. In addition, the mean value of experiment group is higher than the mean of control group. It showed that students understanding skill with learning by utilizing GeoGeobra is better than students understanding skill with conventional learning. This result supports the assistance of GeoGebra to enhance the ability and depth understanding of student on mathematics concept especially in financial mathematics course [17, 18].

5. Conclusion

Based on the result and discussion, it can be concluded that students understanding skill with learning by utilizing GeoGeobra is better than students understanding skill with conventional learning. Modelling activities and utilizing GeoGebra should be taken into consideration to enhance vocational college students understanding skill. This study recommends that utilizing GeoGebra in learning should be continued in the classroom and further research about utilizing GeoGebra in other subjects and populations should be conducted by other researchers.

6. Acknowledgments
Our gratitude goes to all parties who helped researchers in completing this paper. The special thanks go to Directorate Research and Community Service, Ministry Research, Technology and Higher Education of Indonesia for funding support to conduct this study. The author would like to extend gratitude to management of Balikpapan State Polytechnic for facilitating this study to be possible.

References
[1] Fathurrohman M, Porter A L and Worthy A L 2017 Teachers' real and perceived of ICT supported-situation for mathematics teaching and learning *International Journal on Emerging Mathematics Education* **1** 11
[2] Mondal A and Mete J 2012 ICT in higher education: opportunities and challenges *Batter College J. Multidisciplinary Studies* **2** 2
[3] Hennessy S, Harrison D and Wamakote L 2010 Teacher factors influencing classroom use of ICT in Sub-Saharan Africa *Itupale Online J. African Studies* **2** 39
[4] Young J R 2017 Technology integration in mathematics education: Examining the quality of meta-analytic research *International Journal on Emerging Mathematics Education* **1** 71
[5] Oyebolu S and Lemo O 2013 The impact of information and communication technology (ICT) on vocational and technical students' learning *J. Education and Practice* **4** 178
[6] National Council of Teachers of Mathematics 2000 Principles and standards for school mathematics (Reston, VA: NCTM)
[7] Bu L, Spector J M and Haciomeroglu E S 2011 toward model-centered mathematics learning and instruction usinggeogebra*Model-centered learning: Pathways to mathematical understanding using geogebra* (Rotterdam: Sense Publishers)
[8] Ismail Z and Rahman S N A 2017 Learning 2-dimensional and 3-dimensional geometry with Geogebra: Which would students do better? *International Journal on Emerging Mathematics Education* **1** 121
[9] Hall J and Lingefjärd T 2017*Mathematical Modeling: Applications with GeoGebra™* (Sweden: Wiley)
[10] Bu L and Schoen 2011R Geogebra for model-centered learning in mathematics education*Processing Mathematics Through Digital Technologies* (Netherland: Sense Publishers)
[11] Hasek R and Petroskova V 2013 GeoGebra in financial education *North Am. GeoGebra J.* **2** 31
[12] Blum W 2009 Mathematical modelling : can it be taught and learnt ?*Math. Model. Appl.* **1** 45
[13] Delbridge E 2016 Explanation of a loan balance [Online] https://www.thebalance.com/what-is-a-loan-balance-527327. [Accessed: 31-Mar-2017]
[14] Ghozi S 2017 Kesetimbangan hutang*Geogebra Material* [Online] https://ggbm.at/qqCGn7zN.
[15] Apagu V V and Wakili B A 2015 Availability and utilization of ICT facilities for teaching and learning of vocational and technical education in Yobe State Technical College *American J. Engineering Research* **4** 113
[16] Mousoilides N G 2011 Geogebra as a conceptual tool for modelling real world problems*Processing Mathematics Through Digital Technologies* (AW Rotterdam: Sense Publishers)
[17] Sundayana R, Herman T, Dahlan J A and Prahmana R C I 2017 Using ASSURE learning design to develop students' mathematical communication ability *World Transactions on Engineering and Technology Education* **15** 245
[18] Tanujaya B, Prahmana R C I, and Mumu J 2017 Mathematics instruction, problems, challenges, and opportunities: A case study in Manokwari regency, Indonesia *World Transactions on Engineering and Technology Education* **15** 287

IOP Conf. Series: Journal of Physics: Conf. Series **943** (2017) 012011 doi:10.1088/1742-6596/943/1/012011

Construction and reconstruction concept in mathematics instruction

Jeinne Mumu[1], Rully Charitas Indra Prahmana[2] and Benidiktus Tanujaya[1]

[1]University of Papua, Jl. Gunung Salju, Manokwari, West Papua 98314, Indonesia
[2]Universitas Ahmad Dahlan, Jl. Pramuka Kav. 5, Yogyakarta 55161, Indonesia

E-mail: j.mumu@unipa.ac.id

Abstract. The purpose of this paper is to describe two learning activities undertaken by lecturers, so that students can understand a mathematical concept. The mathematical concept studied in this research is the Vector Space in Linear Algebra instruction. Classroom Action Research used as a research method with pre-service mathematics teacher at University of Papua as the research subject. Student participants are divided into two parallel classes, 24 students in regular class, and remedial class consist of 18 students. Both approaches, construct and reconstruction concept, are implemented on both classes. The result shows that concept construction can only be done in regular class while in remedial class, learning with concept construction approach is not able to increase students' understanding on the concept taught. Understanding the concept of a student in a remedial class can only be carried out using the concept reconstruction approach.

1. Introduction

In education system there are various methods and approaches of instruction used by teachers in order to create their teaching more effective [1, 2]. In mathematics instruction, teachers generally use two methods, deductive and inductive methods. Deductive is an instruction method of reasoning by which concrete applications are deducted from general principles, theorems are deduced from definitions and postulates. A deductive approach involves the learners being given a general rule, which is then applied to specific examples and practice exercises [2]. In mathematics instruction, teachers star lesson by writing definitions, followed by presenting examples and counter examples [3]. This method causes students to tend to be inactive in learning.

In contrast to the deductive method, inductive instruction makes use of student "noticing". Instead of explaining a given concept and following this explanation with examples, the teacher presents students with many examples showing how the concept is used [4]. An inductive approach involves the learners detecting, or noticing, patterns and working out a 'rule' for themselves before they practise the concept [5]. This method causes students to tend to be more active in learning. However, not all mathematical concepts can be understood by studying patterns. In an abstract concept, conceptual understanding cannot be done by studying patterns, because patterns cannot be found.

Furthermore, mathematics learning in university should be different from mathematics learning in schools. Mathematics learning in university is more difficult because the concepts studied are abstract. Mathematics learning in school is also abstract but can still be given a real example. One of the abstract mathematics courses in the university is the linear algebra. This is in accordance with the statement of Adams and Elliott [6], which states that one mathematics topic that the majority of preparatory students fear is linear algebra, which is often due to it being perceived as more abstract and thus irrelevant to the 'real world'. Moreover some experts stated that the abstract and formal nature of linear algebra generate two sources of difficulty in its understanding. The nature of linear algebra itself (conceptual difficulties) and the kind of thinking required for the understanding of linear algebra (cognitive difficulties) [7, 8].

One of the most abstract university mathematical concepts in linear algebra course is concept of vector space. On the other hand, Vector Space is one of the important concepts to be well understood by mathematics students. Linear algebra is essentially based on the theory of vector spaces [9]. This concept, as well as the concept of linear independence (or linear dependence), is initially quite challenging for students [10]. Understanding the concept of vector space is an important part in understanding other concepts in other abstract mathematics courses such as algebraic structures and real analysis. Nevertheless vector space learning has been done with a less accurate approach.

However, unlike calculus, that often emphasizes manipulation of symbols in order to solve problems, the focus in linear algebra is on the description of concepts, often through word definitions, and derivation of further concepts from these [11]. Lecturers who teach linear algebra generally start the lesson by defining the vector space and then discuss the example and counter example of the vector space. The learning approach of lecturer causes the students to not understand the concept of vector space comprehensively. Vector Space concepts in the form of definitions containing vector space axioms are only memorized by the students without knowing the process of obtaining them.

Various authors have investigated the learning and teaching of linear algebra from an educational perspective showed that students tend to think about the concepts as the application of techniques [8]. They do not usually understand the meaning of definitions, and they are unable to apply them, even in simple problems. In fact, their findings show that most students manipulate the symbols algebraically without understanding the concepts they refer to Thomas and Stewart [12]. The situation affects the difficulties when the student tries to solve a problem somewhat different from the given example. This situation is different when students understand the concept through a process of constructivism.

Constructivism is a learning theory describing the process of knowledge construction. Knowledge construction is an active, rather than a passive process. Constructivists believe that knowledge should not be just deposited into the learners' minds; instead it should be constructed by the learners through active involvement in the learning process. In the constructivist perspective, knowledge is constructed by the individual through his/her interactions with the environment. Unlike the traditional mode of learning whereby the teacher plays an active role in the teaching/learning environment, and learners passively receive the content, constructivists believe the learning should be centred on the learner [13].

Constructivism in learning mathematics can be done through analogy. Analogy is a sort of similarity [14]. Analogies show how one thing is similar to another. Therefore using analogies in thinking refers to reasoning by using parallels [15]. Analogies are an excellent way to teach mathematics to adults. They enable connections to be forged and assist in removing fear and increase confidence that may have been lost due to boring rote learning. Analogies enable mathematical concepts to be conveyed in a form that students can relate to, thus, increasing their understanding and confidence. The use of analogies enables adult students to form connections between mathematics and real world situations. Connection with the familiar makes mathematical concepts easier to comprehend [6]. Furthermore Mofidi et al [16] stated that one of the methods that teachers apply to assist students with building conceptual knowledge is to use analogies.

The application of constructivism in learning mathematics depends on the initial knowledge that students have. Students who learn a concept should construct their own concepts so that students can understand them well. The process is called a construct. However, in students who have studied the concept, the construction process is different. Those who have already studied a concept, generally have a misconception about the concept. Subanji and Nusantara [17] stated that the error was due to an error in the thinking process, known as pseudo thinking. Pseudo thinking is a thinking process that results in an answer to a problem or construction to a concept "that is not true". The construction concept does not represent the actual thinking. So the analogy process on a student who has studied and new students learn a different concept of analogy approach. Thus in order to improve the understanding of students who have studied a particular concept requires a process called reconstruction.

Reconstruction is an attempt to make corrections to students' misconceptions of the concepts they have learned. Like concept construction, concept reconstruction also uses analogies. The difference between the two lies in the type of analogy used. In reconstruction, the analogy used is based on students' misconception of the concept. Misconceptions can occur in both capable and weak students. According to Harel [18], teacher tends to associate misconception and missing conceptions only with mathematically weak students. But in fact, all students, the weak and the able, in desire to understand and make sense of mathematical concepts, encounter difficulty, and demonstrate as a result behaviors that in many cases are difficult to explain [18].

Based on the previous statement, the following questions can be raised. How do students who study mathematics develop their understanding of a mathematical concept based on the analogy process? Is there a difference in the process of constructing a mathematical concept between students on regular classes and students on remedial classes? Is it true that the construction process is more effectively applied to students in regular classes than applied to students in remedial classes? Is it true that the reconstruction process needs to be done to students in the remedial class? How is the reconstruction process done?

2. Method

The study was conducted using classroom action research. Classroom action research is a wide variety of evaluative, investigative, and analytical research methods designed to diagnose problems in classroom [19]. Furthermore McNiff and Whitehead [20] stated that classroom action research is a method of finding out what works best in classroom so that teacher can improve student learning. This method is a form of enquiry that enable teacher to investigate and evaluate their work [21]. Classroom action research aims to improve the quality of instruction, teaching quality of teacher and learning quality of students [19, 22]. Thus, in this study was conducted evaluation, investigation and analysis of two learning activities, construction and reconstruction concept in order to obtain the appropriate learning activities so that the students' learning achievement who study the concept of Vector Space in Linear Algebra instruction will be better.

Subjects for this research were 42 students of mathematics education from University of Papua, Manokwari West Papua Province, Indonesia. The students are those who contract linear algebra lesson in the 2017/2018 academic year. In this research the subjects, students who participants are divided into two parallel classes, 24 students in regular class, and remedial class consist of 18 students. The remedial class is a class consisting of students who have taken the course, but did not pass the exam.

The grouping of students into two classes aims to acquire groups of students with similar ability to each class. According to Slota [23] teaching a group of students with similar abilities allowed teachers to adjust the pace of instruction to best reach students' needs. A teacher would instruct at a slower pace which provided more repetition and reinforcement with a group of low achieving students than they would have with a group of high achievers.

2.1. Instruction plan

In instruction planning, researchers have to considering the kind of data they will need to collect and the processes involved in the data collection [3, 22]. Instruction plan is prepared through discussion among researchers. The result of the planning establishes that the researcher divided into two responsibilities, as a lecturer and as an observer. There are three instruments produced, namely observation guide, student activity sheet (SAS) and questionnaire. The observation guide is intended to observe student activity during the lesson. SAS consist of some questions used by used to help students understand the concept of vector Space, whereas the questioner was used to measure students' responses to the learning process.

2.2. *Action and observation*
Observations were implemented during the lesson. The classroom action research would be more effective if instruction and observations were hold simultaneously [3, 22]. The learning is carried out separately on two parallel classes, regular class and remedial class. Learning is done by constructing and reconstructing concepts, respectively in each class, through analogy process. Observations were made to students' activities and expressions in learning. Observations are recorded for discussion at the reflection stage.

2.3. *Reflection*
Reflection was carried out after completion of an instruction. Based on some records of observations, conducted by lecturer and observers, reflection is carried out through discussion. Discussion aims to find problems in instruction and tried to find solutions. Criticism and suggestions as a result of reflection are used to rearrange the next instruction plan.

3. Results and discussion
The lecture on Linear Vector Space Concepts was first performed on a regular class. The results of reflection on the regular class are used to rearrange the lesson plans that have been prepared for the remedial class. Based on observations and reflections performed on the regular and remedial class, various problems were found, as well as suggestions that could be used to improve the learning process in the remedial class.

Instruction activities, both in the regular and remedial classes are carried out in three stages of instruction, i.e. preliminary activities, main activities and closing activities. In the preliminary activities, lecturers provide illustrations related to learning objectives and the importance of learning the concept of Vector Space as well as illustration of Vector Space application. Illustrations are also given use a simple example of what analogies are meant in Mathematics. Lecturer then managed the students to sit in groups according to the group that has been prepared. Each group consists of 4 members and is managed by a leader who is a student with a Grade Point Average (GPA) greater than 3.5. Furthermore, the lecturer asks questions to explore students' memories related to the integer system Z, as well as vector operations in the two-dimensional space (R^2) and three-dimensional space (R^3).

The next stage, the lecturer distributes Student Activity Sheets (SAS) for students in groups. Using SAS, students tried to construct the concept of Vector Space. At this stage, the lecturer guides the group in completing SAS. After the appointed time, the lecturer invites each group to write down the results of their investigations on the board. The results were responded by another group, looking for similarities and differences of investigation results. Lecturers also provide opportunities for students who have GPA about 2.00 the challenge to write down the results of his group's investigation. In closing activities, the lecturer guides the students to construct the definition of the Vector Space by reconstructing the results of the appropriate investigation. Thus there is a process experienced by students in understanding the concept of Vector Space. Conceptual knowledge is part of an active process [24]. Conceptual knowledge puts the focus on relationships [25].

The learning activities conducted by students and teachers is known as the process of constructivism. Constructivism focuses upon individuals building up representations of student knowledge, which is tested against experience. When concepts are introduced in school, they are not transmitted to students, but students will attempt to fit them to the models or concepts they currently have [24]. Furthermore, at the end of the lesson, lecturer distributed the questionnaire of smiley face, collected the SAS and the smiley face questionnaire, and gave the personal task.

Some notes about student activities on regular class are described as follows. Some students look to have really learned. This can be seen from their liveliness in groups, they make notes by listening without always looking forward. Indeed, at first some of the students looked confused but long ago they shared a smile with the group members as a sign they understood the results of their group discussion. They are able to ask questions and statements to assure their understanding. This shows that they have understood the concept well. Hopkins [26] states that students asked better quality questions when they had prior knowledge of the study area and had already participated in classroom activities relating to it.

In contrast to the regular class, in remedial class students look confused, which appears in their eyes. They are not able to ask questions well. Their SAS are not done right. This initially confuses the lecturer, but ultimately the lecturer realizes that they need concept reconstruction. This is due to several things, namely students in the remedial class has a low GPA. In addition, these students make mistakes in understanding the concepts learned. The mistake made by students in doing mathematics requires attention [27, 28]. The mistake would seriously affect the subsequent understanding of their mathematical concepts [27]. In order to minimize the impact of the mistake in building the next concept, it is important to track the sources and causes of the mistake [28]. The sources can possibly be found in the formation of student's thinking scheme called the construction process of student's concept.

Based on the results of the reflection of remedial class learning, the SAS reconstructed so-called SAS reconstruction. The SAS reconstruction is used for both regular and remedial classes in the second lecture. SAS reconstruction was prepared after the research team studied what the students knew and did not know in each class, with the aim of reconstructing the students' knowledge of the concepts taught.

Secondary learning activities in regular classes by applying concept reconstruction is not as difficult as teaching in remedial classes. This is because concept errors occur only in a few students. In the remedial class, reconstruction requires considerable activity. Nevertheless both yield satisfactory results. Most students can understand the concept well at the end of the course.

4. Conclusion

Based on the results of the research, it can be concluded that there are significant differences in learning approaches in regular class and remedial class. In regular class students whom construction concept is applied are more active in learning activities than students in remedial class. Students in regular class are more willing to express their opinions, ask questions, answer questions, and working on their SAS. As a result, more percentages of students in regular class who understand the concept of vector space correctly, compared with the percentage of remedial class students who understand the concept of vector space in the first lecture.

Students in regular and remedial classes who have not understood the concept of vector space can be overcome by implementing concept reconstruction. The number of students who did not understand the concept was greatly reduced after analyzing their misconceptions. Based on the results of the analysis is done reconstruction concept to improve understanding of their concept. Thus it can be stated that concept construction is more effectively implemented in the regular class only, while in the remedial class, learning is more effective when implemented by reconstruction of concept. In other words, concept construction can only be done in regular class while in remedial class, learning with concept construction approach is not able to increase students' understanding on the concept taught. Understanding the concept of a student in a remedial class can only be carried out using the concept reconstruction approach. The reconstruction of the concept in the remedial class is preceded by learning what is known and what the student does not know.

5. Acknowledgments

The authors would like to thank the Indonesian ministries of research, technology and higher education that supported and funded this research.

References

[1] Sundayana R, Herman T, Dahlan J A and Prahmana R C I 2017 Using ASSURE learning design to develop students' mathematical communication ability *World Transactions on Engineering and Technology Education* **15** 245
[2] Prahmana R C I, Kusumah Y S and Darhim 2017 Didactic trajectory of research in mathematics

education using research-based learning *J. Phys.: Conf. Ser.* **893** 012001

[3] Tanujaya B, Prahmana R C I and Mumu J 2017 Mathematics instruction, problems, challenges, and opportunities: A case study in Manokwari regency, Indonesia *World Transactions on Engineering and Technology Education* **15** 287

[4] Runisah, Herman T, and Dahlan J A 2017 Using the 5E learning cycle with metacognitive technique to enhance students'mathematical critical thinking skills *International Journal on Emerging Mathematics Education* **1** 87

[5] Lamichhane B R and Belbase S 2017 Images of mathematics held by undergraduate students *International Journal on Emerging Mathematics Education* **1** 147

[6] Adams N and Elliott S 2013 Capture the King: Using Analogies to Teach Mathematics to Adults *International Conference The Future of Education 3rd edition* Retrieved from conference.pixel-online.net/foe2013/common/download/Paper_pdf/209-ITL37-FP-Adams-FOE2013.pdf

[7] Dorier J L and Seirpinska A 2001 *Research into the teaching and learning of linear algebra* The Teaching and Learning of Mathematics at University Level: An ICMI Study (Holton D. ed.) (Dordrecht: Kluwer Academic Publishers)

[8] Meneu M J B, Arcila M M, and Albarracin L 2016 Emphasizing visualization and physical applications in the study of eigenvectors and eigenvalues *Teaching Mathematics and Its Applications* **36** 123

[9] Dorier J L 2002 *Epistemological Analysis of the Genesis of the Theory of Vector Spaces* On the Teaching of Linear Algebra (Dorier, J.L. ed.) (Dordrecht: Kluwer Academic Publishers)

[10] Plan Y and Vershynin R 2013 One- bit compressed sensing by linear programming *Communications on Pure and Applied Mathematics* **66** 1275

[11] Stewart S, and Thomas M O J 2006 Process-object difficulties in linear algebra: Eigenvalues and eigenvectors *Proceedings of the 30th Conference of the International Group for the Psychology of Mathematics Education* Prague, Czech Republic.

[12] Thomas M O J and Stewart S 2011 Eigenvalues and eigenvectors: embodied, symbolic and formal thinking *Math. Edu. Res. J.* **23** 275

[13] Major T E and Mangope B 2012 The constructivist theory in mathematics: the case of botswana primary schools *International Review of Social Sciences and Humanities* **3** 139

[14] Polya G 1954 *Mathematics and Plausible Resoning: Induction and Analogy in Mathematics* (New Jersey: Pricenton University Press)

[15] Leo A M 2008 *On Creativity: Aweking The Creative Mind* (Selangor: Pelanduk Publications)

[16] Mofidi S A, Amiripour P, and Zadeh M H B 2012 Instruction of mathematical concepts through analogical reasoning skills *Indian Journal of Science and Technology* **5** 2916

[17] Subanji and Nusantara T 2016 Thinking process of pseudo construction in mathematics concepts *International Education Studies* **9** 17

[18] Harel G 1999 Students' Understanding of proofs: a historical analysis and implication for the teaching of geometry and linear algebra *Linear Algebra and Its Applications* **302** 601

[19] McKenney S and Reeves T C 2013 *Conducting Educational Design Research* (Abingdon: Routledge)

[20] McNiff J and Whitehead J 2006 *All you need to know about action research: An introduction* (London: Sage Publication)

[21] Koshy V 2005 *Action research for improving practice* (London: Paul Chapman Publishing)

[22] Tanujaya B and Mumu J 2016 *Penelitian tindakan kelas: panduan belajar, mengajar, dan meneliti* (Yogyakarta: Media Akademi)

[23] Slota M J 2011 Evaluating homogenous and heterogenous grouping in a k-12 setting: Do same-sex classrooms affect end-of-grade exam? *Dissertation* (Minneapolis: Capella University)

[24] McCormick R 1997 Conceptual and Procedural Knowledge *International Journal of Technology and Design Education* **7** 141

[25] Hiebert J and Lefevre P 1986 Conceptual and procedural knowledge in mathematics: An

introductory analysis In J. Hiebert (Ed.) *Conceptual and procedural knowledge: The case of mathematics* (Hillsdale, NJ: Lawrence Erlbaum Associates)
[26] Hopkins D 2008 *A Teacher's Guide to Classroom Research* (Berkshire: McGraw-Hill)
[27] Subanji and Nusantara T 2013 Karakterisasi kesalahan berpikir siswa dalam mengonstruksi konsep matematika *Jurnal Ilmu Pendidikan* **19** 208
[28] Prahmana R C I and Kusumah Y S 2016 The hypothetical learning trajectory on research in mathematics education using research-based learning *Pedagogika* **123** 42

IOP Conf. Series: Journal of Physics: Conf. Series **943** (2017) 012012 doi:10.1088/1742-6596/943/1/012012

Modified convolution theorem for one-sided quaternion linear canonical transform

Resnawati and Selvy Musdalifah

Universitas Tadulako, Jl. Soekarno – Hatta, Palu, Indonesia

E-mail: r35n4w4t1@yahoo.com

Abstract. This paper propose a new convolution theorem for one-sided quaternion linear canonical transform (QLCT) assosiated with one-sided quaternion Fourier transform (QFT). We shows some properties of one-sided QLCT and finally we derived the modified convolution theorem for one-sided QLCT.

1. Introduction

Linear canonical transform (LCT) is a linear integral transform with four independent parameters [1], which can be seen as a generalization of many mathematical transforms such as the Fourier transform, fractional Fourier transform and Fresnel transform [2]. LCT has an important role in the analysis of optical systems and signal processing [3]. Particularly, LCT is a very useful tool in the field of optics, because it is able to explain the effects of any quadratic phase system [4].

Generalization of LCT to quaternion algebra is known as quaternion linear canonical transform (QLCT). The noncommutative multiplication of quaternion causes QLCT divided into three types, i.e two-sided, right sided and left sided QLCT [5]. Some important study related to QLCT are asymptotic behaviour of QLCT, two-sided QLCT and its applications and simplified proof of uncertainty principle for two-sided QCLT [6, 7, 8].

The definition of one-sided QLCT, its properties and convolution have been found by [9]. The result showed that the convolution theorem of the two quaternion valued functions is the the sum of multiplication of one-sided QLCT from each function. This paper then will show the relationship between one-sided QLCT and one-sided QFT and derive the new formula of convolution theorem as a sum of multiplication of one-sided QFT.

In this paper, we etablish a modified definition of one-sided QLCT related to one-sided QFT and derive some important properties of one-sided QLCT such as inverse transform, linearity, shift and modulation. We further establish a modified convolution theorem for one-sided QLCT using fundamental relationship between one-sided QLCT and QFT.

The organization of this paper is as follows. Introduction of LCT, QLCT and applications we put on section 1, next in section 2, we introduce the quaternion algebra and give briefly establish our notation. In section 3, we show definition and some important properties of QLCT. In section 4, we derive the new formula of convolution theorem based on relationship between one-sided QLCT and QFT.

2. Quaternion algebra

Quaternion is a linear combination of real scalar and three orthogonal imaginary units (denoted $\boldsymbol{i}, \boldsymbol{j}$ and $\boldsymbol{k}$) with real coefficients, written as

$$\mathbb{H} = \{q = q_0 + \boldsymbol{i}q_1 + \boldsymbol{j}q_2 + \boldsymbol{k}q_3;\ q_0, q_1, q_2, q_3 \in \mathbb{R}\} \tag{1}$$

Which obey Hamiltonian multiplication rules

$$\begin{gathered} \boldsymbol{ij} = -\boldsymbol{ji} = \boldsymbol{k}, \boldsymbol{jk} = -\boldsymbol{kj} = \boldsymbol{i}, \boldsymbol{ki} = -\boldsymbol{ik} = \boldsymbol{j} \\ \boldsymbol{i}^2 = \boldsymbol{j}^2 = \boldsymbol{k}^2 = -1 \end{gathered} \tag{2}$$

For quaternion $q = q_0 + \boldsymbol{i}q_1 + \boldsymbol{j}q_2 + \boldsymbol{k}q_3 \in \mathbb{H}$, q_0 is called the scalar part of q denoted by $q_0 = Sc(q)$ and $\boldsymbol{i}q_1 + \boldsymbol{j}q_2 + \boldsymbol{k}q_3$ is called the vector (pure) part of q, denoted by $\boldsymbol{q}$. the quaternion conjugate of q defined by

$$\overline{q} = q_0 - \boldsymbol{q} = q_0 - \boldsymbol{i}q_1 - \boldsymbol{j}q_2 - \boldsymbol{k}q_3 \tag{3}$$

Multiplication of two quaternion $p = p_0 + \boldsymbol{p}$ and $q = q_0 + \boldsymbol{q}$ yields

$$qp = q_0p_0 + \boldsymbol{q}.\boldsymbol{p} + q_0\boldsymbol{p} + p_0\boldsymbol{q} + \boldsymbol{q} \times \boldsymbol{p} \tag{4}$$

where $\boldsymbol{q}.\boldsymbol{p} = (q_1p_1 + q_2p_2 + q_3p_3)$ and $\boldsymbol{q} \times \boldsymbol{p} = \boldsymbol{i}(q_2p_3 - q_3p_2) + \boldsymbol{j}(q_1p_3 - q_3p_1) + \boldsymbol{k}(q_1p_2 - q_2p_1)$. An anti involution property takes a form

$$\overline{pq} = \overline{q}\overline{p}. \tag{5}$$

From (3), we obtain the norm or modulus of $q \in \mathbb{H}$ defined as the multiplication of quaternion q with the conjugate $\overline{q}$ as

$$|q| = \sqrt{q\overline{q}} = \sqrt{q_0^2 + q_1^2 + q_2^2 + q_3^2}. \tag{6}$$

It is not difficult to see that

$$|qp| = |q||p|, \forall p, q \in \mathbb{H}. \tag{7}$$

Using the conjugate (3) and modulus of q (6), we can define the inverse of $q \in \mathbb{H}\backslash\{0\}$ as

$$q^{-1} = \frac{\overline{q}}{|q|^2}, \tag{8}$$

which is shows that $\mathbb{H}$ is a normed division algebra.

The discussion of quaternion extended into a quaternion valued function, can be expressed as $f: \mathbb{R}^2 \to \mathbb{H}$ where

$$f(x) = f_0(x) + \boldsymbol{i}f_1(x) + \boldsymbol{j}f_2(x) + \boldsymbol{k}f_3(x), \tag{9}$$

And $f_0, f_1, f_2, f_3 \in \mathbb{R}$ are a real valued function. Multiplication of two quaternion valued function $f, g: \mathbb{R}^2 \to \mathbb{H}$ defined as

$$\langle f, g \rangle = \int_{\mathbb{R}^2} f(\boldsymbol{x})\overline{g(\boldsymbol{x})} d^2\boldsymbol{x}, d\boldsymbol{x} = dx_1dx_2 \tag{10}$$

with symmetric scalar part

$$\langle f, g \rangle = \frac{1}{2}[(f, g) + (g, f)] = Sc \int_{\mathbb{R}^2} f(\boldsymbol{x})\overline{g(\boldsymbol{x})}\, d\boldsymbol{x}. \tag{11}$$

In particular, when $f = g$ we obtain the $L^2(\mathbb{R}^2; \mathbb{H})$ norm :

$$\|f\|^2 = (f, f) = \int_{\mathbb{R}^2} |f(\boldsymbol{x})|^2 d\boldsymbol{x}. \tag{12}$$

3. One-sided quaternion linear canonical transform

In this section, we will begin by defining the one-sided QFT (for simplicity of notation, we write QFT instead one-sided QFT) and next we discuss some related and important properties. Furthermore, we defining one-sided QLCT (for simplicity of notation, we also write QLCT instead one-sided QLCT) and derive some important properties.

3.1 QLCT definition and its inverse

Definition 1 *QFT of* $f \in L^1(\mathbb{R}; \mathbb{H})$ *is the transform* $\mathcal{F}_q^{(r)}\{f\}(\boldsymbol{\omega}): \mathbb{R}^2 \to \mathbb{H}$ *defined by*

IOP Conf. Series: Journal of Physics: Conf. Series **943** (2017) 012012 doi:10.1088/1742-6596/943/1/012012

$$\mathcal{F}_q^{(r)}\{f\}(\boldsymbol{\omega}) = \frac{1}{\sqrt{(2\pi)^2}}\int_{\mathbb{R}^2} f(\boldsymbol{x})\mathrm{e}^{-\mu\omega_1 x_1}\mathrm{e}^{-\mu\omega_2 x_2}d\boldsymbol{x}, \tag{13}$$

where $\boldsymbol{x} = x_1e_1 + x_2e_2$, $\boldsymbol{\omega} = \omega_1e_1 + \omega_2e_2$, and $\mathrm{e}^{-\mu\omega_1 x_1}\mathrm{e}^{-\mu\omega_2 x_2}$ are quaternion Fourier kernel.

***Definition** 2. If $f \in L^1(\mathbb{R};\mathbb{H})$ and $\mathcal{F}_q^{(r)}\{f\} \in L^1(\mathbb{R};\mathbb{H})$, then inverse of QFT defined by*

$$f(\boldsymbol{x}) = \mathcal{F}_q^{(r)^{-1}}\left[\mathcal{F}_q^{(r)}\{f\}\right](\boldsymbol{x}) = \frac{1}{\sqrt{(2\pi)^2}}\int_{\mathbb{R}^2} \mathcal{F}_q^{(r)}\{f\}(\boldsymbol{\omega})\mathrm{e}^{\mu\omega_1 x_1}\mathrm{e}^{\mu\omega_2 x_2}d\boldsymbol{\omega}, \tag{14}$$

where $\mathcal{F}_q^{(r)^{-1}}$ is called QFT inverse operator.

***Lemma** 3.* Multiplication of quaternion function $f, g \in L^1(\mathbb{R};\mathbb{H}) \cap L^2(\mathbb{R};\mathbb{H})$ and its QFT can be related by

$$\langle f, g\rangle = \langle \mathcal{F}_q^{(r)}\{f\}, \mathcal{F}_q^{(r)}\{g\}\rangle. \tag{15}$$

***Definition** 4.* Let $A_1 = (a_1, b_1, c_1, d_1)$ and $A_2 = (a_2, b_2, c_2, d_2)$ are matrix parameter satisfying $a_id_i - b_i, c_i = 1, i = 1,2$ and $a, b, c, d \in \mathbb{R}$. QLCT of quaternion function $f \in L^1(\mathbb{R}^2;\mathbb{H})$ defined by

$$\mathcal{L}_A^{(r)}\{f\}(\boldsymbol{\omega}) = \int_{\mathbb{R}^2} f(\boldsymbol{x})\, K_{A_1}(x_1, \omega_1)\, K_{A_2}(x_2, \omega_2)d\boldsymbol{x},\ \ b_1, b_2 \neq 0, \tag{16}$$

where kernel function of the QLCT given by, respectively

$$K_{A_1}(x_1, \omega_1) = \frac{1}{\sqrt{2\pi b_1}}\mathrm{e}^{\mu\frac{1}{2}\left[\frac{a_1}{b_1}x_1{}^2 - \frac{2}{b_1}x_1\omega_1 + \frac{d_1}{b_1}\omega_1{}^2 - \frac{\pi}{2}\right]} \tag{17}$$

and

$$K_{A_2}(x_2, \omega_2) = \frac{1}{\sqrt{2\pi b_2}}\mathrm{e}^{\mu\frac{1}{2}\left[\frac{a_2}{b_2}x_2{}^2 - \frac{2}{b_2}x_2\omega_2 + \frac{d_2}{b_2}\omega_2{}^2 - \frac{\pi}{2}\right]}, \tag{18}$$

with μ is a pure quaternion satisfying $\mu^2 = -1$.
If $A_i = (0,1,-1,0), i = 1,2$, then QLCT in the form (16) reduces to QFT, that is

$$\mathcal{L}_A^{(r)}\{f\}(\boldsymbol{\omega}) = \int_{\mathbb{R}^2} f(\boldsymbol{x})\frac{1}{\sqrt{2\pi}}\mathrm{e}^{-\mu x_1\omega_1}\frac{1}{\sqrt{2\pi}}\mathrm{e}^{-\mu x_2\omega_2}\, d\boldsymbol{x} = \mathcal{F}_q^{(r)}\{f\}(\boldsymbol{\omega}). \tag{19}$$

We see that equation (19) that we established is the same formula of QFT in equation (13). In the following theorem, we establish the the relationship of QLCT and QFT where QLCT can be expressed to QFT formula.

***Theorem** 5.* QLCT of quaternion function $f \in L^1(\mathbb{R}^2;\mathbb{H})$ with matrix parameter $A_1 = (a_1, b_1, c_1, d_1)$ dan $A_2 = (a_2, b_2, c_2, d_2)$ can be reduced to QFT

$$\mathcal{F}_q^{(r)}\left\{g_f\right\}(\boldsymbol{\omega}) = \frac{1}{2\pi}\int_{\mathbb{R}^2} g_f(\boldsymbol{x})\mathrm{e}^{-\mu x_1\omega_1}\mathrm{e}^{-\mu x_2\omega_2}d\boldsymbol{x} \tag{20}$$

Where $\mathcal{F}_q^{(r)}\left\{g_f\right\}(\boldsymbol{\omega}) = \tilde{\mathcal{F}}(\boldsymbol{b\omega})$

$$g_f(\boldsymbol{x}) = \tilde{f}(\boldsymbol{x})\frac{\mathrm{e}^{-\mu\frac{\pi}{4}}}{\sqrt{b_1}}\frac{\mathrm{e}^{-\mu\frac{\pi}{4}}}{\sqrt{b_2}} \tag{21}$$

$$\tilde{f}(\boldsymbol{x}) = \tilde{f}(\boldsymbol{x})\mathrm{e}^{\mu\frac{1}{2}\left(\frac{a_1}{b_1}x_1{}^2\right)}\mathrm{e}^{\mu\frac{1}{2}\left(\frac{a_2}{b_2}x_2{}^2\right)}.$$

with

$$\tilde{\mathcal{F}}(\boldsymbol{\omega}) = \int_{\mathbb{R}^2} g_f(\boldsymbol{x})\frac{1}{\sqrt{(2\pi)^2}}\mathrm{e}^{-\mu x_1\frac{\omega_1}{b_1}}\mathrm{e}^{-\mu x_2\frac{\omega_2}{b_2}}d\boldsymbol{x} = \mathcal{L}_A^{(r)}\{f\}(\boldsymbol{\omega})\mathrm{e}^{-\mu\frac{1}{2}\left(\frac{d_1}{b_1}\omega_1{}^2\right)}\mathrm{e}^{-\mu\frac{1}{2}\left(-\frac{2}{b_2}x_2\omega_2\right)} \tag{22}$$

Proof. From QLCT definition in (16), we have

$$\mathcal{L}_A^{(r)}\{f\}(\boldsymbol{\omega}) = \int_{\mathbb{R}^2} f(\boldsymbol{x})\, K_{A_1}(x_1, \omega_1)K_{A_2}(x_2, \omega_2)d\boldsymbol{x}$$

$$= \int_{\mathbb{R}^2} \left[f(\boldsymbol{x}) \frac{1}{\sqrt{2\pi b_1}} e^{\mu\frac{1}{2}\left(\frac{a_1}{b_1}x_1{}^2\right)} e^{\mu\frac{1}{2}\left(\frac{a_2}{b_2}x_2{}^2\right)} \right] e^{\mu\frac{1}{2}\left(-\frac{2}{b_1}x_1\omega_1\right)} e^{\mu\frac{1}{2}\left(\frac{d_1}{b_1}\omega_1{}^2\right)}$$
$$\cdot \frac{1}{\sqrt{2\pi b_2}} e^{\mu\frac{1}{2}\left(-\frac{\pi}{2}\right)} e^{\mu\frac{1}{2}\left(-\frac{2}{b_2}x_2\omega_2\right)} e^{\mu\frac{1}{2}\left(\frac{d_2}{b_2}\omega_2{}^2\right)} e^{\mu\frac{1}{2}\left(-\frac{\pi}{2}\right)} d\boldsymbol{x}. \qquad (23)$$

Let $\tilde{f}(\boldsymbol{x}) e^{\mu\frac{1}{2}\left(\frac{a_1}{b_1}x_1{}^2\right)} e^{\mu\frac{1}{2}\left(\frac{a_2}{b_2}x_2{}^2\right)}$, then we have

$$\mathcal{L}_A^{(r)}\{f\}(\boldsymbol{\omega}) = \int_{\mathbb{R}^2} \tilde{f}(\boldsymbol{x}) \frac{e^{-\mu\left(\frac{\pi}{4}\right)}}{\sqrt{b_1}} \frac{e^{-\mu\left(\frac{\pi}{4}\right)}}{\sqrt{b_2}} \frac{1}{\left(\sqrt{2\pi}\right)^2} e^{\mu\frac{1}{2}\left(-\frac{2}{b_1}x_1\omega_1\right)} e^{\mu\frac{1}{2}\left(\frac{d_1}{b_1}\omega_1{}^2\right)}$$
$$\cdot e^{\mu\frac{1}{2}\left(-\frac{2}{b_2}x_2\omega_2\right)} e^{\mu\frac{1}{2}\left(\frac{d_2}{b_2}\omega_2{}^2\right)} d\boldsymbol{x} \qquad (24)$$

Multiply both side with $e^{-\mu\frac{1}{2}\left(\frac{d_1}{b_1}\omega_1{}^2\right)} e^{-\mu\frac{1}{2}\left(\frac{d_2}{b_2}\omega_2{}^2\right)}$ then we have

$$\mathcal{L}_A^{(r)}\{f\}(\boldsymbol{\omega}) e^{-\mu\frac{1}{2}\left(\frac{d_1}{b_1}\omega_1{}^2\right)} e^{-\mu\frac{1}{2}\left(-\frac{2}{b_2}x_2\omega_2\right)} = \int_{\mathbb{R}^2} \tilde{f}(\boldsymbol{x}) \frac{e^{-\mu\left(\frac{\pi}{4}\right)}}{\sqrt{b_1}} \frac{e^{-\mu\left(\frac{\pi}{4}\right)}}{\sqrt{b_2}} \frac{1}{\left(\sqrt{2\pi}\right)^2} e^{\mu\frac{1}{2}\left(-\frac{2}{b_1}x_1\omega_1\right)} e^{\mu\frac{1}{2}\left(-\frac{2}{b_2}x_2\omega_2\right)} d\boldsymbol{x}. \qquad (25)$$

Let $g_f(\boldsymbol{x}) = \tilde{f}(\boldsymbol{x}) \frac{e^{-\mu\left(\frac{\pi}{4}\right)}}{\sqrt{b_1}} \frac{e^{-\mu\left(\frac{\pi}{4}\right)}}{\sqrt{b_2}}$, then

$$\mathcal{L}_A^{(r)}\{f\}(\boldsymbol{\omega}) e^{-\mu\frac{1}{2}\left(\frac{d_1}{b_1}\omega_1{}^2\right)} e^{-\mu\frac{1}{2}\left(-\frac{2}{b_2}x_2\omega_2\right)} = \int_{\mathbb{R}^2} g_f(\boldsymbol{x}) \frac{1}{2\pi} e^{\mu\left(-x_1\frac{\omega_1}{b_1}\right)} e^{\mu\left(-x_2\frac{\omega_2}{b_2}\right)} d\boldsymbol{x}$$
$$= \frac{1}{2\pi} \int_{\mathbb{R}^2} g_f(\boldsymbol{x})\, e^{-\mu x_1\left(\frac{\omega_1}{b_1}\right)} e^{-\mu x_2\left(\frac{\omega_2}{b_2}\right)} d\boldsymbol{x}$$
$$= \mathcal{F}_q^{(r)}\{g_f\}\left(\frac{\boldsymbol{\omega}}{\boldsymbol{b}}\right) = \tilde{\mathcal{F}}(\boldsymbol{b\omega}).$$

This ends the proof.

Theorem 6. If $f \in L^1(\mathbb{R}^2;\mathbb{H})$ and $\mathcal{L}_A^{(r)}\{f\} \in L^1(\mathbb{R}^2;\mathbb{H})$, then QLCT inverse can be derived from QFT.
Proof. We have

$$g_f(\boldsymbol{x}) = \frac{1}{2\pi} \int_{\mathbb{R}^2} \mathcal{F}_q^{(r)}\left\{g_f\right\}(\boldsymbol{\omega}) e^{\mu x_1\omega_1} e^{\mu x_2\omega_2} d\boldsymbol{\omega} = \frac{1}{b_1 b_2} \frac{1}{2\pi} \int_{\mathbb{R}^2} \tilde{\mathcal{F}}\left(\frac{\boldsymbol{\omega}}{\boldsymbol{b}}\right) e^{\mu x_1\left(\frac{\omega_1}{b_1}\right)} e^{\mu x_2\left(\frac{\omega_2}{b_2}\right)} d\boldsymbol{\omega}$$
$$= \frac{1}{b_1 b_2\, 2\pi} \int_{\mathbb{R}^2} \mathcal{L}_A^{(r)}\{f\}(\boldsymbol{\omega})\, e^{-\mu\frac{1}{2}\left(\frac{d_1}{b_1}\omega_1{}^2\right)} e^{-\mu\frac{1}{2}\left(\frac{d_2}{b_2}\omega_2{}^2\right)} e^{\mu x_1\left(\frac{\omega_1}{b_1}\right)} e^{\mu x_2\left(\frac{\omega_2}{b_2}\right)} d\boldsymbol{\omega} \qquad (26)$$

It means that

$$f(\boldsymbol{x}) e^{\mu\frac{1}{2}\left(\frac{a_1}{b_1}x_1{}^2\right)} e^{\mu\frac{1}{2}\left(\frac{a_2}{b_2}x_2{}^2\right)} \frac{e^{-\mu\left(\frac{\pi}{4}\right)}}{\sqrt{b_1}} \frac{e^{-\mu\left(\frac{\pi}{4}\right)}}{\sqrt{b_2}} = \frac{1}{b_1 b_2\, 2\pi} \int_{\mathbb{R}^2} \mathcal{L}_A^{(r)}\{f\}(\boldsymbol{\omega})\, e^{-\mu\frac{1}{2}\left(\frac{d_1}{b_1}\omega_1{}^2\right)} e^{-\mu\frac{1}{2}\left(\frac{d_2}{b_2}\omega_2{}^2\right)}$$
$$\cdot e^{\mu x_1\left(\frac{\omega_1}{b_1}\right)} e^{\mu x_2\left(\frac{\omega_2}{b_2}\right)} d\boldsymbol{\omega}$$

Or equivalent with

$$f(\boldsymbol{x}) = \frac{\sqrt{b_1}\sqrt{b_2}}{b_1 b_2\left(\sqrt{2\pi}\right)^2} \int_{\mathbb{R}^2} \mathcal{L}_A^{(r)}\{f\}(\boldsymbol{\omega})\, e^{-\mu\frac{1}{2}\left[\frac{a_1}{b_1}x_1{}^2 - \frac{2}{b_1}x_1\omega_1 + \frac{d_1}{b_1}\omega_1{}^2 - \frac{\pi}{2}\right]} e^{-\mu\frac{1}{2}\left[\frac{a_2}{b_2}x_2{}^2 - \frac{2}{b_2}x_2\omega_2 + \frac{d_2}{b_2}\omega_2{}^2 - \frac{\pi}{2}\right]} d\boldsymbol{\omega}$$
$$= \int_{\mathbb{R}^2} \mathcal{L}_A^{(r)}\{f\}(\boldsymbol{\omega}) \frac{1}{\sqrt{2\pi b_1}} e^{-\mu\frac{1}{2}\left[\frac{a_1}{b_1}x_1{}^2 - \frac{2}{b_1}x_1\omega_1 + \frac{d_1}{b_1}\omega_1{}^2 - \frac{\pi}{2}\right]} \frac{1}{\sqrt{2\pi b_2}} e^{-\mu\frac{1}{2}\left[\frac{a_2}{b_2}x_2{}^2 - \frac{2}{b_2}x_2\omega_2 + \frac{d_2}{b_2}\omega_2{}^2 - \frac{\pi}{2}\right]} d\boldsymbol{\omega}$$
$$= \int_{\mathbb{R}^2} \mathcal{L}_A^{(r)}\{f\}(\boldsymbol{\omega})\, \overline{K_{A_1}(x_1,\omega_1) K_{A_2}(x_2,\omega_2)}\, d\boldsymbol{\omega},$$

Which is inverse of QLCT. This proves the theorem.

3.2 QLCT properties

In this section, we give properties of QLCT due to noncommutative multiplication of quaternion, such as linearity, shifting, and modulation. Due to noncommutative property of quaternion, QLCT has only left linearity with spesific constant defined by

$$\alpha, \beta \in \{q | q = q_0 + \mu q_1, q_0, q_1 \in \mathbb{R}\}, \tag{27}$$

with

$$\mathcal{L}_A^{(r)}\{\alpha f + \beta g\}(\boldsymbol{\omega}) = \alpha \mathcal{L}_A^{(r)}\{f\}(\boldsymbol{\omega}) + \beta \mathcal{L}_A^{(r)}\{g\}(\boldsymbol{\omega}). \tag{28}$$

***Theorem** 7.*Given a quaternion function $f \in L^2(\mathbb{R}^2; \mathbb{H})$, let $\tau_{\boldsymbol{k}} f(\boldsymbol{x})$ denote the shifted function defined by$\tau_{\boldsymbol{k}} f(\boldsymbol{x}) = f(\boldsymbol{x} - \boldsymbol{k})$, where $\boldsymbol{k} \in \mathbb{R}^2$. Then we get

$$\mathcal{L}_A^{(r)}\{\tau_{\boldsymbol{k}} f\}(\boldsymbol{\omega}) = \mathcal{L}_A^{(r)}\{f\}(\omega_1 - a_1 k_1, \omega_2 - a_2 k_2) e^{\frac{-\mu a_1 c_1 k_1^2}{2 + i c_1 k_1 \omega_1}} e^{\frac{-\mu a_2 c_2 k_2^2}{2 + i c_2 k_2 \omega_2}}. \tag{29}$$

Proof. Taking the definition of QLCT (16), we get

$$\mathcal{L}_A^{(r)}\{\tau_{\boldsymbol{k}} f\}(\boldsymbol{\omega}) = \int_{\mathbb{R}^2} \boldsymbol{f}(\boldsymbol{x} - \boldsymbol{k}) \frac{1}{\sqrt{2\pi b_1}} e^{\mu \frac{1}{2}\left[\frac{a_1}{b_1} x_1{}^2 - \frac{2}{b_1} x_1 \omega_1 + \frac{d_1}{b_1} \omega_1{}^2 - \frac{\pi}{2}\right]} \frac{1}{\sqrt{2\pi b_2}} e^{\mu \frac{1}{2}\left[\frac{a_2}{b_2} x_2{}^2 - \frac{2}{b_2} x_2 \omega_2 + \frac{d_2}{b_2} \omega_2{}^2 - \frac{\pi}{2}\right]} d\boldsymbol{x}. \tag{30}$$

By making the change of a variable $\boldsymbol{x} - \boldsymbol{k} = \boldsymbol{m}$, we easily obtain

$$\mathcal{L}_A^{(r)}\{\tau_k f\}(\boldsymbol{\omega}) = \int_{\mathbb{R}^2} f(\boldsymbol{m}) \frac{1}{\sqrt{2\pi b_1}} e^{\mu \frac{1}{2}\left[\frac{a_1}{b_1} m_1{}^2 - \frac{2}{b_1} m_1(\omega_1 - k_1 a_1) + \frac{d_1}{b_1} \omega_1{}^2 - \frac{\pi}{2}\right]} e^{\mu \frac{1}{2}\left[\frac{a_1}{b_1} k_1{}^2\right]} e^{-\mu \frac{1}{2}\left[2 k_1 \frac{\omega_1}{b_1}\right]}$$
$$\cdot \frac{1}{\sqrt{2\pi b_2}} e^{\mu \frac{1}{2}\left[\frac{a_2}{b_2} m_2{}^2 - \frac{2}{b_2} m_2(\omega_2 - k_2 a_2) + \frac{d_2}{b_2} \omega_2{}^2 - \frac{\pi}{2}\right]} e^{\mu \frac{1}{2}\left[\frac{a_2}{b_2} k_2{}^2\right]} e^{-\mu \frac{1}{2}\left[\left(2 k_2 \frac{\omega_2}{b_2}\right)\right]} d\boldsymbol{m}.$$

Therefore, we futher get

$$\mathcal{L}_A^{(r)}\{\tau_k f\}(\boldsymbol{\omega}) = \int_{\mathbb{R}^2} f(\boldsymbol{m}) \frac{1}{\sqrt{2\pi b_1}} e^{\mu \frac{1}{2}\left[\frac{a_1}{b_1} m_1{}^2 - \frac{2}{b_1} m_1(\omega_1 - k_1 a_1) + \frac{d_1}{b_1}(\omega_1 - k_1 a_1 + k_1 a_1)^2 - \frac{\pi}{2}\right]}$$
$$\cdot \frac{1}{\sqrt{2\pi b_2}} e^{\mu \frac{1}{2}\left[\frac{a_2}{b_2} m_2{}^2 - \frac{2}{b_2} m_2(\omega_2 - k_2 a_2) + \frac{d_2}{b_2}(\omega_2 - k_2 a_2 + k_2 a_2)^2 - \frac{\pi}{2}\right]} d\boldsymbol{m}$$
$$\cdot e^{\mu \frac{1 d_1}{2 b_1}\left[2(\omega_1 - k_1 a_1) k_1 a_1 + (k_1 a_1)^2\right]} e^{\mu \frac{1 d_2}{2 b_2}\left[2(\omega_2 - k_2 a_2) k_2 a_2 + (k_2 a_2)^2\right]}$$
$$\cdot e^{\mu \frac{1}{2}\left[\frac{a_1}{b_1} k_1{}^2\right]} e^{-\mu \frac{1}{2}\left[2 k_1 \frac{\omega_1}{b_1}\right]} e^{\mu \frac{1}{2}\left[\frac{a_2}{b_2} k_2{}^2\right]} e^{-\mu \frac{1}{2}\left[\left(2 k_2 \frac{\omega_2}{b_2}\right)\right]}. \tag{31}$$

Applying the definition of the QLCT (16), the above expression can be rewritten in the form

$$\mathcal{L}_A^{(r)}\{\tau_k f\}(\boldsymbol{\omega}) = \mathcal{L}_A^{(r)}\{\tau_k f\}(\omega_1 - a_1 k_1, \omega_2 - a_2 k_2) e^{\mu \frac{1 d_1}{2 b_1}\left[2(\omega_1 - k_1 a_1) k_1 a_1 + (k_1 a_1)^2\right]}$$
$$\cdot e^{\mu \frac{1 d_2}{2 b_2}\left[2(\omega_2 - k_2 a_2) k_2 a_2 + (k_2 a_2)^2\right]} e^{\mu \frac{1}{2}\left[\frac{a_1}{b_1} k_1{}^2\right]} e^{-\mu \frac{1}{2}\left[2 k_1 \frac{\omega_1}{b_1}\right]} e^{\mu \frac{1}{2}\left[\frac{a_2}{b_2} k_2{}^2\right]} e^{-\mu \frac{1}{2}\left[\left(2 k_2 \frac{\omega_2}{b_2}\right)\right]}. \tag{32}$$

We notice that

$$e^{\mu \frac{1 d_1}{2 b_1}\left[2(\omega_1 - k_1 a_1) k_1 a_1 + (k_1 a_1)^2\right]} e^{\mu \frac{1}{2}\left[\frac{a_1}{b_1} k_1{}^2\right]} e^{-\mu \frac{1}{2}\left[2 k_1 \frac{\omega_1}{b_1}\right]} = e^{\mu k_1 \omega_1\left[\frac{d_1 a_1}{b_1} - \frac{1}{b_1}\right]} e^{-\mu \frac{1}{2} k_1{}^2\left[\frac{d_1 a_1}{b_1} - \frac{1}{b_1}\right]},$$
$$e^{\mu \frac{1 d_2}{2 b_2}\left[2(\omega_2 - k_2 a_2) k_2 a_2 + (k_2 a_2)^2\right]} e^{\mu \frac{1}{2}\left[\frac{a_2}{b_2} k_2{}^2\right]} e^{-\mu \frac{1}{2}\left[2 k_2 \frac{\omega_2}{b_2}\right]} = e^{\mu k_2 \omega_2\left[\frac{d_2 a_2}{b_2} - \frac{1}{b_2}\right]} e^{-\mu \frac{1}{2} k_2{}^2\left[\frac{d_2 a_2}{b_2} - \frac{1}{b_2}\right]}.$$

Because $d_i a_i - b_i c_i = 1$, then $\frac{d_i a_i}{b_i} - \frac{1}{b_i} = c_i$, for $i = 1,2$. This means

$$e^{\mu \frac{1 d_1}{2 b_1}\left[2(\omega_1 - k_1 a_1) k_1 a_1 + (k_1 a_1)^2\right]} e^{\mu \frac{1}{2}\left[\frac{a_1}{b_1} k_1{}^2\right]} e^{-\mu \frac{1}{2}\left[2 k_1 \frac{\omega_1}{b_1}\right]} = e^{\mu k_1 \omega_1 c_1} e^{-\mu \frac{1}{2} k_1{}^2 c_1}, \tag{33}$$

and

$$e^{\mu \frac{1 d_2}{2 b_2}\left[2(\omega_2 - k_2 a_2) k_2 a_2 + (k_2 a_2)^2\right]} e^{\mu \frac{1}{2}\left[\frac{a_2}{b_2} k_2{}^2\right]} e^{-\mu \frac{1}{2}\left[\left(2 k_2 \frac{\omega_2}{b_2}\right)\right]} = e^{\mu k_2 \omega_2 c_2} e^{-\mu \frac{1}{2} k_2{}^2 c_2}. \tag{34}$$

Therefore, the equation (32) can be simplify as

$$\mathcal{L}_A^{(r)}\{\tau_k f\}(\boldsymbol{\omega}) = \mathcal{L}_A^{(r)}\{\tau_k f\}(\omega_1 - a_1 k_1, \omega_2 - a_2 k_2) e^{\mu k_1 \omega_1 c_1} e^{-\mu \frac{1}{2} k_1{}^2 c_1} e^{\mu k_2 \omega_2 c_2} e^{-\mu \frac{1}{2} k_2{}^2 c_2}.$$

And this is the desired result.

Theorem 8. Let $\mathbb{M}_{w_0}f$ be modulation operator defined by $\mathbb{M}_{\omega_0}f(x) = f(x)\,\mathrm{e}^{\mu x_1 u_0}\mathrm{e}^{\mu x_2 v_0}$ with $\omega_0 = u_0e_1 + v_0e_2$. Then

$$\mathcal{L}_A^{(r)}\{\mathbb{M}_{\omega_0}f\}(\boldsymbol{\omega}) = \mathcal{L}_A^{(r)}\{f\}(\omega_1 - u_0b_1, \omega_2 - v_0b_2)\mathrm{e}^{\mu\omega_1 u_0 d_1}\mathrm{e}^{-\mu\left(2u_0{}^2 b_1 d_1 - \frac{u_0{}^2 b_1{}^2 d_1}{2}\right)} \cdot \mathrm{e}^{\omega_2 v_0 d_2}\mathrm{e}^{-\mu\left(v_0{}^2 b_2 d_2 - \frac{v_0{}^2 b_2{}^2 d_2}{2}\right)}. \tag{35}$$

Proof. From definition 4, it follows that

$$\mathcal{L}_A^{(r)}\{\mathbb{M}_{\omega_0}f\}(\boldsymbol{\omega}) = \int_{\mathbb{R}^2} f(\boldsymbol{x})\mathrm{e}^{\mu x_1 u_0}\mathrm{e}^{\mu x_2 v_0}\frac{1}{\sqrt{2\pi b_1}}\mathrm{e}^{\mu\frac{1}{2}\left[\frac{a_1}{b_1}x_1{}^2 - \frac{2}{b_1}x_1\omega_1 + \frac{d_1}{b_1}\omega_1{}^2 - \frac{\pi}{2}\right]} \cdot \frac{1}{\sqrt{2\pi b_2}}\mathrm{e}^{\mu\frac{1}{2}\left[\frac{a_2}{b_2}x_2{}^2 - \frac{2}{b_2}x_2\omega_2 + \frac{d_2}{b_2}\omega_2{}^2 - \frac{\pi}{2}\right]}\,d\boldsymbol{x}$$

$$= \int_{\mathbb{R}^2} f(\boldsymbol{x})\frac{1}{\sqrt{2\pi b_1}}\mathrm{e}^{\mu\frac{1}{2}\left[\frac{a_1}{b_1}x_1{}^2 - \frac{2}{b_1}x_1\omega_1 + \frac{d_1}{b_1}\omega_1{}^2 + 2x_1u_0 - \frac{\pi}{2}\right]} \cdot \frac{1}{\sqrt{2\pi b_2}}\mathrm{e}^{\mu\frac{1}{2}\left[\frac{a_2}{b_2}x_2{}^2 - \frac{2}{b_2}x_2\omega_2 + \frac{d_2}{b_2}\omega_2{}^2 + 2e^{\mu x_2 v_0} - \frac{\pi}{2}\right]}\,d\boldsymbol{x}$$

$$= \int_{\mathbb{R}^2} f(\boldsymbol{x})\frac{1}{\sqrt{2\pi b_1}}\mathrm{e}^{\mu\frac{1}{2}\left[\frac{a_1}{b_1}x_1{}^2 - \frac{2}{b_1}x_1(\omega_1 - u_0b_1) + \frac{d_1}{b_1}\omega_1{}^2 - \frac{\pi}{2}\right]} \cdot \frac{1}{\sqrt{2\pi b_2}}\mathrm{e}^{\mu\frac{1}{2}\left[\frac{a_2}{b_2}x_2{}^2 - \frac{2}{b_2}x_2(\omega_2 - v_0b_2) + \frac{d_2}{b_2}\omega_2{}^2 - \frac{\pi}{2}\right]}\,d\boldsymbol{x}.$$

Subsequent calculation reveal that

$$\mathcal{L}_A^{(r)}\{\mathbb{M}_{\omega_0}f\}(\boldsymbol{\omega}) = \int_{\mathbb{R}^2} f(\boldsymbol{x})\frac{1}{\sqrt{2\pi b_1}}\mathrm{e}^{\mu\frac{1}{2}\left[\frac{a_1}{b_1}x_1{}^2 - \frac{2}{b_1}x_1(\omega_1 - u_0b_1) + \frac{d_1}{b_1}\left((\omega_1 - u_0b_1)^2 + 2(w_1 - u_0b_1)u_0b_1 + u_0{}^2b_1{}^2\right) - \frac{\pi}{2}\right]} \cdot \frac{1}{\sqrt{2\pi b_2}}\mathrm{e}^{\mu\frac{1}{2}\left[\frac{a_2}{b_2}x_2{}^2 - \frac{2}{b_2}x_2(\omega_2 - v_0b_2) + \frac{d_2}{b_2}\left((\omega_2 - v_0b_2)^2 + 2(\omega_2 - v_0b_2)v_0b_2 + v_0{}^2b_2{}^2\right) - \frac{\pi}{2}\right]}d\boldsymbol{x}$$

Hence, we get

$$\mathcal{L}_A^{(r)}\{\mathbb{M}_{\omega_0}f\}(\boldsymbol{\omega}) = \mathcal{L}_A^{(r)}\{f\}(\omega_1 - u_0b_1, \omega_2 - v_0b_2)\mathrm{e}^{\mu\omega_1 u_0 d_1}\mathrm{e}^{-\mu\left(2u_0{}^2 b_1 d_1 - \frac{u_0{}^2 b_1{}^2 d_1}{2}\right)}. \cdot \mathrm{e}^{\mu\omega_2 v_0 d_2}\mathrm{e}^{-\mu\left(v_0{}^2 b_2 d_2 - \frac{v_0{}^2 b_2{}^2 d_2}{2}\right)}. \tag{36}$$

This is desired result.

4. Convolution theorem for QLCT

In this section we show that QLCT of convolution quaternionic function. Before doing so, we will define the definition of convolution for quaternion valued function and its properties.

Definition 9. Convolution of quaternion function $f, g \in L^2(\mathbb{R}^2; \mathbb{H})$ denoted by $f * g$, defined by

$$(f * g)(\boldsymbol{x}) = \int_{\mathbb{R}^2} f(\boldsymbol{t})g(\boldsymbol{x} - \boldsymbol{t})\mathrm{e}^{\frac{a_1}{b_1}\mu t_1(t_1 - x_1)}\mathrm{e}^{\frac{a_2}{b_2}\mu t_2(t_2 - x_2)}d\boldsymbol{t}. \tag{37}$$

In the following, we summarize the elementary properties of quaternion convolution for QLCT.

Lemma 10. For quaternion functons f, g and $h \in L^2(\mathbb{R}^2; \mathbb{H})$ and q_1 and q_2 are quaternion constants, we get linearity property defined by

$$(q_1f + q_2g) * h = q_1(f * h) + q_2(g * h), \tag{38}$$

and for real constants r_1 and r_2, we get

$$h * (fr_1 + gr_2) = (h * f)r_1 + (h * g)r_2. \tag{39}$$

IOP Conf. Series: Journal of Physics: Conf. Series **943** (2017) 012012 doi:10.1088/1742-6596/943/1/012012

Proof.

$$(q_1f+q_2g)*h(\boldsymbol{x}) = \int_{\mathbb{R}^2} (q_1f+q_2g)(\boldsymbol{t})h(\boldsymbol{x}-\boldsymbol{t})e^{\frac{a_1}{b_1}\mu t_1(t_1-x_1)}e^{\frac{a_2}{b_2}\mu t_2(t_2-x_2)}d\boldsymbol{t}$$

$$= \int_{\mathbb{R}^2} \left(q_1f(\boldsymbol{t})h(\boldsymbol{x}-\boldsymbol{t}) + q_2g(\boldsymbol{t})h(\boldsymbol{x}-\boldsymbol{t})\right)e^{\frac{a_1}{b_1}\mu t_1(t_1-x_1)}e^{\frac{a_2}{b_2}\mu t_2(t_2-x_2)}d\boldsymbol{t}$$

we easily see that

$$(q_1f+q_2g)*h(\boldsymbol{x}) = q_1\int_{\mathbb{R}^2} f(\boldsymbol{t})h(\boldsymbol{x}-\boldsymbol{t})e^{\frac{a_1}{b_1}\mu t_1(t_1-x_1)}e^{\frac{a_2}{b_2}\mu t_2(t_2-x_2)}d\boldsymbol{t}$$

$$+q_2\int_{\mathbb{R}^2} g(\boldsymbol{t})h(\boldsymbol{x}-\boldsymbol{t})e^{\frac{a_1}{b_1}\mu t_1(t_1-x_1)}e^{\frac{a_2}{b_2}\mu t_2(t_2-x_2)}d\boldsymbol{t}$$

and by the definition of convolution finally we get

$$(q_1f+q_2g)*h(\boldsymbol{x}) = q_1(f*h)(\boldsymbol{x}) + q_2(g*h)(\boldsymbol{x}).$$

This proof the theorem. For the second linearity, we directly see that the theorem is applicable.

Lemma 11. For quaternion function $f, g \in L^2(\mathbb{R}^2;\mathbb{H})$, let $\tau_{\boldsymbol{a}}f(\boldsymbol{x})$ denote the shifted function defined by $\tau_{\boldsymbol{a}}f(\boldsymbol{x}) = f(\boldsymbol{x}-\boldsymbol{a})$, where $\boldsymbol{a} \in \mathbb{R}^2$. The convolution of $\tau_{\boldsymbol{a}}f * g$ defined by

$$(\tau_{\boldsymbol{a}}f * g)(\boldsymbol{x}) = \tau_{\boldsymbol{a}}(f*g). \tag{40}$$

Proof. From convolution definition (37), we see that

$$(\tau_{\boldsymbol{a}}f * g)(\boldsymbol{x}) = \int_{\mathbb{R}^2} \tau_{\boldsymbol{a}}f(\boldsymbol{t})g(\boldsymbol{x}-\boldsymbol{t})\, e^{\frac{a_1}{b_1}\mu t_1(t_1-x_1)}e^{\frac{a_2}{b_2}\mu t_2(t_2-x_2)}d\boldsymbol{t}$$

$$= \int_{\mathbb{R}^2} f(\boldsymbol{t}-\boldsymbol{a})g(\boldsymbol{x}-\boldsymbol{t})\, e^{\frac{a_1}{b_1}\mu t_1(t_1-x_1)}e^{\frac{a_2}{b_2}\mu t_2(t_2-x_2)}d\boldsymbol{t}$$

For $\boldsymbol{s} = \boldsymbol{t} - \boldsymbol{a}$,

$$(\tau_{\boldsymbol{a}}f * g)(\boldsymbol{x}) = \int_{\mathbb{R}^2} f(\boldsymbol{s})g\left((\boldsymbol{x}-\boldsymbol{a})-\boldsymbol{s}\right) e^{\frac{a_1}{b_1}\mu s_1(s_1-x_1)}e^{\frac{a_2}{b_2}\mu s_2(s_2-x_2)}d\boldsymbol{t} = (f*g)(\boldsymbol{x}-\boldsymbol{a})$$

$$= \tau_{\boldsymbol{a}}(f*g)(\boldsymbol{x}).$$

The next two lemmas defined the conjugation and assosiative properties of convolution of two convolution theorem.

Lemma 12. Given quaternion function $f, g \in L^2(\mathbb{R}^2;\mathbb{H})$, conjugation of convolution f, g, defined by

$$\overline{(f*g)(\boldsymbol{x})} = \bar{g} * \bar{f}(\boldsymbol{x}) \tag{41}$$

Lemma 13. Given quaternion function $f, g, h \in L^2(\mathbb{R}^2;\mathbb{H})$, assosiative property of convolution f, g and h defined by

$$(f*g)*h = f*(g*h). \tag{42}$$

Finally, we will derive modified convolution theorem for QLCT using relationship between QLCT and QFT definition. the result compared to [9]. We show that the new formula of convolution theorem for QLCT can be derived from QFT in the following theorem.

Theorem 14. Let $f(\boldsymbol{x}) = f_0(\boldsymbol{x}) + \boldsymbol{i}f_1(\boldsymbol{x}) + \boldsymbol{j}f_2(\boldsymbol{x}) + \boldsymbol{k}f_3(\boldsymbol{x})$ dan $g(\boldsymbol{x}) = g_0(\boldsymbol{x}) + \boldsymbol{i}g_1(\boldsymbol{x}) + \boldsymbol{j}g_2(\boldsymbol{x}) + \boldsymbol{k}g_3(\boldsymbol{x})$ be two quaternion valued function,then one sided QLCT of $f, g \in L^2(\mathbb{R}^2;\mathbb{H})$ is given by

$$\mathcal{L}_A^{(r)}\{f * g\}(\boldsymbol{\omega}) = \mathcal{F}_q^{(r)}\{g_f\}\mathcal{F}_q^{(r)}\{\tilde{f}_0\} + i\mathcal{F}_q^{(r)}\{g_f\}\mathcal{F}_q^{(r)}\{\tilde{f}_1\} + j\mathcal{F}_q^{(r)}\{g_f\}\mathcal{F}_q^{(r)}\{\tilde{f}_2\} + k\mathcal{F}_q^{(r)}\{g_f\}\mathcal{F}_q^{(r)}\{\tilde{f}_3\}, \tag{43}$$

where

$$\tilde{f}_n(\boldsymbol{t}) = f_n e^{\mu\frac{1}{2}\frac{a_1}{b_1}(t_1{}^2)} e^{\mu\frac{1}{2}\frac{a_2}{b_2}(t_2{}^2)}, n = 0,1,2,3.$$

Proof. In this proof, we will decompose the quaternion functions. By the definition of QLCT and its convolution, we obtain

$$\mathcal{L}_A^{(r)}\{f * g\}(\boldsymbol{\omega}) = \int_{\mathbb{R}^2}\int_{\mathbb{R}^2} f(\boldsymbol{t})g(\boldsymbol{x}-\boldsymbol{t})e^{\frac{a_1}{b_1}\mu t_1(t_1-x_1)}e^{\frac{a_2}{b_2}\mu t_2(t_2-x_2)}d\boldsymbol{t}$$
$$\cdot\frac{1}{\sqrt{2\pi b_1}}e^{\mu\frac{1}{2}\left[\frac{a_1}{b_1}x_1{}^2-\frac{2}{b_1}x_1\omega_1+\frac{d_1}{b_1}\omega_1{}^2-\frac{\pi}{2}\right]}\frac{1}{\sqrt{2\pi b_2}}e^{\mu\frac{1}{2}\left[\frac{a_2}{b_2}x_2{}^2-\frac{2}{b_2}x_2\omega_2+\frac{d_2}{b_2}\omega_2{}^2-\frac{\pi}{2}\right]}d\boldsymbol{x}$$

By the change of variables $\boldsymbol{y} = \boldsymbol{x} - \boldsymbol{t}$, the above transform can be written as

$$\mathcal{L}_A^{(r)}\{f * g\}(\boldsymbol{\omega}) = \int_{\mathbb{R}^2}\int_{\mathbb{R}^2} f(\boldsymbol{t})g(\boldsymbol{y})e^{\frac{a_1}{b_1}(-\mu t_1y_1)}e^{\frac{a_1}{b_1}(-\mu t_2y_2)}\frac{1}{\sqrt{2\pi b_1}}\frac{1}{\sqrt{2\pi b_2}}$$
$$\cdot e^{\mu\frac{1}{2}\left[\frac{a_1}{b_1}(y_1+t_1)^2-\frac{2}{b_1}(y_1+t_1)\omega_1+\frac{d_1}{b_1}\omega_1{}^2-\frac{\pi}{2}\right]}e^{\mu\frac{1}{2}\left[\frac{a_2}{b_2}(y_2+t_2)^2-\frac{2}{b_2}(y_2+t_2)\omega_2+\frac{d_2}{b_2}\omega_2{}^2-\frac{\pi}{2}\right]}d\boldsymbol{t}d\boldsymbol{y}$$
$$= \int_{\mathbb{R}^2} f(\boldsymbol{t})\int_{\mathbb{R}^2} g(\boldsymbol{y})\frac{1}{\sqrt{2\pi b_1}}\frac{1}{\sqrt{2\pi b_2}}e^{\mu\frac{1}{2}\left[\frac{a_1}{b_1}y_1-\frac{2}{b_1}(y_1\omega_1)+\frac{d_1}{b_1}\omega_1{}^2-\frac{\pi}{2}\right]}$$
$$\cdot e^{\mu\frac{1}{2}\left[\frac{a_2}{b_2}y_2-\frac{2}{b_2}(y_2\omega_2)+\frac{d_2}{b_2}\omega_2{}^2-\frac{\pi}{2}\right]}d\boldsymbol{y}e^{\mu\frac{1}{2}\frac{a_1}{b_1}(t_1{}^2)}e^{\mu\frac{1}{2}\left(-\frac{2}{b_1}\right)(t_1\omega_1)}$$
$$\cdot e^{\mu\frac{1}{2}\frac{a_2}{b_2}(t_2{}^2)}e^{\mu\frac{1}{2}\left(-\frac{2}{b_2}\right)(t_2\omega_2)}d\boldsymbol{t} \tag{44}$$

from theorem (5) we have

$$\mathcal{L}_A^{(r)}\{f * g\}(\boldsymbol{\omega}) = \int_{\mathbb{R}^2} f(\boldsymbol{t})\mathcal{F}_q^{(r)}\{g_f\}(\boldsymbol{\omega})e^{\mu\frac{1}{2}\frac{a_1}{b_1}(t_1{}^2)}e^{\mu\frac{1}{2}\left(-\frac{2}{b_1}\right)(t_1\omega_1)}e^{\mu\frac{1}{2}\frac{a_2}{b_2}(t_2{}^2)}e^{\mu\frac{1}{2}\left(-\frac{2}{b_2}\right)(t_2\omega_2)}d\boldsymbol{t}. \tag{45}$$

Due to noncommutative property of f and g, we decompose f function and obtain

$$\mathcal{L}_A^{(r)}\{f * g\}(\boldsymbol{\omega}) = \int_{\mathbb{R}^2}\left[f_0\mathcal{F}_q^{(r)}\{g_f\}(\boldsymbol{\omega}) + \boldsymbol{i}f_1\mathcal{F}_q^{(r)}\{g_f\}(\boldsymbol{\omega}) + \boldsymbol{j}f_2\mathcal{F}_q^{(r)}\{g_f\}(\boldsymbol{\omega}) + \boldsymbol{k}f_3\mathcal{F}_q^{(r)}\{g_f\}(\boldsymbol{\omega})\right]$$
$$\cdot e^{\mu\frac{1}{2}\frac{a_1}{b_1}(t_1{}^2)}e^{\mu\frac{1}{2}\left(-\frac{2}{b_1}\right)(t_1\omega_1)}e^{\mu\frac{1}{2}\frac{a_2}{b_2}(t_2{}^2)}e^{\mu\frac{1}{2}\left(-\frac{2}{b_2}\right)(t_2\omega_2)}d\boldsymbol{t}.$$

Because $f_0, f_1, f_2, f_3 \in \mathbb{R}$, then

$$\mathcal{L}_A^{(r)}\{f * g\}(\boldsymbol{\omega}) = \int_{\mathbb{R}^2}\left[\mathcal{F}_q^{(r)}\{g_f\}(\boldsymbol{\omega})f_0 + \boldsymbol{i}\mathcal{F}_q^{(r)}\{g_f\}(\boldsymbol{\omega})f_1 + \boldsymbol{j}\mathcal{F}_q^{(r)}\{g_f\}(\boldsymbol{\omega})f_2 + \boldsymbol{k}\mathcal{F}_q^{(r)}\{g_f\}(\boldsymbol{\omega})f_3\right]$$
$$\cdot e^{\mu\frac{1}{2}\frac{a_1}{b_1}(t_1{}^2)}e^{\mu\frac{1}{2}\left(-\frac{2}{b_1}\right)(t_1\omega_1)}e^{\mu\frac{1}{2}\frac{a_2}{b_2}(t_2{}^2)}e^{\mu\frac{1}{2}\left(-\frac{2}{b_2}\right)(t_2\omega_2)}d\boldsymbol{t}$$

Let $\tilde{f}(\boldsymbol{t}) = f_n e^{\mu\frac{1}{2}\frac{a_1}{b_1}(t_1{}^2)}e^{\mu\frac{1}{2}\frac{a_2}{b_2}(t_2{}^2)}, n = 0,1,2,3$ then

$$\mathcal{L}_A^{(r)}\{f * g\}(\omega) = \int_{\mathbb{R}^2}\left\{\mathcal{F}_q^{(r)}\{g_f\}\tilde{f}_0 + \boldsymbol{i}\left\{\mathcal{F}_q^{(r)}\{g_f\}\tilde{f}_1\right\} + \boldsymbol{j}\left\{\mathcal{F}_q^{(r)}\{g_f\}\tilde{f}_2\right\} + \boldsymbol{k}\left\{\mathcal{F}_q^{(r)}\{g_f\}\tilde{f}_3\right\}\right\}$$
$$\cdot e^{-\mu t_1\frac{\omega_1}{b_1}}e^{-\mu t_2\frac{\omega_2}{b_2}}d\boldsymbol{t}$$

From QFT definition we then have

$$\mathcal{L}_A^{(r)}\{f * g\}(\boldsymbol{\omega}) = \mathcal{F}_q\{g_f\}\mathcal{F}_q\{\tilde{f}\} + i\mathcal{F}_q\{g_f\}\mathcal{F}_q\{\tilde{f}_1\} + j\mathcal{F}_q\{g_f\}\mathcal{F}_q\{\tilde{f}_2\} + k\mathcal{F}_q\{g_f\}\mathcal{F}_q\{\tilde{f}_3\},$$

Where g_f and $\tilde{f}$ defining on theorem 5.

This is the desired result.

5. Result and discussion

Based on QFT relationship with QLCT, we established the new convolution theorem for one-sided QLCT. From the result, we can see the connection between one-sided QLCT and one-sided QFT. Due to noncommutative properties of quaternion, the convolution theorem for two types of QLCT is still an open problem.

6. Acknowledgment

The first and second author are partially supported by *Hibah Penelitian Kerja Sama Perguruan Tinggi* (PEKERTI) 2017 (no. 622.48/UN28.2/PL/2017) Universitas Tadulako, Indonesia.

References

[1] Xiang Q and Kai YQ 2011 On the Relationship Between the Linear Canonical Transform and the Fourier Transform : *International Congress on Image and Signal Processing* IEEE 2214-2217
[2] Gudadhe A S 2012 Linear Canonical Transform on Zemanian Space : *Global Journal of Science Frontier Research Mathematics and Decision Sciences* **12** 16
[3] Pei SC, and Ding JJ 2002 Eigen functions of Linear Canonical Transform : *IEEE Transactions on Signal Processing* **50** 11
[4] Pei SC, Fellow, IEEE, Ding JJ, and Chang JH 2001 Efficient Implementation of Quaternion Fourier Transform, Convolution and Correlation by 2-D Complex FFT : *IEEE Transactions on Signal Processing* **49** 2783
[5] Hitzer E S M 2006 Quaternion Fourier Transform on Quaternion Fields and Generalization: *Advance Applications of Clifford Algebras* **17** 497
[6] Stern A 2006 Why is the Linear Canonical Transform so Little Known : *Workshop on Information Optics* **860** 225
[7] Kou K I, Yu O J, and Morais J 2013 On Uncertainty Principle for Quaternionic Linear Canonical Transform : *Abstract and Applied Analysis Hindawi* **2013** 1
[8] Kou K I and Morais J 2014 "Asymptotic behaviour of the Quaternion linear canonical transformand the Bochner-Minlos theorem :*Applied Mathematics and Computation* **247** 675
[9] Resnawati, Bahri M, and Kusuma J 2013 Transformasi Kanonikal Linear Quaternion Satu Sisi *Thesis* (Makassar: Universitas Hasanuddin)
[10] Bahri M, R. Ashino, and R. Vaillancourt 2015 Convolution theorem for quaternion Fourier transform : properties and applications : *Abstract and Applied Analysis Hindawi* **2013** 1
[11] Bahri M and R. Ashino 2016 A Simplified proof of uncertainty principle for quaternion linear canonical transform : *Abstract and Applied Analysis Hindawi* **2016** 1

IOP Conf. Series: Journal of Physics: Conf. Series **943** (2017) 012013 doi:10.1088/1742-6596/943/1/012013

Implementing thinking aloud pair and Pólya problem solving strategies in fractions

N S H Simpol[1], M Shahrill[1], H-C Li[1] and R C I Prahmana[2]

[1] Sultan Hassanal Bolkiah Institute of Education, Universiti Brunei Darussalam, Bandar Seri Begawan, Brunei Darussalam

[2] Universitas Ahmad Dahlan, Jl. Pramuka Kav. 5, Pandeyan, Yogyakarta, Indonesia

E-mail: masitah.shahrill@ubd.edu.bn

Abstract. This study implemented two pedagogical strategies, the Thinking Aloud Pair Problem Solving and Pólya's Problem Solving, to support students' learning of fractions. The participants were 51 students (ages 11-13) from two Year 7 classes in a government secondary school in Brunei Darussalam. A mixed method design was employed in the present study, with data collected from the pre- and post-tests, problem solving behaviour questionnaire and interviews. The study aimed to explore if there were differences in the students' problem solving behaviour before and after the implementation of the problem solving strategies. Results from the Wilcoxon Signed Rank Test revealed a significant difference in the test results regarding student problem solving behaviour, $z = -3.68$, $p = .000$, with a higher mean score for the post-test ($M = 95.5$, $SD = 13.8$) than for the pre-test ($M = 88.9$, $SD = 15.2$). This implied that there was improvement in the students' problem solving performance from the pre-test to the post-test. Results from the questionnaire showed that more than half of the students increased scores in all four stages of the Pólya's problem solving strategy, which provided further evidence of the students' improvement in problem solving.

1. Introduction

Problem solving is the process of reaching an acceptable solution to a novel problem, which usually involves critical thinking and analytical reasoning skills [1, 2]. Mathematics often requires problem solving skills, and it is not an easy task for teachers to cultivate creative mind and to always design lessons that provide supports in developing problem solving skills [3, 4]. One framework for thinking about problem solving was suggested by Pólya [5], and the strategy is recognised by many people as the steps they undergo during problem solving. He suggested the four stages to problem solving:

1. *Understanding the problem*. The first stage involves comprehending the problem by stating it in our own words, identifying the given information and what is being asked.
2. *Devising a plan*. Once the first stage is done, we can now look for the strategies to solve the problem, for example drawing a diagram or using variables to create an equation.
3. *Carrying out the plan*. This stage includes executing the strategy identified in the previous stage. If the strategy does not work, other strategies will be sought until the problem is solved. Most often, this stage includes the process of trial and error.

4. *Looking back at the process*. Look back at the problem to ensure all parts of the questions are answered and all conditions satisfactorily met. Reflecting back at the problem also involves recognising the undergone process and being able to relate it to other future problems.

1.1 The thinking aloud pair problem solving strategy

Stice [6] suggested that students should be given opportunities to work collaboratively and receive constant feedback on how they are doing in order to improve their learning. One pedagogical technique for problem solving that provides equal experience for the whole class is the 'Thinking Aloud Pair Problem Solving' (hereafter, referred to as TAPPS). Claparède [7] first studied TAPPS, which was later expanded by Whimbey and Lochhead [8]. Their idea was to have students paired up, one as a problem solver, and the other as a listener. The problem solver reads the problem aloud, while the listener listens. The problem solver has to keep talking, expressing all his/her thoughts, and the listener has to prompt the problem solver if he/she stops talking, for example by saying "Tell me what you are thinking". The listener also needs to try to understand what the problem solver has done, so he/she should question the problem solver for further explanation (e.g., "Why do you say that?" "I don't understand. Would you explain that to me?" etc.). The listener should keep in mind not to solve the problem or direct the problem solver to the answer by giving hints. The main purpose of TAPPS is not about getting the right answer, but rather understanding how they reached an answer, even if they are incorrect, which may be achieved through verbalising the process [8].

The TAPPS strategy has been used across disciplines at several secondary levels. For instance, for the application of TAPPS on reading comprehension, it was found that TAPPS increased reading comprehension scores in students with poor reading comprehension ability [9], and even for students with negative attitudes towards reading [10]. Similarly in chemistry, it was found that the students who applied the TAPPS strategy showed greater problem solving ability than those students who applied with other strategies, and that they performed better at recalling mathematical concepts, executing procedures, and conceptual knowledge [11]. Jeon et al. [11] also provided evidence to demonstrate the importance of a strong emphasis on the role of the problem solver and the listener. Their study showed that the listeners exhibited more verbal interactions than the problem solvers. For example, when problem solvers exhibited behaviours of 'requiring agreement', 'providing' responses, or 'modifying' the process or response, the listeners 'agreed', 'asked', and 'pointed out' most of the time.

In the context of mathematics, Kani and Shahrill [12], and Irham and Zainuri [13] examined the application of TAPPS for problem solving on rate, ratio and percentages, and algebra respectively. Kani and Shahrill [12] investigated Year 9 students' problem solving behaviour and perceptions of the TAPPS strategy via questionnaires. They found that the students did not significantly improve their mathematics achievement and conceptual knowledge, but with TAPPS, they were helped to be more conscious of their thinking process and to improve their problem solving skills. Further analysis of the problem solving behaviour questionnaire also showed that the students improved most remarkably in 'understanding the questions'. Kani and Shahrill [12] also pointed out that based on the findings from the TAPPS questionnaire, the students had positive perception and attitudes towards the TAPPS strategy, and they overall agreed that TAPPS could help them in the learning of mathematics.

TAPPS involved collaboration between students, and working collaboratively has become a 21st century trend [14]. Collaborative learning refers to "an instruction method in which students at various performance levels work together in small groups towards a common goal" [15]. Collaborative learning develops social interaction skills, and through it, students will gradually realise the need for positive and supportive interactions in order for the group to succeed. Along the way, they learn to recognise the behaviours that help them work together with their peers, and to identify and reflect their own contributions [16]. Additionally, students could also learn to identify their differences and find solution to disagreements or any social conflicts that may arise when working together [17]. Ultimately, collaborative learning promotes positive societal responses and increases oral communication skills [18, 19].

1.2 Fractions and fractions problem solving

Fractions are one of the essential mathematical concepts that students need to grasp before they can proceed to more advanced mathematics topics [20]. Yet, most students found fractions to be difficult to learn [21, 22]. Some argued that this was due to the confusing relationships between different fractional concepts and arithmetic procedures [23]. For example, the fraction $\frac{3}{8}$ can be seen as three out of eight equal parts (part-whole), as three parts to eight parts (ratio), as three divided by eight (quotient), as a point on a number line (measure), or as three-eighth of a quantity (operator) [24]. On the contrary, it was argued that the concept of fractions begins as a procedural activity [25, 26]. This was consistent with Kerslake [27] who found that many students could perform addition of fractions correctly, but when asked, they were unable to justify the method that they used, suggesting that there is a need for students to learn fractions both procedurally and conceptually.

The sampled students in a Malaysian study found word problems involving fractions to be more difficult compared to the fraction computational problems [28]. According to the experienced teachers they interviewed, students could not relate their mathematical knowledge to real life situation, which resulted in difficulties with solving word problems. The students performed best in addition and subtraction word problems, and worst in multiplication word problems. Division word problems were slightly higher than multiplication, attributing to the fact that students could identify fraction division problem easier than multiplication division. They also observed that students tended to not write down all the necessary steps to solve a problem using the correct mathematical syntax and grammar. Therefore, they believed that students must be trained to write these steps (i.e. their thinking process), so that problem solving can become easier.

2. Method

This study implemented the TAPPS and the Pólya's Problem Solving strategies to support students' learning of fractions. The research question explored was "What differences can be observed in the students' problem solving behaviour before and after the implementation of TAPPS and Pólya's Problem Solving strategies?" An action research framework using a mixed method design was employed with data collected from the pre and post-tests, questionnaires and interviews. A total of 51 mixed ability students (30 male and 21 female) from two Year 7 classes of a government secondary school in Brunei Darussalam took part in this study. At the time of the study, the participants' ages ranged from 11 to 13 years old (M = 12.1, SD = .392). Permission and the approval to conduct the study were obtained from relevant agencies, and the students and the school were ensured that their participations were kept confidential.

2.1 Data collection instruments

The data were collected from the three types of instruments:

1. *Mathematics Achievement Test.* The first author designed this test, which consisted of seven questions: four questions on computational fractions and three questions on fractions word problems. Three experienced teachers were also invited to review the questions to ensure that the test was at an appropriate level of difficulty. The test was used for both of the pre- and post-tests and the duration of the test was 30 minutes.
2. This test consisted of seven questions: four questions on computational fractions and three questions on fractions word problems. The test was used for both the pre- and post-tests and a 30 minutes duration was allowed to complete the test.
3. *Problem Solving Behaviour Questionnaire.* The questionnaire used in this study was adapted from Kani [29]. It was administered to students both before and after the implementation of TAPPS. There were 25 items, and the students were required to rate their answers in a 5-point Likert scale, ranging from 1 (*Never*), 2 (*Seldom*), 3 (*Sometimes*), 4 (*Often*), to 5 (*Always*). All the questionnaire items were based on Pólya's four stages of the problem solving strategy.
4. *Recordings of the lessons and interviews.* Interactions between pairs of students were audio and video recorded in order to analyse students' interactions and problem solving behaviour. In

addition, 17 selected students were interviewed after the post-test to delve further into their problem solving behaviour using the TAPPS strategy. The students were mainly selected for the interviews because they showed great improvement in their pre- to post-test scores, for their notable answers in the questionnaire, and/or for their keen effort during the intervention lessons.

2.2 Procedure

The students were given the pre-test and the problem solving behaviour questionnaire at the start of the first lesson, before being introduced to the TAPPS and the Pólya's problem solving strategies. After the administration of the pre-test and the questionnaire, the students were provided with two instruction sheets in order to help them familiarise with the TAPPS and Pólya's problem solving strategies. The first instruction sheet, which was adopted from Stice [6], provided the guidelines for the problem solver and listener, including a step-by-step procedure on the role of the problem solver and the listener, from sitting comfortably before starting the problem solving to how to first approach the problem. The second instruction sheet provided the students with the guidelines for the Pólya's problem solving strategy [5], which instructed them what they can do in each stage of Pólya's problem solving strategy such as rewriting the information given, or by drawing pictures to solve the given problem.

The students were paired up according to their pre-test scores, where the pairings consisted of one high achiever and one middle achiever, as well as pairs of students with average scores. Subsequently, after the pairings, they were encouraged to apply both of the TAPPS and the Pólya's four-step problem solving strategy to solve the first exercise worksheet. The teacher would facilitate the students when they required her assistance, for example, by guiding the students during the process of understanding the problem through step-by-step questioning. The correct answers were also provided during the lessons to guide the students in their problem solving. To guide the students' use of the TAPPS and Pólya's problem solving strategies, the teacher consistently reminded the students to refer to the disseminated guidelines. In total, the students were given two exercise worksheets during the 3-week intervention lessons, and once they have completed both worksheets, the post-test and the problem solving behaviour questionnaire were then administered. All computational fractions questions from the pre- and post-tests, and the exercise worksheets were required to be solved without using a calculator, and calculators were only permitted for word problem questions. The entire data collection process, including the student interviews took approximately three weeks.

3. Results and discussion

3.1 Differences in students' problem solving behaviour

In order to investigate the differences in students' problem solving behaviour before and after the implementation of TAPPS, the mean scores and standard deviations from the questionnaires (before and after the intervention) were calculated. There was a difference of 6.6, which was then tested for significance level. As the distribution was found to be not normal, the Wilcoxon Signed Rank Test was used instead. According to the test, a significant difference in the student problem solving behaviour was detected, $z = -3.68$, $p = .000$, where the mean score for after ($M = 95.5$, $SD = 13.8$) was higher than the mean score for before ($M = 88.9$, $SD = 15.2$) the implementation of TAPPS, as shown in Table 1. It also showed that the students' problem solving ability have improved in the post-test. And the calculation of effect size ($r = \frac{Z}{\sqrt{N}}$, where, Z = Z-value, N = number of observations i.e. at Time 1 and Time 2) showed a large effect ($r = .36$).

Table 1. Mean scores and standard deviations of problem solving behaviour before and after TAPPS.

(N = 51)	Before	After	Z
Problem solving behaviour	88.9 (15.2)	95.5 (13.8)	-3.68*

Note. *$p < .001$. Standard deviations appear in parentheses next to the mean score.

To explore the abovementioned differences, the data was further analysed by using the Wilcoxon Signed Rank Test. Items that represented each stage of Pólya's problem solving strategy were summed separately, and the mean scores and standard deviations for each stage are presented in Table 2.

Table 2. Mean scores and standard deviations for the four stages of Pólya's problem solving strategy.

(N = 51)	Pre-test	Post-test	Z
Understanding the problem	27.8 (4.36)	30.9 (4.93)	-5.08*
Devising a plan	13.6 (3.33)	14.8 (2.98)	-2.78**
Carrying out the plan	21.6 (4.94)	22.9 (3.67)	-1.84
Looking back or reflecting	22.0 (5.33)	23.1 (4.04)	-1.46

Note. $^*p < .001$. $^{**}p < .01$. Standard deviations appear in parentheses next to the mean score.

The Wilcoxon Signed Rank Test showed that there were statistically significant differences in 'Understanding the problem' and 'Devising a plan' stages, where $z = -5.08$, $p = .000$, and $z = -2.78$, $p = .005$, respectively. The magnitude of differences in the means was large for both stages, according to the effect size ($r = .50$ and $r = .28$). The differences in the other two stages did not differ significantly, $z = -1.84$, $p = .065$ for 'Carrying out the plan' stage and $z = -1.46$, $p = .143$ for 'Looking back or reflecting' stage. Nevertheless, it was also found that there were slight increases in the test scores from the pre-test to the post-test for these two stages.

3.2 Differences in each stage of the Pólya's problem solving strategy

The differences in each stage of Pólya's problem solving strategy were subsequently explored. For a positive impact, the ideal situation would portray a high number of students with increased scores in their problem solving behaviour in the post-test compared to the pre-test. And based on the students' scores obtained from the problem solving behaviour questionnaire in Figure 1, there were more than 27 (out of 51) students whose scores increased in all stages of the Pólya's problem solving strategy.

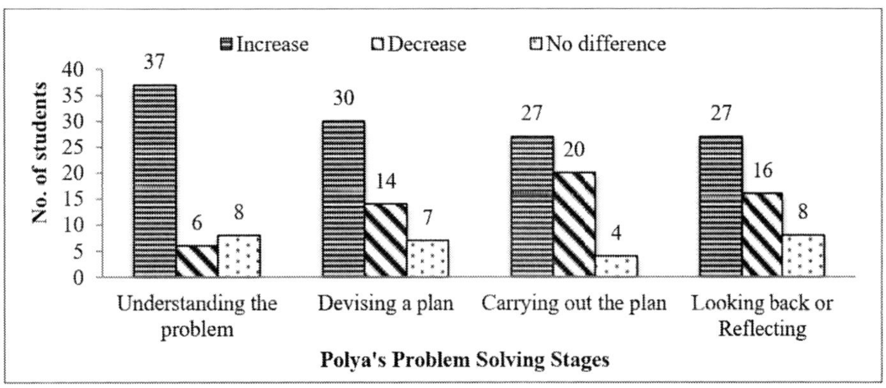

Figure 1. Variations in students' problem solving behaviour (N = 51).

3.2.1 Understanding the problem

The first stage plays an important role in the process of problem solving, since, if the students could not grasp the problem, they would not be able to proceed with the next three stages. As shown in Figure 1, compared to the other stages, this stage showed the most improvement. For example, there were 37 out of 51 students whose scores increased in the 'Understanding the problem' stage after the intervention, while there were relatively fewer students whose total scores decreased (6 students) or remain unchanged (8 students). Figures 2 and 3 show examples of worked solutions from two students, Farah and Fatin (pseudonyms), respectively. It was noticeable that there were changes in their problem solving behaviour before and after the intervention. Both students could be seen writing

IOP Conf. Series: Journal of Physics: Conf. Series **943** (2017) 012013 doi:10.1088/1742-6596/943/1/012013

out their thought processes to help them understand the problem, albeit they did not correctly answer it in the end. Specifically, Figure 2 shows that in the pre-test, Farah only showed the calculations when answering Q4, but after the intervention, her method in understanding the problem involved writing out more important information that she extracted from the question.

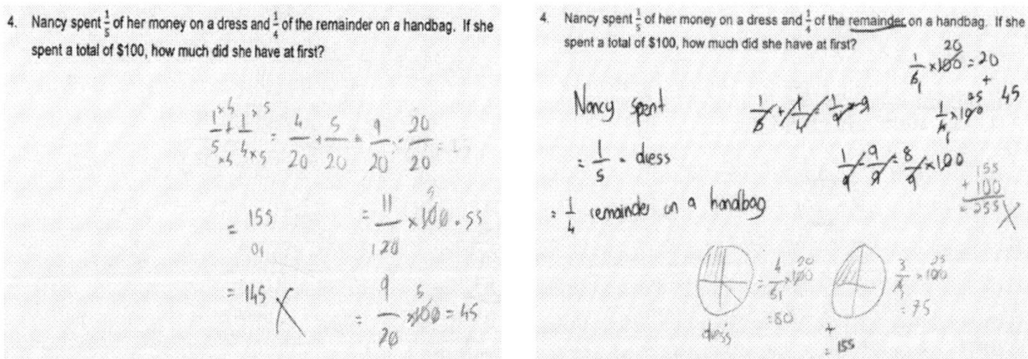

(a) (b)

Figure 2. Examples of Farah's worked solutions for the (a) pre-test and (b) post-test.

Fatin's thought processes were seen in both the pre- and the post-tests, as shown in Figure 3. During the interview, she mentioned that she normally would write down what she understood from the questions. It was also worth noting here that compared to the pre-test, her thought processes appeared to be broader and more detailed in the post-test.

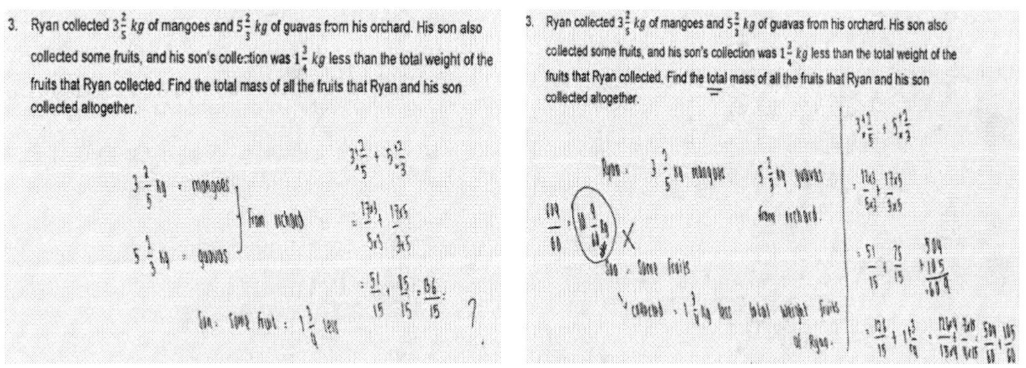

(a) (b)

Figure 3. Examples of Fatin's worked solutions for the (a) pre-test and (b) post-test.

A student's ability in understanding and extracting keywords appeared to determine whether or not they would be able to get the correct answer. This was particularly the case for Q3 and Q4. For instance, this study found that the key to solving Q3 lied on the statements 'less than the total weight' and 'total mass of all the fruits'. As shown in Figure 3, it appeared that Fatin interpreted 'total' by adding up the numbers, but she did not realise what to do with "less than". Meanwhile, the key to understanding Q4 depends on the word 'remainder'. Yet, most students mainly performed operations on $\frac{1}{5}$, $\frac{1}{4}$ and \$100 in which they did not show an understanding of the statement: '$\frac{1}{4}$ of the remainder'. For example, in Figure 2, it was seen that Farah added $\frac{1}{5}$ and $\frac{1}{4}$ and multiplied by 100 when answering Q4. Furthermore, it was also found during the interviews that most of the interviewed students immediately said '$\frac{1}{5}$ and $\frac{1}{4}$', when they were asked what the fractions were for Nancy to spend on a dress and a handbag.

3.2.2 Devising and carrying out a plan

In this study, it was found that more than 40 students showed attempt to devise and carry out a plan when answering the questions both in the pre- and post-test. Notably, they also appeared to become better at devising and carrying out the plans (refer to Figure 1), regardless whether they actually knew how to solve the problems. This demonstrated that most students have, to some extent, improved in problem solving, although not necessarily in achieving the correct answers.

According to Pólya [5], devising a plan includes recalling similar solved problems, and considering whether the method used could also be applied to another problem, which Hatana [30] referred it to 'adaptive expertise'– the ability to apply knowledge and experience to solving problems. In this study, we found that for example, in Q3, there were still 38 students who did not answer the question correctly in the post-test, despite solving several questions of similar concepts in the practice worksheets during the course of the intervention. As mentioned in the section 3.2.1, this may be due to the fact that they did not really understand the problem. Furthermore, it also suggested that many students needed help to develop adaptive expertise, since they were unable to recognise the similarities between the questions that they had done before and the questions in the test and thus, failed to apply the knowledge and experience to solve Q3 in the post-test. In particular, there were nine students who simply made attempts to solve the first or second sentence of Q3 and did not execute their solutions until the end. For example, as shown in Figure 3, Fatin's worked solutions mainly wrote '$3\frac{2}{5}+5\frac{2}{3}$'. Moreover, in the interview, it was found that when the students reviewed their solutions, they realised that they needed to calculate $\left(3\frac{2}{5}+5\frac{2}{3}-1\frac{3}{4}\right)$ to get Ryan's son's collection, however, they remained unaware that they should have also added Ryan's son's collection with Ryan's own collection in order to answer Q3 correctly.

3.2.3 Looking back or reflecting at the process

As shown in Figure 1, there were 27 students who considered themselves to increase their problem solving behaviour in terms of the last stage of Pólya's problem solving strategy: 'looking back or reflecting at the process'. Nevertheless, the analysis of the students' responses to questions in the pre- and post-tests did not show any indications regarding whether or not they looked back at their worked solutions or calculations, both in computational fractions and in word problems. It was also worth noting that probably because calculators were allowed for the word problems, double-checking with the calculation may be done with calculators rather than written computation on the test papers.

3.3 Discussion

The findings of this study showed that out of Pólya's four stages of problem solving process, the students showed the greatest improvement in the 'Understanding the problem' stage, and to some extent, they have also improved in the 'Devising a plan' stage. Yet, as Ahmad, Salim and Zainuddin [28] argued, students tend not to write down all the steps of their solutions; neither in computational problems nor in word problems. Similarly, in this study, it was also found that most of students' development in devising a plan and carrying out the plan could not be seen, since there was not much to be noted down by them on the test papers, especially in word problems where only the final answers were found in most of the students' written responses. This would perhaps be a consequence in allowing the use of calculators, as mentioned earlier. Nevertheless, our study found a consistent result with the literature that students might not be able to explain and recall how to solve the questions, even when they could give the correct answers [27].

Contrary to the claims made by Irham and Zainuri [13] that students from all ability classes (low, intermediate, and high) were able to extract necessary information from the questions, this seemed not to be entirely true in the present study. We found that while the students were able to identify important information such as fractions, they were not able to extract other equally important keywords or statements. For example, in Q4, "Nancy spent $\frac{1}{5}$ *of her money* on a dress and $\frac{1}{4}$ *of the remainder* on a handbag. If she spent a *total of $100*, how much did she *have at first*?" the fractions

IOP Conf. Series: Journal of Physics: Conf. Series **943** (2017) 012013 doi:10.1088/1742-6596/943/1/012013

and the italicised words are equally important in solving the problem. As mentioned in the section 3.2.1, most students were seen either adding or multiplying the two fractions, which they thought that both $\frac{1}{5}$ and $\frac{1}{4}$ were spent from Nancy's money. Meanwhile, it was seen that some students would multiply $\frac{1}{5}$ or $\frac{1}{4}$ with \$100, indicating that they did not have an understanding of the problem as a whole, which prevented them from devising a plan well.

4. Conclusion

The findings of this study demonstrated that the implementation of the TAPPS strategy was helpful in improving students' mathematics problem solving behaviour. One of the improvements in problem solving was in understanding the problem, where after the intervention, some of the responses by the students became more organised. However, in terms of comprehending the entire question, identifying keywords, and executing their solutions fully were still underdeveloped. Many of the students also could not relate and apply a strategy that they have done to other similar questions during the course of the intervention. Moreover, based on the students' performances in the pre- and post-tests, we also observed several errors that the students commonly made, such as performing incorrect order of operations, under-simplification of answers, incorrect inversion, as well as misconceptions in cancelling common factors and common denominators. This is one of the important implications of this study in which the errors and misconceptions discovered would enable teachers to recognise which aspects of the fractions topic that they need to pay more attention to and teach more thoroughly. The participants in this study were Year 7 students, so identifying their conceptual misunderstandings at an early stage before they move on to advanced mathematics were especially important and advantageous.

The TAPPS and Pólya's problem solving strategies were generally well received by the students, and they were eager to participate in the lessons. Amongst the responses were "Grouping and teamwork is good because it becomes easier to understand when he explains", and "Pólya's strategy is helpful because it tells you step by step to solve problems". The present study provided evidence to show that not only did the TAPPS strategy facilitate students to verbalise and organise their thinking, it was also a great medium for students to engage in collaborative learning by trying to understand their partner's thinking. Both strategies could be efficiently utilised either individually or together, but it is recommended that they be used simultaneously since a combination of both would be more effective [31-34]. And both of these strategies could also be used to diversify the teaching of fractions. We believe that if students consistently continue their effort and use both the TAPPS and Pólya's problem solving strategies, they can be helped to eventually become better at problem solving in fractions in particular, and mathematical knowledge in general.

The limitations of this study lie in the intervention lessons. Firstly, there were challenges in conducting these lessons. In the beginning of the first lesson, the students were provided with a detailed guideline on the roles of the problem solver and listener, and they were given explanations on what they can and cannot do as a problem solver or listener. This was repeated continuously in most lessons. Yet, not many students actually followed the rules. The problem solvers tend to exhibit behaviours such as 'requiring agreement', while the listeners 'agreed', 'pointed out', and also to the extent of giving hints to the problem solvers [11]. When tested on what the students did when they were the problem solver and listener, two out of four students understood their roles precisely. They explained that when they were the problem solver, they must say aloud everything and teach the listener unless they no longer know how to explain, and if they were the listener, they must "sit tight, be quiet, and listen" until the problem solver encountered any problems. However, the other two students had some misunderstandings. One student reported to have given some hints and pointed at mistakes when he was the listener, which in fact he should not have done so, while another thought that when he was the problem solver, he should ask if there was anything he could not understand. Therefore, it is recommended that future researches could use visual or physical demonstrations before conducting the lessons, for example, with video demonstrations or through role-play.

Secondly, some pairings might not be as effective as the others [35, 36]. In order to get the most out of the TAPPS and Pólya's problem solving strategies, the students abilities in each pair need to be balanced, so one had a higher score than the other (with a difference of two or three marks between them). However, since there was more number of students whose scores were average, there were also pairs where both had the same scores. Genders and existing seating arrangements were also taken into account. In both classes, the boys and girls were seated separately. In order to minimise movements and unfamiliarity between the students, the girls were paired up with girls, and boys with boys (with the exception of two pairs). During the lessons however, both students in several of the pairings still worked on the problems individually even after changing partners and prompts. Therefore, it is suggested that for each question or lesson, the pairings could also be rotated to allow collaborative partnership in learning.

References

[1] Woods D R 1975 Teaching problem solving skills *Engineering Education* **66** 238
[2] Schoenfeld A H 1983 *Problem Solving in the Mathematics Curriculum: A Report, Recommendations and an Annotated Bibliography (No. 1)* (Washington: Committee on the Teaching of Undergraduate Mathematics, Mathematical Association of America)
[3] Chong M S F and Shahrill M 2016 The use of an emerging framework to explore students' cognitive competency *Indian Journal of Science and Technology* **9**
[4] Yassin N F and Shahrill M 2016 Using Polya's principles as an approach to helping students make sense of problem solving *International Journal of Interdisciplinary Educational Studies* **11** 13
[5] Pólya G 1945 *How to Solve It: A New Aspect of Mathematical Method* (New Jersey: Princeton University Press)
[6] Stice J E 2007 Teaching problem solving *Teachers and Students Sourcebook: Alternative Teaching Methods* (Austin: University of Texas)
[7] Claparède E 1933 La genèse de l'hypothèse *Archives de Psychologie XXIV*(Extrait) (Genève: Librairie Kundig)
[8] Whimbey A and Lochhead J 1999 *Problem Solving & Comprehension* (Mahwah: Lawrence Erlbaum Associates)
[9] Dj M Z and Ali A 2015 Think aloud pair problem solving (TAPPS) strategy in teaching reading *Indonesian EFL Journal: Journal of ELT, Linguistics, and Literature* **1** 131
[10] Rizani F 2015 The influence of TAPPS technique and students' attitude toward students' reading comprehension achievement to the eighth grade students of Air Satan Junior High School *ELTE Journal* **3** 12
[11] Jeon K, Huffman D and Noh T 2005 The effects of thinking aloud pair problem solving on high school students' chemistry problem-solving performance and verbal interactions *Journal of Chemical Education* **82** 1558
[12] Kani N H A and Shahrill M 2015 Applying the thinking aloud pair problem solving strategy in mathematics lessons *Asian Journal of Management Sciences and Education* **4** 20
[13] Irham M and Zainuri Z 2016 Efektivitas pembelajaran think aloud pair problem solving ditinjau dari kemampuan pemecahan masalah *Unnes Journal of Mathematics Education Research* **5** 60
[14] Laal M and Ghodsi S M 2012 Benefits of collaborative learning *Procedia-Social and Behavioral Sciences* **31** 486
[15] Gokhale A A 1995 Collaborative learning enhances critical thinking *Journal of Technology Education* **7** 22
[16] Panitz T 1999 The motivational benefits of cooperative learning *New Directions for Teaching and Learning* **78** 59
[17] Johnson D W and Johnson R T 1985 Relationships between black and white students in intergroup cooperation and competition *The Journal of Social Psychology* **125** 421

[18] Kessler R C and McLeod J D 1985 Social support and mental health in community samples *Social Support and Health* ed S Cohen and L S Syme (San Diego: Academic Press)
[19] Yager S, Johnson D W and Johnson R T 1985 Oral discussion groups-to-individual transfer, and achievement in cooperative learning groups *Journal of Educational Psychology* **77** 60
[20] Siegler R S, Thompson C A and Schneider M 2011 An Integrated theory of whole number and fractions development *Cognitive Psychology* **62** 273
[21] Siegler R S, Fazio L K, Bailey D H and Zhou X 2013 Fraction: The new frontier for theories of numerical development *Trend in Cognitive Sciences* **17** 13
[22] Putri R I I and Zulkardi 2017 Fraction in shot-put: A learning trajectory *Research and Education for Developing Scientific Attitude in Sciences and Mathematics* ed C. Kusumawardani et al. AIP Conference Proceedings 1868 (New York: AIP Publishing)
[23] Li H-C 2011 The development of Taiwanese students' understanding of fractions: A problem-based learning approach *Proceedings of the British Society for Research into Learning Mathematics* **31** 25
[24] Pantziara M and Philippou G 2012 Levels of students' "conception" of fractions *Educational Studies in Mathematics* **79** 61
[25] Charles K and Nason R 2001 Young children's partitioning strategies *Educational Studies in Mathematics* **43** 191
[26] Gray E and Tall D 2007 Abstraction as a natural process of mental compression *Mathematics Education Research Journal* **19** 23
[27] Kerslake D 1986 Fractions: Children's strategies and errors *A Report of the Strategies and Errors in Secondary Mathematics Project* (Windsor, UK: NFER-Nelson)
[28] Ahmad A, Salim S S and Zainuddin R 2004 A study on students' performance in solving word problems involving fractions: Towards the development of a cognitive tool *Proc. in International Conference in Education Melbourne Australia*
[29] Kani N H A 2013 The effects of thinking aloud pair problem solving on students' mathematics performance *Unpublished MTeach Research Exercise* (Brunei Darussalam: Universiti Brunei Darussalam)
[30] Hatano G 1988 *Social and Motivational Bases for Mathematical Understanding* (San Francisco: Jossey Bass)
[31] Bilgin I 2006 The effects of pair problem solving technique incorporating Polya's problem solving strategy on undergraduate students' performance in chemistry *Online Submission* **7** 101
[32] Sundayana R, Herman T, Dahlan J A and Prahmana R C I 2017 Using ASSURE learning design to develop students' mathematical communication ability *World Transactions on Engineering and Technology Education* **15** 245
[33] Widodo S A, Purnami A S and Prahmana R C I 2017 Team accelerated instruction, initials and problem-solves ability in junior high school *International Journal on Emerging Mathematics Education* **1** 193
[34] Tanujaya B, Prahmana R C I, and Mumu J 2017 Mathematics instruction, problems, challenges, and opportunities: A case study in Manokwari regency, Indonesia *World Transactions on Engineering and Technology Education* **15** 287
[35] Othman H R H, Suhaimi Z, Shahrill M and Mahadi M A 2015 To pair or not to pair: Investigating the dynamics of teacher-student interactions in different classroom settings. *Turkish Online Journal of Educational Technology* Special Issue for INTE 2015 675
[36] Othman R, Shahrill M, Mundia L, Tan A and Huda M 2016 Investigating the relationship of student's ability and learning preferences: Evidence from year 7 mathematics' students *The New Educational Review* **44** 125

IOP Conf. Series: Journal of Physics: Conf. Series **943** (2017) 012014 doi:10.1088/1742-6596/943/1/012014

The Van Hiele geometry thinking levels of mild mental retardation students

Z A Shomad, T A Kusmayadi and Riyadi
Universitas Sebelas Maret, Jl.Ir. Sutami No. 36A, Surakarta 57126, Indonesia

E-mail: zahidshomad@gmail.com

Abstract. This research is to investigate the level of mild mental retardation geometry students thinking. This research focuses on the geometry thinking level based on Van Hiele theory. This study uses qualitative methods with case study strategy. Data obtained from observation and tests result. The subjects are 12 mental retardation students. The result show that ability of mild mental retardation students with each other is different but have same level of level thinking geometry. The geometry thinking level of mental retardation students was identified in level 1 of the Van Hiele theory. Based on the level thinking geometry of mental retardation students simplify geometry thinking teachers in selecting appropriate learning methods, choose the materials in accordance with ability, and can modify the material following the geometry thinking level of mental retardation students.

1. Introduction

One of the most important topics in mathematics is geometry [1- 8] who learn about points, lines, figures, the relationships between the line, length, area, volume, and others [9]. Geometry teach the students in the abstract and visually about the size and structure of geometry, such as the characteristics and relationships between one figure with another figures.

The aims of learning geometry is to provide an opportunity for students to have and develop the critical thinking ability, problem-solving ability, and better understanding ability of mathematic by gaining high level of geometry thinking ability [10]. To reach these goals, students need to understand the terms of geometry, spatial problems, and have cause and affect relationship between the features of geometric figures instead of memorizing definitions and theorems [11]. It would be very appropriate to divide the teaching in accordance with the geometry thinking level. However, the increase in the geometry thinking level does not occur naturally, need for learning which stimulates progress level thinking geometry. Geometric thinking levels facilitate students in learning geometry gradually.

One of the important researches about the geometry thinking level is the Van Hiele Geometry Thinking Level. The theory of geometry thinking Levels from the Van Hiele advanced by mathematician Dina Van Hiele and his wife Marie Van Hiele Pieree in 1957. Geometry thinking levels of Van Hiele consists of five levels, namely: 1) Level 0 (Visualization), 2) Level 1 (analysis), 3) Level 2 (Informal Deduction), Level 3 (Deduction), and Level 4 (Rigor) [12, 13]. Geometry thinking levels compiled by Van Hiele have hierarchical stages, making it easier for students in learning high level geometry.

At Level 0, the student identify, resolving names, and operate on geometric figures (e.g., angles, triangles intersecting or parallel lines) according to their appearance [14]. Students can mention the name based on appearance in general. They can mention triangle, square, cube, etc., but they do not explicitly identify the properties of the figures [15], they can't analyze. At Level 1, the student analyzes figures in terms of their components and relationships among components and discovers properties/rules of a class of figures empirically (e.g., by folding, measuring, and using a grid or diagram) [14]. Students analyze the existing components and the relationships between components, and then compile the tools from the figures of empirically. The forms are recognized by their components. They not only look at the visualization, but give attention to the properties. If someone describes a figure with four corners, it is a quadrilateral, though depicted with ugly. However, at this level, properties are not yet an ordered, such as a square is not necessarily identified being a rectangle [16, 17]. At Level 2, the student logically interrelates previously discovered properties/rules by giving or following informal arguments [14]. At this level, students can distinguish the features of figure. Students will understand why each square is a square, but they can't organize necessary lists in order to proof this observation [18, 19]. At Level 3, the student proves theorems deductively and establishes interrelationships among system of theorems [14]. At this level, students are able to explain the relationships between figures using theorems, definitions, and axioms. Students can use their critical thinking ability and problem solving ability. At Level 4, the student establishes theorems in different postulation systems and analyzes/album these systems [14]. They understand that the system is a work area.

Average high students are on level 2-3 [19] and elementary school students are at a level 1-2 [20]. Lately, mathematics and geometry are not only developed for regular students, but also for students with special need. Individuals with Disabilities Education Act (IDEA) stated that it is important to provide access to students with special need in order to get the same curriculum with students normal and expected both get the same instruction on the concepts that are important [21]. One of them is mental retardation. Mental retardation/ intellectual disability is a condition of incomplete development of the mind, characterized by impairment of skills manifested during the developmental period, i.e. cognitive, language, motor and social abilities [22].

According to the American Association on Mental Deficiency (AAMD), there are several types of mental retardation which is measured by IQ, are: 1) Mild mental retardation with IQ score 50-70, 2) Moderate mental retardation with IQ score 36-50, Severe mental retardation 3) with IQ score 20-40, Profound mental retardation 4) with an IQ score below 20, and 5) Unspecified [23]. Elliot further explain some of the difficulties experienced by students studying mental retardation, are: 1) attention, 2) cognitive processes, especially with regard to organizing, classification, and strategy setting, 3) Power Remember, which tends to be bad for short-term memory [24]. The mental age of a student's mental retardation the equivalent age of regular students at the elementary school level, [21]. It implies that although the conditions of the students are weak in the cognitive aspects of mental retardation, but students also have mental retardation geometry thinking level.

Based on the above explanation, geometry thinking level is an important component in learning geometry. Although there has been no research on the level of thinking of students in need special, but there is some literature that describes the geometry for students in need of special students, particularly mental retardation. Therefore, this research aims to analyze the level of geometric thinking about mental retardation students.

2. Method

This study uses qualitative methods with strategy case studies. The qualitative method is a method that aims to find out the events or phenomena experienced by the subject with the description with words and sentences [25]. Case study is an empirical inquiry that purposed to investigates a contemporary phenomenon within its real-life context [26]. Qualitative methods of research with strategy case study is a research aims to know the contextual problem by administering treatment. The problem in this research is a level thought geometry students mild mental retardation.

The research was held at the High School in Sukoharjo in March– May 2017. The subject of this research is the mild mental retardation students chosen by purposive sampling. The subject is used as many as 12 students with mild mental retardation the same capabilities.

Data taken with the awarding of the test on the whole subject.. Data were analyzed with Miles and Huberman steps of analyzing involve data reduction, data display, and verification [27]. The data in the form of the results of the students and then analyzed to find out the level of geometric thinking students.

3. Result and discussion

Table 1 describes the results of level thinking geometry students with mental retardation subject as many as 12 students at a high school in Sukoharjo, Indonesia. In Table 1, there is a code of the student and the geometry thinking level of mild mental retardation student. In the grouping of students in levels is done by analyzing the test results of geometric thinking level students.

Table 1. The task of level thinking geometry students mild mental retardation

Subject	Geometry Thinking Levels
TT01	1
TT02	1
TT03	1
TT04	1
TT05	1
TT06	1
TT07	1
TR08	0
TR09	0
TR10	0
TR11	0
TR12	0

This level think geometry is based on geometric thinking levels test students whohave been validated by experts. The question of the test consists of 10 numbers. Number 1-4 meansures the geometric thinking level 0, number 5-7 measures the geometric thinking level 1, and number 8-10 meansures the geometric thinking level 2. Subject categorized in level thinking 0 if at least 70% correct can work out between the reserved number 1-4. Students are categorized in level 1 can work out if at least 70% correctly on the numbers 5-7, and students are categorized in level 2 may be exercised if at least 70% correctly on the number 8-10.

Table 2. The result of subject geometry thingking test level

Subject	The Question of geometry thinking test level									
	Level 0				Level 1			Level 2		
	1	2	3	4	5	6	7	8	9	10
TT01	F	T	T	T	T	T	T	F	T	F
TT02	T	T	F	T	T	T	T	F	F	F
TT03	T	T	T	F	T	T	T	F	F	T
TT04	T	T	T	T	T	T	T	F	F	F
TT05	T	T	T	F	T	F	T	F	F	F
TT06	T	T	T	T	T	T	T	F	T	F
TT07	T	T	F	T	T	T	F	T	F	F
TR08	T	T	T	F	F	F	T	F	F	F
TR09	T	T	F	T	T	F	F	T	F	F
TR10	T	T	F	T	F	F	F	T	F	F
TR11	T	T	T	F	T	F	F	F	F	F
TR12	T	T	T	F	T	F	F	T	F	F

Based on indicators of the level of thinking of geometry of Van Hiele [28], subject TT01, TT02, TT03, TT04, TT05, TT06, TT07 categorized in level 1 and subject TR08, TR09, TR10, TR11, TR12 categorized in level 0. Subject level 0 able to describe triangle, rectangle, and the rectangle figures appropriately. The subject is not capable of analyzing the characteristics a triangle, square, and rectangular figure, so that the subject is incorrect in choosing a wake which is included in the triangle, square, and rectangular. Based on the analysis, the subject TR08, TR09, TR10, TR11 TR12, and categorized in level 0. While the subject is categorized in level 1 because the subject is able to describe exactly a triangle, square, and rectangular figures. The subject is also capable of analyzing the characteristics of wake up, so you can sort and select the shapes included in the triangle, square, and rectangular figures. The subject of level 1 is unable to associate the relationship figure one with another figure, for example a rectangular including square has four sides of equal length.

3.1 Mild mental retardation student of geometry thinking level 1

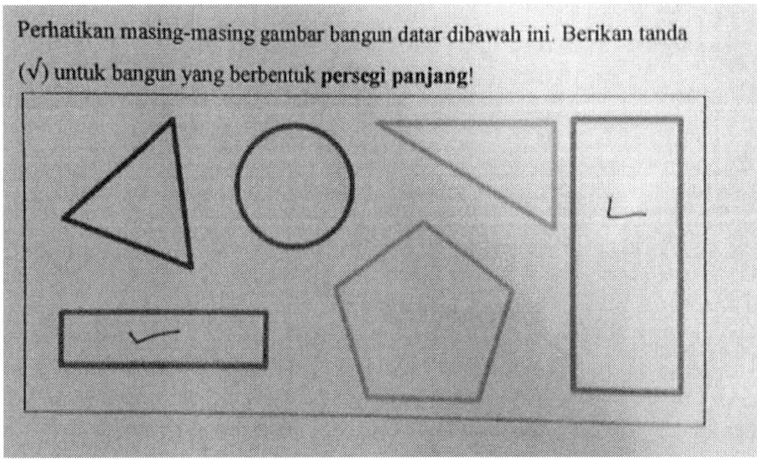

Figure 1. The test result of subject with geometry thinking level 1

Based on Figure 1, the subject can show the image rectangle. The subject of understanding about the characteristics and components of the rectangle. The subject knows the sides that parallel and have the same length. The subject also knows that figure has four right angles, so that students can show the image rectangle. Subject with level 1 can also compare a figure to another figure and describe with precision. However, the subject with level 1 cannot show that a square is a rectangle with a length of the same side.

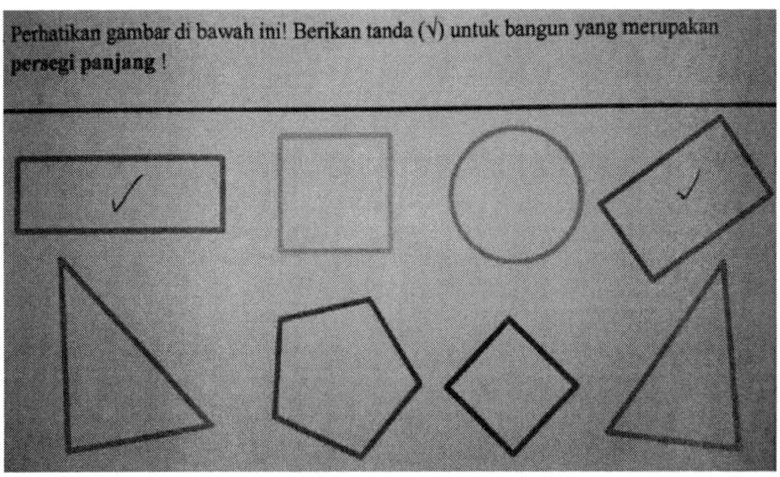

Figure 2. the test result of subject with geometry thinking level 1

Based on Figure 2, the subject of geometric thinking with level 1 can not analyze the relationships between the components of a square and a rectangle. The subjects still think that square is not part of the rectangle [29].

Based on research of Elliot, mental retardation student possible having difficulties with an unusual task they do [24]. Tests and activities from the level of thinking is the new test geometry for mild mental retardation students . However, when compared to the condition of the subject that are experiencing a shortage in the cognitive aspects, then the achievement level thinking geometry students mental retardation level 1 is a good thing. The Van Hiele levels of geometric thinking is a step that the hierarchy, so it is not possible to have the subject thinks the geometry of 3 and 4.

3.2 Mild mental retardation student of geometry thinking level 0

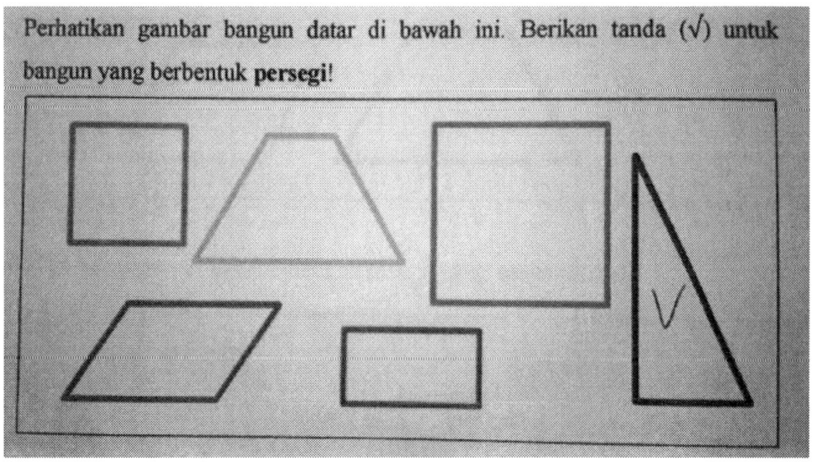

Figure 3. The test result of subject with geometry thinking level 0

Based on the results of the analysis, the subject of geometric thinking with level 0 are having difficulty in naming or identifying figure. The subject was wrong in choosing to figure which is a square. The subject is the difficulty in understanding the characteristics of the figure, so wrong in naming the figure. In Figure 3 above, it is possible to interpret the wrong subject that figure has chosen a right angle so the frontage. Koberlein defines that a square is a quadrilateral wake has a right angle and have a contiguous congruent sides [30, 31]. The subject confuse in comparing the characteristics of figure so that the subject made a mistake in identifying the square. The subject is also unable to identify the characteristics of another figure. This is possible because of the condition of the subject which is a mild mental retardation that students have a less in cognitive aspects.

Based on the above explanation, there is a difference of level thinking geometry students mild mental retardation. It is in accordance with the research conducted by Kemis and Rosnawati, barriers and conditions experienced by each student's mental retardation can be varied, although it has the same type of disability [32]. Although students have mild mental retardation barriers to smarts, but students can still be taught mental retardation. So the thinking level geometry can also be one of the skills that can be developed for students ' mental retardation. Some of the skills that must be mastered by the student's mental retardation are the ability of calculation and mathematics reasoning [33, 34]. Because the stages in thinking level geometry are hierarchical stages, it can help students develop the ability of reasoning in mental retardation.

IOP Conf. Series: Journal of Physics: Conf. Series **943** (2017) 012014 doi:10.1088/1742-6596/943/1/012014

4. Conclusion

The result of this research shows those students who have mild mental retardation level of geometric thinking 0 experience difficulties in identifying and comparing the figure. While students with mental retardation level of geometric thinking 1 has difficulty in connecting the components of a figure. It shows that any individual included in the mild mental retardation may have different geometric thinking levels, although equally included in the category of mental retardation. However, it is important for teachers to know the level of mild mental retardation students thinking for simplicity in explaining the material geometry. If the teacher knows the condition of the student's mental retardation, then handling for student's mental retardation when learning will be easier. The selection of methods, models, strategies, and media in learning geometry are expected to raise the level of mild mental retardation geometry students thinking. One of them with the learning phase geometry developed by Van Hiele.

References

[1] The Ministry of Education of Taiwan (the MET) 1993 *Curriculum Standards for National Elementary School in Taiwan* (Taipei: The Ministry of Education of Taiwan)

[2] The Ministry of Education of Taiwan (the MET) 2000 *The Provisional Guidelines 1-9 Curriculum Grade-Mathematics* (Taipei: The Ministry of Education of Taiwan)

[3] The Ministry of Education of Taiwan (the MET) 2003 *Curriculum Guidelines 1-9-Grade Mathematics* (Taipei: The Ministry of Education of Taiwan)

[4] The Ministry of Education of Taiwan (the MET) 2008 *Curriculum Guidelines 1-9-Grade Mathematics* ((Taipei: The Ministry of Education of Taiwan))

[5] National Council of Teachers of Mathematics (NCTM) 1989 *Curriculum and Evaluation Standards for School Mathematics*, (Reston: NCTM)

[6] The National Council of Teachers of Mathematics (NCTM) 1991 *Professional Standards for Teaching Mathematics* (Reston: NCTM)

[7] National Council of Teachers of Mathematics (NCTM) 1995 *Assessment Standards for School Mathematics* (Reston: NCTM)

[8] National Council of Teachers of Mathematics (NCTM) 2000 *Curriculum and Evaluation Standards for School Mathematics* (Reston: NCTM)

[9] Bieber C, Tuna A, and Korkmaz S 2013 The Mistakes and the Misconceptions of The Eighth Grade Students On The Subject of Angles *European Journal of Science and Mathematical Education* **1** 50

[10] MEB 2000 Ilkögretim okulu matematik dersi programi 5 Siniflar Milli Egitim Basimevi *İlköğretim online*

[11] Battista M T 2002 Learning Geometry in a Dynamic Computer Enviroiment *Teaching Children Mathematics* **8** 333

[12] Crowley, M L 1987 *The Van Hiele Model of The Geometric Development Thought* (Reston: NCTM)

[13] Usiskin, Z P 1982 The van Hiele Levels and Achievement in Secondary School Geometry *Final Report The Cognitive Development and Achievement in Secondary School Geometry Project* Chicago, IL: University of Chicago, Department of Education (ERIC Reproduction Service No. ED 220 228)

[14] Van Hiele P M and van Hiele D 1958 A method of initiation into geometry *Report on Methods of Initiation into Geometry* Groningen: J B Wolters

[15] Lesh R and Landau M 1983 *Acquisition of mathematics concepts and processes* (New York: Academic Press)

[16] Van Hiele P M 1984 English Summary: The Problem of Insight in Connection with School children's Insight into the Subject Matter of Geometry *English translations of selected writings of Dina van Hiele-Geldorf and Pierre M. van Hiele* pp 237-241

[17] Van Hiele P M 1986 *Structure and insight: A theory of mathematics education* (Orlando: Academic Press)
[18] Hoffer A 1981 Geometry is more than proof *The Mathematics Teacher* **74** 11
[19] Yilmaz G K and Koparan T 2015 The Effect of Designed Geometry Teaching Lesson to the Candidate Teachers ' Van Hiele levels of Geometric Thinking *Journal of Education and Training Studies* **4** 129-141
[20] Sayin V and Orbay K 2015 Investigation of 4th Grade Students' Geometric Thinking Levels and Success Scores in Terms of Different Variables *Journal of Studies in Social Sciences* **12** 160
[21] Butler F M, Miller S P, Crehan K, Babbit B and Pierce T 2003 Fraction Instruction for Students with Mathematics Disabilities: Comparing Two Teaching Sequences *Learning Disabilities Research & Practice* **2** 99
[22] World Health Organization 1996 *Guide For Mental Retardation* (Geneva: Division of Mental Helath and Prevention of Subtance Abuse)
[23] The American Association on Mental Deficiency 2009 Classification of Mental Retardation in Washington (American Association on Mental Deficiency (AAMD))
[24] Elliott S N 2000 *Educational Psychology: Third Edition* (Singapore: McGraw-Hill Higher Education)
[25] Moleong L J 2007 *Metodoogi Penelitian Kualitatif* (Bandung: Remaja Rosdakarya)
[26] Zainal Z 2007 Case Study as a Research Method *J Kemanusiaan* **9** 1
[27] Sugiyono 2013 *Metode Penelitian Pendidikan: Pendekatan Kuantitatif, Kualitatif dan R D* (Bandung: Alfabeta)
[28] Fuys D, Geddes D and Tischler R 1988 The Van Hiele Model of Thinking in Geometry Among Adolescents. *Monograph* i-196
[29] Sandipan D S 2014 Area and Perimeter Relation of Square and Rectangle (Relation All Mathematics) *Journal of Mathematics* **10**
[30] Koberlein A 2011 *Elementary Geometry for College Students* (Canada: Nelson Education, Ltd)
[31] Ismail Z and Rahman S N A 2017 Learning 2-dimensional and 3-dimensional geometry with Geogebra: Which would students do better? *International Journal on Emerging Mathematics Education* **1** 121
[32] Kemis and Rosnawati 2013 *Pendidikan Anak Berkebutuhan Khusus Tunagrahita (Peserta didik Berkebutuhan Khusus dengan Hambatan Kecerdasan)* (Jakarta: PT. Luxima Metro Media)
[33] Abdurrahman M 2009 *Pendidikan Bagi Anak Berkesulitan Belajar* (Jakarta: PT Rineka Cipta)
[34] Nurhasanah F, Kusumah Y S and Sabandar J 2017 Concept of triangle: Examples of mathematical abstraction in two different contexts *International Journal on Emerging Mathematics Education* **1** 53

IOP Conf. Series: Journal of Physics: Conf. Series **943** (2017) 012015 doi:10.1088/1742-6596/943/1/012015

Models for financial crisis detection in Indonesia based on M1, M2 per foreign exchange reverse, and M2 multiplier indicators

Sugiyanto[1], Etik Zukhronah[1] and Esteti Sophia Pratiwi[2]

[1]Departement of Statistics, [2]Departement of Mathematics, Faculty of Mathematics and Natural Science, Sebelas Maret University

E-mail: sugiy50@yahoo.co.id, estetipratiwi64@gmail.com

Abstract. Indonesia has been hit by financial crisis in the middle of 1997. The financial crisis that has occurred gives a severe impact to the economy of Indonesia resulting the needs for a detection system of financial crisis. Crisis can be detected based on several indicators such as M1, M2 per foreign exchange reserves, and M2 multiplier. These three indicators can affect the exchange rate stability and may further affect the financial stability so that it can be one of the causes of the financial crisis. This research aims to determine the appropriate model that can detect the financial crisis in Indonesia. Markov switching is an alternative model that can be approach and used often for detecting financial crisis. We can determine the combination of volatility and Markov switching model with AR and volatility model are determined first. The results of this research are that M1 can be modelled by SWARCH (3, 1) while M2 per foreign research exchange reserves and M2 multiplier can be modelled by SWARCH(3,2).

1. Introduction

A country will face the financial crisis if the country's financial system is disrupted then the system may not function efficiently. Financial crisis consists 3 types, one of them is currency crisis. Currency crisis is a situation that there are doubts whether the central bank of a country has sufficient foreign exchange reserves to maintain a country's exchange rate. Currency crisis often leads to the weakening of currency values as a sign of the financial crisis. Financial crisis that hit Asia on the middle of 1997 is begun by the falling of Baht Thailand. The crisis is also happened in Indonesia that has serious effect on the economic stability. After the financial crisis in 1997, International Monetary Fund (IMF) stated the importance of financial crisis detection system [1]. Detection system is a development model by observing the selected economic indicators. According to Kaminsky et al. [2], there are 15 indicators that can be used to detect the crisis; three of them are M1, M2 per foreign exchange reserves, and M2 multiplier. These three indicators can affect the exchange rate stability and may further affect the financial stability so that it can be one of the causes of the financial crisis. Monthly data of M1, M2 per foreign exchange reserves, and M2 multiplier are time series data that are indicated having heteroscedasticity and condition changes. To overcome this condition, SWARCH model is introduced by Hamilton and Susmel [3] on 1994. Some researchers have conducted research on the detection of financial crisis that occured in a country using the combination of volatility and Markov switching models. Chang et al. [4] used SWARCH model to identify the stock volatility foreign and global financial crisis in Korea based on real exchange rate on the period of January 4th 2000 to March 31st 2010. In this study, it will determine the appropriate model for detecting financial crisis in Indonesia based on M1, M2 per foreign exchange reserves, and M2 Multiplier.

2. Experimental Details

2.1. *Autoregressive (AR) Model*

AR model is usually used as a linear time series models. AR model has an order called p that can be determined by partial autocorrelation function (PACF) plot. Model AR (p) can be written as

$$r_t = \phi_1 r_{t-1} + \phi_2 r_{t-2} + \cdots + \phi_p r_{t-p} + a_t, \tag{1}$$

Where r_t log return series is at time-t, $\emptyset_p$ is parameter of AR model (Tsay [7]).

2.2. *Volatility Model*

2.2.1. ARCH Model. Residue of AR model containing heteroscedasticity effect can be modelled using a volatility model where the residue can be expressed as

$$a_t = \sigma_t \epsilon_t \text{ for } \epsilon_t \sim N(0,1) \text{ and } a_t | F_{t-1} \sim N(0, \sigma_t^2), \tag{2}$$

Where F_{t-1} is set of all the information in time, $t-1$. Model ARCH (m) can be written as

$$\sigma_t^2 = \alpha_0 + \sum_{i=1}^{m} \alpha_i a_{t-i}^2, \tag{3}$$

Where m is the order of ARCH model, α_0 is a constant, α_i is a parameter of the ARCH model, and σ_t^2 is the residual variance to the time-t [5].

2.2.2. GARCH Model. High order of ARCH model can be overcome using a GARCH(m, s) which can be written as

$$\sigma_t^2 = \alpha_0 + \sum_{i=1}^{m} \alpha_i a_{t-i}^2 + \sum_{j=1}^{s} \beta_j \sigma_{t-j}^2, \tag{4}$$

Where β_j is parameter of GARCH model [5].

2.2.3. EGARCH Model. Leverage problem on GARCH effect can be overcome by the model EGARCH (m, s) which can be written as

$$\ln \sigma_t^2 = \alpha_0 + \sum_{i=1}^{m} \alpha_i \left(\left| \frac{a_{t-i}}{\sqrt{\sigma_{t-i}^2}} \right| - \sqrt{\frac{2}{\pi}} \right) + \sum_{i=i}^{m} \gamma_i \frac{a_{t-i}}{\sqrt{\sigma_{t-i}^2}} + \sum_{j=1}^{s} \beta_j \ln \sigma_{t-j}^2, \tag{5}$$

Where γ is a parameter Leverage effect [5].

2.3. *SWARCH Model*

According to Hamilton and Susmel [3], Markov switching model for the average conditional can be written as

$$r_t = \mu_{s_t} + \tilde{r}_t \tag{6}$$

where r_t is the log return at time t, $\tilde{r}_t$ is following the process of AR (p) with an average of zero, and $\mu_{s_t}$is the average of state at time t. While SWARCH model can be written as

$$r_t = \mu_{s_t} + a_t, a_t = \sigma_{t,st} \epsilon_t \tag{7}$$

$$\sigma_{t,s_t}^2 = \alpha_{0,s_t} + \sum_{i=1}^{m} \alpha_{i,s_t} a_{t-i}^2, \tag{8}$$

Where μ_{s_t} is the conditional average on a state and σ_{t,s_t}^2 is a variance of residue in a state at time t.

2.4. *Smoothed Probability*

According to Kuan [6], smoothed probability value is defined as

$$P(s_t = i|Z^T;\theta) = \sum_{s_t=1}^{3} P(s_{t+1} = i|Z^T;\theta)P(s_t = i|s_{t+1} = i, Z^T;\theta) \quad (9)$$

Based on Hermosillo and Hesse [7] while the probability of a low-volatility regime has decrease under 0.4 means that the indicator is stabil, the probability of a medium-volatility regime (though declining) still remains at around 0.4-0.6 means that the indicators is prone condition, and if the probability of a high-volatility regime has increased to over 0.6 means that the indicators on a crisis condition.

3. Method

The research used data of M1, M2 per foreign exchange reserves, and M2 multiplier from January 1990 to December 2015 that were obtained from Bank Indonesia and World Bank. The steps of the research are as follows. 1) Create a data plot and then test the stationary of data. If the data are not stationary then transform the data using log return. 2) Make a plot of partial autocorrelation function (PACF) of log return data then form the AR model. 3) Test the effects of heteroscedasticity on the residue of AR model using Lagrange multiplier test. 4) If there is the effects of heteroscedasticity on the residue of AR model, estimates the parameter of ARCH model. 5) Establish the combination of volatility and Markov switching models with the assumption of a three states. 6) Determine the conditions of the crisis based on the smoothed probability.

4. Results and discussion

4.1. Data

Plot data of M1, M2 per foreign exchange reserves, and M2 multiplier can be seen in Figure 1.

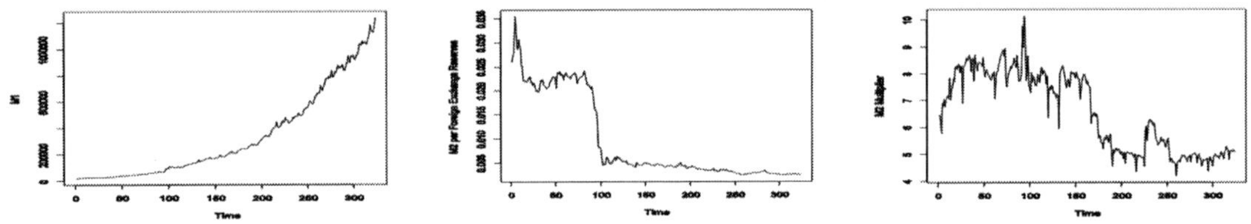

(a) (b) (c)

Figure 1. (a) M1Indicator (b) M2 Per Foreign Exchange Reserves Indicator (c) M2 Multiplier Indicator

Figure 1 indicates that the data are not stationary in mean and variance. It is proven by ADF test with probability value for each data is 0.99, 0.5091, and 0.09072 respectively. Then, the data were transformed using log return. Based on ADF test for log return data, the probability values are 0.01 where the probability values are less than $\alpha = 0.05$ so it can be concluded that the data are stationary. The next step is to form an AR model.

4.2. Establishment of AR Model

AR model can be identified from PACF plot of log return data. Based on M1, it was obtained an AR (1) model i.e. $r_t = -0.179548 + 0.015227 + a_t$. Meanwhile for M2 per foreign exchange, it was obtained an AR (1) model i.e. $r_t = 0.15626r_{t-1} + a_t$ and for M2 multiplier, it was obtained an AR (2) model i.e. $r_t = -0.38173r_{t-1} - 0.21282\, r_{t-2} + a_t$.

Based on Lagrange multiplier test on the residue of model, it was obtained the each probability of M1, M2 per foreign exchange reserves, and M2 multiplier as 2.853×10^{-6}, 5.317×10^{-6}, and 0.00575 respectively. All of probability values are less than 0.05, it means that there are the effect of heteroscedasticity on the residue. To overcome this condition, it was used ARCH model

4.3. Establishment of ARCH Model

For M1, the best model is ARCH (1) which can be written as

$$\sigma_t^2 = 8.483 \times 10^{-4} + 0.2128\, a_{t-1}^2.$$

Meanwhile, the best model for M2 per foreign exchange reserves is ARCH (2) which can be written as

$$\sigma_t^2 = 0.0009494 + 1.1485267\, a_{t-1}^2 + 0.1648690\, a_{t-2}^2,$$

and the best model for M2 multiplier is ARCH(2) which can be written as

$$\sigma_t^2 = 0.0020978 + 0.1821516\, a_{t-1}^2 + 0.0696154\, a_{t-1}^2.$$

Ljung-Box test results the probability values each of M1, M2 per foreign, and M2 multiplier as 0.4703, 0.6893, and 0.7382 respectively. These values are more than 0.05, so it can be concluded that the residue of ARCH models do not contain autocorrelation. Based on the Kolmogorov Smirnov test, the probability values of each indicators are 0.9255, 0.947, and 0.9625. These values are more than 0.05, so it can be concluded that the residue of ARCH models are normally distributed. Based on Lagrange multiplier test, the probability values of each indicators are 0.1612, 0.7343, and 0.8241. These values are more than 0.05, so it can be concluded that the residue of ARCH models do not contain the effect of heteroscedasticity. Based on these tests, it can be concluded that the ARCH models are the appropriate models. To model the changes of condition, it is used Markov switching model.

4.4. Establishment of SWARCH Model

The condition changes in Markov switching model are called states. The condition of intended in this research is low, medium and high volatility. States can be formed by transition probability. The transition probability can be formed as a matrix notation and usually called as matrix of transition probability, for M1 indicator transition probability matrix can be written as the following

$$P_1 = \begin{bmatrix} 0.07676399 & 0.28453180 & 0.6986911025 \\ 0.38344527 & 0.53779923 & 0.3011849593 \\ 0.53979073 & 0.17766900 & 0.0001239382 \end{bmatrix}.$$

Based on P_1 obtained that the probability to survive on low volatility state is 0.07676399. Probability changes from low to medium volatility state is 0.28453180. Probability changes from low to high volatility state is 0.6986911025. Probability changes from medium to low volatility state is 0.38344527. Probability survive in medium volatility state is 0.53779923. As well as the probability changes from medium to high volatility state is 0.3011849593. Probability changes from high to low volatility state is 0.53979073. Probability changes state from high to medium volatility state is 0.17766900. Probability survive in high volatility state is 00.00012393829.The matrix of transition probability for M2 per foreign exchange reserves and M2 multiplier stated in P_2 and P_3 as follows

$$P_2 = \begin{bmatrix} 0.3296875 & 0.29289357 & 0.5331779 \\ 0.5149849 & 0.67176968 & 0.1895393 \\ 0.1553276 & 0.03533675 & 0.2772827 \end{bmatrix} \text{ and}$$

$$P_3 = \begin{bmatrix} 0.0004510231 & 0.4594832 & 0.3716147 \\ 0.7649777181 & 0.4446863 & 0.2884468 \\ 0.2345712588 & 0.0958305 & 0.3399384 \end{bmatrix}.$$

Based on P_1, the parameter estimates of SWARCH(3,1) model for M1 can be written as follows

$$\mu_t \begin{cases} 0.029641, state\ 1, \\ 0.013654, state\ 2, \\ -0.009820, state\ 3, \end{cases} \quad \sigma_t^2 \begin{cases} 0.000077 + 0.359388 a_{t-1}^2, state\ 1, \\ 0.000192 + 0.019840 a_{t-1}^2, state\ 2, \\ 0.000008 + 0.280558 a_{t-1}^2, state\ 3, \end{cases}$$

Based on P_2, the parameter estimates of SWARCH(3,2) model for M2 per foreign exchange reserves can be written as follows

$$\mu_t \begin{cases} -0.018669, & state\ 1, \\ 0.001943, & state\ 2, \\ -0.022587, & state\ 3, \end{cases}$$

$$\sigma_t^2 \begin{cases} 0.00000004 + 0.02538471 a_{t-1}^2 + 0.04343017 a_{t-2}^2, state\ 1, \\ 0.00000003 + 0.00526216 a_{t-1}^2 + 0.00178388 a_{t-2}^2, state\ 2, \\ 0.00006397 + 0.30972356 a_{t-1}^2 + 0.05636983 a_{t-2}^2, state\ 3. \end{cases}$$

Based on P_3, the parameter estimates of SWARCH(3,2) model for M2 multiplier can be written as follows

$$\mu_t \begin{cases} 0.01936, state\ 1, \\ -0.00300, state\ 2, \\ -0.02996, state\ 3, \end{cases}$$

$$\sigma_t^2 \begin{cases} 0.000047 + 0.059475 a_{t-1}^2 + 0.078826 a_{t-2}^2, state\ 1, \\ 0.000001 + 0.116317 a_{t-1}^2 + 0.015176 a_{t-2}^2, state\ 2, \\ 0.000116 + 0.169118 a_{t-1}^2 + 0.000192 a_{t-2}^2, state\ 3. \end{cases}$$

where μ_t and σ_t^2 is the conditional mean and variance of SWARCH models.

4.5. Crisis Detection

Crisis detection using SWARCH(3,1) and SWARCH(3,2) can be seen by value of smoothed probability. Figure 2 show the smoothed probability plots of SWARCH (3.1) modeled by M1 and SWARCH(3,2) modeled by M2 per foreign exchange reserves and M2 multiplier.

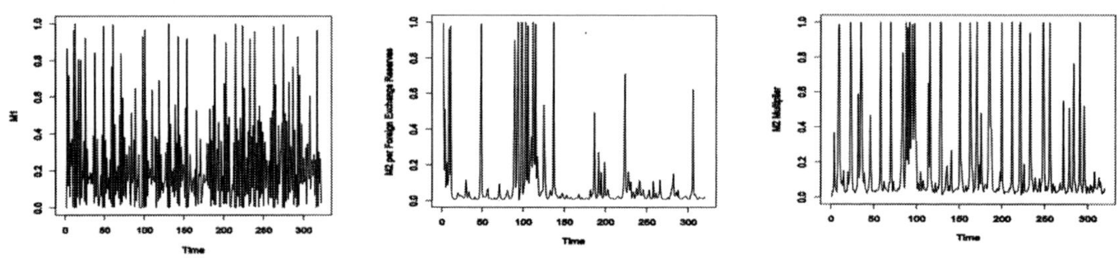

(a) (b) (c)

Figure 2. (a) Smoothed Probability of M1 (b) Smoothed Probability of M2 per Foreign Exchange Reserves (c) Smoothed Probability of M2 Multiplier.

Crisis condition signed with value of smoothed probability that greater than 0.6 for each indicators as shown in Figure 2. Table 1 showed the crisis period that has been detected on 1997, 1998, and 2008 based on the value of the smoothed probability that greater than 0.6 by M1. M2 per foreign exchange reserves, and M2 multiplier.

Table 1. The crisis detection based on smoothed probability

Year	M1	M2 per Foreign Exchange Reserves	M2 Multiplier
1997	July	September	August, October, December
1998	April, July	January, May, June, October	March, June
2008	January, October, December	November	September, October

Table 1 showed that smoothed probability for M1, M2 per foreign exchange reserves, and M2 multiplier can detect the financial crisis that hit Indonesia in 1997, 1998, and 2008.

5. Conclusion

The results of this research are M1 can be modelled by SWARCH(3,1) while M2 per foreign exchange reserves and M2 multiplier can be modelled by SWARCH(3,2) that can be used to detect the financial crisis in Indonesia in 1997, 1998, and 2008.

References

[1] Abimanyu A and M H Imansyah 2008 *Sistem Pendeteksian Dini Krisis Keuangan di Indonesia,* (Yogyakarta: Universitas Gajah Mada)
[2] Kaminsky G, S Lizondo and C M Reinhart 1998 *Leading Indicators of Currency Crisis*, IMF Working Paper **45**
[3] Hamilton J D and R Susmel 1994 Autoregressive Conditional Heteroscedasticity and Changes in Regime, *Journal of Econometrics* **64** 307
[4] Chang K, K Y Cho and M Hong 2010 Stock Volatility, Foreign Exchange Rate Volatility and The Global Financial Crisis, *Journal Of Economic Research* **15** 249
[5] Tsay R S 2005 *Analysis of Financial Time Series* (Canada: John Wiley and Sons)
[6] Kuan C M 2002 The Markov Switching Model, *Journal of Economics Academia Sinica* **1** 1
[7] Hermosillo B G and Hesse H 2009 *Global Market Conditions and Systematic*, Risk IMF Working Paper **230**

Mathematics learning on geometry for children with autism

F E Widayati[1,2], B Usodo[1] and I Pamudya[1]

[1]Universitas Sebelas Maret, Jl. Ir. Sutami No. 36A Kentingan, Surakarta, Indonesia
[2]Sekolah Menengah Al Firdaus, Jl. Al Kautsar, Kartasura, Sukoharjo, Indonesia

E-mail: faridaesti27@gmail.com

Abstract. The purpose of this research is to describe: (1) the mathematics learning process in an inclusion class and (2) the obstacle during the process of mathematics learning in the inclusion class. This research is a descriptive qualitative research. The subjects were a mathematics teacher, children with autism, and a teacher assistant. Method of collecting data was observation and interview. Data validation technique is triangulation technique. The results of this research are : (1) There is a modification of lesson plan for children with autism. This modification such as the indicator of success, material, time, and assessment. Lesson plan for children with autism is arranged by mathematics teacher and teacher assistant. There is no special media for children with autism used by mathematics teacher. (2) The obstacle of children with autism is that they are difficult to understand mathematics concept. Besides, children with autism are easy to lose their focus.

1. Introduction

Education is a basic need of every human being in order to ensure its survival is more dignified so that education has an important role in community development. Without education, human will be difficult to develop and even going backward. A state has an obligation to provide high-quality education to every citizen. In Indonesia, the Law number 20 comment 2003 about the National Education System on Article 5 states that every citizen has the same right to obtain a high-quality education [1]. This implies that all children in all circumstances, including children with special needs, are entitled to an equal opportunity to obtain a proper education. Statement of the *Individuals with Disabilities Education Act* (IDEA) suggests that it is important to provide access to the crew in order to get the same curriculum as normal and expected students both get the same instruction on concepts that are important [2].

As a form of government concerns about the special education for children with special needs, the government shall provide an education program suitable for all conditions of children. The new paradigm for education for students with special needs is putting them together with the other children to access education for all without distinction as to the rich, poor, with or without special needs. This is consistent with the results of the Dakar Declaration in 2000 which resulted in political and strategic decisions on education for all. The contents of the declaration, among others, stated that ensuring human learning needs of all young and adults are met through equitable access to programs of learning and life skills as appropriate [3].

Inclusive education is a regular education in which there are children with special needs. Through inclusive education, the children with special needs are educated alongside the other children (without special needs) to foster their potential. However, not all schools can develop the skills, the knowledge, the programs, and the resources to ensure all students reach their full potential [3]. Inclusive education prevents the discrimination against the children with special needs such that they can learn to live in the society. The society here means a society consisting of people with and without special needs live together, not separated, as a community. It starts with the school community. Similarly, this also applied to children with special needs of autism.

The Center for Disease Control and Prevention found that 1% of children aged 8 years in the United States meet the criteria for Autism Spectrum Disorder (ASD) in 2006. That is only for children aged 8 years there have been 40,000 individuals diagnosed with ASD. After 2000 there was a real increase in just 12 years, the number of children with autism has tripled compared with the previous decade. The remarkable number of suspected children with autism increases Indonesia. In 2015, it is estimated there are approximately 12,800 children with autism and 134,000 people with autism spectrum in Indonesia [4].

Autism is an organic developmental disorder that affects the ability of children to interact with people [5]. Children with autism experience the typical developmental disorder that includes perception, linguistic, cognitive, communication from mild to severe, and like living in his own world, characterized by the inability to communicate verbally and non-verbally with the external environment. Children with autism cannot relate to the others. It means that the ability to build relationships with other people disturbed by their inability to communicate and to understand the feelings of others.

Children with autism have very heterogeneous capabilities which can be leveled into high, average and below average [5]. The parents of children with autism generally have the assumption that children with autism have many advantages and they understand that children with autism have a chance to learn better than the other children with special needs. In the reality, most of the children with autism have constraints in terms of communication skills, social interaction, behavior, sensory, and learning styles which are very different from one child to another. Seeing this condition, it can be understood that children with autism have different characteristics from one another. The unique conditions of autistic children will greatly affect the education services, the program model, the learning strategies, and the resources needed for their learning activities, especially, as the focus of this article, in the mathematics learning.

Based on a description of the autistic child condition, obstacles probably occurs in the mathematics learning process when those abilities impaired. The teacher needs to concern with this condition by designing the mathematics learning to enable all children to learn according to their capacities. However, the actual learning services for children with autism tend to be generalized classical, no adoption of autism needs nor individual needs. One of the schools that organize inclusive education is SMP Al Firdaus. It is a junior high school in Sukoharjo, Central Java, Indonesia. The preliminary observation in SMP Al Firdaus gave data that there are three students with autism. Two students in the 7th grade and one student in the 8th-grade students. All the three students with autism are assisted by teacher assistant.

One of the subjects that must be taught to students with special needs is mathematics. Schools have to teach mathematics and students should learn mathematics. Indirectly, the statement explained that the inclusion school should also continue to teach mathematics, but the material presented to each student's special needs may be different between individuals from one another, depending on the condition of the student [6]. The success of student learning is not only influenced by the factors of students, but also of factors of teachers, especially in the inclusion classroom. According to the result of preliminary interview towards a mathematics teacher at SMP Al Firdaus, the difficulty in learning at children with autism happened because of the limitations of knowledge in understanding autism and how to handle a student with special needs. Identification of effective mathematics learning strategies for students with autism can facilitate success in both standardized testing and inclusion in a regular classroom setting. The success of mathematics learning in the classroom cannot be separated from the learning process and the interactions that occur in the classroom. To optimize the learning process, the readiness of the students and the teachers are required to establish the planning of the activities during the learning. The preparation of individual learning concerning to the needs of students with and without disabilities is more important than learning outcomes [7]. Based on the explanation, the purpose of this study was to determine (1) how the process of mathematics learning in the inclusive classroom, (2) the obstacles experienced by teachers during the learning process in the inclusive classroom of SMP Al Firdaus Sukoharjo.

2. Method

This study is a qualitative research which aims to understand the phenomenon of what is experienced by the subject of the study such as behavior, perception, motivation, action, and others, in a holistic manner, and by way of description in the form of words and language, in a specific context naturally and by utilizing a variety of natural methods [8]. The research was conducted at SMP Al Firdaus Sukoharjo.

The data in this study are the information about the mathematics learning process including the readiness of teachers before the learning process, the implementation and the evaluation, the student's activity during the learning, and the information about the difficulties experienced by the children of autism during the mathematics learning process in the inclusive classroom. The information obtained through observations in the mathematics learning as well as interviews with the mathematics teacher, the children with autism, and the teacher assistant. There are three sources of data in this study, namely: the key informants, places, and events, as well as documents. We were supported by instruments of observation towards the mathematics learning process and the guidelines for the interview.

3. Result and discussion

3.1. Mathematics Learning in Inclusive Classroom

The readiness of the teacher and the students with and without autism are very necessary before starting the mathematics learning. In the SMP Al Firdaus, mathematics teacher prepares the learning resources and media necessary for the learning process in the classroom inclusion before the learning begins. There are no special media prepared for students with autism. In order to help the concentration of students with autism, the teacher place students with autism to sit in the front row. In addition, there is a communication advance especially between mathematics teacher and teacher assistant regarding the material to be taught and to be adjusted to the student's ability.

There are three students with autism at SMP Al Firdaus. Two of them are in the 7th grade and one of them is in the 8th grade. Each student with autism in SMP Al Firdaus has different abilities. This makes the mathematics learning delivered differently. A student with autism in the 7th grade joined regular mathematics learning with partial modification. It means that the student still follows the regular teaching in classrooms, but with a modified curriculum that covers most of the materials and the assessments. The other one student with autism in 7th grade obtain mathematics learning with full modifications. It means that the student does not join the mathematics learning in a regular classroom but in a separate room and taught by a teacher assistant. Specifically, the material obtained was also different and not equal to the 7th grade but was equal to the 3rd grade.

A student with autism in the 8th grade gets mathematics learning by joining the regular classroom, using the regular curriculum, and do not get a modification of the curriculum. During the process of mathematics learning in the inclusive classroom, teacher assistant is ready in the classroom to help mathematics teachers in guiding students with autism when experiencing difficulties during the learning so that students with autism can still follow the material given by the teacher of mathematics well.

There is a modification of lesson plan for children with autism. This modification such as the indicator of success, material, time, and assessment. A lesson plan for children with autism is arranged by mathematics teacher and teacher assistant. The lesson plan was arranged using contextual approach. Thus, the students were introduced by the geometry material related to everyday life. This is in line with a research states that the ultimate goal of educating children with autism is to prepare students to achieve independence in the functioning of life and become a successful member of society [9]. This is because students with autism are difficult to gain knowledge of mathematical concepts. Thus, it is important that teacher use the most effective methods to teach students mathematics skills.

At the beginning of the learning activity, mathematics teacher give the students a motivation to focus more on teaching and conduct apperception by repeating the material and give the material-related questions without distinguishing students with and without autism. Questions for students with autism sometimes have to be repeated more than once to help students concentrate and focus.

Mathematics teacher connects the material and the context in daily life so that students with autism are easier to understand, for example, on the material of solid geometry, to help students understand the cube teachers carry teaching aid such as toy section, and to facilitate understanding of the beams, teachers using wrap toothpaste. The assessment stage performed almost as they both monitor the progress of students during the learning takes place, differing only in providing feedback on the success of students, especially students with autism, since they have different capabilities. Assessment instruments for each student with autism are also different, tailored to the abilities of each student, and has been agreed by the mathematics teacher and teacher assistant special. Assessment tools for students with autism with the regular curriculum prepared by the teachers of mathematics, whereas for autistic students with curriculum modifications either partially or fully drawn up by a teacher assistant with guidance and approval of mathematics teacher.

3.2. Experienced teacher in the classroom during the process of inclusion

At times during the learning process of mathematics, there are factors that experienced by mathematics teacher at SMP Al Firdaus Sukoharjo. Teachers have difficulty in providing materials to students with autism considering students with autism have difficulty in planting concept. The biggest obstacle for teachers in the mathematics learning process in the inclusive classroom is that the concentration of student with autism was less focused on learning as well as the lack of learning media or teaching aid to support the students with autism in learning mathematics.

4. Conclusion

Based on the results and discussion, it can be concluded that: (1) The process of mathematics learning in the inclusive classroom of SMP Al Firdaus Sukoharjo has a good preparation both the teacher and the students; (2) At the beginning of the learning activities, the mathematics teacher conducts apperception stage by repeating the previous material and provide related question; (3) At its core activities, mathematics teacher connects the material and daily context; (4) In the assessment phase, it is conducted by monitoring the progress of students during the learning takes place, differing only in the provision of feedback on student success; (5) Obstacles experienced by teacher in the learning process is on how to maintain the concentration of students with autism in learning mathematics.

References

[1] MOEC 2003 *Law number 20 comment 2003 about the national education system* (Jakarta: Ministry of Education and Culture)
[2] Butler F M, Miller S P, Crehan K, Babbit N and Pierce Y 2003 Fraction instruction for students with mathematics disabilities: comparing two teaching sequences *The Division for Learning Disabilities of the Council for Exceptional Children* **2** 99
[3] Forbes F 2007 Towards inclusion: An Australian perspective *Support for Learning* **22**
[4] Sastry A and Aguirre B 2014 *Parenting anak dengan autisme* (Yogyakarta: Pustaka Pelajar)
[5] Hadis H A 2006 *Pendidikan anak berkebutuhan khusus-autistik* (Bandung: Alfabeta)
[6] Steedy K 2008 Effective mathematics instruction *National Dissemination Center for Children with Disabilities (NICHCY)* **3**
[7] Babra M and Nyakudzuka J 2014 The inclusion of children with mental disabilities: A teacher's perspective *International Journal of Humanities Social Science and Education* (IJHSSE) **1** 65
[8] Moleong L J 2013 *Metodologi penelitian kualitatif* (Bandung: Remaja Rosdakarya)
[9] Gillies R M and Carrington S 2004 Inclusion: culture, policy and practice: A Quessland Perspective *Asia Pacific Journal of Education Strategies* **75** 175

IOP Conf. Series: Journal of Physics: Conf. Series **943** (2017) 012017 doi:10.1088/1742-6596/943/1/012017

Teaching materials of algebraic equation

S A Widodo[1,2], R C I Prahmana[3], A S Purnami[2] and Turmudi[1]

[1]Universitas Pendidikan Indonesia, Jl. DR. Setiabudi, Bandung, Indonesia
[2]Universitas Sarjanawiyata Tamansiswa, Jl. Batikan UH III/1043, Yogyakarta, Indonesia
[3]Universitas Ahmad Dahlan, Jl. Pramuka Yogyakarta, Indonesia

E-mail: sriadi@ustjogja.ac.id

Abstract: The purpose of this paper is to know the effectiveness of teaching materials algebraic equation. This type of research used experimental method. The population in this study is all students of mathematics education who take numerical method in sarjanawiyata tamansiswa of university; the sample is taken using cluster random sampling. Instrument used in this research is test and questionnaire. The test is used to know the problem solving ability and achievement, while the questionnaire is used to know the student's response on the teaching materials. Data Analysis technique of quantitative used Wilcoxon test, while the qualitative data used grounded theory. Based on the results of the test can be concluded that the development of teaching materials can improve the ability to solve problems and achievement.

1. Introduction

One of the reasons for the low students' achievement caused by the less ability in solving math. It can be shown that learners to solve math problems steps used to solve problems are incomplete/uncompleted and unsystematic, usually students simply write down the steps to implement the plan while the step of understanding, planning and looking back answers rarely written [1]. Though Ability to solve the problem is a basic ability of a person in solving a problem that involves critical thinking, logical, systematic, formulate concepts and capital success for students in solving mathematical problems, develop their ideas, and prepare yourself to face a problem in daily life –day

To solve a mathematical problem requires a systematic way or steps to make the process of solving the problem become easy and focused. One way to solve the problem is from Polya principle ie understand the problem, devise a plan make a plan, carry out our plan, and look back at the completed solution [2]. The Understand the problem stage refers to an understanding of what is known, what is asked, or whether sufficient, insufficient, exaggerated or contradictory conditions to seek out are asked [2-4]. In the make a plan stage or make a plan refers to how the related settlement strategy. A problem cannot be solved properly without good planning [4, 5]. The planning to solve the problem depends largely on the creative experience of students in making the settlement of a problem, the more varied their experience, the more creative the students tend to be in developing the problem solving plan [6]. At the stage of making plans, students can formulate mathematical problems or develop mathematical models, implement strategies to solve problems, find connections between concepts or theories that support each other and look for the formulas needed to solve problems, in addition students need to know first some math problem solving strategies to solve the problem can be done more effectively [2], students can take Polya's heuristics to see if, by modifying the conditions in the problem, new, more accessible problems may lead to which can be used as a springboard to break the original [5].

Carry out our plan stage, which referred to this stage is students are ready to do the calculation with all kinds of necessary data including the concepts and formulas or equations as appropriate [2, 4, 5]. Students will examine each of the steps outlined in the plan and write in detail to make sure each step is correct, the student must be able to form a more standard problematic system, in the sense that the formulas to be used are ready-to-use formulas according to what is used in the matter. In the last stage of the look back at the completed solution, the student will look back at the answer to make sure that the answer to the problem is correct. This step is important to check whether the results obtained are in accordance with the provisions and there is no contradiction to the question [1, 2, 4, 5].

In solving the problem, learners sometimes do not write down what is known and asked of the problem faced. So educators are difficult to guess, whether these learners have understood the problems faced or not. If the educator states that the learner has not understood the problem, the learner is able to solve the problem correctly. But if the educator declares that the learner understands the problem, the learner has not written down what is known and what is being asked. This implies that there are student habits such as not writing down what is known and what is being asked in understanding the problem. In addition, learners are sometimes unable to tell what steps to take to resolve the problem. Learners do not understand the steps to be taken to plan to solve the problem. Sufficient requirement and sufficient conditions in resolving already can illustrate the plan is sometimes not done by the investigators. In the step of re-examining the answers, learners almost entirely do not perform the process. Learners assume that this step makes the time to solve the problem is wasting time. At the re-examination stage is the lowest-weighted stage in the classification of thinking levels, so most learners do not perform the check-back stage in solving the problem [7].

Teaching materials have a very important position in learning, namely as a representation of the explanation of educators in front of the class [8]. On the other hand, teaching materials are positioned as tools or means to achieve competence [9]. So the preparation of teaching materials should be guided Competence to be achieved. Teaching materials are prepared without a guided competence, certainly will not provide many benefits to learners. The making of teaching materials is part of the innovation development process in education. The teaching materials used should not always be conventional but as educators, at least take action to improve the paradigm, perspective, thinking, attitudes, habits, professionalism, and behavior in teaching. Such as the development of teaching materials CELMI (Cellular Explorer Learning Movie Instruction) is a mobile-based learning media, where the development of this media is background by the misuse of the mobile telephone which is increasingly rampant done by students [10]. Thus the educators provide innovation in making teaching materials because it will impact on the smooth activities of learning students so as not to tend to feel bored.

Along with the increasingly modern education system and the growing demands, not infrequently schools still use conventional ways in implementing the learning process. Learning in a conventional way is usually done in solitary, meaning that the learning process that starts from planning, implementation, to the evaluation of student learning done by one teacher [11]. Planning is done by the teacher is usually in the form of preparing books or materials that almost the same as the previous year. This is can be showed by lesson plan that will be used during the learning process are still the same as in previous years and teachers do not have the motivation to prepare a good lesson plan so that in the preparation the lesson plan is not serious [12]. Whereas the preparation of lesson plan should at least adjust the characteristics of learners. So if the teacher is still using the same RPP with previous years then the teacher is indirectly learners are considered the same characteristics, whereas humans born into the world have unique differences [11].

There are several weakness of mathematics learning in Indonesian schools. The first is the low ability of teachers to use varied teaching methods. The second is the teaching ability of teachers only limited to answer the questions. The third is the teacher does not want to change the learning that is already considered correct and effective. And the last is teacher uses only conventional learning regardless the thinking aspect of learners [13].

To produce an active, easy to understand, and fun learning for the students, a learning model is needed that enables students to participate actively in the learning process, a learning process that enables students to participate actively during the learning process and interested students can be created by learning using the device [14]. One of the learning tools are that often used during the learning process such as teaching materials or textbooks.

Based on the result of previous research, it is found that the result of the validation test of algebra equation obtained by the average of 4.38 with very good category [11], the result of the validation is necessary to conduct experimental research or limited research in order to know the effectiveness of teaching materials equation algebra that has been developed. Based on these problems, the purpose of this article is to find out the material algebraic equations that have been developed to improve the ability to solve problems and learning achievements.

2. Method

The method used in this research is experimental research model, as a follow-up research from RND in the previous year. Population is the whole subject of research [15] , so the population in this study is all students of mathematics education who take numerical method in sarjanawiyata tamansiswa of university. The sampling technique used in this study is simple random, where the opportunity to take students who take the courses of numerical methods as research samples are the same. Of the students who took the numerical method taken as many as 58 students who are used as research samples.

The instrument used in this study is test. The test is used to obtain data about the ability to solve problems and learning achievement. Scoring capability problem solving refers to predetermined indicators that students can understand the problem, students can plan to solve problems, students can implement plans to solve problems, and students do the examination of the answers to the matter [16, 17]. The type of evaluation is described in table 1, and scoring learning achievement refers to the predefined ideal answer model.

Table 1. Description of the problem solving ability score

Polya Principle	Score	Indicators
Understand the problem	0	Student does not write anything so the student does not understand the meaning of the proposed problem
	1	Student writes the data/concept/knowledge that is not related to the proposed problem so that the student does not understand the proposed problem
	2	Student only write or reveal what is known or what is asked only
	3	Student is able to write down or reveal what is known and asked of the problem posed clearly
Devise a plan make a plan	0	Students do not tell or write steps to solve the problem
	1	Students tell or write step steps to solve the problem but not coherent
	2	Students write down sufficient terms and conditions necessary or formula of problem submitted and use all information collected.
Carry out our plan	0	Students are not able to implement the plan that has been made,
	1	Students implement the plan that has been made, but the error of the procedure and error algorithm
	2	Students implemented the plan that has been made, but there are procedural errors and algorithm errors
	3	Students do the plan that has been made, using the correct troubleshooting steps and no procedural errors occur, but an algorithm error occurs
	4	Students do the plan that has been made, using the steps to solve the problem correctly, no procedural errors and algorithm errors.
Look back at the completed solution	0	Student does not conduct a re-examination of the answers,
	1	Student re-examines the answers.

To test the proposed hypothesis using non-parametric test statistic with Wilcoxon test. The hypothesis proposed in this study is that there is an increase in learning achievement and problem-solving skills in the students after using algebra equation development result materials.

3. Result and discussion

3.1. Result

The result of previous research is obtained that the average of validation result of teaching material is 4.38 with very good category, so that the developed teaching material needs to be tested [11]. Then next learning materials development result is called algebraic equation.

Table 2. Output statistic non paramateric

	Problem-solving skills	Achievement
Man-Whitney U	569.000	1285.000
Wilcoxon W	2280.000	2996.000
Z	–6.159	–2.194
Asymp. Sig	0.000	0.028

Based on the result of statistical calculation (Table 2), there are two findings. The first is about student achievement on algebraic equation it is found that Mann-Whitney U is 1285.000; Wilcoxon W of 2996.000; Z_{obs} of -2.194 with a significance of 0.028. Based on these results it can be decided that there are differences in student achievement on algebraic equation before and after using the development result of teaching materials. The second is about students' ability to solve algebraic equations, Mann-Whitney U is 569.000; Wilcoxon W for 2280,000; Z_{obs} of -6.112 with a significance of 0.000. Based on these results it can be decided that there are differences in the ability of students in solving algebraic equation problems before and after using the development result of teaching materials.

Table 3. Descriptive statistic

		Minimum	Maximum	Mean	Variance
Problem solving skills	Before	10	32	21,97	24,74
	After	21	35	27,84	12,27
Achievement	Before	22	80	58,52	257,17
	After	16	100	67,48	651,38

Based on Table 3, there are two conclusions. The first is the average score of student achievement before and after using the teaching materials of the development result, it can be concluded that student achievement after using the better teaching materials when compared before using the materials development results. Based on these results it can be decided that teaching materials algebraic equations of development results can improve student achievement. The second is the average score of students' ability in solving algebraic equation problems before and after using teaching materials of development result, it can be concluded that the ability of students in solving algebraic equation problem after using better teaching materials when compared before using development result material. Based on these results it can be decided that the teaching materials algebraic equations of development results can improve students' ability in solving algebraic equation problems

3.2. Discussion

In learning mathematics, problem solving is an importance thing to be instilled to learners. There are several reasons that problem solving becomes important and becomes one of the basic skills of a

IOP Conf. Series: Journal of Physics: Conf. Series **943** (2017) 012017 doi:10.1088/1742-6596/943/1/012017

person in solving math problems. First, problem solving cannot be separated in everyday life, with the ability to solve problems that can be used to provide solutions or answers to problems faced more analytical so that someone can be a problem solver. In other words, when students are trained to solve problems, learners will be able to make decisions, because learners have become skilled about how to gather relevant information, analyze information, and be aware of the need to re-examine the results that have been obtained. Second, in mathematics learning, problem solving can be used to formulate concepts, develop ideas or ideas, and capital success for students in solving mathematical problems, because a concept or principle will be meaningful if it can be applied in problem solving. Third, the mathematical standards in schools should include standard content and process standards. Process standards include problem solving, reasoning and verification, interconnection, communication, and representation [18]. Fourth, one of the objectives of learning mathematics in Indonesia is that students are expected to foster critical thinking skills, logical, systematic, accurate, effective, and efficient in solving problems [19]. Achieving or not the purpose of mathematics learning one of them can be seen from the success learners in understanding mathematics and utilizing this understanding to solve mathematical problems as well as other sciences.

Improving the ability to solve problems for learners in learning mathematics is one of the tasks of an educator. One of the factors that influence the ability of learners in solving mathematical problems is the factor from outside the students such as the model of learning used by educators during the learning process. To produce active learning, easy to understand, and fun for students needed a model of learning that makes students actively participate in the learning process, the learning process that makes students actively participate during the learning process and interested students can be created by learning using the learning tools [14].

Students who are interested and actively participate in the learning process leads to better learning achievement when compared with passive learners and not interested in the learning process [11]. At least the learning achievement of learners who are interested and active will increase, when compared with passive learners. This is because learners increasingly understand the concepts of material delivered by educators if the student is not passive in receiving material submitted by the teacher [20].

Achievement and problem solving skills are largely determined by the quality of learning. The quality of learning is determined by the teaching materials used by teachers and learners [21]. Development of teaching materials is one form of learning process activities to improve or improve the quality of ongoing learning. The use of teaching materials in the learning process is one way to overcome the passive learning situation. Learners will be helped in the learning process by using teaching materials, helping educators to reduce material presentation time and increase mentoring time for learners, help educational institutions in completing the curriculum and achieve instructional goals with the time available [22].

The results of previous research indicate that learning tools in the form of teaching materials that have been developed are able to provide potential (positive) impact on the achievement of learning mathematics [23], and positively influence the activity and process skill to learner's achievement and the influence of activity and process skill together equal to achievement of 39% [13].

Test can be used to evaluate various aspects of teaching. Measurement of the effectiveness of learning should always be associated with the achievement of learning objectives. One of the indicators that can be used to define the effectiveness of learning is the carefulness of behavioral acquisition [24, 25]. Conscientious mastery of learned behavior, also often referred to as a performance error rate, can be used as an indicator to define the effectiveness of learning. The more accurately the students master the learned behavior, the more effective the lesson has been, the smaller the error rate, the more effective the lesson will be.

Referring to that opinion, the effectiveness of learning can be measured by providing a set of tests to learners. The test is used to measure the extent of learners' skills or mastery of the learner's material in understanding mathematics. Because one of the objects of this study is the ability of students in solving problems and achievement of learning, the skills or competencies that must-have is that students are able to solve or solve mathematical problems faced by learners so that learners have a good learning achievement.

Learners who are interested and actively participate in the learning process leads to better learning achievement when compared with passive elatif learners and not interested in the learning process. At least the learning achievement of learners who are interested and active will increase, when compared with passive learners. This is because learners increasingly understand the concepts of material delivered by educators.

4. Conclusion

In the previous research it was found that the teaching materials of algebraic equations that have been developed are validated by the experts, while the results of the learning material obtained the average score of 4.38 in the very good category. Based on limited trial, it found that teaching materials of algebraic equations can improve the students' problem solving ability and learning achievement.

References

[1] Pardimin and Widodo S A 2015 Increasing skills of student in junior high school to problem solving in geometry with guided *Journal of Education and Learning (EduLearn)* **10** 390

[2] Melis E and Ullrich C 2003 How to teach it-Polya-inspired scenarios in ActiveMath In *Proceedings of the International Conference on AI in Education AIED* pp 141-147

[3] Polya G 2014 *How to solve it: A new aspect of mathematical method* (New Jersey: Pricenton University Press)

[4] Ortiz E 2016 The problem-solving process in a mathematics classroom *Transformations* **1** 4

[5] El Sayed R A E 2002 Effectiveness of problem posing strategies on prospective mathematics teachers' problem solving performance *Journal of Science and Mathematics Education in Southeast Asia* **25** 56

[6] Indarwati D, Wahyudi, and Ratu N 2014 Peningkatan kemampuan pemecahan masalah matematika melalui penerapan problem based learning untuk siswa kelas V SD *Satya Widya: Jurnal Penelitian Pengembangan Kependidikan* **30** 17

[7] Utomo D U 2012 Pembelajaran lingkaran dengan pendekatan pemecahan masalah versi polya pada kelas VIII di SMP PGRI 01 Dau *Widya Warta* **XXXVI** 145

[8] Sumartana I W E, Sujana I W, Wiyasa I K N, and Kes M 2014 Pengaruh pembelajaran kontekstual yang berbasis tutor sebaya berbantuan bahan ajar tehadap hasil belajar IPS siswa kelas V SD gugus 8 abiansemal Badung tahun ajaran 2013/2014 *Mimbar PGSD* **2**

[9] Fidiana L, Bambang S, and Pratiwi D 2012 Pembuatan dan implementasi modul praktikum fisika berbasis masalah untuk meningkatkan kemandirian belajar siswa kelas XI *Unnes Physics Education Journal* **1** 38

[10] Susanto B, Dwijanto, and Siskandar 2012 Celmi media pembelajaran berbantuan telephone celluler untuk meningkatkan kompetensi penyolderan elektronika *Innovative Journal of Curriculum and Educational Technology* **1** 74

[11] Widodo S A 2017 Development of teaching materials algebraic equation to improve problem solving *Infinity* **6** 61

[12] Setiyasih R M 2016 Kesesuaian rencana pelaksanaan pembelajaran (RPP) dengan pelaksanaan pembelajaran bahasa jawa di SMP se-kecamatan Pulosari kabupaten Pemalang *Piwulang Jawa: Javanese Learning and Teaching Journal* **4**

[13] Nugroho A A 2011 Pengembangan perangkat pembelajaran matematika berbasis smart dengan strategi TAI pada materi segitiga kelas VII *Jurnal Aksioma* **2**

[14] Supriyono S, Setiawan T B, and Trapsilasiwi D 2014 Pengembangan perangkat pembelajaran matematika model student facilitator and explaining setting contextual teaching and learning (CTL) pada subpokok bahasan prisma dan limas kelas VIII semester genap *Pancaran Pendidikan* **3** 53

[15] Creswell J W 2012 *Educational research: planning, conducting, and evaluating quantitative* (Boston: Pearson)
[16] Widodo S A, Purnami A G, and Prahmana R C I 2017 Team accelerated instruction, initials and problem-solves ability in junior high school *International Journal on Emerging Mathematics Education* **1** 193
[17] Widodo S A 2015 Efektivitas pembelajaran team accelerated instruction terhadap kemampuan memecahkan masalah matematika siswa kelas VIII SMP kota Jogjakarta *AdMathEdu* **5** 183
[18] National Council of Teachers of Mathematics (NCTM) 2000 *Principles and standards for school mathematics* (Virginia: NCTM)
[19] Badan Standar Nasional Pendidikan (BSNP) 2006 *Lampiran peraturan menteri pendidikan nasional no 22 tahun 2006* (Jakarta: Depdiknas)
[20] Yeni E M 2011 Pemanfaatan benda-benda manipulatif untuk meningkatkan pemahaman konsep geometri dan kemampuan tilikan ruang siswa kelas v sekolah dasar *Jurnal Edisi Khusus* **1** 63
[21] Winarni D S 2017 Analisis kesulitan guru paud dalam membelajarakan ipa pada anak usia dini *Edu Sains: Jurnal Pendidikan Sains & Matematika* **5** 12
[22] Permana Y and Sumarmo U 2007 Mengembangkan kemampuan penalaran dan koneksi matematik siswa sma melalui pembelajaran berbasis masalah *Educationist* **1** 116
[23] Nizarwati, Hartono Y and Aisyah N 2013 Pengembangan perangkat pembelajaran berorientasi konstruktivisme untuk mengajarkan konsep perbandingan trigonometri siswa kelas X SMA *Jurnal Pendidikan Matematika* **3** 57
[24] Reigeluth C M and Merrill M D 1979 Classes of instructional variables *Educational Technology* **19** 5
[25] Sundayana R, Herman T, Dahlan J A and Prahmana R C I 2017 Using ASSURE learning design to develop students' mathematical communication ability *World Transactions on Engineering and Technology Education* **15** 245

IOP Conf. Series: Journal of Physics: Conf. Series **943** (2017) 012018 doi:10.1088/1742-6596/943/1/012018

An analysis on Kendall Lee queueing system with non-preemptive priority at BRI Ahmad Dahlan Yogyakarta

S D Ayuningtyas and N Binatari

Department of Mathematics Education, Faculty of Mathematics and Natural Science, Yogyakarta State University, Jl Colombo No.1 Depok Karangmalang Sleman Yogyakarta Indonesia

E-mail: nikenasih@uny.ac.id

Abstract. This research was aimed to determine the Kendall Lee model, measure the Performances and analyze the optimal condition of queue at tellers line at BRI Unit K.H Ahmad Dahlan Yogyakarta. First, the model was constructed based on Kendall Lee in which the parameter values were obtained from the fitness test of the data distribution. Next, the Performances measurements were counted using Little's Formula while the optimal condition was hold based on the total cost. Since the addition of the number of servers will influence the total cost, the optimal condition is reached for a number of servers with lowest total cost and satisfying steady state condition. The result shows that the queue system in teller counter at BRI Ahmad Dahlan Yogyakarta uses M/M/2 model with non-preemptive service distribution. The optimum number of the server due to its total cost was found to be two servers. By these, it could be concluded that the system has been using optimal condition.

1. Introduction

In our everyday life, the phenomenon of the queue is very common. A queue is a part of the circumstances that occur in a series of random operational activities in a service facility [1]. The queue occurs because the need for a service exceeds the capacity available for the service. Therefore, to increase the system's performance, the service providers provide multiple servers. Moreover, some system differs the customers into different priorities such as in customer service centers, airport security checkpoints, hospital emergency rooms, cloud computing systems or processor management in certain computer operating systems [2].

In a priority queue, each customers has a relative priority to be served [3]. Based on its priority level, there are two types of priority service discipline. The first one is the non-preemptive priority. If during the service time, one with higher priority customer arrives, it must wait until the item of lower priority has been served [4]. The second one is the preemptive priority. If during the service time, one with higher priority level arrives, it immediately displaces the lower priority customer. In this case, priority service discipline brings new problem on the queueing system.

The queue theory is one of the branches of applied mathematics that can be used to solve queuing problems. According to Heizer and Render, this has been very usefull for a service or manufacturing company [5]. One model of queue system was introduced by Kendall, which later was developed by Lee. The line model of Kendall Lee is a model line that is used to specify the characteristics of a queue includes the distribution of arrival and departure of customers, the number of servers, service

IOP Conf. Series: Journal of Physics: Conf. Series **943** (2017) 012018 doi:10.1088/1742-6596/943/1/012018

discipline, the capacity of the system and the number of customers who want to enter the queue system as the source. The used Kendall Lee line model is denoted by $(a\ /\ b\ /\ c)$: $(d\ /\ e\ /\ f)$ [6], where if the system capacity and calling source are both infinite then it is simplified by $a\ /\ b\ /\ c\ /\ d$.

A research study on the system with priority service discipline has been done by numerous researchers. For the single server, it has been discussed in [3]. Specifically, a system with two priority classes has been solved in [4] while the waiting time for markovian multiserver with non-preemptive priority $M\ /\ M\ /\ c$ was discussed in [7]. Research on the system of service priority queue with non-preemptive multi-server was done by Gail, Hantler, and Taylor [8]. From the research, they have derived the Performances measurements of the system and the waiting time for each priority class. A research conducted by Kao and Wilson [9] also discusses the customer waiting time for the system of non-preemptive priority queue multiple servers with two priority classes. Another study that discusses queuing system with a priority of service was done by Ni'amah and Sugito [10].

A queue with non-preemptive priority was implemented by a service provider namely BRI KH Ahmad Dahlan Yogyakarta at teller's line. The abundance of customers from the customer service's line and surveyor's line to teller's line has a higher priority to be served than the direct customers. By these, they found their direct customer service's delays are lining up to get service. The service delays occur whenever there is an abundance of customer service. Therefore, an analysis of this queueing system should be done.

In this research, it will be discussed the Kendall Lee model with non-preemptive priority occurred in BRI Unit KH Ahmad Dahlan Yogyakarta. The purpose of this study was to analyze the model line for Kendall Lee, to measure the Performances of the service system at teller line, and to find the number of server which optimize the total cost.

2. Method

The following will explain the methods and research design, location and time of the study, data collection techniques and design, and data analysis techniques.

2.1. Method and research design

The method used in this research was descriptive research method. Descriptive research method aims to describe the character of a variable, group or social phenomena that occur in society [11]. The design used in this research was the case and field research design.

2.2. Data collection technique

The research was conducted at PT Bank Rakyat Indonesia Unit KH Ahmad Dahlan Yogyakarta which is located at KH Ahmad Dahlan st No. 08 Yogyakarta. The process of retrieval data was obtained in the customer's waiting room, figure 1. The data was collected using the primary data collection technique in two ways, by using interview and field observation. The data are the interarrival time of abundance customers from customer service, the average rate of customer arrival of the queue directly to the teller and the service time.

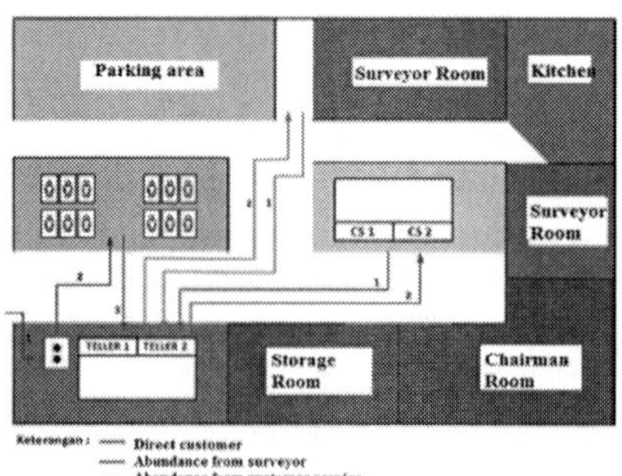

Figure 1. Queue system.

2.3. Data analysis technique

Customer data analysis techniques in teller line BRI Unit KH Ahmad Dahlan in Yogyakarta is as follows:

1. The primary data for the number of direct customer arrival, abundance customer arrival time and the service rate are grouped per 15 minutes and obtained for 4 hours of research per days.
2. Poisson distribution test for the data was conducted using the Kolmogorov-Smirnov test. According to Siegel [12], the calculation steps for Kolmogorov-Smirnov test are (a) define the - theoretical cumulative function, ie the expected cumulative distribution under H_0, (b) observe the values by pairing each interval $S_{(N)}(X)$, the cumulative frequency distribution which is obtained from a random sample of N observations, with intervals $F_{(0)}(X)$, the cumulative distribution function under the H_0 assumption, and then subtract $F_{(0)}(X)$ with $S_{(N)}(X)$ for each levels. (c) count the test statistic, $D = \max\left|F_{(0)}(X) - S_{(N)}(X)\right|$. (d) determine the Dtable value which is determined by referring to the critical values from the one sample Kolmogorov-Smirnov table.
3. The prosedure to conduct a one sample Kolmogorov-Smirnov test was:
   Null hypothesis H_0 : the data follows the Poisson distribution,
   Alternative hypothesis H_1 : the sata do not follows the Poisson distribution.
   The level of significance : α
   Critical value : the null hypothesis, H_0, is rejected if the test statistic D is greater or equals than the critical value obtained from the table, $D \geq D_{tabel}$.
4. Determine the queue model of the customers queueing system in teller line at Bank Rakyat Indonesia (Persero) Branch Office Unit KH Ahmad Dahlan Yogyakarta. According to Taha [6], Kendall Lee's Queue Model is denoted as follows:
   $$(a/b/c):(d/e/f).$$
   where
   a : Distribution of the arrival rate
   b : Distribution of services or departures time rate
   c : Number of server in parallel, $c = 1, 2, 3, \ldots$
   d : Service disciplines, such as FCFS, LCFS, SIRO
   e : The maximum capacity of the system
   f: The calling source
5. The traffic intensity for the i-th priority customer is $\rho_i = \dfrac{\lambda_i}{c\mu}$, where μ is the average rate of customer services, λ_i is the average of customer arrival rate on *i*–th priority, and c is the number of tellers to serve. Examine the steady-state solution, $\rho = \rho_1 + \rho_2$ It is exist if the value is less than one. In other words, the system has reached a stable condition.
6. Calculate the Performances of the queuing system. The measures of Performances are the expected value of customer waiting time on all priorities (W_q) and the expected number of i-th customers in the queue system $\left(W_s^{(i)}\right)$. According to Gross et al [13], W_q and $W_s^{(i)}$ can be calculated successively using equation (1) and equation (2).

$$W_q = \sum_{i=1}^{2} \frac{\lambda_i}{\lambda} W_q^{(i)} \quad (1)$$

and

$$W_s^{(i)} = W_q^{(i)} + \frac{1}{c\mu}, \tag{2}$$

where

$$W_q^{(i)} = \frac{\left[c!(1-\rho)(c\mu)\sum_{n=0}^{c-1} \frac{(c\rho)^{(n-c)}}{n!} + c\mu \right]^{-1}}{(1-\sigma_{i-1})(1-\sigma_i)} \text{ and } \sigma_i = \sum_{k=1}^{i} \rho_k, \sigma_0 = 0 \tag{3}$$

6. Optimize the queue system by considering its total cost. The cost analysis has taken a significant place in the queueing theory [14]. The total cost was defined as the sum total of the total cost of services per unit time, and the cost of the waiting *customers* per unit time [5]. In mathematical symbol, the cost model can be counted using the following formula:

$$ETC(c) = ESC(c) + EWC(c) \tag{4}$$

where

c	: the number of servers
$ETC(c)$	: the total costs for c servers per unit time.
$ESC(c)$	: the expected service cost per unit time for c servers.
$EWC(c)$	: the expected waiting cost per unit time for c servers.

According to Gross et al [13], to calculate the value of the expectations of many customer types are in the queue system, it can be using the equation 5:

$$L_s^{(i)} = \frac{\lambda_i \left[c!(1-\rho)(c\mu)\sum_{n=0}^{c-1} \frac{(c\rho)^{(n-c)}}{n!} + c\mu \right]^{-1} + \rho_i (1-\sigma_{i-1})(1-\sigma_i)}{(1-\sigma_{i-1})(1-\sigma_i)} \tag{5}$$

3. Research and Discussion

The queueing in teller line at BRI Unit KH Ahmad Dahlan follows priority discipline. The abundance customer from the customer service line and the surveyor line to the teller line has higher priority to be served. Therefore, the direct customer in the teller line should wait until the abundance customer are acompletely serviced. In otherwise, if the arrival abundance customer occurs when all the teller are being serving the customers, then the abundance customer should wait until the service is completely done. This type of priority discipline is called a nonpreemptive priority service discipline.

The study of this research deals only for teller service. Once customer completely served from the teller line, it is assumed that the customer left the queueing system. Therefore, the queue follows single phase multi channel model. The queue and service flow model can be seen from figure 2.

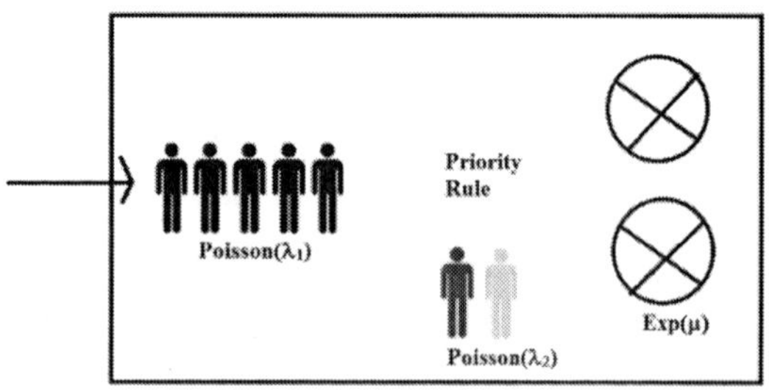

Figure 2. Queue model.

3.1. Distribution Match Test

Distribution suitability test was used to test the data directly teller line customer arrival, arrival overflow customers, and customer service is the Kolmogorov-Smirnov test. It was obtained that the direct customer arrival per 24 minutes was Poisson distributed with parameter $\lambda_1 = 8$ the abundance customers arrival was following Poisson distribution with $\lambda_2 = 1$ and the service was exponentially distibuted, $\mu = 9$.

3.2. Kendall Lee's Queue Model

Teller line systems in PT *Bank Rakyat Indonesia* (Persero) Branch Office Unit KH Ahmad Dahlan Yogyakarta has a model line of Kendall Lee $(M/M/2):(NPD/\infty/\infty)$. This model means arrivals and customer service Poisson distributed with 2 servers. Service discipline on the model $(M/M/2):(NPD/\infty/\infty)$ contains non-preemptive priority rule with unlimited system capacity and calling source.

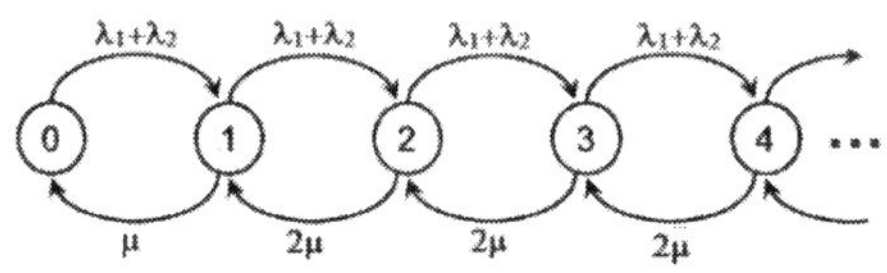

Figure 3. State transition diagram.

3.3. Steady State Check

The steady state condition is reach if the traffic intensity is less than one.

$$\rho = \frac{\lambda_1 + \lambda_2}{2 \cdot \mu} = \frac{1}{2} < 1$$

From this result, the queuing system with two server has reached steady state condition.

3.4. Performances measurements

The system characteristics could be recognised by looking at the expected rate of waiting time of the customer in queue line for both priority, W_q, and the expected rate of time spent by i-th type of customer on the system, $W_s^{(i)}$. By substituting the parameter's values into equation (1) and equation (2), the performances measurements from the system are as presented in table 1.

Table 1. Performance's measurements

Measurements	Value
$W_q^{(1)}$ (minute)	1.4084
$W_q^{(2)}$ (minute)	1.9746
W_q (minute)	1.0614
$W_s^{(1)}$ (minute)	2.1584
$W_s^{(2)}$ (minute)	2.7246

3.5. Cost Analysis Queue

In order to provide a better service, management could perform three ways, add the number of servers, decrease the service time or reduce the system capacity. This research only considered with the number of servers by using the cost model. The cost model can be used to evaluate and determine the optimum number of servers in the queue system [15]. The operating costs to be considered in this research were the cost of waiting for the customer, the salary and the logistics costs for adding a number of tellers. Further information, *teller* in PT Bank Rakyat Indonesia (Persero) Branch Office Unit KH Ahmad Dahlan Yogyakarta work for seven hours each day and five days in a week.

There was an extra cost for servicing if there is an additional teller. The expected service cost per unit time is equals to the number of server which is multiplied with the service cost for one server. In other word, the formula for the expected service cost can be written as equation (6).

$$ESC(c) = c \times C_1 \tag{6}$$

Now, we considered to the value of the service cost for each server. To serve the customer, the management should pay for the salary and the logistic cost. The salary for a teller per month is Rp 3,500,000 and the logistic cost to buy a computer, desk, chairs, and printer are Rp 4,000,000, Rp 1,100,000, Rp 700,000 and Rp 1,700,000, respectively. Assumed that the life period of the logistic was five years, then the service cost per minute is given in table 2.

Table 2. Cost of services

Cost Type	Unit	Cost per Month (Rp)	Cost per Day (Rp)	Cost per Hour (Rp)	Cost per Minute (Rp)
Salary	Person	3.500.000	175,000	25,000	417
Computer	Set	67,000	3,350	479	8
Table	Unit	19,000	950	136	3
Chairs	Unit	12,000	600	86	2
Printer	Unit	29,000	1.450	208	4
Total		3,627,000	181,350	25909	434

From the table, it can be concluded that the service cost per minute for one server is Rp 434.

Next, the expected waiting cost is considered to the customer's disadvantage for being in the system. Suppose that in order to be served, the customer should leave from work. In other words, the cost of waiting equals the cost of the customer if it is not going to work.Therefore, the waiting costs for customers are related to the per capita income of Indonesia. According to the Central Bureau of Statistics per capita income of Indonesia in 2016 was Rp 47.96 million. Therefore, the cost of waiting customer per minute, C_2, was Rp 476. The expected waiting cost per unit time

$$EWC(c) = L_s \times C_2. \tag{7}$$

Table 3 presents the three scenario of number of tellers, which are one, two and three, and the traffic intensity, ρ.

Table 3. The expected number of customer

Number of Teller (c)	λ	μ	$\rho = \frac{\lambda}{c\mu}$	***Steady State***	L_s **(customer)**
1	$\frac{77}{240}$	$\frac{77}{240}$	1	No	-
2	$\frac{77}{240}$	$\frac{77}{240}$	$\frac{1}{2}$	Yes	1,4203
3	$\frac{77}{240}$	$\frac{77}{240}$	$\frac{1}{3}$	Yes	1,0988

Based on the table 3, for one teller, the system will not reach a stable condition. It means that the teller will have no time for idle or rest. Since the optimal condition is reach if the traffic intensity is less than 1, a one teller queue system is not an optimal condition. Next, the system's stability was found to be in minimum two tellers. The greater number of servers, the less expected number of customers in the system. Therefore, we determined the optimum number of servers regarding to the number satisfied steady state condition but also minimize the total cost. Table 4 shows the total needed cost if there is an additional teller.

Table 4 Total cost per minute

Total Teller	Service Cost	Total Service Cost	Waiting cost for Customers	Number of Customers in the System	Total Waiting Cost	Total Cost
c	C_1	$c \times C_1$	C_2	L_s	$L_s \times C_2$	$(c \times C_1) + (L_s \times C_2)$
2	Rp 434	Rp 868	Rp 476	1.4203	Rp 677	Rp 1.545
3	Rp 434	Rp 1.302	Rp 476	1.0988	Rp 524	Rp 1.826

Based on the calculation of the total cost per minute in table 4, the optimal number of tellers was two tellers because it requires the lowest total cost, accounted to Rp 1,545. Since the number of tellers *at the moment wass* 2 systems, we can conclude that the queue system was already under optimal conditions.

4. Conclusion

Based on the results, the queue system for teller service at PT Bank Rakyat Indonesia (Persero) Branch Office Unit KH Ahmad Dahlan in Yogyakarta follows *multiple channels single-phase* line models Kendall Lee $(M / M / 2 / NPD)$. It means that the arrivals and service are Poisson distributed, two *servers* are available and the applied service's discipline is a *non-preemptive* priority. In addition, the capacity of queuing systems and calling sources is unlimited. The size of the Performances of the queue system in the company includes the average expected value of the customer's waiting time during 1.0614 minutes with an average value of the expected timing of the overflow customers and clients direct line tellers are in the queue until the system had been served consecutively for 2.1584 min and 2.7246 minutes. Teller queue system in the enterprise can achieve optimal conditions currently served by two tellers. Therefore, the queue system is already under optimal conditions.

References

[1] Kakiay T J 2004 *Dasar teori antrian untuk kehidupan nyata* (Yogyakarta: Andi)
[2] Brandwajn A and Begin T 2017 Multi-server preemptive priority queue with general arrivals and service times *Performance Evaluation* (Amsterdam: Elsevier)
[3] Miller R G 1960 Priority Queues. *Ann. Math. Statist.* **31** 86
[4] Chang W 1965 Queuing with nonpreemptive and preemptive-resume priorities. *Operations Research* **13** 1020-22
[5] Heizer J and Render B 2011 *Operations Management:* **10** (New Jersey: Pearson).
[6] Taha H A 2007 *Operation Research and Introduction* (Upper Saddle River: Pearson Education, Inc)
[7] Kella O and Yechiali U 1985 Waiting times in the non-preemptive priority M/M/C queue. *Stochastic Models,* **1** 257-62
[8] Gail H R, Hantler S L *et al* 1988 Analysis of a non-preemptive priority multiserver queue *Advances in Applied Probability* **20** 852-79
[9] Kao P C and Wilson S D 1999 Analysis of non-preemptive priority queues with multiple servers and two priority classes. *European Journal of Operational Research* **118** 181-93
[10] Ni'amah D and Sugito 2011 Queuing system with priority service *Proceedings of the National Seminar on Statistics* (Semarang: UNDIP)
[11] Martono N 2011 *Quantitative research methods* (Jakarta: Rajawali Pers).
[12] Siegel S 1956 *Nonparametric Statistics for the Behavioral Sciences* (New York: Mc Graw-Hill)
[13] Gross D 2008 *Fundamentals of Queuing Theory:* **4**. (New Jersey & Sons, Inc: John Wiley).

[14] Mishra S S, Yadav D K 2010 Computational approach to cost and profit analysis of clocked queueing networks *Contemporary Engineering Sciences* **3** 365-70
[15] A Muhajir, N Binatari 2017 Queueing System Analysis of Multi Server Model at XYZ Insurance Company in Tasikmalaya City *AIP Conference Proceedings* **1868**

IOP Conf. Series: Journal of Physics: Conf. Series **943** (2017) 012019 doi:10.1088/1742-6596/943/1/012019

Behaviour analysis of students in solving mathematics word problem

A T Wahyudi, S Subanti and B Usodo

Master Program of Mathematics Education, Faculty of Teacher Training and Education, Universitas Sebelas Maret, Jalan Ir. Sutami 36A Surakarta 57126, Indonesia

E-mail: asad3.w@gmail.com

Abstract. The purpose of this research was to identify the behaviour of students with auditory learning style in solving mathematics word problem on geometry material. The procedure was giving a test of mathematics word problem to students and interviews to representatives of students have an auditory learning style. The result of the research found that behaviours of students with auditory learning style demonstrated in solving mathematics word problems on geometry material showed the category of problem solving behaviour namely DTA-not proficient, DTA-Proficient and DTA-Limited of context. Thus, it can be said that the students with auditory learning style in solving mathematics word problem have behaviour solving problem still not optimal. Therefore, it is important for the teachers to consider auditory learning style of students in solving mathematics word problem to design and to plan better mathematics learning.

1. Introduction

Mathematics is one of the sciences which play the important role in supporting advancement and development in science and technology. Mathematics existence in this life is very significant because it is a universal science which can be conducted in another discipline. It is in line with the explanation that the study of mathematics is considered to be very important in each and every country of the world. Students are required to learn mathematics which is considered as a basic education since the skill of mathematics computation is essential in every walk of life [1]. In addition, mathematics is at the heart of many successful careers and successful lives for societal development, particularly in the extraordinary and accelerating change circumstances [2].

Indonesia is one of the countries recognized how the important mathematics in education. Thus, it has become one of the obligatory lessons in every education level. It must be mastered from basic level (elementary school) until high level (university). One of the signs whether mathematics learning process succeeds or not in a class showed from students' achievement in applying concept and formula in solving a task. However, based on the report of Trends in International Mathematics and Science Study (TIMSS) 2011, Indonesia ranked 38 of 42 countries with the overall score mean is 386 [3]. The result of Programme for International Student Assessment (PISA) 2015 showed that Indonesia ranked 63 out of 70 countries regarding with mathematics problem solving ability [4]. Also, more than 60% Indonesian students were only able to complete the problem under level 2. It indicates the low of mathematics problem solving ability of Indonesian students.

The low students' ability in solving mathematics problem also supported based on the result of observation we have conducted in SMP 3 Kebakkramat. It is a junior high school in Karanganyar, Central Java, Indonesia. The observation showed the low students' ability in solving mathematics' problem. Most of them acquired difficulty in mathematics learning in solving word problem include geometry lesson. Thus, they often get confused for comprehending what is the aim of the task and find difficulties to declare mathematics model from a word problem. It is in line with the explanation that one of the difficulties which are mostly obtained by them in mathematics problem is in solving mathematics word problem [5, 6]. The difficulties which have been obtained by them is not only in solving the problem requiring interpretation of the sentence but also the difficulties in mathematics model which have a meaning related to a problem.

The difficulties in solving mathematics word problem given are probably influenced by some factors, one of them is kind of students' learning style in the classroom. Students' learning style is a manner which is chosen by someone to receive information from the environment and to process them continually which each of students has their students' learning style selves. Thus, it influences students to determine how to study mathematics effectively. Manners of study effectively help them to catch and understand this lesson. A research conducted by Makhlouf, Martinez and Dahawy found that kinesthetic learning style owned by male and female students made the learning result statistically different which male students have better achievement than female [7]. But, for visual learning style and auditory style, the results showed that students' learning style give the different result for each student [7]. It is also supported by Halim [8], Zahro and Assyar [9] which shows that students' learning style influenced significantly by the students' behaviour in solving word problems.

In learning process at the SMP 3 Kebakkramat, the teacher often uses conventional learning model. The teacher gives a speech to the students who have an auditory learning style. Sometimes, the teacher also addresses problems to the students and they usually use three steps: writing the given information, writing the purpose, and writing the solution. Furthermore, they are given a mathematics word problem. The teacher explains what is given in the problem. Then, the teacher writes what the purpose is, and finally, the teacher writes its answer. Hereby, the students are given the explanation about it clearly. In order to memorize the steps, they are given mathematics word problem by the teacher with the same model with the example provided. Following these steps, most of the students in the classroom still could not solve the word problems.

In mathematics learning, especially problem solving, one of the very important things is the process which the students should have a custom and curiosity in solving the problems. In this condition, students' behaviour in problem solving is very vital. Pape [10] observed and described the students' behaviour when solving mathematics word problem from the perspectives of the reading comprehension process, including the kind of reading strategy. They read repeatedly, conclude, ask a question and interpret the structure of the text to improve reading comprehension. Moreover, Pape [10] classified kind of behaviours in solving mathematics word problem into two parts. They are direct translation approach and meaning-based approach. In the direct translation approach, it was divided into sub-categories. They are direct translation approach proficient, direct translation approach not proficient and direct translation approach limited context. On the other hand, the meaningful based approach was divided into two sub-categories. They are meaning-based approach full context and meaning-based approach justification.

Based on the research conducted by Pape [10], direct translation approach category was categorized by the students which had the lack of interpreting information from the problem or using problem context include all the sources in every element and relation between problems element. Based on this consideration, the researcher can conclude that the characteristic of direct translation approach includes the students who could not relate the problem elements yet in solving problems process. Furthermore, based on research conducted by Pape [10], meaningful-based approach category was divided behaviours changed include read repeatedly in acquiring an information with problem context, and explain the correct answer to operate calculation system. Based on the consideration the researcher can conclude that the characteristic of meaningful based approach includes information recording, using problem context which has description and assessment in mathematics operation.

We are interested to identify and to classify the students' behaviour in solving mathematics word problem in geometry lesson, especially the material of prism which is considered an auditory learning style. The purpose of the research is to identify and to classify kind of students' behaviour in solving mathematics word problem in geometry lesson especially in prism material considered by auditory learning style.

2. Method

The design of this research is descriptive qualitative using with case study method. Qualitative research was a research aimed to understand phenomena regarding subject's experience holistically through the description in the form of word and language [11]. Yin [12] defined case study as an empirical inquiry that investigates a contemporary phenomenon within its real-life context. Thus, our research tried to investigate and to identify deeply and holistically the contextual phenomena experienced by the subject. In this research, the contextual phenomena are the behaviour of students in solving mathematics word problem.

Before selecting the subject, we gave a learning style questionnaire to obtain students with auditory learning style. Then, we gave mathematics word problem and interviewed the research subjects. The research was conducted at SMP 3 Kebakkramat in 2017. The subjects in this research were three students from the 8th grade chosen by using purposive sampling taken from 32 students with auditory learning style considered to communicate their idea or their process in problem solving. Through this procedure, we obtained data of the answer sheet in solving the mathematics word problem and the result of interviewing the research subjects.

In this research, the data have been obtained through written tests and interviews on the subject of research. Furthermore, to verify the credibility of the data, we used triangulation method. We also used analysis steps by Miles and Huberman, including the data reduction, presentation, and verification [13].

3. Result and discussion

Based on the result acquired from the three research subjects with auditory learning style is followed by an interview to find out the subject' behaviour in solving the word problems. The result of data analysis of written test is provided in Figure 1.

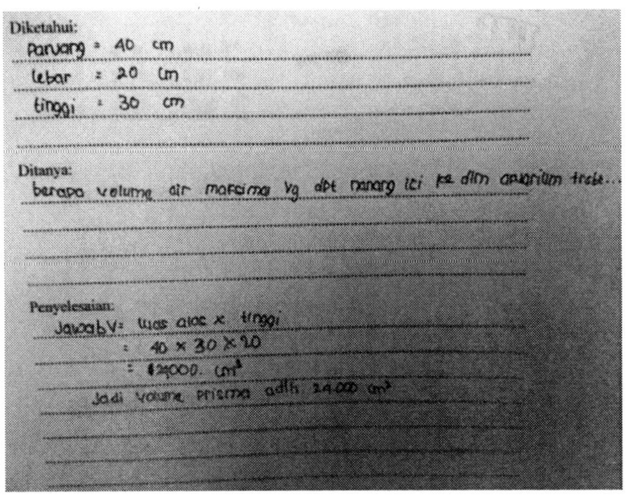

Figure 1. The answer provided by Subject 1 indicated by DTA-Limited Context

Based on the answer provided by Subject 1, Subject 1 could mention the given information in the problem context and declared what is the task question. Subject 1 was able to write the length of aquarium, the wide of aquarium and the height of aquarium in the answer sheet. Furthermore, Subject 1 was able to write what they inquire in the problem in writing how much volume of the water which is able to be fulfilled into the aquarium. However, the calculation and description were limited that appeared in the lack of calculation process. In the calculation of Subject 1, there is no writing in length and wide before changing into a number. In the answer sheet, Subject 1 has provided the conclusion indicated in the last writing of the answer. The result of interview towards Subject 1 suggests that Subject 1 used limited problem context. There is a limitation in the mathematical calculation, and read repeatedly followed with a calculation which probably includes problem context, and writes the last sentence completion directly.

Further, we got an answer sheet from the Subject 2 as shown in Figure 2.

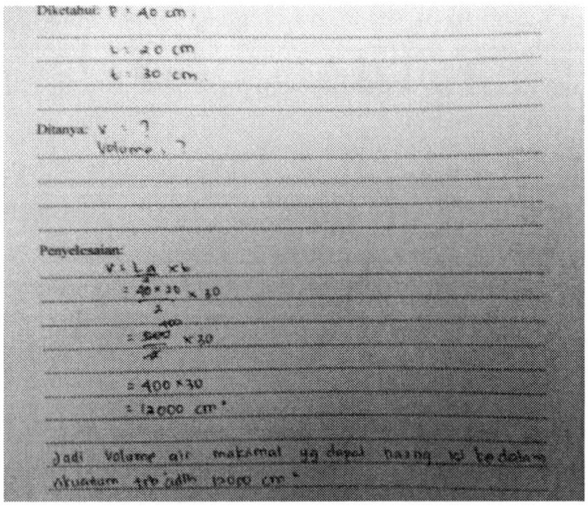

Figure 2. The answer provided by Subject 2 indicated by DTA-Proficient

Based on the answer provided by Subject 2, Subject 2 was able to declare what it was found in the problem. It appeared on the answer sheet that Subject 2 wrote the length, the width, and the height of the aquarium correctly. However, in this condition, Subject 2 did not use the problem context in calculation process correctly. Because, whether it is wrong way in interpretation formula in the aquarium with prism shape, the effect appeared inappropriate context with the problem.

Based on the interview towards Subject 2, Subject 2 did not read repeatedly the problem. In this condition, Subject 2 did not use the problem context in the process of the calculation. Furthermore, a description of mathematical calculation is not provided. Subject 2 provided the conclusion of the solution, but it was inappropriate with problem context.

The last answer sheet was obtained from Subject 3 as shown in Figure 3.

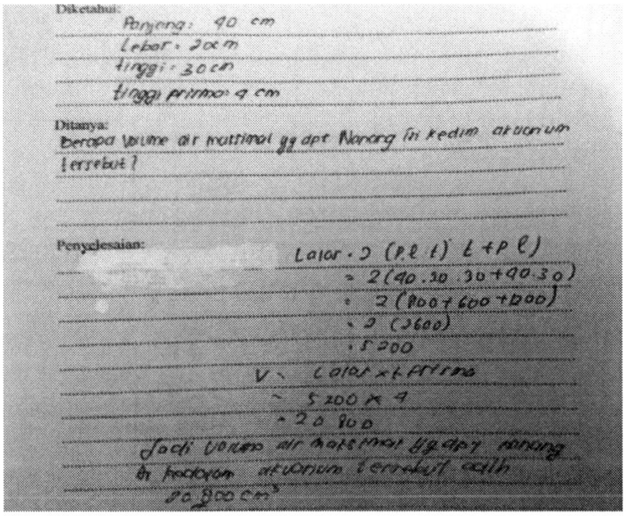

Figure 3. The answer provided by Subject 3 indicated by DTA-not Proficient

Based on the answer provided by Subject 3, Subject 3 has used problem context. However, Subject 3 provided an incorrect height of aquarium at prism size. Subject 3 did not use problem context in the process of the calculation and then provided the wrong result. Subject 3 incorrectly used the formula for the base area of the aquarium. As the result, the last answer is incorrect. The result of interview towards Subject 3 suggests that Subject 3 did not read repeatedly the problems. Subject 3 seems to use the problem context in the calculation process, but it is inappropriate with what it should be.

According to the data analysis in written test and interviews, we can acquire the classification and identification of students' behaviour in solving mathematics word problem in prism material with auditory learning style as follows.

Subject 1's behaviour in solving mathematics word problem described that the behaviour category is Direct Translation Approach (DTA) with sub-category Direct Translation Approach Limited Context, include: (1) using problem context in calculation was limited, (2) in this condition it describes about limitation in mathematical calculation, and (3) read repeatedly is directly followed by calculation which probably includes using problem context, (4) write the last answer completion.

Subject 2's behaviour in solving mathematics word problem described behaviour category for Direct Translation Approach (DTA) for sub-category Direct Translation Approach Proficient, include: (1) not read repeatedly the test, (2) not using problem context in the process even though in the calculation, (3) not describing mathematical calculation, (4) in this condition, the last answer was inappropriate with problem context.

Subject 3's behaviour in solving mathematics word problem described behaviour category for Direct Translation Approach (DTA) for sub-category Direct Translation Approach Not Proficient, include: (1) not read repeatedly the test, (2) not use problem context in the process even though in the calculation, (3) not explain the mathematical calculation, (4) the difficulties in calculation on problem solving, (5) there is the last answer, but it was inappropriate with problem context.

Based on the result of the test and interview for the Subject 1, it is acquired that Subject 1 has behaviour characteristic in problem solving as follows: (1) used problem context in the calculation, but it was limited. It is able to be found from the capability of subject 1 to write an to mention what is detected, is questioned, and used problem context in the calculation; (2) Those are limited explanation in the mathematical calculation; it seemed a few descriptions for problem completion; (3) read repeatedly which is directly followed by the calculation which is probably including used problem context. It is acquired from the reason of Subject 1 when solving the problem; and (4) write the last answer completion which seemed from the capability of subject 1 in answering the last result with problem context.

Moreover, based on consideration of written test and interview towards Subject 2, it is acquired that Subject 2 has behaviour category in solving the problem, including: (1) not read repeatedly the test. It is acquired from the reason when Subject 2 was solving the problem, (2) not using problem context in the process even though in the calculation. Subject 2 is able to be detected by capability in explaining and answering the problem which is wrong in describing what is detected in the used of formula in problem solving, (3) not explaining for mathematical calculation which appeared from the answer of Subject 2 without giving any description in every step in problem solving; (4) in this condition, the last inappropriate answer with problem context appeared from the subject 3 capability is failed in writing the last answer. And then, based on the consideration of the written test and interview for Subject 3, it is acquired that Subject 3 has behaviour characteristics in problem solving, including: (1) not read repeatedly in the test acquired from the reason when it was solving the problem; (2) not using problem context in the process even though in calculation which is found from the capability in explaining and answering the fault which is still fault in describing what is detected and is used the formula in problem solving; (3) not describing mathematical calculation which appeared from the answer of Subject 3 without giving description in every step in problem solving; (4) the difficulties in calculation process in solving problem. And also, it has been found difficulties to operate multiplication and addition; (5) in this condition, the last inappropriate answer with problem context appeared from the Subject 3 capability is failed in writing the last answer.

Based on the consideration and description of discussion, it is acquired that the characteristic of Subject 1 is classified as Direct Translation Approach (DTA) for sub-category Direct Translation Approach Limited Context, Subject 2 is classified as Direct Translation Approach Proficient category, and Subject 3 is classified as Direct Translation Approach Not Proficient. In this condition, it is appropriate with the behaviour in problem solving which is divided the behaviour in solving mathematics word problem when they solve the word problem into two parts [10, 14]. They are direct translation approach and meaningful based approach. Direct translation approach is divided into three sub-categories. They are direct translation approach proficient, direct translation approach not proficient and direct translation approach limited context. Whereas, the meaningful based approach is divided into two parts. They are meaningful based approach full context and meaningful based approach justification. Based on consideration from the data analysis, this behaviour is being in the kind of problem solving by Pape. Based on the three consideration, it is appropriate with the result of several research which classified students' behaviour in solving mathematics word problem, include: meaning-based approach full context, meaning-based approach limited context, direct translation approach limited context, direct translation approach proficient and direct translation approach not proficient [15-17]. Therefore, it can be said that the students' behaviour in solving the mathematics word problem with auditory learning style is not optimal yet which appeared in kind of behaviours in direct translation approach (DTA).

4. Conclusion

Based on the result and discussion the research concludes that classification of students' behaviour in solving mathematics word problem be considered from auditory learning style is Direct Translation Approach (DTA) for sub-category Direct Translation Approach Limited Context, Direct Translation Approach Proficient and Direct Translation Approach Not Proficient.

It shows that the students with auditory learning style in solving the mathematics word problem have behaviour solving problem still not optimal. Thus, the teacher should guide when the students read the task, they must write the information from the problem context which has been known. After that, they write a method used include the reason. And when they are writing the solving procedure, they must write the sequences process and the end of the result. Besides, the teachers must select of models, methods, strategies, and mathematics learning media to optimize the behaviour of students in solving the mathematics word problem. And it is really important for mathematics teachers to consider more attention toward students' auditory learning style in designing better mathematics learning.

References

[1] Mahanta D 2012 Achievement in Mathematics: Effect of Gender and Positive/Negative Attitude of Students *International Journal of Theoretical & Applied Sciences* **4** 157
[2] Amirali M 2010 Students' Conceptions of the Nature of Mathematics and Attitudes towards Mathematics Learning *Journal of Research and Reflections in Education* **4** 27
[3] Mullis I V S, Martin M O, Foy P and Arora A 2012 *TIMMS 2011 International Result in Mathematics* (Chestnut Hill: TIMSS & PIRLS International Study Center)
[4] PISA 2016 *Results Excellence and Equity in Education Volume I* (Paris: OECD Publishing)
[5] Marlina L 2013 Penerapan Langkah Polya dalam Menyelesaikan Soal Cerita Keliling dan Luas Persegi Panjang *Jurnal Elektronik Pendidikan Matematika Tadulako* **1** 1
[6] Maharani H R, Sukestiyarno, and Waluya B 2017 Creative thinking process based on Wallas model in solving mathematics problem *International Journal on Emerging Mathematics Education* **1** 177
[7] Makhlouf, Martinez and Dahawy 2012 A Comparison of Preferred Learning Styles between Vocational and Academic Secondary School Students in Egypt *Institute for Learning Styles Journal* **5** 1
[8] Halim A 2012 Pengaruh Strategi Pembelajaran dan Gaya Belajar Terhadap Hasil Belajar Fisika Siswa SMPN 2 Secanggang Kabupaten Langkat. *Jurnal Tabularasa PPS Unimed* **9** 141
[9] Zahroh U and Assyar B 2014 Kecenderungan Gaya Belajar Mahasiswa dalam Menyelesaikan Masalah Fungsi Bijektif *Jurnal Kebijakan dan Pengembangan Pendidikan* **2** 1
[10] Pape S J 2004 Middle School Children Problem Solving Behaviour A Cognitive Analysis *Journal for Research in Mathematics Education.* **135** 187
[11] Moleong L J 2007 *Metodologi Penelitian Kualitatif* (Bandung: Remaja Rosdakarya)
[12] Zainal Z 2007 Case Study as a Research Method *Jurnal Kemanusiaan* **9** 1
[13] Sugiyono 2013 *Metode Penelitian Pendidikan: Pendekatan Kuantitatif, Kualitatif dan R&D* (Bandung: Alfabeta)
[14] Tanujaya B, Prahmana R C I, and Mumu J 2017 Mathematics instruction, problems, challenges, and opportunities: A case study in Manokwari regency, Indonesia *World Transactions on Engineering and Technology Education* **15** 287
[15] Pratiwi E 2013 Analisis Perilaku Siswa Kelas IV SD dalam Memecahkan Soal Cerita Matematika Berdasarkan Tahapan Analisis Kesalahan Newmann *Jurnal Cakrawala Pendidikan* **15** 67
[16] Widodo S A, Purnami A S, and Prahmana R C I 2017 Team accelerated instruction, initials and problem-solves ability in junior high school *International Journal on Emerging Mathematics Education* **1** 193
[17] Syamsuri, Purwanto, Subanji, and Irawati S 2017 Using APOS theory framework: Why did students unable to construct a formal proof? *International Journal on Emerging Mathematics Education* **1** 135

IOP Conf. Series: Journal of Physics: Conf. Series **943** (2017) 012020 doi:10.1088/1742-6596/943/1/012020

The integration of technology in teaching mathematics

D Muhtadi[1,2], Wahyudin[2], B G Kartasasmita[2] and R C I Prahmana[3]

[1]Universitas Siliwangi, Jl. Siliwangi No. 24, Tasikmalaya, Indonesia
[2]Universitas Pendidikan Indonesia, Jl. Setiabudi No. 229 Bandung, Indonesia
[3]Universitas Ahmad Dahlan, Jl. Pramuka Kav. 5, Pandeyan, Yogyakarta, Indonesia

Email: dedimuhtadi@unsil.ac.id

Abstract. This paper presents the Transformation of Technological Pedagogical and Content Knowledge (TPACK) of three pre-service math teacher. They participate in technology-based learning modules aligned with teaching practice taught school and became characteristic of teaching method by using the mathematical software. ICT-based learning environment has been the demands in practice learning to build a more effective approach to the learning process of students. Also, this paper presents the results of research on learning mathematics in middle school that shows the influence of design teaching on knowledge of math content specifically.

1. Introduction

The challenge of accelerating change in the field of education in the 21st century is to find and develop tools that can make the efficiency and value of teaching and learning [1]. Technology is having an impact and change in efficiency as a tool of knowledge that can change the way the subject matter taught in teaching practice facilitates teacher and student learning in using new technology as a tool for collecting, organising and evaluating information in solving problems and innovate the practical ideas in the reality of life [2, 3, 4].

Many math teachers are aware of the opportunities of technology, such as interactive whiteboards, graphics calculators, software mathematics dynamic, program graphics, the system of computer algebra and others, too because in learning [5, 6]. However, the quality of the use of technology not only use the technology itself but how the technology selected can be integrated into content particular in learning. A teacher with the knowledge TPACK assumed to be able to figure out how to integrate technology into specific content and can apply the pedagogy most appropriate. Teachers in the 21st century are expected to be able to find out how to integrate technology into every aspect of education such as curriculum design, implementation, management and evaluation [7]. Therefore, it is very important for a teacher to come with the knowledge TPACK to survive in the educational system in the future. Recommendations from the Associated Mathematics Teacher Educators [8], states that the teacher education programme should provide the opportunity, for teachers, gained the knowledge and experience necessary to incorporate the technology in context teaching and learning of Mathematics.

IOP Conf. Series: Journal of Physics: Conf. Series **943** (2017) 012020 doi:10.1088/1742-6596/943/1/012020

The theoretical framework of TPACK, built on the framework of PCK (Pedagogic Content Knowledge) belongs to Schulman [9, 10, 11, 12]. Basic assumptions of this framework are that the teacher side of knowledge about technology and optimization of combined in the classroom is a balanced combination of content, pedagogy, and technology [11, 12, 13]. Harris and Hofer [14] mention the TPACK as something special, the application of knowledge-based content that supports the high integration of technology.

Based on Figure 1, TPACK is the center of the model, a representation of the slices of the following Knowledge:

1. PCK: how to teach specific content-based materials;
2. TCK: how to choose and use technology to communicate with the knowledge content;
3. TPK: how to use a particular technology when teaching.

TPACK ability of a teacher can be achieved when the teacher knows: (a) how technology tools can change the pedagogical strategies and representations of content to teach a particular topic, and (b) how the technology tools and impact on representation understanding of students against the topic [15]. Therefore it is important for teachers to learn how to have the ability of particular content domain TPACK for learning to be more effective, and the need to develop a design-based guide to assist him in the TPACK the process of teaching.

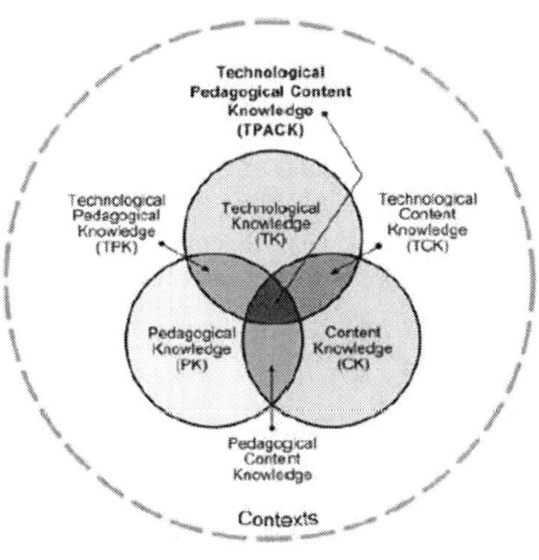

Figure 1. Components of the TPACK framework [11]

TPACK framework concept can give a new direction to describe the instructional idea in solving problems related to the integration of technology into teaching and learning practices as the theoretical basis for developing curriculum in teacher education programme [3, 16]. Research on teacher education reported that the TPACK model could be used as a potentially useful framework for preparing and developing teacher competence in school teaching [17, 18].

The purpose of this research is to explore technology-based learning approach in the context of the preparation and development of the competence of pre-service mathematics teacher to face the challenges of the changing flow of information and communication technology move more quickly. This article presents an exploration of the transformation competence TPACK of pre-service math teacher and how to integrate technology into teaching and learning mathematics of high school students.

2. Method

2.1. Participants

Two hundred and seventeen pre-service math teachers at a University in the town of Tasikmalaya have been taking computer courses. For research, three pre-service math teachers in this program are invited to participate in a case study. Participants in the study this case study is one female and two males, and those aged 22 to 23 years. All of them have satisfactory basic ICT skills, but they do not have the experience of using ICT for experiments in and teaching mathematics in the past.

2.2. The use of mathematical software in learning

Participants are introduced to the use of mathematical software such as GeoGebra, Cabri, and Geometers Sketchpad as a tool and medium for learning mathematics. The module created presents examples of material that can be taught by using the software.

After completing the learning presentation using mathematical software, participants are encouraged into critical open discussion forums to consider the potential impacts of technology-based teaching methods based on the TPACK framework. The critical open discussions on this case are consistent with the criteria for ICT-TPACK consisting of (a) the identification of the appropriate topic to teach with technology; (b) identification of appropriate representation to change content; (c) identification of teaching strategies that are hard to do in the traditional way; (d) the selection of the right tools and pedagogical use of their capabilities; and (e) the identification of the right integration strategies [19].

2.3. Data collection

To investigate the TPACK transformation from three pre-service math teachers who participated in this research, two instructional design tasks by using mathematical software into practice teaching and the learning process of students. The task of designing the first given to a participant in the first week with learning materials using mathematical software, before the first case study presentations. At the end of the learning materials, participants are assigned to complete the task of designing the two using mathematical software into practice teaching and the learning process of students. Also, the results of the study conducted with high school students put on the units of analysis are related to present experimental results the use of mathematical software in teaching practices.

2.4. Data analysis

For analyzing transformation competence of TPACK three pre-service math teachers, using the analysis of the contents in writing systematic design tasks. The coding system of the rubric with the TPACK framework developed and then encoded based on the design of the tasks the seven categories defined by the TPACK framework, namely CK, PK, TK, PCK, TCK, TPK, and TPACK.

3. Result

3.1. TPACK transformation of pre-service math teacher

3.1.1. The 1st Pre-service Math Teacher (PMT-1)

PMT-1 is male 23 years old, has no teaching experience, has no work experience, proficient using excel, and can use GeoGebra. Mathematical instructional design tasks with the TPACK framework can be summarized in Table 1.

Table 1. Summary of TPACK component for the 1st pre-service math teacher

Design of Learning Mathematics (DLM)	TPACK Components															
	CK		PK		TK		PCK		TPK		TCK		TPACK			
	Pn	Cn	Pn	Cn	Pn	Cn	Pn	Cn	Pn	Cn	Pn	Cn	Nn	Pn	Cn	An
DLM-1 (cube)	x		x			x	x		x		x		x			
DLM-2 (surface area of a cube)	x		x			x		x	x			x			x	

Note: Pn = perceptional levels, Cn = conceptional level, An = action level, Nn = no perception

For the preparation of instructional design, PMT-1 chooses flat-side geometry topics with the material properties of the cube for DLM-1 and the surface area of the cube for instructional DLM-2. On DLM-1, PMT-1 identifies content from the properties of the cube, such as the shape of the side of the cube, the positions of the sides, and the position of the ribs. PMT-1 describes learning cube properties by observing the shape of the side of the cube, the positions of the sides of rib and positioning objects that exist in the classroom or instructional media are presented. The use of software GeoGebra also shown to have more details about the properties of the cube especially avoid misconceptions students at space position of divination orthogonal. However, the PMT-1 did not explain the link between content and presentation, how to use GeoGebra software in learning. The third component of this given the PMT-1 alternative nature functioned as a supporting idea (*supporting the idea*) in justifying the previous explanation.

On DLM-2, the PMT-1 describe the surface area of a cube (CK-Pn), how to find the surface area of the cube (PK-Pn) as well as the use of GeoGebra to explain the steps of determining the surface area of the cube (TK-Pn). The relationship between the content and pedagogical knowledge explained that to be able to find the surface area of the cube, the students should understand the breadth of each side of the cube (PCK-Cn). The area on each side of the cube it will easily be understood when it is displayed in the net-mesh cubes served in GeoGebra (TCK-Cn). However, the PMT-1 did not explain in detail about the steps the presentation of this topic in GeoGebra. Transformation on the use of technology emerging from the results of students' difficulties in the identification of the observation of the media presented (TPK-Pn). The PMT-1 conceptual relationship between mention content, pedagogical and technological in DLM-2 (TPACK-Cn), i.e., by revealing the need for technology to represent the shape of the surface of the cube, so that facilitate students in identifying the area of each side of the cube.

3.1.2. The 2nd Pre-service Math Teacher (PMT-2)

PMT-2 is a 22-year-old female, private tutor, does not have the experience of teaching in schools, can use excel, and know the software GeoGebra but have never used. Mathematical instructional design tasks with the TPACK framework can be summarized in Table 2.

Table 2. Summary of TPACK component for the 2nd pre-service math teacher

Design of Learning Mathematics (DLM)	TPACK Components															
	CK		PK		TK		PCK		TPK		TCK		TPACK			
	Pn	Cn	Pn	Cn	Pn	Cn	Pn	Cn	Pn	Cn	Pn	Cn	Nn	Pn	Cn	An
DLM-1 (Integer Operations)		x	x		x		x		x		x		x			
DLM-2 (Linear equations of one variable)		x	x		x		x			x		x			x	

Note: Pn = perceptional levels, Cn = conceptional level, An = action level, Nn = no perception

On DLM-1, PMT-2 identify materials about calculating integer operations (CK-Cn) and how to transform the operations of subtraction, multiplication, and division into the summation operation, as well as using the inductive way to indicates that the negative multiplication by negative is positive (PK-Pn). CK needs to be delivered through the use of computer technology (TCK-Pn), like GeoGebra, Cabri or GPS, so the shape of the operation of subtraction, multiplication, and division and does not appear as though it has been superseded by the operation of addition (TPACK-Pn). The use of computer technology is meant to visualize forms of positive numbers and negative numbers (TCK-Pn) so that the students can distinguish between the sign of the number with the mark operation (PCK-Pn).

On DLM-2, the concept of the linear equations of one variable (LEOV) identified by PMT-2 about the use of the concept of operations to calculate an integer (CK-Pn). Because the term segment is a new concept for the students classes VII, a representation of a segment must be changed into another concept that is easy to understand students (PCK-Pn), i.e. through a visual representation (TCK-Cn) so that the term variables and constants are perceived by the students as if real (TK-Pn). Students must have the perception that the addition and subtraction of a variable are essentially the same with the addition and subtraction of constants (PK-Pn). Computer animation is very supportive towards learning this because representation variables and constants are easy to modify (TPK-Pn). By showing the relationship between CK, PK, and TK, PMT-2 has a good understanding of how the integration of content and technology, pedagogy (TPACK-Cn) so that the delivery of the material linear equations one variable (LEOV) the nature of *mind-on* can be a *hands-on activity*.

3.1.3. The 3rd Pre-service Math Teacher (PMT-3)

PMT-3 is a man of 23, have teaching experience, be able to use excel, and GeoGebra. Mathematical instructional design tasks with the TPACK framework can be summarized in Table 3.

Table 3. Summary of TPACK component for the 3rd pre-service math teacher

Design of Learning Mathematics (DLM)	TPACK Components															
	CK		PK		TK		PCK		TPK		TCK		TPACK			
	Pn	Cn	Pn	Cn	Pn	Cn	Pn	Cn	Pn	Cn	Pn	Cn	Nn	Pn	Cn	An
DLM-1 (geometry)		x	x				x		x		x				x	
DlM-2 (ball)		x	x		x			X		x		x				x

Note: Pn = perceptional levels, Cn = conceptional level, An = action level, Nn = no perception

On DLM-1, PMT-3 identifies the difficulty students on the material geometry in determining the corresponding position, either side of a corresponding or corresponding angle (CK-Cn). The result triggered the use of computer technology to a clip-right position between waking up so that the corresponding position (TK-Pn). Thus students can easily observe a corresponding side (TCK-Pn) to form the equation geometry is right (PK-Pn). Students can easily change the position of some images being mutual position corresponds to the way some wake that will be adjusted (TPK-Pn). The way is a technique that is good especially for students with low space divination (PCK-Pn). PMT-3 has a good understanding of the relationship between the PK, CK, TK, and delivery of TPACK (TPACK-Cn).

On DLM-2, PMT-3 identify the concept of balls (CK-Cn) is an abstract concept compared to wake up other spaces. PMT-3 looked at that space is a vast awake pedestal x-height (when the sockets and its roof congruent) and 1/3 wide x base x height (when only has the base and had no roof) (PK-Pn). The ball appears to be different from the other room, but PMT-3 has the perception that the concept could be applied to other room wake up to find the formula volume sphere (PCK-Cn). When the bottom quadrilateral is input pine on blankets and its vertex at the point the ball is placed in the center of the ball through the virtual image (TK-Pn), then students will easily that pedestal of n is a square blanket pine fruit balls (TCK-Cn). Furthermore, students will immediately be able to find the volume of the fruit n pine quadrilateral, i.e., $4\pi r^2$ x 1/3 x r (TPK-Cn). PMT-3 identifies the relationship between CK, PK, and TK which is actionable in the real implications in learning (TPACK).

3.2. Implementation of learning at school (the result of a research class of pre-service math teacher)

Implementation of learning at the school of the three prospective mathematics teacher is teaching practice in experiment results of applying learning by using the mathematical software [20]. The findings in the field show that understanding TPACK pre-service math teachers affect the practice of teaching mathematics in the classroom and instructional practices affect the conceptual understanding of middle school students in the field of mathematics.

4. Discussion

Empirical findings provide evidence that understanding TPACK conceptual of the third pre-service math teachers have undergone a transformation and improvement of knowledge that is generated by the following training using learning software-math. Also, the findings also show that math teacher candidates on TPACK task design both have a higher level of competence than the first design task. In more detail, it is interesting to note that prospective teachers of mathematics one have resulted in the perception of the PCK, the conception of TPK and TCK, for the development of competencies at the level of the conception of TPACK. Also, evidence that TPK and TCK PMT-2 has changed the perception of the level of conception similar to the PMT-3 who has changed PCK, TPK, and TCK at the perception level to the conception level. According to the transformation of their knowledge, the level of competence of the TPACK for PMT-2 and PMT-3 need to be developed to a higher level, from perception level to conception and action respectively.

The results of this research show that the technology-based learning has a positive impact on the development of the competence TPACK of a pre-service mathematics teacher. The result is because the use of technology-based learning materials for model TPACK pre-service teacher something to do with the knowledge content with changing shape significantly TPACK [16, 21, 22]. Moreover, consistently, research findings stating that technology-based teaching for teacher professional competency development led to a significant change in the learning activities of high quality.

5. Conclusion

This paper reports the use of technology-based learning approach to developing the competence of pre-service math teacher TPACK and transforming knowledge in integrating technology into practice teaching mathematics after technology-based learning training. In an effort to better serve the needs of the high-quality mathematics teachers, the result of this case study illustrates that competence TPACK can, in particular, be regarded as core attributes for future math teachers. The result of learning model with using the tools of technology can help students develop an understanding of mathematics and mathematics teaching method practices affect for teachers. Also, the competency of the TPACK mathematics teacher can have a direct impact on student learning in mathematics lessons. The results of this study show that the technology-based learning approaches can be an effective role in preparing and enhancing competency TPACK for teachers of mathematics.

Acknowledgments

The researcher thanks to supervisors and also the research fund, BPPDN KEMENRISTEK DIKTI. Hopefully, the results of this study can have a positive impact on the author in particular and the world of education in general.

References

[1] Prahmana R C I and Kusumah Y S 2016 The hypothetical learning trajectory on research in mathematics education using research-based learning *Pedagogika* **123** 42

[2] Edelson D C 2001 Learning-for-use: A framework for the design of technology-supported inquiry activities *Journal of Research in Science Teaching* **38** 355

[3] Jimoyiannis A 2010 Designing and implementing an integrated technological pedagogical science knowledge framework for science teacher professional development *Computer & Education* **55** 1259

[4] Young J R 2017 Technology integration in mathematics education: Examining the quality of meta-analytic research *International Journal on Emerging Mathematics Education* **1** 71

[5] Tanujaya B, Prahmana R C I, and Mumu J 2017 Mathematics instruction, problems, challenges, and opportunities: A case study in Manokwari regency, Indonesia *World Transactions on Engineering and Technology Education* **15** 287

[6] Fathurrohman M, Porter A L, and Worthy A L 2017 Teachers' real and perceived of ICT supported-situation for mathematics teaching and learning *International Journal on Emerging Mathematics Education* **1** 11

[7] Jang S J and Tsai M F 2012 Exploring the TPACK of International Society for Technology in Education *Computers & Education* **59** 327

[8] Association of Mathematics Teacher Educator (AMTE) 2006 Preparing teachers to use technology to enhance the learning of mathematics: A position of the Association of Mathematics Teacher Educators

[9] Shulman L S 1986 Those who understand: Knowledge growth in teaching *Educational Researcher* **15** 4

[10] Shulman L S 1987 Knowledge and teaching: Foundations of the new reform *Harvard Educational Review* **57** 1

[11] Mishra P and Koehler M 2006 Technological pedagogical content knowledge: A framework for teacher knowledge *The Teachers College Record* **108** 1017

[12] Mishra P, Koehler M J, and Kereluik K 2009 The song remains the same: Looking back to the future of educational technology *Tech Trends* **53** 48

[13] Cox S 2008 A conceptual analysis of technological pedagogical content knowledge *Unpublished doctoral dissertation* Provo, UT: Brigham Young University

[14] Harris J and Hofer M 2009 Instructional Planning Activity Types as Vehicles for Curriculum-Based TPACK Development. In I. Gibson et al. (Eds.) *Proceedings of Society for Information Technology & Teacher Education International Conference* 2009 (Chesapeake, VA: AACE) pp 4087-4095

[15] Graham C R, Burgoyne N, Cantrell P, Smith L, St. Clair L, and Harris R 2009 TPACK development in science teaching: Measuring the TPACK confidence of in-service science teachers *TechTrends* **53** 70

[16] Chai C S, Koh J H L, Tsai C C, and Tan W L 2011 Modeling primary school preservice teachers' Technological Pedagogical Content Knowledge (TPACK) for meaningful learning with information and communication technology (ICT) *Computers & Education* **57** 1184

[17] Doering A, Veletsianos G, Scharber C, and Miller C 2009 Using the technological, pedagogical, and content knowledge framework to design online learning environments and professional development *Journal of Educational Computing Research* **41** 319

[18] Voogt J, Tilya F, and Akker JvD 2009 Science teacher learning for MBL-supported student-centered science education in the context of secondary education in Tanzania *Journal of Science Education and Technology* **18** 429

[19] Angeli C, and Valanides N 2009 Epistemological and methodological issues for the conceptualization, development, and assessment of ICT TPCK: Advances in technological pedagogical content knowledge (TPCK) *Computers & Education* **52** 154
[20] Lee M H, Chang C Y, and Tsai C C 2009 Explore in the earth science classroom with relation to their attitudes and achievement *International Journal of Science Education* **31** 1811
[21] Prahmana R C I, Kusumah Y S, and Darhim 2017 Didactic trajectory of research in mathematics education using research-based learning *J. Phys.: Conf. Ser.* **893** 012001
[22] Sundayana R, Herman T, Dahlan J A, and Prahmana R C I 2017 Using ASSURE learning design to develop students' mathematical communication ability *World Transactions on Engineering and Technology Education* **15** 245

IOP Conf. Series: Journal of Physics: Conf. Series **943** (2017) 012021 doi:10.1088/1742-6596/943/1/012021

Stability of C_0-quasi semigroups in Banach spaces

Sutrima[1,2], Ch.Rini Indrati[1], and Lina Aryati[1]
[1]Mathematics Department of Universitas Gadjah Mada, Yogyakarta, Indonesia
[2]Mathematics Department of Universitas Sebelas Maret, Surakarta, Indonesia

E-mail: sutrima@mipa.uns.ac.id

Abstract. We concern on the non-autonomous abstract Cauchy problems $\dot{x} = A(t)x(t)$ on Banach spaces X. If $A(t)$ is the infinitesimal generator of a C_0-quasi semigroup $R(t,s)$ on X and $x_0 \in \mathcal{D}$, domain of $A(t)$, then the solution of the equation has uniquely representation $x(t) = R(0,t)x_0$. This representation shows that the stability of the quasi semigroup $R(t,s)$ influences the stability of the solution. In this paper, we investigate the stabilities of C_0-quasi semigroups following the existing theory of stabilities of C_0-semigroups $T(t)$ and bounded evolution operators $U(t,s)$. We devote the uniform, exponential, and strong stability of C_0-quasi semigroups in Banach spaces. The results are applicable for a large class of the time-dependent differential equations with unbounded coefficients in Banach spaces.

1. Introduction

Consider the non-autonomous abstract Cauchy problems

$$\begin{cases} \dot{x}(t) & = A(t)x(t), \quad t \geq 0 \\ x(0) & = x_0, \qquad\quad x_0 \in X \end{cases} \tag{1}$$

on Banach spaces X (see [1] and [2]). The x is an unknown function from the real interval $[0,\infty)$ into a Banach space X and $A(t)$ is a closed linear operator in X with domain $\mathcal{D}(A(t)) = \mathcal{D}$, independent of t and dense in X. The restrictive assumptions guarantee the existence of solutions of (1) (see [1] and [2]). Let $U(t,s)$ be an evolution family of bounded linear operators on X. The solution of (1) is induced by evolution operators $U(t,s)$,

$$x(t) = U(t,0)x_0. \tag{2}$$

This representation shows that the stability of the family $U(t,s)$ influences the behavior of the state $x(t)$ as time t evolves.

The problem (1) can be solved using another approach called quasi semigroup $R(t,s)$ on Banach space X. In this context $A(t)$ is the infinitesimal generator of $R(t,s)$ with domain $\mathcal{D}$ and for $x_0 \in \mathcal{D}$ the solution of (1) is given by $x(t) = R(0,t)x_0$. Definition of an infinitesimal generator and properties of quasi semigroups can be reviewed in detail in [3], [4], [5], [6], [7] and the references therein. By the following definition we verify that every C_0-semigroup on X is a C_0-quasi semigroup on X, but it is not conversely.

Let X be a Banach space and $\mathcal{L}(X)$ be the set of all bounded linear operators on X. A two-parameter commutative family $\{R(t,s)\}_{s,t\geq 0}$ in $\mathcal{L}(X)$ is called a strongly continuous quasi semigroup, in short C_0-quasi semigroup, on X if for each $r,s,t\geq 0$ and $x\in X$:

(a) $R(t,0)=I$, identity operator on X,
(b) $R(t,s+r)=R(t+r,s)R(t,r)$,
(c) $\lim_{s\to 0^+}\|R(t,s)x-x\|=0$,
(d) there exists a continuous increasing function $M:[0,\infty)\to[0,\infty)$ such that

$$\|R(t,s)\|\leq M(s).$$

If in (d) $M(s)=1$ for all $s\geq 0$, then $\{R(t,s)\}_{s,t\geq 0}$ is called contraction. Here we note that the norm $\|\cdot\|$ of (c) and of (d) acts on X and on $\mathcal{L}(X)$, respectively. In the sequel for simplicity we denote the quasi semigroup $\{R(t,s)\}_{s,t\geq 0}$ by $R(t,s)$. Next, by (d) of the definition of C_0-quasi semigroup if $R(t,s)$ is a C_0-quasi semigroup on a Banach space X and $R'(t,s)$ is the dual of $R(t,s)$ on the dual space X', then for every $x\in X$ and $x'\in X'$ the map

$$t\mapsto\|R(t,s)x\|\quad\text{and}\quad s\mapsto\|R'(t,s)x'\|$$

are measurable. These properties facilitate in characterizing of stability of C_0-quasi semigroups.

The most noted theory of stability of linear evolution operators has been proven by Datko [8]. In a few years later Pazy [9] completed the Datko's result and this result is well-known by Datko-Pazy Theorem. Generalizations of this result are obtained in [10], [11], [12], and [13] for exponential stability, in [14] and [15] for exponential instability, in [16] and [17] for exponential dichotomy, and in [18] the case of polynomial stability. In particular for autonomous case, the role of the family $U(t,s)$ is played by a C_0-semigroup $T(t)$. Some developments of the stability of C_0-semigroups are founded in [19], [20], [21], [22], [23], [24], and [25]. The developments of concepts of stability include uniformly stable, exponentially stable, strongly stable, weakly stable, and polynomially stable. Some results use the characterization of spectrum or resolvent of the infinitesimal generator of C_0-semigroup in proofs. These approaches are different from the Datko-Pazy Theorem approach. The Datko's proof use idea of a Lyapunov functional in Hilbert Space while the Pazy's proof base on the integral of norm of evolution operator.

By a quasi semigroup approach Megan and Cuc [6] generalized the Datko's result to investigate the exponential stability. Megan and Cuc [6] construct the sufficient and necessary conditions for exponential stability using an admissible increasing function as upper bound of integral of norm of the quasi semigroup either using an admissible increasing sequence. Moreover, Cuc [26] achieved to construct the necessary and sufficient conditions for uniform exponential dichotomy of C_0-quasi semigroup in Banach spaces. The obtained results generalize the similar theorem which are obtained by Datko [8] and Pazy [9] for the exponential stability and Preda and Megan [17] for the exponential dichotomy of C_0-quasi semigroups, respectively. Currently there has been no research that addresses the strong stability of C_0-quasi semigroups in Banach spaces. Besides that the results of Datko [8] and Megan and Cuc [6] give a possibility of alternative proof for the sufficiency and necessity for uniform and exponential stability.

In this paper we characterize uniform stability, exponential stability, and strong stability of C_0-quasi semigroups in Banach spaces. Specially, we investigate the sufficient and necessary conditions for uniform and exponential stability with a different approach to [6]. These are the main objects of this paper that is organized into the following three sequential sections. The stability and uniform stability are characterized in Section 2. Section 3 deals with the exponential stability in various types. The strong stability is characterized in Section 4.

2. Uniformly Stable

We begin with the non-uniform stability or stability of the C_0-quasi semigroups on a Banach space. This is the simplest stability.

Definition 2.1. *A C_0-quasi semigroup $R(t,s)$ on a Banach space X is said to be:*

(a) stable if there exists an increasing function $N : \mathbb{R}^+ \to (0,\infty)$ such that

$$\|R(t,s)x\| \leq N(t)\|x\|$$

for all $t,s \geq 0$ and $x \in X$;

(b) uniformly stable if there exists a positive constant $N > 0$ such that

$$\|R(t,s)x\| \leq N\|x\|$$

for all $t,s \geq 0$ and $x \in X$.

The Definition 2.1 states that if $R(t,s)$ is uniformly stable, then $R(t,s)$ is stable. The converse is not valid.

Example 2.2. The C_0-quasi semigroup $R(t,s)$ defined by

$$R(t,s)x = \frac{1+t}{1+t+s}x$$

for all $t,s \geq 0$ and $x \in \mathbb{R}$, is uniformly stable on $\mathbb{R}$.

The following theorem is the main result of this section.

Theorem 2.3. *Let $R(t,s)$ be a C_0-quasi semigroup on a Banach space X. The following statements are equivalent.*

(a) $R(t,s)$ is uniformly stable.

(b) There are $p,q > 1$ with $\frac{1}{p}+\frac{1}{q} = 1$, N_p, and N_q' such that

$$\sup_{t\geq 0, s>0} \left(\frac{1}{s}\int_0^s \|R(t+v,s-v)x\|^p dv\right)^{\frac{1}{p}} \leq N_p\|x\| \tag{3}$$

for all $x \in X$, and

$$\sup_{t\geq 0, s>0} \left(\frac{1}{s}\int_0^s \|R'(t,v)x'\|^q dv\right)^{\frac{1}{q}} \leq N_q'\|x'\| \tag{4}$$

for all $x' \in X'$.

Proof. $(a) \Rightarrow (b)$. By hypothesis there is a positive constant $N > 0$ such that

$$\|R(t,s)x\| \leq N\|x\|$$

for all $t,s \geq 0$ and $x \in X$. Since the inequality holds for every $t,s \geq 0$, then for $p > 1$ and $s \geq v$ it follows

$$\left(\frac{1}{s}\int_0^s \|R(t+v,s-v)x\|^p dv\right)^{\frac{1}{p}} \leq N\|x\|$$

for all $x \in X$. Hence $R(t,s)$ satisfies (3) with $N_p = N$. Since $\|R(t,s)\| = \|R'(t,s)\|$, analogously we can prove that $R(t,s)$ satisfies (4) with $N_q' = N$.

$(b) \Rightarrow (a)$. The Hölder inequality implies

$$\begin{aligned} s|x'(R(t,s)x)| &= \int_0^s |R'(t,v)x'(R(t+v,s-v)x)|dv, \quad s \geq v \\ &\leq \int_0^s \|R'(t,v)x'\|\|(R(t+v,s-v)x)\|dv \\ &\leq \left(\int_0^s \|R'(t,v)x'\|^q dv\right)^{\frac{1}{q}} \left(\int_0^s \|R(t+v,s-v)x\|^p dv\right)^{\frac{1}{p}}. \end{aligned}$$

Hence

$$|x'(R(t,s)x)| \leq \left(\frac{1}{s}\int_0^s \|R'(t,v)x'\|^q dv\right)^{\frac{1}{q}} \left(\frac{1}{s}\int_0^s \|R(t+v,s-v)x\|^p dv\right)^{\frac{1}{p}}.$$

By hypothesis there are $N_q', N_p > 0$ such that

$$|x'(R(t,s)x)| \leq N_q' N_p \|x'\|\|x\|. \tag{5}$$

On the other hand we have

$$|x'(R(t,s)x)| \leq \|x'\|\|R(t,s)\|\|x\|.$$

By definition of $\|R(t,s)\|$ it follows

$$\|R(t,s)\| \leq N_q' N_p := N < \infty$$

for all $t, s \geq 0$. This states that $R(t,s)$ is uniformly stable. ∎

The implication (b) $\Rightarrow$ (a) in Theorem 2.3 is not valid whenever (3) either (4) does not hold. The modifying of Example 2 of [11] with $R(t,s)x(\xi) = T(s)x(\xi)$ describes this situation.

3. Exponentially Stable

In this section we observe the exponential stability of C_0-quasi semigroups which is more special than the uniform stability. We begin with term of uniformly exponentially bounded of quasi semigroups.

Definition 3.1. *A C_0-quasi semigroup $R(t,s)$ on a Banach space X is said to be uniformly exponentially bounded if there are $\omega \in \mathbb{R}$ and $N_\omega \geq 1$ such that*

$$e^{\omega s}\|R(t,s)\| \leq N_\omega \tag{6}$$

for all $t, s \geq 0$.

There exists a C_0-quasi semigroup $R(t,s)$ which is not uniformly exponentially bounded, see Remark 1 of [11]. Specially, if $R(t,s) = T(s)$, where $T(s)$ is a C_0-semigroup on Banach space X, then $R(t,s)$ is uniformly exponentially bounded on X. Example 1 of [11] shows that the converse is not valid.

Definition 3.2. *A C_0-quasi semigroup $R(t,s)$ on a Banach space X is said to be:*

(a) exponentially stable on X if there are constant $\alpha > 0$ and increasing function $N : \mathbb{R}^+ \to [1,\infty)$ such that

$$e^{\alpha s}\|R(t,s)x\| \leq N(t)\|x\|$$

for all $t, s \geq 0$ and $x \in X$;

(b) uniformly exponentially stable on X if there are constant $\alpha > 0$ and $N \geq 1$ such that

$$e^{\alpha s}\|R(t,s)x\| \leq N\|x\|$$

for all $t, s \geq 0$ and $x \in X$;

(c) exponentially stable in the Barreira-Valls sense if there are constant $N \geq 1$, $\beta \geq \alpha > 0$ such that

$$e^{\alpha s}\|R(t,s)x\| \leq Ne^{(\beta-\alpha)t}\|x\|$$

for all $t, s \geq 0$ and $x \in X$.

From Definition 3.2 it is obvious that if $R(t,s)$ is uniformly exponentially stable, then $R(t,s)$ is uniformly stable and exponentially stable in the Barreira-Valls sense. If $R(t,s)$ is exponentially stable in the Barreira-Valls sense, then $R(t,s)$ is exponentially stable. If $R(t,s)$ is exponentially stable, then $R(t,s)$ is stable.

If $R(t,s)$ is uniformly exponentially bounded, then we can choose the positive constant ω such that (6) holds. Therefore, if $R(t,s)$ is uniformly exponential stable, then $R(t,s)$ is uniformly exponentially bounded. Example 2.2 describes that there exists a uniformly stable quasi semigroup but it is not exponentially stable.

Example 3.3. The C_0-quasi semigroup $R(t,s)$ defined by

$$R(t,s)x = e^{-\beta s}x, \quad \beta > 0$$

for all $t, s \geq 0$ and $x \in \mathbb{R}$, is uniformly exponentially stable on $\mathbb{R}$ (with $0 < \alpha < \beta$).

Example 3.4. Given a C_0-quasi semigroup $R(t,s)$ by

$$R(t,s)x = e^{(t+s)\sin(t+s)-t\sin t-3s}x$$

for all $t, s \geq 0$ and $x \in \mathbb{R}$. The quasi semigroup is exponentially stable in the Barreira-Valls sense (with $\alpha = 2$ and $\beta = 4$), but it is not uniformly exponentially stable on $\mathbb{R}$.

Supposed that $R(t,s)$ is uniformly exponentially stable on $\mathbb{R}$. There are constants a $\alpha > 0$ and $N \geq 1$ such that

$$e^{\alpha s}|R(t,s)x| = e^{\alpha s}e^{(t+s)\sin(t+s)-t\sin t-3s}|x| \leq N|x|$$

for all $t, s \geq 0$ and $x \in \mathbb{R}$. However, if we set $t = 2n\pi - \frac{\pi}{2}$ and $s = \frac{\pi}{2}$,

$$e^{2(n-1)\pi} \leq Ne^{\alpha(\pi/2)},$$

which is a contradiction for sufficiently large n.

Example 3.5. Given a function $u : \mathbb{R}^+ \to [1,\infty)$ where $u(n) = e^{n2^n}$ and $u\left(n + \frac{1}{2^n}\right) = 1$ for non-negative integer n. Let $R(t,s)$ be a C_0-quasi semigroup defined

$$R(t,s)x = \frac{u(t)}{u(t+s)e^s}x$$

for all $t, s \geq 0$ and $x \in \mathbb{R}$. The $R(t,s)$ is exponentially stable but it is not exponentially stable in the Barreira-Valls sense on $\mathbb{R}$.

If we suppose that $R(t,s)$ is exponentially stable in the Barreira-Valls sense, then there are constants $N \geq 1$ and $\beta \geq \alpha > 0$ such that

$$e^{\alpha s}\|R(t,s)x\| \leq Ne^{(\beta-\alpha)t}\|x\|$$

for all $t, s \geq 0$ and $x \in \mathbb{R}$. However, for $t = n$ dan $s = \frac{1}{2^n}$ we have

$$e^{n(2^n-(\beta-\alpha))}e^{(\alpha-1)\frac{1}{2^n}} \leq N$$

which is a contradiction for sufficiently large n.

Next we investigate the sufficient and necessary conditions for the uniform exponential stability of C_0-quasi semigroup. The following results are generalization of the exponential stability of C_0-semigroups and family of bounded evolution operators. The results are also the alternative sufficient conditions of [6].

Lemma 3.6. *If C_0-quasi semigroup $R(t,s)$ is uniformly exponential bounded on a Banach space X, then $R(t,s)$ is uniformly exponentially stable on X.*

Proof. By the definition of C_0-quasi semigroup there is a function $g : \mathbb{R}^+ \to \mathbb{R}^+$ such that

$$\inf_{t,s\geq 0} g(t+s) < 1 \quad \text{dan} \quad \|R(t,s)\| \leq g(s),$$

for all $t, s \geq 0$. Lemma 1 of [11] implies that $R(t,s)$ is uniformly exponentially stable on X. ■

Theorem 3.7. *Let $R(t,s)$ be a uniformly exponentially bounded C_0-quasi semigroup on a Banach space X. The following statements are equivalent.*

(a) $R(t,s)$ is uniformly exponentially stable.

(b) There exist $p, q > 1$ with $\frac{1}{p} + \frac{1}{q} = 1$, N_p, and N_q' such that

$$\sup_{t\geq 0, s>0}\left(\frac{1}{s}\int_0^s \|R(t+v, s-v)x\|^p dv\right)^{\frac{1}{p}} \leq N_p\|x\| \tag{7}$$

for all $x \in X$, and

$$\sup_{t,s\geq 0}\left(\int_0^s \|R'(t,v)x'\|^q dv\right)^{\frac{1}{q}} \leq N_q'\|x'\| \tag{8}$$

for all $x' \in X'$.

Proof. Analogously to the proof of Theorem 2.3, the implication $(a) \Rightarrow (b)$ is trivial.

$(b) \Rightarrow (a)$. By hypothesis there are $\omega \in \mathbb{R}$ and $N_\omega \geq 1$ such that

$$e^{\omega s}\|R(t,s)\| \leq N_\omega$$

for all $t, s \geq 0$. If $s \geq 1$, then for every $x' \in X'$ implies

$$\left(\frac{1}{s}\int_0^s \|R'(t,v)x'\|^q dv\right)^{\frac{1}{q}} \leq \left(\int_0^s \|R'(t,v)x'\|^q dv\right)^{\frac{1}{q}} \leq N_q'\|x'\|.$$

Conversely, if $0 < s < 1$, then

$$\left(\frac{1}{s}\int_0^s \|R'(t,v)x'\|^q dv\right)^{\frac{1}{q}} \leq \left(\frac{1}{s}\int_0^s N_\omega^q e^{\omega q}\|x'\|^q dv\right)^{\frac{1}{q}} \leq N_\omega e^{\omega}\|x'\|.$$

From the both we conclude that if the condition (8) is valid, then the inequality

$$\sup_{t\geq 0, s>0}\left(\frac{1}{s}\int_0^s \|R'(t,v)x'\|^q dv\right)^{\frac{1}{q}} \leq N_q^{'}\|x'\|$$

for all $x' \in X'$ is. Hence, Theorem 2.3 implies that the C_0-quasi semigroup $R(t,s)$ is uniformly stable on X. Therefore, there is a constant $N_1 > 0$ such that

$$\|R(t,s)\| \leq N_1 \tag{9}$$

for all $t, s \geq 0$.

On the other hand for $s > v \geq 0$ we have

$$\begin{aligned} s|x'(R(t,s)x)| &\leq \left(\int_0^s \|R'(t,v)x'\|^q dv\right)^{\frac{1}{q}}\left(\frac{1}{s}\int_0^s \|R(t+v,s-v)x\|^p dv\right)^{\frac{1}{p}} s^{\frac{1}{p}} \\ &\leq N_p\|x\|s^{\frac{1}{p}}\left(\int_0^s \|R'(t,v)x'\|^q dv\right)^{\frac{1}{q}} \leq N_p N_q^{'}\|x\|\|x'\|s^{\frac{1}{p}}. \end{aligned}$$

Consequently, there is a constant $N_2 > 0$ such that

$$s^{\frac{1}{q}}\|R(t,s)\| \leq N_2 \tag{10}$$

for all $t, s \geq 0$. By adding (9) and (10) we have

$$\|R(t,s)\| \leq \frac{N_1 + N_2}{1 + s^{\frac{1}{q}}} := g(s).$$

According to Lemma 3.6, the $R(t,s)$ is uniformly exponentially stable on X. ∎

Corollary 3.8. *Let $R(t,s)$ be a uniformly stable C_0-quasi semigroup on a Banach space X. The $R(t,s)$ is exponentially stable on X if and only if $R(t,s)$ satisfies (8).*

By implementing the principle duality to Theorem 3.7 we obtain the following theorem.

Theorem 3.9. *Let $R(t,s)$ be a uniformly stable C_0-quasi semigroup on a Banach space X. The following statements are equivalent.*

(a) $R(t,s)$ is uniformly exponentially stable.

(b) There exist $p, q > 1$ with $\frac{1}{p} + \frac{1}{q} = 1$, N_p, and $N_q^{'}$ such that

$$\sup_{t,s\geq 0}\left(\frac{1}{s}\int_0^s \|R'(t,v)x'\|^q dv\right)^{\frac{1}{q}} \leq N_q^{'}\|x'\| \tag{11}$$

for all $x' \in X'$, and

$$\sup_{t\geq 0, s>0}\left(\int_0^s \|R(t+v,s-v)x\|^p dv\right)^{\frac{1}{p}} \leq N_p\|x\| \tag{12}$$

for all $x \in X$.

The following theorem shows that the uniformly stable assumption in Corollary 3.8 can be replaced by the exponentially bounded assumption.

IOP Conf. Series: Journal of Physics: Conf. Series **943** (2017) 012021 doi:10.1088/1742-6596/943/1/012021

Theorem 3.10. *Let $R(t,s)$ be uniformly exponentially bounded C_0-quasi semigroup on a space Banach X. The following statements are equivalent.*

(a) $R(t,s)$ is uniformly exponentially stable.

(b) There exist $q \geq 1$ and N_q' such that

$$\sup_{t,s\geq 0}\left(\int_0^s \|R'(t,v)x'\|^q dv\right)^{\frac{1}{q}} \leq N_q'\|x'\|$$

for all $x' \in X'$.

Proof. Analogously to proof of Theorem 2.3, the implication $(a) \Rightarrow (b)$ is trivial.

$(b) \Rightarrow (a)$. By hypothesis there exist $\omega > 0$ and $N_\omega \geq 1$ such that

$$\|R(t,s)\| \leq N_\omega e^{\omega s} \tag{13}$$

for all $t, s \geq 0$. For $q > 1$ is fixed. If $s \geq 1$, then for every $x' \in X'$

$$\begin{aligned}\int_0^1 N_\omega^{-q} e^{-\omega q u} du |x'(R(t,s)x|^q &\leq \left(\int_0^s \|R'(t,v)x'\|^q \|R(t+v, s-v)\|^q N_\omega^{-q} e^{-\omega q v} dv\right)\|x\|^q \\ &\leq \left(\int_0^s \|R'(t,v)x'\|^q dv\right)\|x\|^q \leq \left(N_q'\right)^q \|x'\|^q\|x\|^q.\end{aligned}$$

On the other hand we have

$$|x'(R(t,s)x|^q \leq \|R(t,s)\|^q\|x'\|^q\|x\|^q.$$

Therefore

$$\|R(t,s)\| \leq H(\omega)N_\omega N_q'$$

for all $t \geq 0$, where $H(\omega) = \left(\int_0^1 e^{-\omega q u} du\right)^{-1}$.

Conversely, if $0 \leq s < 1$, then by (13) we have

$$\|R(t,s)\| \leq N_\omega e^{\omega}$$

for all $t \geq 0$. Thus, $R(t,s)$ is uniformly stable on X. Corollary 3.8 implies that $R(t,s)$ is uniformly exponentially stable on X.

For $q = 1$ and the fact that $R(t,s)$ is a quasi semigroup, then for $s > v \geq 0$ we have

$$\begin{aligned} s|x'(R(t,s)x)| &\leq \int_0^s \|R'(t,v)x'\|\|R(t+v,s-v)x\|dv \\ &\leq \left(\int_0^s \|R'(t,v)x'\|dv\right) M(s)\|x\| \leq N_1'\|x'\|M(s)\|x\|.\end{aligned}$$

Hence

$$\|R(t,s)\| \leq \frac{N_1' M(s)}{s} := g(s)$$

for all $t, s \geq 0$. Lemma 3.6 implies that $R(t,s)$ is uniformly exponentially stable on X. ∎

IOP Conf. Series: Journal of Physics: Conf. Series **943** (2017) 012021 doi:10.1088/1742-6596/943/1/012021

4. Strongly Stable

In this section we consider a strong stability concept which is weaker than the uniform exponential stability.

Definition 4.1. *A C_0-quasi semigroup $R(t,s)$ on a Banach space X is said to be:*

(a) strongly stable if

$$\lim_{s\to\infty} \|R(t,s)x\| = 0$$

for all $t \geq 0$ and $x \in X$;

(a) uniformly strongly stable if

$$\lim_{t,s\to\infty} \|R(t,s)x\| = 0$$

for all $x \in X$.

From the definition it is obvious that every uniformly strongly stable C_0-quasi semigroup is a strongly stable C_0-quasi semigroup but the converse is not valid.

Example 4.2. Given a C_0-quasi semigroup by

$$R(t,s)x = \frac{1+t}{1+t+s}x$$

for all $t, s \geq 0$ and $x \in \mathbb{R}$. The $R(t,s)$ is strongly stable, but it is not uniformly strongly stable on $\mathbb{R}$.

It is obvious that $R(t,s)$ is strongly stable on $\mathbb{R}$. However, for $t = s$ and $x \neq 0$ we have

$$\lim_{t,s\to\infty} |R(t,s)x| = \lim_{s\to\infty} \frac{1+s}{1+2s}|x| = \frac{1}{2}|x| \neq 0.$$

So, $R(t,s)$ is not uniformly strongly stable on $\mathbb{R}$.

Example 4.3. Given a C_0-quasi semigroup $R(t,s)$ by

$$R(t,s)x = e^{-(s^2+2st)}x$$

for all $t, s \geq 0$ and $x \in \mathbb{C}$. The $R(t,s)$ is uniformly strongly stable on $\mathbb{C}$.

It follows that

$$\lim_{t,s\to\infty} |R(t,s)x| = \lim_{t,s\to\infty} \frac{1}{e^{s^2+2st}}|x| = 0$$

for all $x \in \mathbb{C}$. So, $R(t,s)$ is uniformly strongly stable on $\mathbb{R}$.

Lemma 4.4. *If $R(t,s)$ is a uniformly strongly stable C_0-quasi semigroup on a Banach space X, then $R(t,s)$ is uniformly stable on X.*

Proof. By hypothesis and fact the function $s \mapsto \|R(t,s)\|$ is bounded on every bounded interval $[0,\alpha]$, the Uniformly Boundedness Theorem (Theorem A.3.19) of [27] implies that $R(t,s)$ is uniformly stable on X. ■

Theorem 4.5. *Let $R(t,s)$ be a uniformly stable C_0-quasi semigroup on a Banach space X and $x \in X$.*

(a) If there exists an unbounded sequence $(s_n) \subseteq \mathbb{R}^+$ such that $\|R(t,s_n)x\| \to 0$ for $n \to \infty$, then $\lim_{s\to\infty} \|R(t,s)x\| = 0$ for every $t \geq 0$.

(b) If $R(t,s)$ is a C_0-quasi semigroup of contraction, then $\lim_{s\to\infty}\|R(t,s)x\|$ exists, for every $t \geq 0$.

Proof. (a) By Lemma 4.4, $R(t,s)$ is uniformly stable. Set $N = \sup_{t,s\geq 0}\|R(t,s)\|$. By boundedness of (s_n), for every $n \in \mathbb{N}$ there is s such that $v_n = s - s_n > 0$. In this case we can choose (v_n) to be an increasing sequence. The hypothesis implies that $\|R(t,v_n)x\| \to 0$ for every $t \geq 0$ as $n \to \infty$. Given any $\epsilon > 0$. There is $K \in \mathbb{N}$ such that for $n \geq K$ we have

$$\|R(t,v_n)x\| < \epsilon.$$

Consequently, for all $n \geq K$ we have

$$\|R(t,s)x\| \leq \|R(t+v_n,s_n)\|\|R(t,v_n)x\| \leq N\epsilon$$

for every $x \in X$ and $t \geq 0$. So, $\|R(t,s)x\| \to 0$ as $s \to \infty$ for every $t \geq 0$ and $x \in X$.

(b) For $t \geq 0$ fixed and $x \in X$. The function $s \mapsto \|R(t,s)x\|$ is decreasing. In facts, if $s_1 < s_2$, where $s_2 = s_0 + s_1, s_0 > 0$ and $R(t,s)$ is quasi semigroup of contraction, then

$$\|R(t,s_2)x\| = \|R(t+s_1,s_0)R(t,s_1)x\| \leq \|R(t+s_1,s_0)\|\|R(t,s_1)x\| \leq \|R(t,s_1)x\|.$$

Therefore, there is an unbounded sequence $(s_n) \subseteq \mathbb{R}^+$ such that $(\|R(t,s_n)x\|)$ is a bounded decreasing sequence of real numbers. The Convergence Monotone Theorem implies that for every $t \geq 0$, $\lim_{s\to\infty}\|R(t,s)x\|$ exists. ■

Corollary 4.6. *Let $R(t,s)$ be a uniformly stable C_0-quasi semigroup on a Banach space X. The following statements are equivalent.*

(a) $R(t,s)$ is strongly stable.

(b) For all $x \in X$, $t \geq 0$, and $p \geq 1$

$$\lim_{s\to\infty}\frac{1}{s}\int_0^s \|R(t,v)x\|^p dv = 0$$

Proof. $(a) \Rightarrow (b)$. This necessity is trivial.

$(b) \Rightarrow (a)$. By uniform stability we have

$$\begin{aligned} s\|R(t,s)x\|^p = \int_0^s \|R(t,s)x\|^p dv &\leq \int_0^s \|R(t+s-v,v)R(t,s-v)x\|^p dv \\ &\leq N^p \int_0^s \|R(t,u)x\|^p du. \end{aligned}$$

Therefore,

$$\|R(t,s)x\|^p \leq N^p\left(\frac{1}{s}\int_0^s \|R(t,u)x\|^p du\right). \tag{14}$$

It follows that the right hand side of (14) converges to zero as $s \to \infty$, so is the left hand side. Thus, $R(t,s)$ is strongly stable on X. ■

The following theorem is a generalization of Corollary 4.6 neglecting the uniform stability.

Theorem 4.7. *Let $R(t,s)$ be a C_0-quasi semigroup on a Banach space X. The following statements are equivalent.*

(a) $R(t,s)$ is strongly stable.

(b) There exist $p, q > 1$ with $\frac{1}{p} + \frac{1}{q} = 1$ and a function $N_q^{'} : \mathbb{R}^+ \to (0, \infty)$ such that

$$\lim_{s\to\infty} \frac{1}{s} \int_0^s \|R(t+v, s-v)x\|^p dv = 0$$

for all $x \in X$ and $t \geq 0$, and

$$\sup_{s>0} \left(\frac{1}{s} \int_0^s \|R'(t,v)x'\|^q dv \right)^{\frac{1}{q}} \leq N_q^{'}(t)\|x'\|$$

for all $x' \in X'$ and $t \geq 0$.

Proof. $(a) \Rightarrow (b)$. This necessity is a direct consequence of Corollary 4.6.

$(b) \Rightarrow (a)$. By the Hölder inequality we have

$$\begin{aligned} s|x'(R(t,s)x)| &\leq \int_0^s \|R'(t,v)x'\|\|(R(t+v,s-v)x)\|dv, \quad s \geq v \\ &\leq \left(\int_0^s \|R'(t,v)x'\|^q dv \right)^{\frac{1}{q}} \left(\int_0^s \|R(t+v,s-v)x\|^p dv \right)^{\frac{1}{p}}. \end{aligned}$$

Hence

$$|x'(R(t,s)x)| \leq \left(\frac{1}{s} \int_0^s \|R'(t,v)x'\|^q dv \right)^{\frac{1}{q}} \left(\frac{1}{s} \int_0^s \|R(t+v,s-v)x\|^p dv \right)^{\frac{1}{p}}.$$

Furthermore

$$\|R(t,s)x\| \leq N_q^{'}(t) \left(\frac{1}{s} \int_0^s \|R(t+v,s-v)x\|^p dv \right)^{\frac{1}{p}}. \tag{15}$$

Since the right hand side of (15) converges to zero as $s \to \infty$, $R(t,s)$ is strongly stable on X. ■

By inspecting the proof of Theorem 4.7 we have the sufficiency for the uniform strong stability.

Corollary 4.8. *Let $R(t,s)$ be a C_0-quasi semigroup on a Banach space X. The following statements are equivalent.*

(a) $R(t,s)$ is unformly strongly stable.

(b) There exist $p, q > 1$ with $\frac{1}{p} + \frac{1}{q} = 1$ and a positive constant $N_q^{'} > 0$ such that

$$\lim_{s\to\infty} \frac{1}{s} \int_0^s \|R(t+v, s-v)x\|^p dv = 0$$

for all $x \in X$ and $t \geq 0$, and

$$\sup_{t\geq 0, s>0} \left(\frac{1}{s} \int_0^s \|R'(t,v)x'\|^q dv \right)^{\frac{1}{q}} \leq N_q^{'}\|x'\|$$

for all $x' \in X'$.

Remark 4.9. We have characterized the various types of the stability for C_0-quasi semigroups on Banach spaces which is similar to C_0-semigroups and evolution operators. The results are applicable for the time-dependent differential equations with unbounded coefficients in Banach spaces. In particular, these results can also be applied to the investigation of the stability, controllability, and observability of the time-dependent linear systems. The next works, we shall investigate the polynomial stability of C_0-quasi semigroups in Banach spaces which is more general than the exponential stability.

Acknowledgments
The authors are grateful to the reviewer for helpful comments.

References

[1] Curtain R F and Pritchard A J 1978 Infinite-Dimensional Linear Systems Theory (New York: Springer-Verlag)
[2] Pazy A 1983 Semigroups of Linear Operators and Applications to Partial Differential Equations (New York: Springer)
[3] Barcenas D, Leiva H and Moya A T 2006 The Dual Quasi-Semigroup and Controllability of Evolution Equations *Journal of Mathematical Analysis and Applications* **320** 691-702
[4] Janfada M 2010 On Regularized Quasi-Semigroups and Evolution Equations *Abstract and Applied Analysis* Article ID 785428 13
[5] Leiva H and Barcenas D 1991 Quasi-semigroups, Evolution Equation and Controllability (Merida: Facultad De Ciencias Departemento De Matematica)
[6] Megan M and Cuc V 1999 On Exponential Stability of C_0-Quasi-Semigroups in Banach Spaces *Le Matematiche* Liv 229-41
[7] Sutrima, Indrati C R Aryati L and Mardiyana 2017 The Fundamental Properties of Quasi-Semigroups, *Journal of Physics: Conference Series* article ID 855 9
[8] Datko R 1972 Extending A Theorem of A.M Lyapunov to Hilbert Space *Journal of Mathematical Analysis and Applications* **32** 610-16
[9] Pazy A 1972 On The Applicability of Lyapunov's Theorem in Hilbert Space *SIAM Journal on Mathematical Analysis* **3** 291-94
[10] Buşe C 1994 On Nonuniform Exponential Stability of Evolutionary Process *Rendiconti del Seminario Matematico Universit e Politecnico di Torino* **52** 396406
[11] Buşe C Megan M Prajea M and Preda P 2007 The Strongly Variant of A Barbashin's Theorem on Stability of Solution for Non-Autonomous Differential Equations in Banach Spaces *Integral Equations and Operator Theory* **59** 491500
[12] Megan M and Stoica C 2008 On Uniform Exponential Trichotomy of Evolution Operators in Banach Spaces *Integral Equations and Operator Theory* **60** 499506
[13] Minda A A and Megan M 2011 On (h,k)-Stability of Evolution Operators in Banach Spaces *Applied Mathematics Letters* **24** 44-48
[14] Megan M Sasu A L and Sasu B 2004 Exponential Stability and Exponential Instability for Linear Skew-Product Flows *Matematica Bohemica* **129** 225-43
[15] Megan M and Stoica C 2008 Exponential Instability of Skew-Evolution Semiflows in Banach Spaces *Studia Universitatis Babe-Bolyai, Mathematica* **53** 17-24
[16] Minh N V Räbiger F and Schnaubelt R 1998 Exponential Stability, Exponential Expansiveness, and Exponential Dichotomy of Evolution Equation on the Half-Line *Integral Equations and Operator Theory* 332-53
[17] Preda P and Megan M 1985 Exponential Dichotomy of Evolutionary Processes in Banach Spaces *Czechoslovak Mathematical Journal* **35** 312-23
[18] Megan M Ceausu T and Ramneantu M L 2011 Polynomial Stability of Evolution Operators in Banach Spaces, *Opuscula Matematica* **31** 279-88
[19] Bátkai A Engel K J and Schnaubelt H 2006 Polynomial Stability of Operator Semigroups *Mathematische Nachrichten* **279** 1425-40
[20] Batty C J K Chill R and Tomilov Y 2002 Strong Stability of Bounded Evolution Families and Semigroups *Journal of Functional Analysis* **193** 116-39
[21] Elkoutri A and Taoudi M A 2003 Spectral Inclusions and Stability Results for Strongly Countinuous Semigroups *International Journal of Mathematics and Mathematical Sciences* PII.S0161171203202088 2379-87
[22] Eisner T 2007 Stability of Operator and C_0-Semigroups Disertation (der Mathematischen Fakultät der Eberhard-karls-Universitat Tübingen)
[23] Eisner T and Farkas B 2007 Weak Stability for Orbits of C_0-Semigroups on Banach Spaces *Functional Analysis and Evolution Equations* 201-08
[24] deLaubenfels H Phóng V Q and Wang S 2002 Stability of Semigroups of Operator and Spectral Subspaces *Semigroup Forum* **64** 337-54
[25] Paunonen L 2011 Pertubation of Strongly and Polynomially Stable Riesz-Spectral Operator *Systems and Control Letters* **60** 234-48
[26] Cuc V 2001 A Generalization of A Theorem of Datko and Pazy *New Zealand Journal of Mathematics* **30** 31-9

[27] Curtain R F and Zwart H J 1995 Introduction to Infinite-Dimensional Linear Systems Theory (New York: Springer)

Scaffolding in geometry based on self regulated learning

A S Bayuningsih, B Usodo and S Subanti
Universitas Sebelas Maret, Jl. Ir. Sutami No. 36 A, Kentingan, 57126, Jebres, Surakarta, Jawa Tengah, Indonesia

E-mail: anisasepti415@gmail.com

Abstract. This research aim to know the influence of problem based learning model by scaffolding technique on junior high school student's learning achievement. This research took location on the junior high school in Banyumas. The research data obtained through mathematic learning achievement test and self-regulated learning (SRL) questioner. Then, the data analysis used two ways ANOVA. The results showed that scaffolding has positive effect to the mathematic learning achievement. The mathematic learning achievement use PBL-Scaffolding model is better than use PBL. The high SRL category student has better mathematic learning achievement than middle and low SRL categories, and then the middle SRL category has better than low SRL category. So, there are interactions between learning model with self-regulated learning in increasing mathematic learning achievement.

1. Introduction

Mathematics is one of the most important and influential lessons for someone's future. According to *National Council of Teachers of Mathematics* (NCTM), in this fluctuate era, someone who can understand and master Mathematics, has wider opportunity to determine his future [1]. This is because by mastering mathematical competencies, someone can develop themselves in several of life optimally, otherwise someone who does not master mathematical competence, their future will be limited. According to survey held by *Programme for International Student Assessment* (PISA) at 2015 in collaboration with 72 countries, showed that Indonesia has increased to previous year, from 375 point in 2012 become 386 in 2015. However, this achievement is still low compared to the average of *Organization for Economic Cooperation and Development* (OECD)'s amount 490; and Indonesia's position is ranked sixth from the last. Mathematics is a branch of abstract science; means that the learning process requires understanding and problem-solving that are not real [2]. One of abstract branch is geometry. A branch concerned with point, straight line, plane figures, space, spatial figure and the relations between them was called geometry [3]. Geometry is a part of mathematics which related to student's life, because almost all of the objects around them are geometry [4]. The study of geometry helped students develop the skills of visualization, critical thinking, intuition, problem-solving, deductive reasoning, logical argument and proof [5]. So, one of the important branch of mathematics was geometry.

Based on observation and interview to junior high school teacher said that for students, geometry is one of difficult lesson. Moreover, most of the mathematics learning process is still teacher-center; teacher is the only one source. The teacher is dominant and the students do not required be independent and active during the learning process. Generally, the application of mathematics learning process gain from teacher, it will be makes the students passive [6]. The learning process in Indonesia is still dominated by conventional learning or direct, so it approved that this is a serious problem whence most of the students did not understand what they learned because they were passive during the learning process so their achievement very unlikely to develop [7].

In this case, the teacher is required to revise learning strategies, technique and media that support the success of learning process. The learning process focuses on solving problem then teachers are recommended to use problem-based learning. The key components of the problem-based learning is the act of reflection where students are asked to apply what they have learned in other situations which refers to transfer knowledge. There are five components in the learning process in the classroom which include students, teachers, content, methods and the environment. Williams [8] Stated that teachers must understand the context of mathematics in order to apply the suitable technique. One of technique that can be applied in learning process is scaffolding [9]. Scaffolding is one of technique teacher can choose to help students who face difficulties [10]. Difficulties in learning must be experienced by students especially when faced new materials or information. Scaffolding can be applied to every teacher, adults, or experts in their daily life. However, in order to process scaffolding properly and effectively they need to get enough training and experience.

Furthermore, the combination of Problem Based Learning and scaffolding technique will benefit the students in formulating and organizing students' mathematical learning. Scaffolding is success combined with PBL on student learning [11]. The exact and appropriate technique is feasible to influence students in mathematic learning. Based on those reasons, technique scaffolding is used for Problem Based Learning model expected to increase the mathematic learning achievement. Much of the students thought that mathematics is difficult and boring. It has impact to student's mental in doing something that is depending on other people. The complex problem of mathematics is actually from the students' mind itself. That cognitive problem influenced their academic achievement [12] Academic achievement have role into some life aspects such as anxiety, self-esteem, and optimism [13]. The student with high academic achievement tended to have a strong competitive motivation than whose with low academic achievement [14]. One of cognitive aspect which influenced achievement is self-regulated learning. Thus, learning process nowadays considered that one of its aims is to free the students from their need of teacher, so they can learn independently in their whole [12,15,16]; and to learn independently the students should be a self-regulated learner [17]. Self-regulated learning that used in this research are autonomy, creative and initiative, responsible, and able to refrain [18]. The aim of this research is to know whether PBL-Scaffolding can achieve better result of study than PBL without Scaffolding and direct learning. Beside that to know the student's achievement with self-regulated learning in high category, achievement with self-regulated learning in middle category, achievement with self-regulated learning in low category, and also to know the interaction between learning model and self-regulated learning to the student's learning achievement.

2. Method

This research was a quasi-experimental research, because the researcher cannot control all the relevant variables. The research design is 2X3 factorial that aims to know the influence of two variables independent to dependent. The independent variable for this research is PBL model with scaffolding technique for experimental class and PBL model to class control. Next, for other independent variable is self-regulated learning which rated to high, middle, and low categories. Population for this research is all of the students at SMP 2 Sokaraja VII grade class 2016/2017. The sample was class VII B (experimental class) 36 students and class VII F (control class) 36 students. The mathematic learning achievement obtained by using multiple choice test with geometry material, whereas SRL obtained by using questioner data based on the indicator made by SRL aspect.

3. Results and discussion

Mathematic test for students was 25 multiple choice given to experimental class and control class. Based on the result, the average learning mathematic achievement for experimental class 1 was 36, and control class was 36. Before the ANOVA two ways test, normality test held on the data. The mathematic learning achievement normality test is held by using SPSS software with 5% level of significance. The result of normality test of student's mathematics learning shows that in the experimental class the sig value is 0.395 > 0.05 and control class the sig value is 0.199 > 0.05. Therefore from the statement above can be take conclusion that the sample of achievement tests comes from a normally distributed population. The homogeneity test of student achievement test shows that the value of sig. = 0.521 > 0.05. Therefore, the conclusion can take from the population has the same or homogenous variance. In this study, data analysis used two-way ANOVA shown in table 1 with the following results.

Table 1. The result of hypothesis test

Source	Type III Sum of Squares	df	Mean Square	F	Sig.
Corrected Model	6210.098[a]	5	122.020	17.410	.000
Intercept	315708.574	1	315708.574	4.425E3	.000
Model	385.301	1	385.301	5.401	.023
SE	4974.635	2	2487.317	34.865	.000
Model*SE	542.667	2	271.333	3.803	.027
Error	4708.513	66	71.341		
Total	390820.000	72			
Corrected Total	10918.611	71			

Table 2. The result of post hoc tests anava

(I) SE	(J) SE	Mean Difference (I-J)	Std. Error	Sig.	95% Confidence Interval	
					Lower Bound	Upper Bound
High	middle	13.26*	2.486	.000	7.30	19.22
	low	24.17*	2.820	.000	17.41	30.93
Middle	high	-13.26*	2.486	.000	-19.22	-7.30
	low	10.92*	2.395	.000	5.17	16.66
Low	high	-24.17*	2.820	.000	-30.93	-17.41
	middle	-10.92*	2.395	.000	-16.66	-5.17

Table 3. Homogeneous subsets result

SE	N	Subset		
		1	2	3
Low	19	61.47		
Middle	36		72.39	
High	17			85.65
Sig.		1.000	1.000	1.000

Based on the calculation Table 1, can be seen that the influence of Problem Based Learning model with scaffolding technique toward mathematics learning achievement has the value of F arithmetic = 5.401 with sig = 0.023 < 0.05, so H0 rejected and Ha accepted. It can take the conclusion that there are differences between the students' mathematics learning achievement in the experimental class used

Problem Based Learning model with scaffolding technique with the control class used PBL. Whereas, based on the average of the students' mathematic learning achievement, so the group of students used scaffolding technique in PBL bigger than the students with PBL non technique scaffolding.

This is because during the process of learning mathematics using scaffolding students become more understand the material. When the discussion process takes place, the problem will be solved by the students. The provision of assistance usually provided by teachers in the PBL process is minimal so that students are not fully understood. The scaffolding was provide by teachers to students due to the learning process can develop a conceptual in children and help the learning process [19]. Teachers only provided scaffolding with sufficient so that students will not feel disturbed and feel neglected. Hence, the mathematics learning outcome of students can be improved.

The influence of self-regulated learning towards the mathematic learning achievement has a value of F arithmetic of 34.865 and sig = 0.000 <0.05, so can take the conclusion that there is a difference between categories in self-regulated learning. Based on table 2 it can be seen that there is a difference between self-regulated learning of high category students toward middle and low category. While based on table 3 it shows that the self-regulated learning of the high category is better than the middle and low category self-regulated learning, and then the self-regulated learning of the middle category is better than the low self-regulated learning category. Based on Table 1, On column Class*SRL obtained sig. = 0.027 <0.05, so can take the conclusion that there is interaction between Problem Based Learning model with scaffolding technique and self-regulated learning (independence learn) student to student's mathematics learning achievement.

The use of learning model aims to overcome the problems in teaching and learning process so the material can be delivered maximally. Most of the students in learning the geometry material based on the understanding, thus, when given the problem students do not understand it. A lack understanding of concepts can lead to poor mathematical achievement. One of causes students lack of understanding because teacher uses a less precise learning model. Learning model that enables for learners to solve daily problems and depth understand the concept is the learning model in which learners faced with a problem and seek solutions. Giving problems in everyday life is very important for students' thinking, so the use of PBL models is the right solution to use in learning. PBL is a promising approach not only to build mathematics understanding but also to test students' conceptual knowledge [20]. PBL uses problems to improve student achievement, so it can stimulate students to build the right solution for the given problem [21]. Based on the results of this research indicate that the model of learning and self-regulated learning affect mathematics learning achievement. The use of learning strategies helps students learn activities and apply what they already know to be effective in learning [22]. The effectiveness of PBL in mathematics learning can improve creative thinking and provide great opportunities for students to actively participate in the learning process [23].

One of the benefits that can be obtained from PBL is the emergence of students' attitude, objective and curiosity in solving the problem. Student curiosity causes students to become eager to learn and know their knowledge. Student curiosity provides motivation for students to seek answers of questions or problems that arise and deal with them. The presentation of geometry problems in a way that stimulates curiosity and encourages exploration, can improve their learning and attitudes toward mathematics [24]. Learning with scaffolding techniques helps students in solving the problem. The use of scaffolding is a support given to the student when the teacher gave a problem and the student cannot solve the problem [25].

In this research, it can be seen that when students are given scaffolding technique in mathematics learning process, it makes students become easier in understanding and confidence about the result obtained to solve the problem. Unlike a class which without using scaffolding techniques, to solve problems the students are unsure with the result. If students understand the material that obtained, then when students find the same problem will be able to solve it well. In the process of learning mathematics using PBL many students are lazy to do, because teachers guide them only in general and demands them to solve the problems based on existing sources. Unlike the case with PBL which is using scaffolding techniques, students who ask to the teacher was guided by giving instructions to solve the problem. Assistance is not only given at the beginning of problem solving, but help or support can be given in the middle or at the end of the problem solving. Any help will be reduced and given opportunities to learners along with the ability to complete a job independently.

Scaffolding in learning is a teaching technique that consists of teaching a new skill by inviting students together to accomplish tasks that are too difficult if the student solve their own, and help the student in building an understanding of knowledge and new processes [26-29]. It shows that providing support to learners at the right time will create a better learning process of mathematics, otherwise if does not right, then it will interference. Scaffolding should be provided sufficiently and timely so that learners can be responsible and independent of the learning they are doing [19].

When the learning model combined with scaffolding techniques will establish student's confidence in solving a problem, so they can find the right solution to overcome the existing shortcomings. Through confidence, it will make them express opinions and independent in learning so that they can understand easily and able to solve the problems. In this case the effective scaffolding technique is used in problem based learning process (PBL). Providing appropriate scaffolding techniques in the learning of mathematics can improve the intellectual development of learners in achieving the goals of mathematics learning.

4. Conclusion

Based on the results of research above, the conclusion that can be taken is the difference of student achievement between students using PBL model with scaffoding technique and PBL. The achievement of learning mathematics using PBL with scaffolding technique gives better result than the achievement of learning mathematics without scaffolding. There is a difference in learning achievement between high, middle and low category students. However, the high SRL category student has better mathematic learning achievement than middle and low SRL categories, and then the middle SRL category has better than low SRL category. Based on the research discussion, the results showed that the influence of Scaffolding can increase the achievement of higher learning achievement of mathematics in PBL than the achievement of learning achievement not using scaffolding.

5. Acknowledgments

The authors thank to the Department of Magister Mathematics Education, Faculty of Teacher Training and Education, Universitas Sebelas Maret.

References

[1] National Council of Teachers of Mathematics 2000 *Principles and Standards for School Mathematics* (United State: Nasional Council of Teachers of Mathematics, Inc)
[2] Syah M 2011 *Psikologi Pendidikan dengan Pendekatan Baru* (Bandung: Remaja Rosdakarya)
[3] Biber C, Tuna A and Korkmaz S 2013 The mistakes and the misconceptions of the eighth grade students on the subject of angles *European Journal of science and mathematics education* **1** 50
[4] Safrina K, Ikhsan M and Ahmad A 2014 Peningkatan Kemampuan Pemecahan Masalah Geometri melalui Pembelajaran Kooperatif Berbasis Teori Van Hiele *Jurnal Didaktik Matematika* **1** 355
[5] Jones K 2002 *Issues In The Teaching and Learning of Geometry* in Linda Haggarty (Ed), *Aspects of Teaching Secondary Mathematics: perspectives on practice* (London: RoutledgeFalmer) Chapter 8 121
[6] Ruseffendi E T 1991 *Pengantar kepada membantu guru mengembangkan kompetensinya dalam pengajaran matematikan untuk meningkatkan CBSA* (Bandung: Tarsito)
[7] Komalasari K 2011 *Pembelajaran Kontekstual Konsep dan Aplikasi* (Bandung: PT Refika Aditama)
[8] Williams K C and Williams C C 2011 Five key ingredients for improving student motivation *Research in Higher Education Journal 12* 1

[9] Zakaria E, Chin L C and Daud Y 2010 The Effects of Cooperative Learning on Students' Mathematics Achievement and Attitude toward Mathematics *Journal of Social Sciences* **6** 272
[10] Sutiarso S 2009 Scaffolding dalam Pembelajaran Matematika *Proc Seminar Nasional Penelitian, Pendidikan dan Penerapan MIPA Fakultas MIPA* (Yogyakarta: Universitas Negeri Yogyakarta)
[11] Greening T 1998 *Scaffolding for Succes in Problem Based Learning* (Australia: The University of Ballarat) Med Educ Online Series
[12] Sabornie E J, Cullinan D, Osborne S S and Brock L B 2005 Intellectual academicy and behavioral functioning of students with high-incidence disabilities: A cross-categorical meta-analysis *Council for Exceptional Children* **72** 47
[13] El-Anzi F O 2005 Academic achievement and its relationship with anxiety, self-esteem, optimism, and pessimissm in kuwaiti students *Social Behavior and Personality* **33** 95
[14] Lens W, Lacante M, Vansteenkiste M and Herrera D 2005 Study persistence and academic Achievement as a function of the type of competing tendencies *European Journal of Psychology ofEducation* **20** 275
[15] Slavin 2009 *Educational Psychology* (New Jersey: Prentice-Hall, Inc)
[16] Jacobsen R R and Harris S 2002 Does the type of campus influence self-regulated learning as measured by the motivated strategies for learning questionnaire (MSLQ) *Education* **128** 412
[17] Woolfolk 2008 *Educational Psychology Active Learning Edition Tenth Edition* (Boston: Allyn and Bacon)
[18] Desmita 2009 *Psikologi Perkembangan Peserta Didik* (Bandung: PT Remaja Rosdakarya)
[19] Walqui A 2006 Scaffolding Instruction for English Language Learners: A Conceptual Framework *The International Journal of Bilingual Education and Bilingualism* **9** 159
[20] Macmath S, Wallace J and Chi X 2009 Problem-Based Learning in Mathematics A Tool for Developing Students' Conceptual Knowledge *The Literacy and numeracy*
[21] Ricon T, Rosenblum S and Schreuer N 2010 Using problem based learning in training health professionals: should it suit the individual's learning style *Journal Scientific Research Publishing* **25** 143
[22] Mahmoud A and Kamel A 2014 The Effect of Using Discovery learning Strategy in Teaching Grammatical Rules to First Year General Secondary Student on Developing Their Achievement and Metacognitive Skills International *Journal of Innovation and Scientific Research* **5** 146
[23] Padmavathy R D 2013 Effectiveness of problem based learning in mathematics *International Multidisciplinary e-Journal* **2**
[24] Jones K 2002 *Issues in the Teaching and Learning of Geometry* In: Linda Haggarty (Ed) Aspects of Teaching Secondary Mathematics: perspectives on practice. (London: RoutledgeFalmer) Chapter 8 pp 121-139
[25] Van de Pol J, Volman M and Beishuizen J 2010 Scaffolding in Teacher–Student Interaction: A Decade of Research *Educ Psychol Rev* **22** 271
[26] Kurniasih AW 2012 Scaffolding sebagai Alternatif Upaya Meningkatkan Kemampuan Berpikir Kritis Matematika *Jurnal Kreano* **3** 113
[27] Sundayana R, Herman T, Dahlan J A and Prahmana R C I 2017 Using ASSURE learning design to develop students' mathematical communication ability *World Transactions on Engineering and Technology Education* **15** 245
[28] Ismail Z and Rahman S N A 2017 Learning 2-dimensional and 3-dimensional geometry with Geogebra: Which would students do better? *International Journal on Emerging Mathematics Education* **1** 121
[29] Nurhasanah F, Kusumah Y S, and Sabandar J 2017 Concept of triangle: Examples of mathematical abstraction in two different contexts *International Journal on Emerging Mathematics Education* **1** 53

Characteristics of manipulative in mathematics laboratory

A Istiandaru, V Istihapsari, R C I Prahmana, F Setyawan and A Hendroanto

Mathematics Education Department of Universitas Ahmad Dahlan, Jalan Prof. Dr. Soepomo, S.H. Warungboto, Yogyakarta, 55164 Indonesia.

E-mail: afit.istiandaru@pmat.uad.ac.id

Abstract. A manipulative is a teaching aid designed such that students could understand mathematical concepts by manipulating it. This article aims to provide an insight to the characteristics of manipulatives produced in the mathematics laboratory of Universitas Ahmad Dahlan, Indonesia. A case study was conducted to observe the existing manipulatives produced during the latest three years and classified the manipulatives based on the characteristics found. There are four kinds of manipulatives: constructivism manipulative, virtual manipulative, informative manipulative, and game-based manipulative. Each kinds of manipulative has different characteristics and impact towards the mathematics learning.

1. Introduction

The definition of a good mathematics teaching practice is oftenly related to the achievement of the learning goals. If the students could achieve the learning goals through a particular teaching approach, then the approach can be regarded as good [1]. One of the mathematics leaning goals is to create what the students learn become meaningful [2, 3]. In this case, representation becomes important since mathematics teachers cannot bring the mathematics object in front of their students [4, 5]. Furthermore, the idea of reducing the abstract sense in mathematics learning has been massively promoted to help students understand the mathematics concepts better [1, 2, 6, 5, 7–9]. Therefore, they need manipulatives to show what the objects look like and what properties attached to the objects.

Manipulative is a kind of teaching aid designed to represent abstract mathematical ideas concretely. It can be in a form of visual and physical appearance which can be manipulated by students through hands-on-activities [8]. Manipulative is believed to play important role in encouraging students to learn mathematics. Some studies report that manipulatives, both concrete and virtual, have significant impact in students'achievement [10–12]. Manipulative provides opportunity for the students, especially in the thinking stage of operational concrete to understand a concept by supporting the process of assimilation and accommodation [6].

On the other hand, simply using manipulative does not guarantee that the learning would be effective [10]. There are many factors which must be considered in using the manipulatives. Teacher needs to consider the relevancy with material or concept [13], the goal of using the manipulative (game for fun or finding a concept) [5], the learning approach, and the students' characteristics [10]. In some cases, one or two considerations above are failed to be well understood by teacher such that the learning goals could not be achieved optimally.

Despite on the challenge of the effectiveness, the use of manipulative is still promising to help students construct their knowledge through meaningful activities. That is why mathematics teachers and pre-service teachers need to develop the appropriate manipulatives to support their professional skill.

The mathematics education department of Universitas Ahmad Dahlan, Indonesia, is one that have a huge attention in developing manipulatives to support mathematics learning. During the last three years, there are many manipulatives produced in its laboratory based on the need analysis of the material and the current curriculum. This article aims to provide an insight to the characteristics of manipulatives produced in the mathematics laboratory which is then can be used to track the potential effectiveness and challenge for the use of the manipulatives at school.

2. Method

This research was an exploratory case study towards the characteristics of manipulatives produced by the mathematics education laboratory in Universitas Ahmad Dahlan. As the nature of exploratory case study which allows data collection conducted prior to the definition the case [14, 15], we explored the manipulatives before we get down to the characteristics as the focus topic.

All manipulatives produced in the latest three years were documented, classified, and then their characteristics were observed upon the appropriate use for mathematics learning. The documentation result is also confirmed by a secondary data taken from the laboratory. It was a record of manipulative register in which all manipulatives produced should be registered before displayed and used in the future. The data was then classified based on the characteristics found. The classification also considered what kind of learning approach is compatible with the use of the manipulatives. Furthermore, interviews to the corresponding lecturers who adviced the production of the manipulatives were also conducted to gain more explanation towards the findings.

3. Results and discussion

The laboratory record gave an early information that there are 127 manipulative products registered during the latest three years. Further documentation was conducted and successfully found all of the products with various conditions. It was only 74.8% of them which can be accessed and used by students or lecturers in appropriate condition, while the rest are damaged. Most of the damaged products are the ones produced three years ago, having paperboard or styrofoam as its material. It leads to the first finding that the manipulative products with such materials only last no more than three years for appropriate use. An interview addressed to the lecturer who adviced for the production, we refer to as Subject 1, confirming that the durability of the products is an important issue they concern currently.

> "I myself admit that three years ago I did not concern in limiting the spesification of material. Thus, some students used fragile material and thus it was not durable. For now and then, I have set the spesification of material which could maintain the durability of the products, such as plastic coated paperboard instead of paperboard only" (Subject 1 testimony).

The manipulatives in the laboratory are products of two subjects. They are the Mathematics Learning Media Development and the Learning Multimedia. The former produces concrete products while the later produces virtual manipulatives (CDs containing Macromedia Flash files). The documentation of the data can be seen in Table 1.

Table 1. Manipulative products based on its type

Type	Percentage Number of Products
Concrete product	36.2%
Virtual product	63.8%

Based on the Table 1, most of manipulatives produced were the virtual ones, as the product of the Learning Multimedia subject. Furthermore, the observation towards the products' characteristics found

IOP Conf. Series: Journal of Physics: Conf. Series **943** (2017) 012023 doi:10.1088/1742-6596/943/1/012023

that the products can be categorized into three types of manipulatives, (1) the constructivist manipulative, (2) the informative teaching aid/model, and (3) the mathematics fun games. The result of classification can be seen in Table 2.

Table 2. Manipulative products based on its characteristics

Type	Percentage Number of Products
Constructivist manipulative	10.2%
Informative teaching aid/model	81.9%
Mathematics fun games	7.9%

Based on the Table 2, most of the manipulative products can be categorized as informative teaching aid, while the other two categories share the rest percentage. Each type of manipulative has its own characteristics. The characteristics also determine the compatibility of its use in the mathematics learning.

3.1. Constructivist manipulative

Constructivist manipulative, or usually just called manipulative, refers to manipulative definition in general. Manipulative material is defined as concrete object representing mathematical ideas which is designed such that students could understand mathematical concepts by manipulating it [8]. It can be touched, moved around, and appeals to several senses including the socio-cultural needs of the students [7]. By using the keywords of definition, we can understand the important atribution that the object shall be able to be manipulated by the students. On the contrary, if the students can not do any manipulation towards the model, then it is not categorized in this classification.

In the mathematics education laboratory, there is only 10.2% of thc manipulatives produced. We tried to find out whether this number is a "by design" phenomena by interviewing the lecturers. Two lecturers were interviewed, the Subject 1 and Subject 2, and they gave quite similar answer. They said that they let the students analyze the material and design the manipulative based on their creativity.

> "First, I instructed the students (mathematics education department) to work in group, analyzing the material, and making assumption if particular material is considered difficult by students (primary or secondary school). Then, they designed the relevant manipulative and my role is controling that they will not mislead the mathematical concept" (Subject 1).

> "Sometimes I ask them to pay attention on some materials which are usually considered problematic for secondary school students. Then, they work in group designing what manipulative would be useful for learning" (Subject 2).

From the interview above, there is no control on what kind of manipulative the students will produce. This phenomena will lead to further question about what makes the manipulative effective for learning, and do they aware of the students' spesific behavior while learning by ustilizing manipulative.

A simple example of the manipulative in this category is presented in Figure 1.

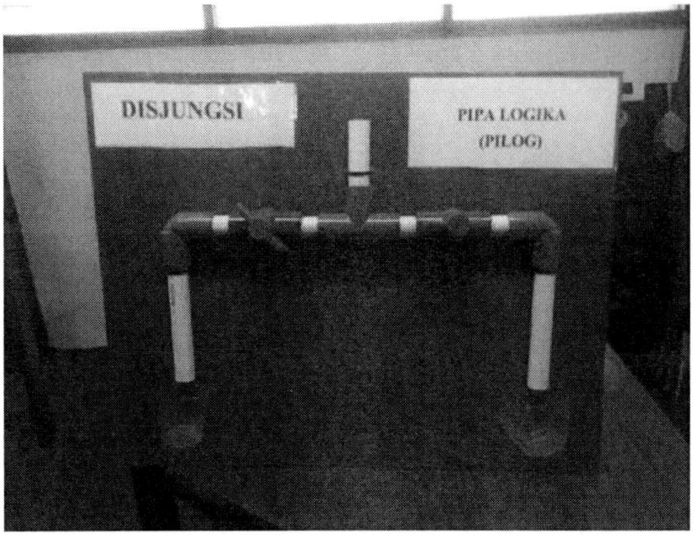

Figure 1. Logic pipe as model for disjunction

If we look at the Figure 1, the logic pipe can be categorized as constructivist manipulative since the student could manipulate the model in some designed conditions. The knop of the pipe represents the true or false of statement. If it is “ON”, the water flows, which means “TRUE” statement, while “OFF” means “FALSE” statement. The compound sentence represented by Figure 1 is disjunction. The combination of “ON” and “OFF” conditions will lead to the truth value of the disjunction. If the water flows, then the disjunction is “TRUE”, and vice versa. Using this logic pipe, students can find the table of truth value of a disjunction.

The use of this manipulative indeed requires attention from teachers to aware of the goals and approach of learning. Many studies refers the use of manipulative to the idea of constructivism addressed by Piaget and Vigotsky, who suggest that students need to construct knowledge by themselves through questioning activities and the teacher provides assistance which the students need in a form of scaffolding [5]. In here, manipulative is suitable with constructivism learning and contextual learning where the students could find the concepts through discussion and interaction with the manipulative [3]. Furthermore, manipulative also support the heuristic of problem solving which is also problematic in the context of Indonesian secondary school students as reported by some studies [16]. The other issue is that it is essential for teachers to make the students realize the manipulatives as tools instead of as toys. If manipulatives are regarded as "toys", students will think that all they can do is playing with the manipulative rather than learning something which the manipulatives bring to them.

3.2. Informative teaching aid/model

This type of manipulative often create misunderstanding among educators. By definition in Section 3.1, this type cannot be categorized as manipulative. Therefore, we can use the term of informative teaching aid. The idea of the teaching aid is simply giving information to the students. It is a ready-made object which the students learn the concept as it is delivered.

However, some studies still refers this teaching aid, especially the virtual one, as manipulative though there is no significant manipulation conducted by students [17]. A virtual manipulative is an interactive visual representation of a mathematical object which allow students to construct knowledge [7]. If students just click to see the next material in a smartphone or giving a response to get a “Yes” or “No” feedback, then the media cannot be categorized as virtual manipulative. These media mainly adopt the expository method and provide a virtual presentation of material which then could substitute conventional book. This media invites many sceptical idea that it does not accommodate the idea of constructivism attributed in manipulative [7].

This phenomena is also found in the multimedia produced by mathematics laboratory of Universitas Ahmad Dahlan, most of the multimedia are categorized as informative virtual teaching aid. They used Flash application to just deliver the material and the information is given by clicking some buttons provided. In fact, some applications can be utilized to develop appropriate virtual manipulative, such as Geogebra and Cabri 3D which allow students to manipulate figures [17].

An example of informative teaching aid is presented in Figure 2.

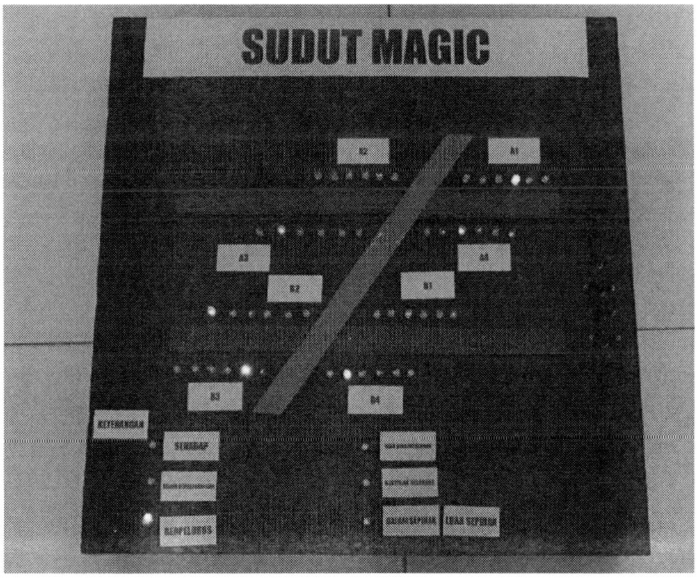

Figure 2. Magic angle as an informative teaching aid

The teaching aid presented in Figure 2 is used to give ready-made information to the students. If the students turn on the button, then various LED lamp in some indicated angles will light up. A legend information of the light meaning is provided, such as green light represent supplementary angles, or blue light means alternate interior angles, and so on. There is no possibility for students to manipulate the teaching aid.

The reasons why manipulative use is suggested by many educators is that it is promising to engage the students with constructivism which make sure the retention of information. However, inapropriate design or implementation of manipulative will cause the use of manipulative become ineffective since the students do not really understand that they construct a concept [9]. They are just being told and are risky to forget the material.

3.3. Mathematics fun games

The third type of manipulative produced in mathematics laboratory of Universitas Ahmad Dahlan is mathematics fun games. Yet, according to the definition, this type cannot be categorized as manipulative, either. It is because the students only play and follow certain rule without doing any manipulation.

This fun games is useful for drilling and promoting joyful learning. However, it has challenge for the concept retention. Students should not lost focus that they learn something (concept) [11, 18, 19]. Teachers who believe manipulatives are just used for fun are not going to genuinely incorporate manipulatives and the concepts they were meant to convey into their instructional lessons [2]. Thus, the use of mathematics fun games should also consider the learning goals.

An example of mathematics fun games in mathematics laboratory is presented in Figure 3.

Figure 3. Roulette games for quadratic equation material

The Figure 3 shows a roulette games for quadratic equation material. Students turn the roulette and get topic pointed by the arrow. They have to do problems or challenge related to the topic and then confirmed by a key answer provided in an envelope. This is a drilling method by using a fun game media.

4. Conclusion

Manipulative shall be used properly according to the goal and the approach of learning. The success of manipulative use depends on the appropriate use in the mathematics learning. Therefore, we have to consider the characteristics of the manipulative which will be developed. In the mathematics education laboratory of Universitas Ahmad Dahlan, there are various manipulatives produced every year. We found that there are types found, namely: (1) the constructivist manipulative, (2) the informative teaching aid, and (3) the mathematics fun games. The constructivist manipulative allows the students to do manipulation on the dynamic model to construct knowledge by themselves. The informative teaching aid simply gives a ready-made information to the students. It is suitable for expository learning. The mathematics fun games allow the students to play and have a drilling upon certain exercise or problems. For future research, it is potential to test the effectiveness of the three categories above to confirm the findings and also the state of the art of this research among the other relevant studies.

References

[1] Pepin B, Xu B, Trouche L and Wang C 2017 Developing a deeper understanding of mathematics teaching expertise: an examination of three Chinese mathematics teachers' resource systems as windows into their work and expertise *Educ. Stud. Math.* **94** 257

[2] Furner J M and Worrell N L 2017 The importance of using manipulatives in teaching math today *Transformations* **3** 1

[3] Fuadi R, Johar R and Munzir S 2016 Peningkatkan kemampuan pemahaman dan penalaran matematis melalui pendekatan kontekstual *Didakt. Mat.* **3** 47

[4] Jitendra A K, Nelson G, Pulles S M, Kiss A J and Houseworth J 2016 Is mathematical representation of problems an evidence-based strategy for students with mathematics difficulties? *Except. Child.* **83** 8

[5] Kelly C A 2006 Using manipulatives in mathematical problem solving: a performance-based analysis *Math. Enthus.* **3** 184
[6] Baroody A J 2017 What role should concrete experiences play in early childhood mathematics instruction? *Interventions in Early Childhood Mathematics Education* ed J Sarama and D Clements (Champaign: University of Illinois at Urbana-Champaign) pp 1-73
[7] Durmus S and Karakirik E 2006 Virtual manipulatives in mathematics education: a theoritical framework *Turkish Online J. Educ. Technol.* **5** 117
[8] Moyer P 2001 Are we having fun yet? how teachers use manipulatives to teach mathematics *Educ. Stud. Math.* **47** 175
[9] Uttal D H, Scudder K V and DeLoache J S 1997 Manipulatives as symbols: A new perspective on the use of concrete objects to teach mathematics *J. Appl. Dev. Psychol.* **18** 37
[10] Tucker S I, Lommatsch C W, Moyer-Packenham P S, Anderson-Pence K L and Symanzik J 2017 Kindergarten children's interactions with touchscreen mathematics virtual manipulatives: An innovative mixed methods analysis *Int. J. Res. Educ. Sci.* **3** 646
[11] Tucker S I and Johnson T N 2017 I thought this was a study on math games: Attribute modification in children's interactions with mathematics apps *Educ. Sci.* **7** 50
[12] Schacter J and Jo B 2017 Improving preschoolers' mathematics achievement with tablets: a randomized controlled trial *Math. Educ. Res. J.* **29** 313
[13] Hidayah I, Pujiastuti E and Chrisna J E 2017 Teacher's stimulus helps students achieve mathematics reasoning and problem solving competences *J. Phys. Conf. Ser.* **824** 11001
[14] Tellis W M 1997 Introduction to case study *Qual. Rep.* **3** 1
[15] Rahardjo M 2017 *Studi Kasus dalam Penelitian Kualitatif: Konsep dan Prosedurnya* (Malang: Islamic University of Maulana Malik Ibrahim)
[16] Kusdinar U, Sukestiyarno S, Isnarto I and Istiandaru A 2017 Krulik and Rudnik model heuristic strategy in mathematics problem solving *International Journal on Emerging Mathematics Education* **1** 205
[17] Collett P, Elizabeth P and Steyn C 2017 Using African designs in virtual manipulatives for geometrical concept development *Bridges 2017 Conf. Proc.* (Port Elizabeth: Nelson Mandela University) pp 455-8
[18] Sundayana R, Herman T, Dahlan J A and Prahmana R C I 2017 Using ASSURE learning design to develop students' mathematical communication ability *World Transactions on Engineering and Technology Education* **15** 245
[19] Tanujaya B, Prahmana R C I and Mumu J 2017 Mathematics instruction, problems, challenges, and opportunities: A case study in Manokwari regency, Indonesia *World Transactions on Engineering and Technology Education* **15** 287

IOP Conf. Series: Journal of Physics: Conf. Series **943** (2017) 012024 doi:10.1088/1742-6596/943/1/012024

Models for financial crisis detection in Indonesia based on bank deposits, real exchange rate and terms of trade indicators

Sugiyanto, Etik Zukhronah and Anis Nur Aini

Department of Statistics, Sebelas Maret University, Jl. Ir. Sutami no. 36-A, Kentingan, Surakarta, 57126, Indonesia

E-mail : sugiy50@yahoo.co.id

Abstract. Several times Indonesia has experienced to face a financial crisis, but the crisis occurred in 1997 had a tremendous impact on the economy and national stability. The impact of the crisis fall the exchange rate of rupiah against the dollar so it is needed the financial crisis detection system. Some data of bank deposits, real exchange rate and terms of trade indicators are used in this paper. Data taken from January 1990 until December 2016 are used to form the models with three state. Combination of volatility and Markov switching models are used to model the data. The result suggests that the appropriate model for bank deposit and terms of trade is SWARCH (3,1), and for real exchange rates is SWARCH (3,2).

1. Introduction

Impaired financial system stability can lead to financial crisis. In order to maintain this stability, it is necessary to monitor the indicators that contribute to the occurrence of the crisis. Financial crisis that hit Asia on the middle of 1997 is begun by the falling of Baht Thailand. The crisis is also happened in Indonesia that have seriously effect on the economic stability. Detection of crisis in a country could be seen through many indicators. According to Kaminsky et al. [1], there are 15 indicator that can be used to detect the crisis, three of them are bank deposits, real exchange rates, and terms of trade.

Monthly data of bank deposits, real exchange rates, and terms of trade are a time series data that are indicated having heteroschedasticity and condition changes. To overcome this condition, it is used a combination of volatility and Markov switching models. Such models include SWARCH (switching autoregressive conditional heteroscedasticity), MS-GARCH (Markov switching generalized autoregressive conditional heteroscedasticity), and MS-EGARCH (Markov switching exponential generalized autoregressive conditional heteroscedasticity). SWARCH model is introduced by Hamilton and Susmel [2]. Some researchers have conducted research on the detection of financial crisis that occured in a country using the combination of volatility and Markov switching models. Chang et al. [3] used SWARCH model to identify the stock volatility foreign and global financial crisis in Korea based on real exchange rate on the period of January 4th 2000 to March 31st 2010 with three states assumption.

Gray [4] introduced the MS-GARCH model that have the same characteristics with SWARCH but involve more simple parameters, to model the rate data of the United States from January 1970 to April 1994. Mwamba and Majadibodu [5] identify a currency crisis on South Africa based on indicators of foreign currency using the MS-GARCH (1,1) model.

IOP Conf. Series: Journal of Physics: Conf. Series **943** (2017) 012024 doi:10.1088/1742-6596/943/1/012024

Henry [6] modeled short-term rates data in the UK on period of January 2nd 1980 until August 29th 2007 using the MS-EGARCH model. The result shows that the model MS-EGARCH able to capture the volatility asymmetries and changing conditions on the data. Shojaei [7] investigated the influence of oil price crisis in Tehran Stock Exchange. The model used is the MS-EGARCH (1,1) with two states, namely a low state (recession) and high state (expansion).

2. Experimental Details

2.1. *Autoregressive (AR) Model*

AR model is usually used as a linear time series models. AR model has an orde called p that can be determined by partial autocorrelation function (PACF) plot. Model AR(p) can be written as

$$r_t = \phi_1 r_{t-1} + \phi_2 r_{t-2} + \cdots + \phi_p r_{t-p} + a_t,$$

where r_t is log return series at time-t, $\emptyset_p$ is parameter of AR model (Tsay [8]).

2.2. *Volatility Model*

2.2.1. ARCH Model. Residue of AR model containing heteroscedasticity effect can be modeled using a volatility model where the residue can be expressed as

$$a_t = \sigma_t \epsilon_t \text{ for } \epsilon_t \sim N(0,1) \text{ and } a_t | F_{t-1} \sim N(0, \sigma_t^2),$$

where F_{t-1} is set of all the information in time $t-1$. Model ARCH(m) can be written as

$$\sigma_t^2 = \alpha_0 + \sum_{i=1}^{m} \alpha_i a_{t-i}^2,$$

where m is the order of ARCH model, α_0 is a constant, α_i is a parameter of the ARCH model, and σ_t^2 is the residual variance to the time-t (Tsay [8]).

2.2.2. GARCH Model. High order of ARCH model can be overcome using a GARCH(m,s) which can be written as

$$\sigma_t^2 = \alpha_0 + \sum_{i=1}^{m} \alpha_i a_{t-i}^2 + \sum_{j=1}^{s} \beta_j \sigma_{t-j}^2,$$

where β_j is parameter of GARCH model (Tsay [8]).

2.2.3. EGARCH Model. Leverage problem on GARCH effect can be overcome by the model EGARCH (m, s) which can be written as

$$\ln \sigma_t^2 = \alpha_0 + \sum_{i=1}^{m} \alpha_i \left(\left| \frac{\alpha_{t-i}}{\sqrt{\sigma_{t-i}^2}} \right| - \sqrt{\frac{2}{\pi}} \right) + \sum_{i=i}^{m} \gamma_i \frac{\alpha_{t-i}}{\sqrt{\sigma_{t-i}^2}} + \sum_{j=1}^{s} \beta_j \ln \sigma_{t-j}^2,$$

where γ is a parameter Leverage effect (Tsay [8]).

IOP Conf. Series: Journal of Physics: Conf. Series **943** (2017) 012024 doi:10.1088/1742-6596/943/1/012024

2.3. *SWARCH Model*

According to Hamilton and Susmel [2], Markov switching model for the average conditional can be written as

$$r_t = \mu_{s_t} + \tilde{r}_t$$

where r_t is the log return at time t, $\tilde{r}_t$ is following the process of AR (p) with an average of zero, and $\mu_{s_t}$is the average of state at time t. While SWARCH model can be written as

$$r_t = \mu_{s_t} + a_t, a_t = \sigma_{t,st}\epsilon_t$$

$$\sigma^2_{t,s_t} = \alpha_{0,s_t} + \sum_{i=1}^{m} \alpha_{i,s_t} a^2_{t-i},$$

where μ_{s_t} is the conditional average on a state and σ^2_{t,s_t} is a variance of residue in a state at time t.

2.4. *Smoothed Probability*

According to Kuan [9], smoothed probability value is defined as

$$P(s_t = i|Z^T;\theta) = \sum_{s_t=1}^{3} P(s_{t+1} = i|Z^T;\theta)P(s_t = i|s_{t+1} = i, Z^T;\theta)$$

Based on Hermosillo and Hesse [10] while the probability of a low-volatility regime has decrease under 0.4 means that the indicator is stabil, the probability of a medium-volatility regime (though declining) still remains at around 0.4-0.6 means that the indicators is prone condition, and if the probability of a high-volatility regime has increased to over 0.6 means that the indicators on a crisis condition.

3. Research Method

The research used data of bank deposits, real exchange rate, and terms of trade from January 1990 to December 2016 that were obtained from Bank Indonesia and World Bank. The steps of the research are as follows.

1) Create a data plot and then test the stationary of data. If the data are not stationary then transform the data using log return.
2) Make a plot of partial autocorrelation function (PACF) of log return data then form the AR model.
3) Test the effects of heteroscedasticity on the residue of AR model using Lagrange multiplier test.
4) If there is the effects of heteroscedasticity on the residue of AR model, estimates the parameter of ARCH model.
5) Establish the combination of volatility and Markov switching models with the assumption of a three states.
6) Determine the conditions of the crisis based on the smoothed probability.

4. Results and Discussion

4.1. *Data*

Plot data of bank deposits, real exchange rate, and terms of trade can be seen in Figure 1.

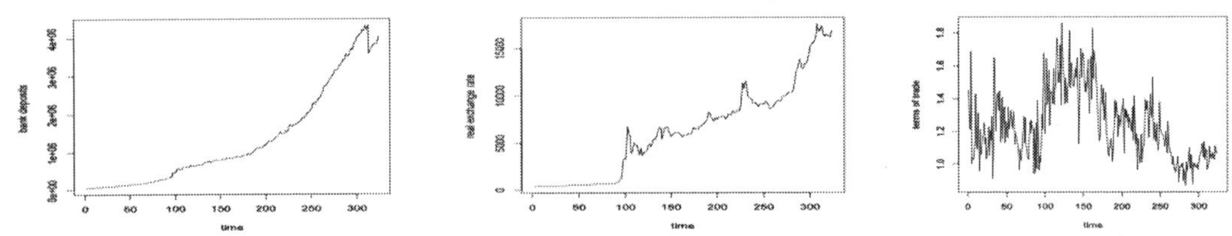

(a) (b) (c)

Figure 1. (a) Bank Deposits Indicator (b) Real Exchange Rate Indicator (c) Terms of Trade Indicator

Figure 1 indicates that the data are not stationary in mean and variance. It is proven by ADF test with probability values for each data are 0.9322, 0.4678, and 0.3247 respectively. Then, the data were transformed using log return. Based on ADF test for log return data, the probability value is 0.01 where it is less than $\alpha = 0.05$ so it can be concluded that the data are stationary. The next step is to form an AR model.

4.2. Establishment of AR Model

AR model can be identified from PACF plot of log return data. Based on bank deposits, it was obtained an AR(1) model i.e. $r_t = 0.22527r_{t-1} + a_t$. Meanwhile for real exchange rate, it was obtained an AR(3) model i.e. $r_t = 0.41724r_{t-1} - 0.17960r_{t-2} + 0.14351r_{t-3} + a_t$ and for terms of trade, it was obtained an AR(2) model i.e. $r_t = -0.61933r_{t-1} - 0.31982\, r_{t-2} + a_t$.

Based on Lagrange multiplier test on the residue of model, it was obtained the each probability of bank deposits, real exchange rate, and terms of trade as 0.003321 , 7.752×10^{-13}, and 0.0002142 respectively. All of probability values are less than 0.05, it means that there are the effect of heteroscedasticity on the residue. To overcome this condition, it was used ARCH model

4.3. Establishment of ARCH Model

For bank deposits, the best model is ARCH(1) which can be written as

$$\sigma_t^2 = 0.0003955 + 0.5610426\, a_{t-1}^2.$$

Meanwhile, the best model for real exchange rate is ARCH(2) which can be written as

$$\sigma_t^2 = 1.515 \times 10^{-4} + 1.834\, a_{t-1}^2 + 0.03617\, a_{t-2}^2,$$

and the best model for terms of trade is ARCH(1) which can be written as

$$\sigma_t^2 = 0.0074505 + 0.2724708\, a_{t-1}^2.$$

Ljung-Box test results the probability values each of bank deposits, real exchange rate and terms of trade as 0.1007, 0.2514, and 0.8663 respectively. These values are more than 0.05, so it can be concluded that the residue of ARCH models do not contain autocorrelation. Based on the Kolmogorov Smirnov test, the probability values of each indicators are 0.4969, 0.7335, and 0.972. These values are more than 0.05, so it can be concluded that the residue of ARCH models are normally distributed. Based on Lagrange multiplier test, the probability values of each indicators are 0.9241, 0.7629, and 0.2167. These values are more than 0.05, so it can be concluded that the residue of ARCH models do not contain the effect of heteroscedasticity. Based on these tests, it can be concluded that the ARCH models are the appropriate models. To model the changes of condition, it is used Markov switching model.

4.4. Establishment of SWARCH Model

The condition changes in Markov switching model are called states. The condition of intended in this research is low, medium and high volatility. States can be formed by transition probability. The transition probability can be formed as a matrix notation and usually called as matrix of transition probability, for bank deposits indicator transition probability matrix can be written as the following

$$P_1 = \begin{bmatrix} 0.42029 & 0.943161 & 0.057943 \\ 0.573877 & 0.000863 & 0.582474 \\ 0.005834 & 0.055976 & 0.359583 \end{bmatrix}.$$

Based on P_1 obtained that the probability to survive on low volatility state is 0.42029. Probability of transition from low to medium volatility state is 0.573877. Probability of transition from low to high volatility state is 0.005834. Probability of transition from medium to low volatility state is 0.943161. Probability survive in medium volatility state is 0.000863. As well as the probability of transition from medium to high volatility state is 0.055976. Probability of transition from high to low volatility state is 0.057943. Probability of transition from high to medium volatility state is 0.582474. Probability survive in high volatility state is 0.359583.The matrix of transition probability for real exchange rate and terms of trade stated in P_2 and P_3 as follows

$$P_2 = \begin{bmatrix} 0.455456 & 0.230373 & 0.127314 \\ 0.488802 & 0.753937 & 0.038933 \\ 0.055742 & 0.01569 & 0.833752 \end{bmatrix} \text{ and}$$

$$P_3 = \begin{bmatrix} 0.512377 & 0.603143 & 0.082906 \\ 0.464982 & 0.259947 & 0.45517 \\ 0.022641 & 0.13691 & 0.461925 \end{bmatrix}.$$

Based on P_1, the parameter estimates of SWARCH(3,1) model for bank deposits can be written as follows

$$\mu_t \begin{cases} 0.015223, for\ state\ 1, \\ -0.00163, for\ state\ 2, \\ 0.048424, for\ state\ 3, \end{cases} \sigma_t^2 \begin{cases} 1.911 \times 10^{-6} + 0.010015 a_{t-1}^2, for\ state\ 1, \\ 0.0001058 + 0.3710515 a_{t-1}^2, for\ state\ 2, \\ 7.722 \times 10^{-5} + 0.8681867 a_{t-1}^2, for\ state\ 3, \end{cases}$$

Based on P_2, the parameter estimates of SWARCH(3,2) model for real exchange rate can be written as follows

$$\mu_t \begin{cases} 0.004737,\ for\ state\ 1, \\ 0.003718, for\ state\ 2, \\ 0.026658,\ for\ state\ 3, \end{cases}$$

$$\sigma_t^2 \begin{cases} 3.133 \times 10^{-6} + 0.410228 a_{t-1}^2 + 0.031234 a_{t-2}^2,\ for\ state\ 1, \\ 4.052 \times 10^{-6} + 0.137679 a_{t-1}^2 + 0.001046 a_{t-2}^2, for\ state\ 2, \\ 1.221 \times 10^{-5} + 0.007305 a_{t-1}^2 + 0.013346 a_{t-2}^2,\ for\ state\ 3. \end{cases}$$

Based on P_3, the parameter estimates of SWARCH(3,1) model for terms of trade can be written as follows

$$\mu_t \begin{cases} -0.00152, for\ state\ 1, \\ -0.00289, for\ state\ 2, \\ 0.008067, for\ state\ 3, \end{cases}$$

$$\sigma_t^2 \begin{cases} 6.57 \times 10^{-6} + 0.224834 a_{t-1}^2, for\ state\ 1, \\ 3.536 \times 10^{-6} + 0.050367 a_{t-1}^2, for\ state\ 2, \\ 7.489 \times 10^{-5} + 1.542798 a_{t-1}^2, for\ state\ 3. \end{cases}$$

where μ_t and σ_t^2 is the conditional mean and variance of SWARCH model.

4.5. Crisis Detection

Crisis detection using SWARCH(3,1) and SWARCH(3,2) can be seen by value of smoothed probability. Figure 2 show the smoothed probability plots of SWARCH (3,1) modeled by bank deposits and terms of trade while SWARCH(3,2) modeled by real exchange rate.

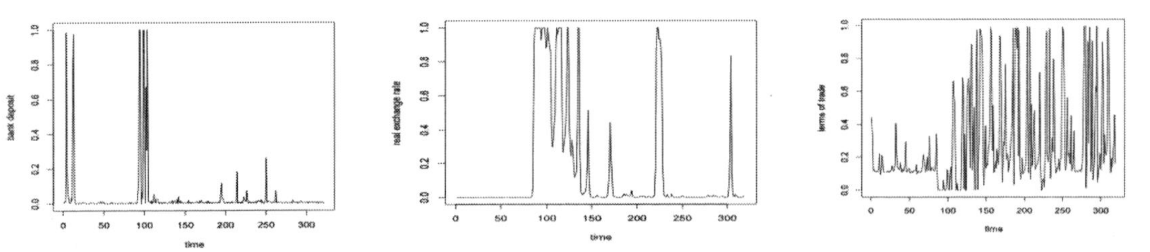

(a) (b) (c)

Figure 2. (a) Smoothed Probability of bank deposits (b) Smoothed Probability of real exchange rate (c)

Smoothed Probability of terms of trade.

Crisis condition signed with value of smoothed probability that greater than 0.6 for bank deposits and real exchange rate indicators, that greater than 0.8 for terms of tradeindicators as shown in Figure 2. Table 1 showed the crisis periode that has been detected on 1997, 1998, and 2008 based on the value of the smoothed probability that greater than 0.6 by bank deposits and real exchange rate, that greater than 0.8 by terms of trade.

Table 1. The crisis detection based on smoothed probability

Year	Bank Deposits	Real Exchange Rate	Terms of Trade
1997	December	July, August, September, October, November, December	January, August, September, December
1998	January, May, June, August, October	January-December	February, March, April, June, August
2008		October, November, December	February, April

Table 1. showed that smoothed probability for real exchange rate and terms of trade can detect the financial crisis that hit Indonesia in 1997, 1998, and 2008. That smoothed probability for bank deposits can not detect the financial crisis that hit Indonesia in 2008

5. Conclusion

The results of this research are bank deposit and terms of trade can be modelled by SWARCH(3,1) while real exchange rate can be modelled by SWARCH(3,2) that can be used to detect the financial crisis in Indonesia in 1997, 1998, and 2008.

References

[1] Kaminsky G, Lizondo S, and Reinhart CM 1998 Leading Indicators of Currency Crises *IMF Working Paper* **45**

[2] Hamilton J D, and Susmel R 1994 Autoregressive Conditional Heteroscedasticity and Changes in Regime, *Journal of Econometrics* **64** 307-333

[3] Chang K, Cho K Y, and Hong M 2010 Stock Volatility Foreign Exchange Rate Volatility and The Global Financial Crisis *Journal of Economic Research* **5** 249-72

[4] Gray, S F 1996 Modeling the Conditional Distribution of Interest Rates as A Regime-Switching Process *Journal of Finance Economics* **42** 27-62

[5] Mwamba, J M and Majadibodu, T 2012 Implied Volatility of Foreign Exchange Option: A Leading Indicator for Currency Crisis Identification, *African Journal of Business Management* **6** 10766-74

[6] Henry, TO 2007 *Between The Rock and a Hard Place: Regime Switching in the Relationship Between Short-Term Interest Rates and Equity Returns in the UK* (Victoria : The University of Melbourne)

[7] Shojaei A, Khezri M, and Samadi, S Z 2013 The Asymmetrical Effect of Oil Market Shocks on Tehran Stock Exchange: A Regime Switching Mode *Journal of Basic and Applied Scientific Research* **3** 1149-55

[8] Tsay R S 2005 *Analysis of Financial Time Series* (Canada : John Wiley and Sons)

[9] Min K C 2002 Lecture On The Markov Switching Model *Institute of Economics Academia Sinica*

[10] Hermosillo B. G, and Hesse H 2009 Global Market Conditions and Systematic *Risk IMF Working Paper* **230**

IOP Conf. Series: Journal of Physics: Conf. Series **943** (2017) 012025 doi:10.1088/1742-6596/943/1/012025

Conjecturing via analogical reasoning constructs ordinary students into like gifted student

Supratman, N Ratnaningsih and S Ryane

Universitas Siliwangi, Jl. Siliwangi No. 24 Tasikmalaya, West Java, Indonesia

E-mail: supratman@unsil.ac.id

Abstract. The purpose of this study is to reveal the development of knowledge of ordinary students to be like gifted students in the classroom based on Piaget's theory. In exposing it, students are given an open problem of classical analogy. Researchers explore students who conjecture via analogical reasoning in problem-solving. Of the 32 students, through the method of think out loud and the interview was completed: 25 students conjecture via analogical reasoning. Of the 25 students, all of them have almost the same character in problem solving/knowledge construction. For that, a student is taken to analyze the thinking process while solving the problem/construction of knowledge based on Piaget's theory. Based on Piaget's theory in the development of the same knowledge, gifted students and ordinary students have similar structures in final equilibrium. They begin processing: assimilation and accommodation of problem, strategies, and relationships.

1. Introduction

Today education emphasizes the need for students to develop the ability to think and solve problems. There is much hope to promote this thinking by using constructivist lessons or problem-based learning in the classroom [1]. Also in the learning process; learners are aware of the cognition process and can control and manage their cognition [2]. Despite the fact that the process of teaching and learning in Indonesia, there are many math teachers who teach the procedure without explaining why a particular procedure is used. As a result, students believe that in solving the problem, it is enough to choose the solution procedure according to the given problem. In this case, the focus of learning is not why certain procedures are used to solve problems. However, students choose the procedure to solve the problem, in addition, students argue: how to solve the problem using the procedure. The process of emphasizing "learning" leads to student behavior only "copying" the procedures undertaken by the teacher, without understanding why they should use this procedure. Therefore, if the problem is slightly changed or slightly modified, students cannot solve the problem. In learning, emphasis on the use of procedures without giving the right reasons is wrong. Teacher actions like this are the beginning of the formation of the pseudo-thinking process in students. Teachers sometimes do not realize that students already have the basic knowledge that can be used to build new knowledge. In addition, teachers sometimes forget that one student with another has a different conceptual understanding process [3]. Such learning will produce students in the lowest level of thinking below is seen from the hierarchy of thinking [4].

To avoid the occurrence of pseudo-thinking, then Students need to be given the opportunity to conjecture in learning when delivering new material through problem-solving. By reason of conjecturing in solving problems using current knowledge to build new knowledge for him [1] Conjecturing is one of the activities of students in active learning. However, it is possible to solve mathematical problems in which students find hunches and guesses. As stated [2] to solve problems, students do one of three possibilities, namely through hunches, guessing, and conjecturing.

Furthermore, Polya says that conjecturing is a part of mathematics and suspect's one of the branches of mathematics. Similarly [3] explains that when learning mathematics in a classroom, students are encouraged to create and test conjectures. In addition, students are helped to develop their reasoning skills. In the evaluation standard [5], the core of students doing mathematics is conjecturing by demonstrating the logical validity of conjecturing. J. Mason, at all [4] says that basically, the students' competence in mathematics includes two (2) things i.e. conjecturing and convincing.

In the implementation of the student, estimation cannot be separated using existing knowledge, both to solve problems and for construction of new knowledge. This suggests that students conjecturing via analogical reasoning as revealed [6]. Conjecturing via analogical reasoning is a learning process that emphasizes a scientific approach to active student learning in mathematics. Analogical reasoning according to [7] refers to the ability to understand and build similar structural similarities to objects whose surface features are always the same. This is called conjecture through analogy as revealed [8].

Analogical reasoning for the construction of knowledge [6] analogical reasoning in classroom math learning [9] analogical reasoning refers to the ability to understand and build structural similarities according to objects whose surface features are not always the same [9], conjecturing research through analogical reasoning in mathematical discovery [10], occurs in the math class although this is more the exception than the rule [6], building creative thinking [8]. Thus, mathematical education will be a very useful study of the use of conjecturing through analogical reasoning as a learning device.

Conjecturing via analogical reasoning can be used in learning for a scientific approach [6]. In addition, conjecturing via analogical reasoning is necessary to enhance the learning of active students in mathematics. Thus, avoid the pseudo thinking in constructing new knowledge [3]. Many junior high schools in Indonesia are learning gifted students and regular students are in one class, so classroom learning makes it difficult for teachers to apply learning models and approaches. Seeing these conditions requires wise treatment from the teacher so that all students are served to construct new knowledge/problem-solving.

The problem is "can ordinary students be constructed like gifted students?"

In this section, we combine several main theories that have examined conjecturing through analogical reasoning, as well as those who have investigated the process of abstraction in mathematical thinking and learning [2, 11-16] used more sociocultural and contextual systems approach to abstraction. We know that reification is the dominant mode to explain the process of abstraction, in which the process becomes an object of manipulation in itself as terminology [12]. In a sociocultural and contextual approach [17-18] reveal a shift in the construction of concepts occurring in diverse environmental discourses, through figures and reflections in the classroom facilitates reflecting abstractions. Radford [19] reveals despite the inherent tension in both abstractions, namely logical-deductive, and socio-discursive, but both can be seen as complementary components of the theory. That is, in turn can provide different meanings, connections, or explanations of the phenomenon under study. Our conceptual framework consists of Piaget and contextual (socio-cultural) views, analogical thinking, and abstraction. Specifically using conjecturing via analogical reasoning as instructional tools for use in constructing new knowledge can be developed so on.

2. Literature review

Some of the theories underlying this research are as follows:

2.1. Conjecturing via analogical reasoning

According to [20] conjecturing can be done in five ways, including conjecturing by analogy. In addition, according to English [21], the analogy consists of a pedagogical analogy, a problem of classical analogy and analogy (using A: B: C: D). Krulik in [22] argue that hierarchical thinking consists of: (1) recall, (2) basic thinking, (3) critical thinking, and (4) creative thinking. Furthermore, Krulik et al. say that reasoning is part of the thinking process. This thinking included in reasoning is basic thinking, critical thinking, and creative thinking. Thus conjecturing through analogical reasoning is a conjecture to solve the "analogy target" based on "basic analogy" (knowledge already possessed/mastered). The invention [14] solves a number of interior corners of the tetrahedron and pentagon (the target analogies to the C: D relationship) before the student has mastered the 180^0 triangle's interior angle (the analogy/knowledge base already possessed by A: B relationship). Furthermore, the discovery of Supratman [8] for the construction of a cone equation (as the analogy target) is based on the construction of the angle bisector equation (as the basis of the analogy).

2.2. *Conjecturing via analogical reasoning of students*

Findings of Supratman [8] the students in solving the problem are divided into 6 groups, ie (1) the students do not do conjecture, (2) the students do conjectures, but the wrong conjecturing, and not via analogy reasoning, (3) the students do conjecture and the results of conjecturing they are true but not via analogical reasoning, because some students have learned from books and some are from the internet, (4) students make conjectures via analogical reasoning but their conjectures are wrong, (5) students make conjectures via analogical reasoning and their conjecture results true, but did not develop their conjecture. so that, they produce only one conjecture, and (6) the student does conjecture via analogy reasoning and his conjecture results correctly and the students are able to develop their conjecturing. So, they produce some conjecturing.

2.3. *Construction of new knowledge*

When a person learns, They will be even disequilibrium. That leads to assimilation and accommodation processes. Suppose the structure of the problem is not shared in depend of the student's cognitive structure. With this process, the cognitive structure evolves through a process of altering, combining, or forming a new scheme until this final equilibrium condition is said to be the construction of new knowledge or adaptation to the environment [23] Adaptation process begins to absorb the structure of the problem (intelligent behavior) which is the input to the cognitive structure in the initial level equilibrium condition, which goes beyond the initial cognitive structure. It results in disequilibrium between the problem structure and the cognitive structure. Therefore, there is an adaptation of the cognitive structure through assimilation and accommodation. After assimilation and accommodation, there will be a cognitive structure in the new level equilibrium, as it is constructed by new knowledge. Illustration of Piaget's theory of modification of [24] as shown in Figure1.

2.4. *Ordinary student*

In Indonesia, many schools that do not appreciate gifted students are supported by the low pedagogical mastery of teachers [1] so that the learning of one student with the other is the same. This resulted in learning; students only touched the recall and basic thinking only. In order not to be the wrong perception then we distinguish between ordinary students with mental students (mental retardation / mentally retarded). As for the meaning of ordinary children does not mean belonging to children with IQ under 84 as expressed in American Association on Mental Deficiency (AAMD). The regular students in this study were officially registered students and participated in regular classes affiliated with the Ministry of Education at its level [25] Students in the same grade but for one reason or another, students are not able to solve problems or are unable to construct new knowledge directly for complex problems.

2.5. *Gifted students*

Although gifted students can cause problems for teachers because they are not a homogeneous group. Gifted students according to [26] students who have unlimited problem-solving abilities. an image of a gifted student according to the National Association for Gifted Children (NAGC) as "someone who shows, or has the potential to perform, a remarkable level of performance in one or more areas of expression" [27]. Students who are gifted mathematically include those who have a high ability or have a high interest. Some students in this group may be gifted with intuitive knowledge of mathematical concepts, while others have a passion for the subject although they may have to work hard to learn it. Many gifted students make themselves obvious to parents, guardians, and teachers by understanding and articulating math concepts at an earlier age than expected [3]. They are often found to easily make connections between study topics and often cannot explain how they get answers quickly [28] can be seen in Figure 1.

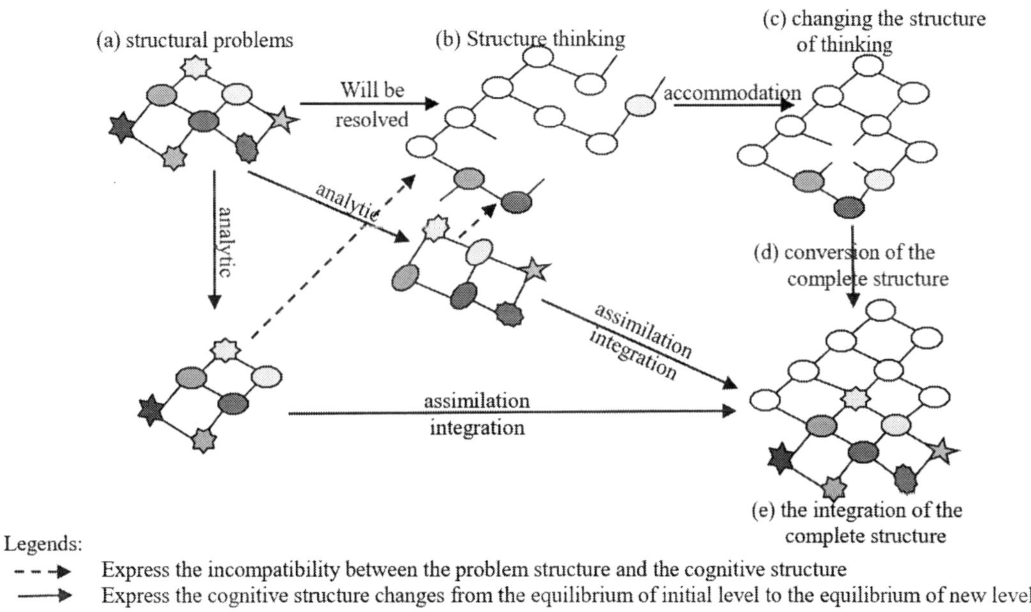

Figure 1. Occurrence of Assimilation process and accommodation (modification of Subanji and Supratman [24])

3. Method

3.1. Research subject

The authors conducted a study of 52 students of Junior High School from the district of Ciamis, which is considered to have an ordinary ability. Comparative subject (S1) was taken from a class of accelerated students from SMP in Tasikmalaya, to solve the problem solved by the research subject (S2).

3.2. Research methodology

This research is a qualitative-explorative research in taking verbal data, because the data to be collected and used is verbal data resulting from the exploration of students who are able to solve problems such as gifted students. The study was conducted at SMP in district of Ciamis. To see the construction of new knowledge is seen based on Piaget Theory. Piaget theory consists of assimilation and accommodation end then assisted premises by Think out loads. The research instrument is a researcher as an instrument, supported by a sheet of instrument task and completed by interview in accordance with [29] opinion. The reason of the researcher is as an instrument, because the researcher is the research manager as well as one of the instruments in data collection that cannot be replaced by other instruments. The initial task sheet instrument is as follows.

- The first problem if you have two points let A and B, A not equal B. What would you do with those two points?
- Next, there is another point suppose C, and $A \neq B \neq C$. What would you do with that point C? And what do you find?

- Next, there is another point let D, and $A \neq B \neq C \neq D$. What would you do with that point D? And what do you find?
- Next, there is another point suppose E, and A ≠ B ≠ C ≠ D ≠ E. What would you do with that E point? And what do you find?
- Next, there is another point suppose F, and $A \neq B \neq C \neq D \neq E \neq F$. What would you do with that F point? And what do you find?
- Next, there is another points suppose G, H, and $A \neq B \neq C \neq D \neq E \neq F \neq G \neq H$. What would you do with those points of G and H? and what do you find?

The problem is to construct the concept of the number of interior of corners-n = $(n-2) \times 180^0$

3.3. Data collection procedures
The study examines conjecturing via analogical reasoning to build students of ordinary ability to become like gifted students in constructing new knowledge based on Piaget's adaptation framework. Data collection, researchers provide problems related to the analogy to a student to complete. While doing the conjecturing process through analogy reasoning, the students express out loud what he is thinking. Researchers record verbal expressions and record the expressions (behaviors) and unique things that students do when conjecturing through analogical reasoning [30]. If you have finished one student, then do the same thing to the other students, to get the subject in accordance with the desired. Such data retrieval, according to [31] is said in the Think Out Loud method, though [27] says think aloud (think aloud). In this method is done to ask the research subject to solve the problem as well as tell the thinking process.

4. Result and Discussion
4.1. Result
Of the 32 students who participated, found the following: 2 students did not conjecture only able to make a straight line and did not want to continue. 5 students do not conjecture via analogical reasoning, because they have read books related to the problem. The remaining 25 students were able to find the concept of the number of interior angles for the corner-n is $(n-2)180^0$. From the 25 students initially: 3 students are only able to make various triangles and the number of interior corners of the triangle. They are named students type A (StA); 7 students are only able to create rectangles and the number of rectangular interior corners. They are named students type B (StB); 8 students are only able to create pentagonal and the number of pentagonal interior angles. they are named students type C (StC); 5 students are only able to make hexagons and the number of interior corners. they are named students type D (StD); The remaining 2 students are able to make the octagonal and the number of interior corners, even able to construct the concept of the number of n-corner interior angles is $(n-2)180^0$. They are named students type E (StE). StA, StB, StC and StD were able to develop their conjecture like StE. Before that, the researcher reminds StA, StB, StC and StD, that you can develop such conjecture as you did before. After that, they can to the construction of concept the number of an interior of $corner - n = (n-2)180^0$. Because they all share the same characteristics students are taken to see the new knowledge construction process, who more consistently explains a question and is named S2. While for the results of interviews with students as follows:

Researcher (R): Why do not you continue your work?
StA: Ehm (while he toyed with his pencil). oh yes I can sir, that is making a rectangle. Then They did the next conjecture to find the number of interior of n-angles is $2 \times 180^0 = 360^0$

then researchers combine like StB

R : do you give up making a rectangle? I added a point. What will you make?

StB: he paused, and then he replied I can make a pentagon and the number of angles in it. eeh eeh (while he made a line connecting point A to point D and point B to point D) oooh there are 3 triangles. Means the number of angles is $3 \times 180^0 = 540^0$.

In fact, they make a line segment by connecting two points so they get 3 triangles. As for the form as can be seen in Figure 2.

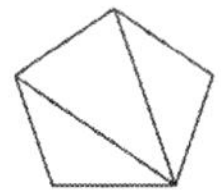

Figure 2. The students' work in drawing the pentagon and making it into several triangles

Then researchers combine like StC

R : Can you make further conjectures if I give one more point?
StC : I will try it sir. (They make a hexagon and divide into four triangles) as soon as he answers the result is $4 \times 180^0 = 720^0$

In fact, they make a line segment by connecting two points so they get 4 triangles. As for the form as can be seen in Figure 3.

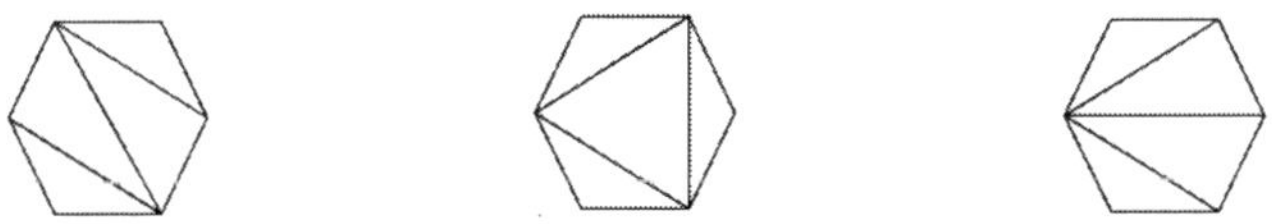

(a) (b) (c)

Figure 3. The students' work in drawing the pentagon and making it into several triangles

Then researchers combine like StD

R : if I add two points. Can you expect it?
StD : I will make an octagon so that the number of interior corners is $6 \times 180^0 = 1.080^0$

so the StD as S1 to find the number of interior angles for the n-segment is $(n-2)180^0$. As for researchers interviewed S1 as follows.

R : Can you do this (while the researcher shows the task sheets instrument)?
S1 : oh sir I do not think I can (while he was scratching his head)
R : Have you got this problem before?
S1 : Not yet, sir
R :what do you know with two points A and B?
S1 : these are the two points representing the line or segment AB
R : after added point C why do you make a triangle?
S1 : because if I put on the elongation of line AB, point C is only a point located on line AB which can be before A, can be between A and B, and can after B. For that, I make triangle, and triangle the number of angles is 180^0 *(while he shows the number of angles from the three angles to the angle of the straightener by tearing the triangle into an alignment angle.*
R : After added point D. Why do you make a rectangle?
S1 : Actually I want to create a line for line, line and line. But I did not bring a run so I made things easier. Actually I want to create a line for line, line and line. But I did not bring a run so I made things easier. The number of inner corners I can guess is 360^0 *because the rectangle is twice the triangle.*
R : After added point E. Why do you make a quadrilateral?

S1 : *Actually I want to make a line for a triangle and the other. But I did not bring a run for drawing, finally I made a rectangle. The number of angles in a quadrilateral I can guess is* 360^0 *because the rectangle is twice the triangle.*
R : *After added point F. Why do you make a hexagonal?*
S1 : *I draw a hexagon has an interior corner is* $4 \times 180^0 = 620^0$.
R : *After added point G and H. Why do you make a octagonal?*
S1 : *I draw a octagonal has an interior corner is* $(8-2) \times 180^0 = 1080^0$.
R : *What if added point again so as to form "n-rectangle". What is the n-corner interior angle?*
S1 : *the number of n-corner interior angles is* $(n-2) \times 180^0 = 180^0 n - 360^0$

When researchers interviewed the S2 as follows.

R : *Can you do this? and what is your conclusion?*
S2 : *yes I try first sir (as well as when the researcher offered to S2.*
R : *Can you do this? and what is your conclusion?*
S2 : *yes I try first Pak (he did it straight away and within 23 minutes have found the number of interior corners in terms of -n =* $(n-2) \times 180^0$*)*

The conjecture of S1 is the same as the conjecture rather than S2, but for S1 must be step by step different with S2. Because S2 directed conjecture for solve problem and his conjecture is thru. Process of thinking of S1 and S2 can be look at in Figure 4.

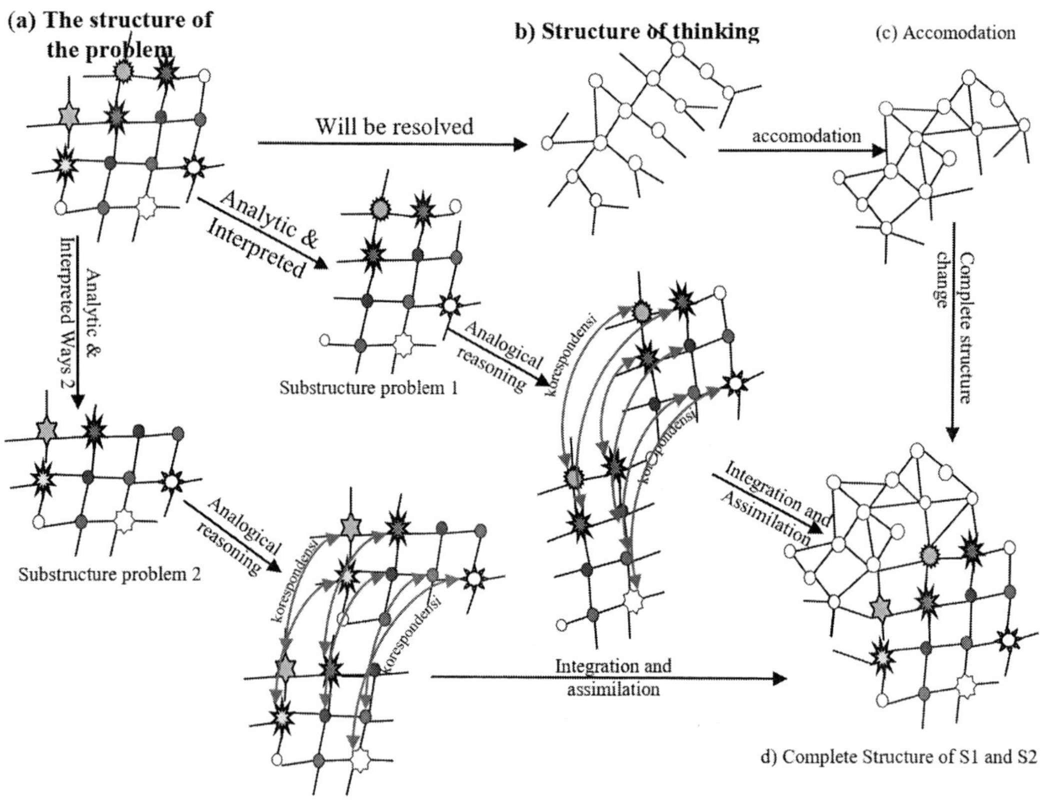

Figure 4. Process of thinking of S1 and S2

4.2. Discussion

S1 guessed through analogical reasoning to solve problems with the division of the problem structure. This shown when the problem presented simultaneously to the S1, S1 looks confusion. But after the question is given little by little as the stage of building the concept of looking for the number of interior corners of an aspect-n = (n-2) $\times 180^0$. But S1 does conjecture via analogical reasoning. The reason the problem has never been S1 found, but S1 uses the knowledge that he previously knew.

Why say S1 and S2 do conjecture? Because S1 and S2 do not know the problem before, so really natural conjecturing occur. Why are S1 and S2 supposed to be conjecturing via analogical reasoning? because S1 and S2 do allegations through the use of concepts and events that have been previously owned. For example: S1 and S2 to estimate the number of interior angles by tearing the triangle into 3 parts . Then he suspects the number of interior angles of the triangle is to form a straight angle. S1 and S2 never saw that straight angle was 180^0. Further, the quadrilateral consists of two triangles; the pentagon consists of three segments and so on so that S1 and S2 have the same conjecture for the number of n-corner interior angles (n-2) x180^0.

As Supratman pointed out in [31] that the events experienced by S1, who experienced category stages of StA, StB, StC, and StD , beginning with accommodation of problem , accommodation of strategic and accommodation of relationships. And then followed by assimilation of problems, assimilation of strategy and assimilation of relations. As for the S2, begins with assimilation of the problem, assimilation of strategy and assimilation of the relationship. Then it followed by accommodation of the problem back to the assimilation of the strategy and assimilation of the relationship [32-33]. But in the construction of new knowledge has a common structure of thinking between ordinary students with gifted students.

5. Conclusion

Conjecturing via analogical reasoning constructs ordinary students into like gifted student, but this study required further verification of other materials.

References

[1] Supratman, Ryane S, and Rustina R 2016 Conjecturing via Analogical Reasoning in Developing Scientific Approach in Junior High School Students *J. Phys. Conf. Ser. 693 012017*

[2] Pólya G 1954 *Induction and Analogy in Mathematics and Plusible Reasoning* (London: Princeton University Press)

[3] Van de Walle J A, Karp K S and Bay-Williams J M 2010 *Elementary and Middle School Mathematics: Teaching Developmentally* (Boston New York San Francisco: Allyn & Bacon Mexico)

[4] Mason J, Burton L, and Stacey K 2010 Thinking mathematically *SECOND* **10** PEARSON

[5] National Council of Teachers of Mathematics 1989 Commission on Standards for School Mathematics 1989 Curriculum and evaluation standards for school mathematics (NCTM)

[6] Alexander P A and Buehl M 2004 Seeing The Possibilies: Contructing and Validating Measures of Mathematical and Analogical reasoning for Young Children, *in Mathematical and Analogical Reasoning of Young Learners, L. D. English, Ed* (New Jersey: Lawrence Erlbaum Associates) 23

[7] Sriraman B 2006 An ode to imre lakatos: Quasi-thought experiments to bridge the ideal and actual mathematics classrooms *Interchange* **37** 155

[8] Ahmanmaedi S 2013 Conjecturing via Analogical Reasoning to Explore Creative Thinking. 한국수학교육학회 학술발표논문집 305-316

[9] Richland L E, Holyoak K J and Stigler J W 2004 Analogy Use in Eighth-Grade Mathematics Classrooms *Cogn. Instr* **22** 37

[10] Worral J and Zahar E 1977 *Proofs and refutations the logic of mathematical discovery IMRE LAKATOS* (London, New York, Melbourne: CAMBRIDGE UNIVERSITY PRESS)

[11] Dubinsky E 1991 Reflective abstraction in advanced mathematical thinking *Adv. Math. Think* 95–

121
[12] Gray E M and Tall D O 1994 Duality, ambiguity, and flexibility: A 'Proceptual' view of simple arithmetic *J. Res. Math. Educ. J. Res. Math. Educ* **25** 116
[13] Gomez P and Canadas M C 2016 Difficulties in Mathematics Teachers Training In Learning Phenomenological Analysis *Rev Latinoam. Investig. en Matemática Educ* **10** 311
[14] Lee K H and Sriraman B 2011 Conjecturing via reconceived classical analogy *Educ. Stud. Math* **76** 123
[15] Ron G, Hershkowitz R and Dreyfus T 2008 The Identification of Partially Correct Constructs," *Proc. 31st Annu. Conf. Matheamtics Educ. Res. Gr. Australas* 421–427
[16] White P and Mitchelmore M C 2010 Teaching for Abstraction: A Model *Math. Think. Learn* **12** 205
[17] Cappetta R W and Zollman A 2013 Agents of Change in Promoting Reflective Abstraction : A Quasi-Experimental Study on Limits in College Calculus," REDIMAT *J. Res. Math. Educ* **2** 343
[18] Hershkowitz I, Lanes O and Lamb M E 2007 Exploring the disclosure of child sexual abuse with alleged victims and their parents *Child Abus. Negl* **31** 111
[19] Radford A N 2008 Duration and outcome of intergroup conflict influences intragroup affiliative behaviour. *Proc of the Royal Society of London B: Biological Sciences* 275(1653) 2787-2791
[20] Cañadas M C, Deulofeu J, Figueiras L, Reid D A and Yevdokimov O 2007 The conjecturing process: Perspectives in theory and implications in practice *Journal Teach. Learn* **5** 55–72
[21] English L D 2004 *Mathematical and Analogical Reasoning in Early Childhood," in Mathematical and Analogical Reasoning of Young Learners* (New Jersey: Lawrence Erlbaum Associates)
[22] Krulik S, Rudnick J and Milau E 2003 *Problem Solving and Reasoning," in Teaching Mathematics in Middle School* (Mexico : T. Mueller, Ed. Allyn & Bacon Mexico)
[23] Supratman A M 2013 Piaget's Theory in the Development of Creative Thinking. *Research in Mathematical Education, J. The Korean Society of Mathe. Educ/Series D* **17** 291
[24] Subanji R and Supratman A M 2015 The Pseudo-Covariational Reasoning Trought Processes in Constructing Graph Function of Reversible Event Dynamics Based on Assimilation and Accomodation Frameworks *Research in Mathematical Education. J. The Korean Society of Mathe. Educ/Series D* **19** 61
[25] Awamleh H, Al Assaf J A F. Borini I S, and Rahman A S M B A 2013 The Intelligence's Level of Gifted and Ordinary Students in Seventh and Eighth Grades , in Accordance with the Raven's Advanced Matrices Test in Irbid Governorate in Relation to Some Variables (Sex , Mothers' Qualification)," Am. Int *J. Contemp. Res* **3** 26
[26] Gagné F 1995 From giftedness to talent: A developmental model and its impact on the language of the field *Roeper Rev* **18** 103
[27] Van Tassel-Baska J 2007 Leadership for the future in gifted education: Presidential address, NAGC 2006 *Gifted Child Quarterly* **5** 5
[28] Rotigel, J V and Fello, S 2005 *Mathematically Gifted Students: How Can We Meet Their Needs?* In S. K. Johnsen & J. Kendrick (Eds.), A gifted child today reader. Math education for gifted students (Waco, TX: Prufrock Press) 3-16
[29] Yin R K 2016 Case Study Research Design and Methods (5th ed.) *J. Progr. Eval* **1** 108
[30] Sundayana R, Herman T, Dahlan J A and Prahmana R C I 2017 Using ASSURE learning design to develop students' mathematical communication ability *World Transactions on Engineering and Technology Education* **15** 245
[31] Olson G M, Mack R L, and Duffy S A 1981 Cognitive aspects of genre *Poetics* **10** 283
[32] Syamsuri, Purwanto, Subanji, and Irawati S 2017 Using APOS theory framework: Why did students unable to construct a formal proof? *International Journal on Emerging Mathematics Education* **1** 135
[33] Maharani H R, Sukestiyarno, and Waluya B 2017 Creative thinking process based on Wallas model in solving mathematics problem *International Journal on Emerging Mathematics Education* **1** 177

IOP Conf. Series: Journal of Physics: Conf. Series **943** (2017) 012026 doi:10.1088/1742-6596/943/1/012026

The students' mathematical argumentation in geometry

Sukirwan[1,2], Darhim[2], T Herman[2] and R C I Prahmana[3]

[1] Universitas Sultan Ageng Tirtayasa, Jl. Jakarta Km. 4, Serang 42121, Indonesia
[2]Universitas Pendidikan Indonesia, Jl. Dr. Setiabudi 229, Bandung 40154, Indonesia
[3]Universitas Ahmad Dahlan, Jl. Pramuka Kav. 5, Yogyakarta 55161, Indonesia

E-mail: sukirwan@untirta.ac.id

Abstract. The main objective of this research is to analyze the student's mathematical argumentation when dealing with geometry. The method is used qualitative method with grounded theory to know how the students provide an explanation or an answer against claims so that the quality of the vernacular students will be drawn up with clear from how students compose a series of arguments. The results showed that there were still many students basically experiencing constraints in argumentation, but the quality of the reasoning appears to be a variation of the argument appeared, include: inductive, algebra, visual and perceptual. In addition, the starting point of the students composes a series of arguments generally starts from claims that arise in a matter. Proof of claim further builds upon the relationship between the characteristics of data with mathematical objects that appear in the acquired mathematical knowledge from previous students. Relationship spelled out in a series of statements and reasons which support the claims through the fourth argument.

1. Introduction

The reasoning is the ability that is always inherent in any mathematical activity. In curriculum documents [1, 2], it is said that reasoning is one of the fundamental mathematical capabilities. Its implications not only on the basic capabilities in understanding mathematics [3], but remains a strategic issue in the development of mathematical learning challenges in the future.

The reasoning is one of the very complex mathematical abilities. This complexity makes reasoning as an ability that is not easy to achieve. Some studies report a constraint faced by students when exposed to reasoning [4, 5, 6]. These studies present the constraints in relation to the quality of mathematical reasoning in which students generally give an explanation of the difficulty in using the model or mathematical statement as well as making a statement that supports or refuting arguments [5].

More recently, a study of the quality of mathematical reasoning is much in touch with the argumentation [7]. Argumentation is seen as the study of effective reasoning [8], and becomes a fundamental tool that helps students improve the competency of reasoning [9]. The study generally uses on qualitative approaches [7, 10]. The study of argumentation has many researchers with extremely variable themes, among them: the structure of general argument with Toulmin Argumentation Pattern (TAP) [11, 12], the taxonomy of proof scheme [13, 14]; representation [15]; and verification [7, 15].

TAP [11] is seen as the starting point for the development of the study of argumentation. According to Toulmin [11], the validity of logical is not the main thing in an argument, but the orientation of the more important is the content and the structure of semantic matching. The Toulmin analysis is called informal logic [10], which later inspired many researchers to come up with the term informal argumentation [16]. Some other researchers who distinguish the term formal and informal argumentation [17, 18] in which the first term refers to mathematical proof [18, 19]. Other researchers agreed on the difference between proof and argument [20, 21] and placing proof as a series of stages/logic chains organizing process called cognitive unity [20]. In this study, the term proof is distinguished with the argument, that is, the more degrading aspects of semantic rather than syntactic [22].

This paper only wants to reveal an aspect of the study of argumentation with regard to the representation of a mathematical argument emerges from the students of the junior high school. Study on these aspects is viewed essentially given the students of the junior high school lies on the transition of concrete thinking to formal. Overview of aspects of the representation of this argument will give valuable illumination to the development of learning or certain approaches are needed to improve students' mathematical argumentation. Selected topics in this study were the topic of geometry. The study of the argument on this topic has many experts earlier especially deal with a taxonomy of the proof scheme [9, 14, 23]. Study of van Hiele [23] including the most phenomenal by creating a taxonomy of geometric, include visual, descriptive/analytical, deductive, informal, formal, deductive rigor. Nurhasanah, Sabandar, and Kusumah [24] adopted the van Hiele geometry to investigate the emerge of mathematical abstraction to the study of the geometry of the students. Although not mentioned explicitly about argumentation, an investigation into the mathematical abstraction of the variations that occur in students with regards to the student experience in drawing up arguments. The third researcher this does not compile the taxonomy as well as the van Hiele, but a variation of how students are doing abstraction is the main goal to be achieved in the research. In this way, it has been done before by Liu [15]. Study Liu [15] focusing on variations of answers students in solving geometry problems with 4 gave rise to a form of argumentation, namely: inductive, algebra, visual and perceptual.

2. Method

The research was carried out on the even semester on 2016/2017. The research approach is qualitative use the method of grounded theory. Research subjects are assigned 33 students in the class VIII in one of the junior high school in Tangerang, Indonesia. The main data obtained from test results on the topics of geometry, especially cube and block. The test results are analyzed through three stages of grounded theory, namely: *open coding, selective coding,* and *theoretical coding* [25]. At the first stage, test results were analyzed through two steps: (1) identification of the structure of mathematical argumentation, and (2) grouping the arguments based on the representation of argumentation that appears. The arguments are then grouped by similarity of the characteristics and the representation of argumentation in the form of ARG 1, ARG 2 and so on. ARG 9 is categorized as irrelevant argumentation, while ARG 10 is categorized as arguing that simply does not show up or no categorized. At the second stage, this is done by considering the sub-categories that appear to define a core category. The measures undertaken are (1) doing analysis of category or subcategory that appears by specifying the dominant symptoms from each of the categories (subcategories), (2) assigns a category to the core by means of linking between categories that have been defined in the previous step, and (3) reviewing deepening against each category core that has been established through the interview against the selected sample theoretically [26]. And the last stage, theory or conjecture is built upon synchronization and triangulation of the data. The measures taken are (1) conducting analysis and synchronization of data obtained in the previous stage, (2) perform a triangulation of data through analysis of the work students and in-depth interview against the respondent elected, (3) devise theories (conjectures based on the results of the analysis), synchronization and triangulation.

The following are the instruments used in a mathematical argument test students.

Koko claims that the areas of the diagonal fields of a cube are equal.
How about your opinion about the revelation of Koko? Make an argument in favor of denying the Koko opinion!

Firman claims that two blocks of the equal volume have a different surface area if the length, width, and height of each block are different.
How about your opinion about the question of Firman? Make an argument in favor of denying the Firman opinion!

3. Result and discussion

3.1. Open coding

The results of studies generally describe the existence of variations in themathematical argument. This is reflected from the arguments that appear on the stage of open coding, as shown in table 1.

Table 1. The Variation of Student's Mathematical Argumentation

Argumentation	Question 1		Question 2	
	N	Percentage	N	Percentage
ARG 1	5	15	22	67
ARG 2	4	12	2	6
ARG 3	3	9	3	9
ARG 4	2	6	0	0
ARG 5	3	9	0	0
ARG 9	6	18	2	6
ARG 10	10	30	4	12

In table 1, the percentage of the arguments on the question 1 for category irrelevant argumentation (ARG 9) and argumentation which do not appear (ARG 10) goes far beyond other arguments. This shows that students are still experiencing constraints to produce argumentation. Instead, on the question 2, the percentage of ARG 1 is high compared with most other arguments. This gives you an idea that the trend of mathematical argumentation that appears on the question 2 is ARG 1. Likewise, although the arguments in the question 1 dominating in ARG 9 and ARG 10, the argument for the category appears, still the highest argument 1 compared to the other arguments.

3.2. Selective coding

This step begins by identifying the arguments appear appropriate representation. On the argumentation 1, students use some examples to support claims are filed. On the question 1, students must first define a few cubes with a certain size; for example, 4cm. Students then calculate the length of the field diagonals of cube ($4\sqrt{2}$cm), then the area of the diagonals field of the cube ($16\sqrt{2}cm^2$). Students who understand the properties of the cube, it is generally claimed that areas of the diagonal fields that opposite sides of 4cm are equal, namely $16\sqrt{2}cm^2$. As the basis of the conclusions drawn, the students claim that area of the diagonal fields will always be the same on other cubes which has a rib 6cm ($36\sqrt{2}cm^2$), 8cm ($64\sqrt{2}$ cm^2), ... and so on. On the question 2, students choose a pair of the block that has the same volume but the area of the same, for example, the pair of the block with a size 12cm x 2cm x 2cm and 4cm x 3cm x 4cm. Students claim that the two blocks in the same volume ($48cm^3$) but the area of the different ($104cm^2$ and $90cm^2$). Students also have tried a couple of other blocks and the block pairs are claiming that volume is the same, but the size of the corresponding ribs are different, the breadth is definitely different.

On the argumentation 2, students use symbolic representation to assert that claim in each reserved was accepted or not. Students generally bring up any variables to determine a specific value from the element of the unknown material in question. On the question 1, students establish specific variables (s, r, or x) to represent the value of the ribs on any cube. These variables are then used to identify the elements of the diagonal field of any cube last until the conclusion of that vast symbolic field of diagonal fields of a cube is equal. On the question 2, the students set two blocks with different sizes.

The students then claimed that both the block in the same volume and must be pointed out that they are not equal. At this stage, no one can demonstrate that the claims involved proved to be right. Of course, students can use authentication with contra-example, but no one is able to do so.

On the argument 3, students build their arguments on the picture to support the proposed claim. On the question 1, students set up a cube. Students then determine two areas of diagonal fields and claim that the wide diagonal both fields are equal. Student outlines one of the broad fields of the diagonal equation. With the properties of the cube is known by previous, students then get the broad field of the other diagonal. This way is done to get the other diagonal fields so that the obtained conclusions that the sixth field of diagonal cube breadth equal. On the question 2, students create an illustration picture of two blocks that have length, width and height are different. A measure of length, width, and height of each block are the length of the line segments representing length, width, and height of each rib block. All students in this category are indeed could not propose further conjecture, so the outcome of this argument is not up to the expected conclusion.

On the argument 4, students base their arguments on the context is known. On the question 1, student imagined a pair of cuboid objects. Both cubes are congruent. If each cube sliced through one of the diagonal fields, then by altering their respective pieces of the cube, a new cube shape is obtained which remains congruent. Furthermore, students can also do other slices on the diagonal and keep generating more new cubes are congruent. The arguments 4 including arguments that at least show-up and even in the reserved question 2 do not appear at all. Basically, students need a lot of experience to interact with the real world that presents contextual problems so arguments 4 can appear in the imagination of the mind of the students.

On the argumentation 5, students base their arguments on the connections between mathematical objects based on the characteristics of the objects that have been previously understood. On the question,1 students begin making arguments by mentioning cube properties. For example, a cube has the properties, among others: the six sides congruent squares, mutually perpendicular sides, ribs-congruent opposite sides and perpendicular to each other. Student directly claims that the area of the diagonal fields of the cube is equal because the sides of the cube are congruent.

3.3. Theoretical coding

From some of the argumentation that appears that students have different ways of making explanations supporting against argumentation. On the argumentation 1, students use some special cases to check the truth of the statement asked. The conclusion is drawn when each case demonstrates the truth of the statement asked. This way is an indication of inductive argumentation. On the argumentation 2, students expressed any case through symbolic representation. Because the case took any, then the truth of the statement that is submitted is a foregone conclusion. This way is an indication of the algebraic argument. On the argument 3, students must first define an image as a starting point for investigation. Students then create math sentences based on the facts contained in the image. Conclusions are drawn when the facts demonstrate the truth of the statement. Because the proof of the truth of the statement is sourced from the image, then this way is an indication of visual argumentation. On the argument 4, the students started the idea of proof from a context that is known to object that resemble the existing context on the statement. Conclusions are drawn based on the perception that the truth of the statement appears in the context of the imagined or manipulated. This way is an indication of perceptual argumentation. On the argumentation 5, students base on characteristics of mathematical objects that exist in the statement. Although this argument brings up new arguments, all students in this group are never completely on the expected conclusions. Basically, new students bring up the initial idea to create an interconnection between the properties owned this object with the statement that will be proven. Thus students are using characteristics of mathematical objects on a statement that would then be applied by means of inductive argumentation, visual, perceptual and algebra. The results of the following interview give an explicit regarding the next student actions based on arguments 5.

Teacher : How are the diagonal field of this and that?
Student 5 : Equal, Sir?
Teacher : Why?
Student 5 : Similarly rectangular? This line and it's is equal, the properties of the cube.
Teacher : How do I calculate the areas?
Student 5 : (Confused), but the length of the rib is not known, Sir? For Example, I take 4 cm.
Teacher : Student 11 mentions that the length of field diagonal of the cube is equal, what that means?
Student 11 : I don't know, but, for example, in this classroom? (While point to the roof of the class). If I pull the line from there to the end here just the same with the draw a line from the tip of next door here to there.
Teacher : Next?
Student 11 : Lines on the wall on your right and left, right Sir?
Teacher : Why?
Student 11 : Properties of the cube.
Teacher : How about Student 24?
Student 24 : From this picture, I know Sir, but not sure?
Teacher : Why?
Student 24 : Since the cube sides of a square, then the diagonal field is equal. Right, Sir?

The results of the interview give an overview that students use the characteristics of mathematical objects arise in planning to make a statement in making an explanation of the claim lodged. The claim is a phenomenon that appears on every student's argumentation in which claims are used as a *starting point* to justify the statement in question. Referring to this claim, students then create an explanation so that the data obtained can come to the statement of claim filed. This data is connected to students' prior knowledge gained so that students utilize the characteristics of mathematical objects that appear to elaborate on claims based on several cases (inductive), a symbolic representation (algebra), images (Visual), and imagination (perceptual).

4. Conclusion

The student's mathematical argumentation include inductive, algebra, visual and perceptual argumentation. The future of argumentations emerge from the fourth argument is *inductive*, the conclusion is based on the truth of some cases; *algebraic*, the conclusion is based on the general case with representation symbolic; *visual*, the conclusion is based on geometry picture; *perceptual*, the conclusion is based on a known context. In the justification of a statement, students generally bring up claims as a starting point. Students then elaborate data by connecting on mathematical object characteristics in question. Students use inductive, algebra, visual, and perceptual and to make a series of statements and reasons that support against claims filed.

References

[1] National Council of Teachers Mathematics 2000 *Principles and standards for school mathematics* (Reston V A : National Council of Teachers of Mathematics)
[2] Sundayana R, Herman T, Dahlan J A, and Prahmana R C I 2017 Using ASSURE learning design to develop students' mathematical communication ability *World Transactions on Engineering and Technology Education* **15** 245
[3] Ball D and Bass H 2003 Making mathematics reasonable in school In J Kilpatrick, G Martin and D Schifter eds *A Research Companion to Principles and Standards for School Mathematics* (Reston V A : National Council of Teachers of Mathematics) p 27
[4] Wardhani S and Rumiati 2011 *Instrument penilaian Hasil belajar matematika SMP: Belajar dari PISA dan TIMSS* (Yogyakarta: P4TK Matematika)
[5] Rizta A, Zulkardi, and Hartono Y 2013 Pengembangan soal penalaran model TIMSS matematika

SMP *Jurnal Penelitian dan Evaluasi Pendidikan* **17** 230
[6] Tanujaya B, Prahmana R C I, and Mumu, J 2017 Mathematics instruction, problems, challenges, and opportunities: A case study in Manokwari regency, Indonesia *World Transactions on Engineering and Technology Education* **15** 287
[7] Bergqvist T 2005 How students verify conjectures: Teachers' expectations *Journal of Mathematics Teacher Education* **8** 171
[8] Zarebski Toulmin's model of argument and the "logic" of science discovery *Studies in Logic, Grammar and Rhetoric* **16** 267
[9] Lee T-N 2015 Developing a theoretical framework to assess Taiwanese primary students' geometric argumentation In M Marshman, V Geiger and A Bennison eds *Proceedings of the 38th Annual Conference of the Mathematics Education Research Group of Australia* (Sunshine Coast: MERGA) p 365
[10] Inglis M, Mejia-Ramos J P, and Simpson A 2007 Modelling mathematical argumentation: The importance of qualification *Educational Studies Mathematics* **66** 3
[11] Toulmin S E 1958 *The use of argument* (Cambridge: Cambridge University Press)
[12] von Aufschnaiter C, Erduran S, Osborne J, and Simon S 2008 Arguing to learn and learning to argue: Case studies of how students' argumentation relates to their scientific knowledge *Journal of Research in Science Teaching* **45** 101
[13] Pedemonte B 2008 Argumentation and algebraic proof *ZDM, The International Journal of Mathematics Education* **40** 385
[14] Stylianides A J 2007 Proof and proving in school mathematics *Journal for Research in Mathematics Education* **38** 289
[15] Liu Y 2013 Aspects of mathematical arguments that influence eighth-grade students' judgment of their validity *Dissertation* (Ohio: The Ohio State University)
[16] Brem S K and Rips L J 2000 Explanation and evidence in informal argument *Cognitive Science* **24** 573
[17] Aberdein A 2005 The uses of argument in mathematics *Argumentation* **19** 287
[18] Yackel E 2002 What we can learn from analyzing the teacher's role in collective argumentation *Journal of Mathematics Behavior* **21** 423
[19] Banegas J A 2003 Argumentation in mathematics In An Aberdeen and I J Dove eds *The Argument of Mathematics* (Netherland: Springer) p 47
[20] van Bendegem J P and van Kerkhove B 2009 Mathematical arguments in context *Foundations of Science* **14** 45
[21] Prahmana R C I and Kusumah Y S 2016 The hypothetical learning trajectory on research in mathematics education using research-based learning *Pedagogika* **123** 42
[22] Larios-Osorio V and Acuña-Soto C 2009 Geometrical proof in the institutional classroom environment In F-L Lin, F-J Hsieh G Hanna, and M de Villiers eds *Proceedings of the International Commission on Mathematical Instruction* (Taipei: National Taiwan Normal University, Taiwan) p 59
[23] van Hiele P M 1986 *Structure and insight: A theory of mathematics education* (New York: Academic Press)
[24] Nurhasanah F, Kusumah Y S, and Sabandar J 2017 Concept of triangle: Examples of mathematical abstraction in two different contexts *International Journal on Emerging Mathematics Education* **1** 53
[25] Jones M and Alony I 2011 Building the use of grounded theory in doctoral studies *International Journal of Doctoral Studies* **6** 95
[26] Creswell J W 2010 *Reseach design pendekatan qualitative, quantitative dan mixed* A Fawaid (Translator) (Yogyakarta: Pustaka Pelajar)

IOP Conf. Series: Journal of Physics: Conf. Series **943** (2017) 012027 doi:10.1088/1742-6596/943/1/012027

On locating independent domination number of amalgamation graphs

Dwi Agustin Retno Wardani [1,4], **Dafik**[1,2], **Ika Hesti Agustin**[1,3], **Elsa Yuli Kurniawati** [1,3]

[1] CGANT University of Jember, Indonesia
[2] Mathematics Edu. Depart. University of Jember, Indonesia
[3] Mathematics Depart. University of Jember, Indonesia
[4] Mathematics Edu. Depart. IKIP PGRI Jember, Indonesia

E-mail: 2i.agustin@gmail.com, d.dafik@unej.ac.id, ikahesti.fmipa@unej.ac.id

Abstract. An independent set or stable set is a set of vertices in a graph in which no two of vertices are adjacent. A set D of vertices of graph G is called a dominating set if every vertex $u \in V(G) - D$ is adjacent to some vertex $v \in D$. A set S of vertices in a graph G is an independent dominating set of G if D is an independent set and every vertex not in D is adjacent to a vertex in D. By locating independent dominating set of graph G, we mean that an independent dominating set D of G with the additional properties that for $u, v \in (V - D)$ satisfies $N(u) \cap D \neq N(v) \cap D$. A minimum locating independent dominating set is a locating independent dominating set of smallest possible size for a given graph G. This size is called the locating independent dominating number of G and denoted $\gamma_{Li}(G)$. In this paper, we analyze the locating independent domination number of graph operations.

1. Introduction

Let G be a nontrivial, finite, simple, undirected and connected graphs, with vertex set $V(G)$, edge set $E(G)$ and with no isolated vertex, for more detail definition of graph see [1, 2]

A set D of vertices of a graph $G = (V, E)$ is dominating if every vertex in $V(G) - D$ is adjacent to some vertex in D. The domination number of G, denoted by $\gamma(G)$, is the minimum cardinality of a dominating set in G. A locating-dominating set is a dominating set D that locates all the vertices in the sense that every vertex not in D is uniquely determined by its neighborhood in D. The locating domination number of G, denoted by $\gamma_L(G)$, is the minimum cardinality of a locating dominating set in G. A locating-dominating set of order $\gamma_L(G)$ is called an $\gamma_L(G)$-set The concept of a locating dominating set was introduced and first studied by Slater [3, 4, 5, 6] and also Waspodo *et. al.* [8] studied the bound of distance domination number of edge comb product.

For definition and notation of locating dominating set in [7] explained that the open neighborhood of a vertex $v \in V(G)$ is $N_G(v) = \{u \in V(G); uv \in E(G)\}$ and its closed neighborhood is the set $N_G[v] = N_G(v) \cup \{v\}$. For a set D of vertex set of G, $N_G[D]$ is the union of all closed neighborhoods of vertices in D. The degree of v is $d_G(v) = |N_G(v)|$. If the graph G is a connected graph, we simply write $V(G), E(G), N(v), N[v], N[D]$ and $d(v)$ rather than $V(G), E(G), N_G(v), N_G[v], N_G[D]$ and $d_G(v)$, respectively.

A dominating set of G and denoted by D, if $N[v] \cap D \neq \emptyset$ for all vertex $v \in G$, or equivalently, $N[D] = V(G)$. Any two vertices u and $v \in V(G)\ D$ are located by D if they have distinct neighbors in D that is, $N(u) \cap D \neq N(v) \cap D$. If a vertex $u \in V(G)\ D$ is located from every other vertex in $V(G)\ D$, we simply say that u is located by D. A set D is a locating set of G if any two distinct vertices outside D are located by D. In particular, if S is both a dominating set and a locating set, then S is a locating dominating set. We remark that the only difference between a locating set and a locating dominating set in G is that a locating set might have a unique non-dominated vertex.

A set D of vertices in a graph G is an independent dominating set of G if D is an independent set and every vertex not in D is adjacent to a vertex in D. The definition and notation of locating independent dominating set of graph G similarly the definition and notation of locating dominating set, we mean that a locating independent dominating set D of G with the additional properties that D is an independent set and every vertex not in D is adjacent to a vertex in D. The locating independent dominating number of a graph G, denoted by $\gamma_{LI}(G)$, is the minimum cardinality of a locating independent dominating set of graph G. A locating independent dominating set of order $\gamma_{LI}(G)$ is called an $\gamma_{Li}(G)$-set.

2. Main Results

The definition of amalgamation of graph is taken from [9]. Let G_i be a simple connected graph, for $i \in \{1, 2, \ldots, t\}$ and $t \in \mathcal{N}$ and $|V(G_i)| = k_i \geq 2$ for some $k_i \in \mathcal{N}$. For $t \geq 2$ let $\{G_1, G_2, \ldots, G_t\}$ be a finite collection of graphs and each G_i, $i \in \{1, 2, \ldots, t\}$, has a fixed vertex v_{oi} called a *terminal*. The amalgamation denote by $Amal(G_i, v_{oi})$.

In this section, we determine the exact values of locating independent dominating number of some special graphs and its operations namely star graph S_n, $Amal(S_n, v, m)$, path graph P_n, $Amal(P_n, v, m)$, wheel graph W_n, $Amal(W_n, v, m)$, ladder graph L_n, $Amal(L_n, v, m)$.

Lemma 2.1. *For any graph G of order n, the lower bound of locating independent domination number of amalgamation graph $Amal(G, v, m)$ is $\gamma_{Li}\ (Amal(G, v, m)) \geq m(\gamma_{Li}(G) - 1) + 1$.*

Proof. The graph $Amal(G, v, m)$ is a connected graph of order $|V(Amal(G, v, m))| = (p(G) - 1)m + 1$ and size $|E(Amal(G, v, m))| = (q(G))m$.

To prove the lemma above, we claim that $\gamma_{Li}\ (Amal(G, v, m)) \geq m(\gamma_{Li}(G) - 1) + 1$. To convince this, assume that $\gamma_{Li}\ (Amal(G, v, m)) < m(\gamma_{Li}(G) - 1) + 1$. The intersection between the neighborhood $N(v)$ with $v \in V(G) - D$ will be empty set. Thus, it is a contradiction. See Figure 1 for illustration. □

Theorem 2.2. *For $n \geq 3$, the locating independent domination number of S_n is $\gamma_{Li}(S_n) = n$.*

Proof. Star graph S_n is a connected graph with vertex set $V(S_n) = \{A\} \cup \{x_i; 1 \leq i \leq n\}$ and edge set $E(S_n) = \{Ax_i; 1 \leq i \leq n\}$. The order and size of S_n are $|V(S_n)| = n + 1$ and $|E(S_n)| = n$.

We claim that $\gamma_{Li}\ (S_n) \geq n$. To convince the proof, assume that $\gamma_{Li}(S_n) < n$. Let the dominator vertex set of S_n, for $n \geq 3$, be $D = \{x_i; 1 \leq i \leq n - 1\}$, thus $|D| = n - 1$, and let non-dominator vertex set of S_n, for $n \geq 3$, be $V - D = \{A\} \cup \{x_n\}$. Then we get the intersection of the neighborhood $N(v)$ with $v \in V(G) - D$ and dominator set D, in the following.

$$N(A) \cap D = \{x_i; 1 \leq i \leq n - 1\}$$
$$N(x_n) \cap D = \emptyset$$

It can be seen that the intersection between the neighborhood $N(v)$ with $v \in V(G) - D$, for $N(x_n) \cap D = \emptyset$. Thus, the dominator set D do not dominate all vertices in $V(S_n)$. It concludes that, by assuming $\gamma_{Li}(S_n) < n$, it will not comply the condition of locating independent dominating set. Therefore, the lower bound of locating independent domination number of

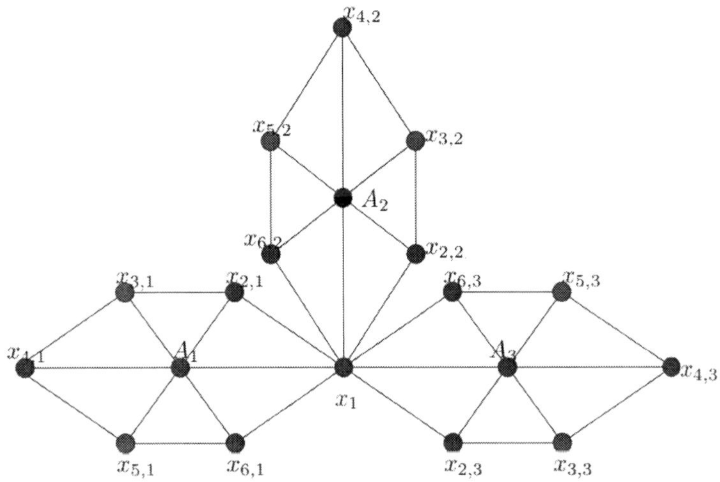

Figure 1. Amalgamation of Wheel Graph $Amal(W_n, v, m)$

S_n is $\gamma_{Li}(S_n) \geq n$. Furthermore, we will show that the upper bound of locating independent domination number of S_n is $\gamma_{Li}(S_n) \leq n$. Choose $D = \{x_i; 1 \leq i \leq n\}$ as the dominator set of S_n, for $n \geq 3$, thus $|D| = n$. Choose $V - D = \{A\}$ as the non-dominator set of S_n for $n \geq 3$. We will get the intersection between the neighborhood $N(v)$ with $v \in V(G) - D$ and dominator set D, in the following.

$N(A) \cap D = \{x_i; 1 \leq i \leq n\}$

It can be seen that the intersection between the neighborhood $N(v)$ with $v \in V(G) - D$ are all different, and it is not empty set. The dominator set D does not dominate all vertices in $V(S_n)$. It can be concluded that, for $\gamma_{Li}(S_n) \leq n$, it will comply the condition of locating independent dominating set. Thus $\gamma_{Li}(S_n) \leq n$. Hence, then the locating independent domination number of S_n is $\gamma_{Li}(S_n) = n$. □

Theorem 2.3. *Let G be an amalgamation graph of star S_n with $n \geq 3$ and $m \geq 3$. Then locating independent domination number of $Amal(S_n, v, m)$ is $\gamma_{Li}(Amal(S_n, v, m)) = m(n-1) + 1$.*

Proof. The graph $Amal(S_n, v, m)$ is a connected graph with $V(Amal(S_n, v, m)) = \{x\} \cup \{A_i; 1 \leq i \leq m\} \cup \{x_{i,j}; 1 \leq i \leq m; 1 \leq j \leq n-1\}$ and $E(Amal(S_n, v, m)) = \{xA_i; 1 \leq i \leq m\} \cup \{A_i x_{i,j}; 1 \leq i \leq m; 1 \leq j \leq n-1\}$. The order of this graph is $|V(Amal(S_n, v, m))| = nm + 1$ and the size is $|E(Amal(S_n, v, m))| = nm$. To prove the above theorem $\gamma_{Li}(Amal(S_n, v, m)) = m(n-1) + 1$, we will show that the lower bound $\gamma_{Li}(Amal(S_n, v, m)) \geq m(n-1) + 1$ and the upper bound $\gamma_{Li}(Amal(S_n, v, m)) \leq m(n-1) + 1$.

Firstly, we will show that $\gamma_{Li}(Amal(S_n, v, m)) \geq m(n-1) + 1$. By Lemma 2.1, we have $\gamma_{Li}(Amal(S_n, v, m)) = m(\gamma_{Li}(S_n) - 1) + 1$. Since by Theorem 2.2 we have $\gamma_{Li}(S_n) = n$, thus we get so $\gamma_{Li}(Amal(S_n, v, m)) \geq m(n-1) + 1$. Furthermore, we will show that the upper bound of locating independent domination number of $\gamma_{Li}(Amal(S_n, v, m)) \leq m(n-1) + 1$. We consider $D(Amal(S_n, v, m)) = \{x\} \cup \{x_{i,j}; 1 \leq i \leq m; 1 \leq j \leq n-1\}$ as the dominator set of $Amal(S_n, v, m)$ for $n \geq 3$ and $m \geq 3$. It is clearly to see that $|D| = m(n-1) + 1$, and the dominator set D dominates all vertices of $G = Amal(S_n, v, m)$. By definition, we have the non-dominator set of $Amal(S_n, v, m)$ for $n \geq 3$ and $m \geq 3$ is $V(Amal(S_n, v, m)) - D(Amal(S_n, v, m)) = \{A_i; 1 \leq i \leq m\}$. The intersection between the neighborhood $N(v)$ with $v \in V(G) - D(G)$ and dominator set $D(G)$ is as follows.

$N(A_i) \cap D = \{x\}, \{x_{i,j}; 1 \leq i \leq m; 1 \leq j \leq n-1\}$

It can be seen intersection between the neighborhood $N(v)$ with $v \in V(G) - D(G)$ and the obtained dominator set D are uniques and it is not empty set. Thus, it can be concluded that $\gamma_{Li}(Amal(S_n, v, m)) \leq m(n-1)+1$. D also complies the condition of locating independent dominating set. Hence, the lower bound and upper bound of locating independent domination number respectively, are $\gamma_{Li} \geq m(n-1)+1$ and $\gamma_{Li} \leq m(n-1)+1$. It conludes the locating independent domination number of $Amal(S_n, v, m)$ is $\gamma_{Li}(Amal(S_n, v, m)) = m(n-1)+1$. □

Theorem 2.4. *For $n \geq 4$, locating independent domination number of P_n is $\gamma_{Li}(P_n) = \lceil \frac{2n}{5} \rceil$.*

Proof. Path graph P_n is a connected graph with vertex set $V(P_n) = \{x_i; 1 \leq i \leq n\}$ and edge set $E(P_n) = \{x_i x_{i+1}; 1 \leq i \leq n-1\}$. The order and size of P_n are $|V(P_n)| = n$ and $|E(P_n)| = n-1$.

We claim that $\gamma_{Li}\ (P_n) \geq \lceil \frac{2n}{5} \rceil$. To convince the proof, assume that $\gamma_{Li}(P_n) < \lceil \frac{2n}{5} \rceil$. Let the dominator vertex set of P_n, for $n \geq 4$, $D = \{x_i; i \equiv 0 \bmod 2; i > 2\}$ those $|D| = \lceil \frac{2n}{5} \rceil$ and non-dominator vertex set of P_n for $n \geq 4$ is $V - D = \{x_2\} \cup \{x_i; i \equiv 1 \bmod 2\}$. Then we get the intersection of the neighborhood $N(v)$ with $v \in V(G) - D$ and dominator set D, in the following.

$N(x_i) \cap D = \{x_i; i \equiv 0 \bmod 2; i > 2\}$
$N(x_1) \cap D = \emptyset$
$N(x_2) \cap D = \emptyset$

It can be seen that the intersection between the neighborhood $N(v)$ with $v \in V(G) - D$, for $N(x_1), N(x_2) \cap D = \emptyset$. Thus, the dominator set D do not dominate all vertices in $V(P_n)$. It concludes that, by assuming $\gamma_{Li}(P_n) < \lceil \frac{2n}{5} \rceil$, it will not comply the condition of locating independent dominating set. Therefore, the lower bound of locating independent domination number of P_n is $\gamma_{Li}(P_n) \geq \lceil \frac{2n}{5} \rceil$. Furthermore, we will show that the upper bound of locating independent domination number of P_n is $\gamma_{Li}(P_n) \leq \lceil \frac{2n}{5} \rceil$. Choose $D = \{x_i; i \equiv 0 \bmod 2\}$ as the dominator set of P_n, for $n \geq 4$, thus $|D| = \lceil \frac{2n}{5} \rceil$. Choose $V - D = \{x_i; i \equiv 1 \bmod 2\}$ as the non-dominator set of P_n for $n \geq 4$. We will get the intersection between the neighborhood $N(v)$ with $v \in V(G) - D$ and dominator set D, in the following.

$N(x_i) \cap D = \{x_i; i \equiv 0 \bmod 2\}$

It can be seen that the intersection between the neighborhood $N(v)$ with $v \in V(G) - D$ are all different, and it is not empty set. The dominator set D does not dominate all vertices in $V(P_n)$. It can be concluded that, for $\gamma_{Li}(P_n) \leq \lceil \frac{2n}{5} \rceil$, it will comply the condition of locating independent dominating set. Thus $\gamma_{Li}(P_n) \leq \lceil \frac{2n}{5} \rceil$. Hence, then the locating independent domination number of P_n is $\gamma_{Li}(P_n) = \lceil \frac{2n}{5} \rceil$. □

Theorem 2.5. *Let G be a amalgamation graph of path (P_n) with $n \geq 4$ and $m \geq 3$. Then locating independent domination number of $Amal(P_n, v, m)$ is $\gamma_{Li}\ (Amal(P_n, v, m)) = m(\lceil \frac{2n}{5} \rceil - 1) + 1$.*

Proof. The graph $Amal(P_n, v, m)$ is a connected graph with $V(Amal(P_n, v, m)) = \{x\} \cup \{x_{i,j}; 1 \leq i \leq m; 1 \leq j \leq n-1\}$ and $E(Amal(P_n, v, m)) = \{xx_{i,1}; 1 \leq i \leq m\} \cup \{x_{i,j}x_{i,j+1}; 1 \leq i \leq m; 1 \leq j \leq n-2\}$. The order of this graph is $|V(Amal(P_n, v, m))| = nm - m + 1$ and the size is $|E(Amal(P_n, v, m))| = nm - m$. To proof the above theorem $\gamma_{Li}(Amal(P_n, v, m)) = m(\lceil\frac{2n}{5}\rceil - 1) + 1$, we will show that the lower bound $\gamma_{Li}(Amal(P_n, v, m)) \geq m(\lceil\frac{2n}{5}\rceil - 1) + 1$ and the upper bound $\gamma_{Li}(Amal(P_n, v, m)) \leq m(\lceil\frac{2n}{5}\rceil - 1) + 1$.

Firstly, we will show that γ_{Li} $(Amal(P_n, v, m)) \geq m(\lceil\frac{2n}{5}\rceil - 1) + 1$. By Lemma 2.1, we have $\gamma_{Li}(Amal(P_n, v, m)) = m(\gamma_{Li}(P_n) - 1) + 1$. Since by Theorem 2.4 we have $\gamma_{Li}(P_n) = \lceil\frac{2n}{5}\rceil$, thus we get so $\gamma_{Li}(Amal(P_n, v, m)) \geq m(\lceil\frac{2n}{5}\rceil - 1) + 1$. Furthermore, we will show that the upper bound of locating independent domination number of $\gamma_{Li}(Amal(P_n, v, m)) \leq m(\lceil\frac{2n}{5}\rceil - 1) + 1$. We consider $D(Amal(P_n, v, m)) = \{x\} \cup \{x_{i,j}; 1 \leq i \leq m; j \equiv 0 \bmod 2\}$ as the dominator set of *Amal* (P_n, v, m) for $n \geq 4$ and $m \geq 3$. It is clearly to see that $|D| = m(\lceil\frac{2n}{5}\rceil - 1) + 1$, and the dominator set D dominates all vertices of $G = Amal(P_n, v, m)$. By definition, we have the non-dominator set of $Amal(P_n, v, m)$ for $n \geq 4$ and $m \geq 3$ is $V(Amal(P_n, v, m)) - D(Amal(P_n, v, m)) = \{x_{i,j}; 1 \leq i \leq m; j \equiv 1 \bmod 2\}$. The intersection between the neighborhood $N(v)$ with $v \in V(G) - D(G)$ and dominator set $D(G)$ is as follows.

$$N(x_{i,j}) \cap D = \{x\}, \{x_{i,j}; 1 \leq i \leq m; j \equiv 0 \bmod 2\}$$

It can be seen intersection between the neighborhood $N(v)$ with $v \in V(G) - D(G)$ and the obtained dominator set D are uniques and it is not empty set. Thus, it can be concluded that $\gamma_{Li}(Amal(P_n, v, m)) \leq m(\lceil\frac{2n}{5}\rceil - 1) + 1$. D also complies the condition of locating independent dominating set. Hence, the lower bound and upper bound of locating independent domination number respectively, are $\gamma_{Li} \geq m(\lceil\frac{2n}{5}\rceil - 1) + 1$ and $\gamma_{Li} \leq m(\lceil\frac{2n}{5}\rceil - 1) + 1$. It conludes the locating independent domination number of $Amal(P_n, v, m)$ is $\gamma_{Li}(Amal(P_n, v, m)) = m(\lceil\frac{2n}{5}\rceil - 1) + 1$. □

Theorem 2.6. *For $n \geq 3$, the locating independent domination number of L_n is $\gamma_{Li}(L_n) = n$.*

Proof. Ladder graph L_n is a connected graph with vertex set $V(L_n) = \{x_i; 1 \leq i \leq n\} \cup \{y_i; 1 \leq i \leq n\}$ and edge set $E(L_n) = \{x_ix_{i+1}; 1 \leq i \leq n-1\} \cup \{y_iy_{i+1}; 1 \leq i \leq n-1\} \cup \{x_iy_i; 1 \leq i \leq n\}$. The order and size of L_n are $|V(L_n)| = 2n$ and $|E(L_n)| = 3n - 2$.

We claim that γ_{Li} $(L_n) \geq n$. To convince the proof, assume that $\gamma_{Li}(L_n) < n$. Let the dominator vertex set of L_n, for $n \geq 3$, be $D = \{x_i; 1 \leq i \leq n-1; i = \text{odd}\} \cup \{y_i; 1 \leq i \leq n-1; i = \text{even}\}$, thus $|D| = n - 1$, and let non-dominator vertex set of L_n, for $n \geq 3$, be $V - D = \{x_i; 1 \leq i \leq n; i = \text{even}\} \cup \{y_i; 1 \leq i \leq n; i = \text{odd}\} \cup \{x_n; n = \text{odd}\} \cup \{y_n, n = \text{even}\}$. Then we get the intersection of the neighborhood $N(v)$ with $v \in V(G) - D$ and dominator set D, in the following.

$$N(x_i) \cap D = \{x_{i-1}, x_{i+1}, y_i\};\ 2 \leq i \leq n-2; i = \text{even}$$
$$N(x_n) \cap D = \{x_{n-1}\};\ n = \text{even}$$
$$N(x_{n-1}) \cap D = \{x_{n-2}, y_{n-1}\};\ n = \text{odd}$$
$$N(x_n) \cap D = \emptyset;\ n = \text{odd}$$
$$N(y_1) \cap D = \{x_1, y_2\}$$
$$N(y_i) \cap D = \{x_i, y_{i-1}, y_{i+1}\};\ 3 \leq i \leq n-2; i = \text{odd}$$

$N(y_n) \cap D = \{y_{n-1}\}$; $n =$ odd
$N(y_{n-1}) \cap D = \{x_{n-1}, y_{n-2}\}$; $n =$ even
$N(y_n) \cap D = \emptyset$; $n =$ even

It can be seen that the intersection between the neighborhood $N(v)$ with $v \in V(G) - D$, for $N(x_n) \cap D = \emptyset$ for n odd and $N(y_n) \cap D = \emptyset$ for n even. Thus, the dominator set D do not dominate all vertices in $V(L_n)$. It concludes that, by assuming $\gamma_{Li}(L_n) < n$, it will not comply the condition of locating independent dominating set. Therefore, the lower bound of locating independent domination number of L_n is $\gamma_{Li}(L_n) \geq n$. Furthermore, we will show that the upper bound of locating independent domination number of L_n is $\gamma_{Li}(L_n) \leq n$. Choose $D = \{x_i; 1 \leq i \leq n; i = \text{odd}\} \cup \{y_i; 1 \leq i \leq n; i = \text{even}\}$ as the dominator set of L_n, for $n \geq 3$, thus $|D| = n$. Choose $V - D = \{A\}$ as the non-dominator set of S_n for $n \geq 3$. We will get the intersection between the neighborhood $N(v)$ with $v \in V(G) - D$ and dominator set D, in the following.

$N(x_i) \cap D = \{x_{i-1}, x_{i+1}, y_i\}$; $2 \leq i \leq n-2; i =$ even
$N(x_n) \cap D = \{x_{n-1}, y_n\}$; $n =$ even
$N(y_1) \cap D = \{x_1, y_2\}$
$N(y_i) \cap D = \{x_i, y_{i-1}, y_{i+1}\}$; $3 \leq i \leq n; i =$ odd
$N(y_n) \cap D = \{x_n, y_{n-1}\}$; $n =$ odd

It can be seen that the intersection between the neighborhood $N(v)$ with $v \in V(G) - D$ are all different, and it is not empty set. The dominator set L does not dominate all vertices in $V(L_n)$. It can be concluded that, for $\gamma_{Li}(L_n) \leq n$, it will comply the condition of locating independent dominating set. Thus $\gamma_{Li}(L_n) \leq n$. Hence, then the locating independent domination number of L_n is $\gamma_{Li}(L_n) = n$. □

Theorem 2.7. *Let G be a amalgamation graph of ladder (L_n) with $n \geq 2$ and $m \geq 2$, then locating independent domination number of $Amal(L_n, v, m)$ is γ_{Li} $(Amal(L_n, v, m)) = nm$.*

Proof. The graph $Amal(L_n, v, m)$ is a connected graph with $V(Amal(L_n, v, m)) = \{y_1\} \cup \{x_i^j; 1 \leq i \leq n; 1 \leq j \leq m\} \cup \{y_{i+1}^j; 1 \leq i \leq n-1; 1 \leq j \leq m\}$ and $E(Amal(L_n, v, m)) = \{y_1 x_1^j; 1 \leq j \leq m\} \cup \{y_1 y_2^j; 1 \leq j \leq m\} \cup \{x_{i+1}^j y_{i+1}^j; 1 \leq i \leq n-1; 1 \leq j \leq m\} \cup \{x_i^j x_{i+1}^j; 1 \leq i \leq n-1; 1 \leq j \leq m\} \cup \{y_{i+1}^j y_{i+2}^j; 1 \leq i \leq n-2; 1 \leq j \leq m\}$. The order of this graph is $|V(Amal(L_n, v, m))| = 2nm - m + 1$ and the size is $|E(Amal(L_n, v, m))| = 3nm - 2m$. To prove the above theorem $\gamma_{Li}(Amal(L_n, v, m)) = nm$, we will show that the lower bound $\gamma_{Li}(Amal(L_n, v, m)) \geq nm$ and the upper bound $\gamma_{Li}(Amal(L_n, v, m)) \leq nm$.

Firstly, we will show that γ_{Li} $(Amal(L_n, v, m)) \geq nm$. By Lemma 2.1, we have $\gamma_{Li}(Amal(L_n, v, m)) = m(\gamma_{Li}(L_n) - 1) + 1$. Since by Theorem 2.6 we have $\gamma_{Li}(L_n) = n$, thus we get so $\gamma_{Li}(Amal(L_n, v, m)) \geq m(n-1) + 1$. Furthermore, we will show that the upper bound of locating independent domination number of $\gamma_{Li}(Amal(L_n, v, m)) \leq nm$. We consider $D(Amal(L_n, v, m)) = \{x_i^j; 1 \leq i \leq n; 1 \leq j \leq m; i = \text{odd}\} \cup \{y_i^j; 1 \leq i \leq n; 1 \leq j \leq m; i = \text{even}\}$ as the dominator set of $Amal$ (L_n, v, m) for $n \geq 2$ and $m \geq 2$. It is clearly to see that $|D| = nm$, and the dominator set D dominates all vertices of $G = Amal(L_n, v, m)$. By definition, we have the non-dominator set of $Amal(L_n, v, m)$ for $n \geq 2$ and $m \geq 2$ is $V(Amal(L_n, v, m)) - D(Amal(L_n, v, m)) = \{y_1\} \cup \{x_i^j; 1 \leq i \leq n; 1 \leq j \leq m; i = \text{even}\} \cup \{y_i^j; 3 \leq i \leq n; 1 \leq j \leq m; i = \text{odd}\}$. The intersection between the neighborhood $N(v)$ with $v \in V(G) - D(G)$ and dominator set $D(G)$ is as follows.

$N(y_1) \cap D = \{x_1^j, y_2^j; 1 \leq j \leq m\}$
$N(x_2^j) \cap D = \{x_1^j, x_3^j, y_2^j\}$; $1 \leq j \leq m$

$N(x_i^j) \cap D = \{x_{i-1}^j, x_{i+1}^j, y_i^j\}$; $4 \le i \le n-2; 1 \le j \le m; i = \text{even}$
$N(x_n^j) \cap D = \{x_{n-1}^j, y_n^j\}$; $1 \le j \le m; n = \text{even}$
$N(y_i^j) \cap D = \{x_i^j, y_{i-1}^j, y_{i+1}^j\}$; $3 \le i \le n; 1 \le j \le m; i = \text{odd}$
$N(y_n^j) \cap D = \{x_n^j, y_{n-1}^j\}$; $1 \le j \le m; n = \text{odd}$

It can be seen intersection between the neighborhood $N(v)$ with $v \in V(G) - D(G)$ and the obtained dominator set D are uniques and it is not empty set. Thus, it can be concluded that $\gamma_{Li}(Amal(L_n, v, m)) \le nm$. D also complies the condition of locating independent dominating set. Hence, the lower bound and upper bound of locating independent domination number respectively, are $\gamma_{Li} \ge nm$ and $\gamma_{Li} \le nm$. It conludes the locating independent domination number of $Amal(L_n, v, m)$ is $\gamma_{Li}(Amal(L_n, v, m)) = nm$. □

3. Concluding Remarks

In this paper, we have determined the exact values of locating independent dominating number of some graph operations, namely ladder graph S_n, $Amal(S_n, v, m)$, path graph P_n, $Amal(P_n, v, m)$, wheel graph W_n, $Amal(W_n, v, m)$, ladder graph L_n, $Amal(L_n, v, m)$. As we have mentioned in introduction, to prove weather an locating independent dominating number is a hard problem. Thus, it still gives the following open problem.

Open Problem 3.1. *Let G be any connected graph, determine sharper lower bounds of $\gamma_{Li}(G)$ in term of the degrees of the graph?*

3.1. Acknowledgments

We gratefully acknowledge the support from CGANT University of Jember of year 2017.

References

[1] Gross J L, Yellen J and Zhang P 2014 *Handbook of graph Theory* Second Edition CRC Press Taylor and Francis Group
[2] Chartrand G and Lesniak L 2000 *Graphs and digraphs 3rd* ed (London: Chapman and Hall)
[3] P.J. Slater 1987 Dominating and location in acyclic graphs, Networks **17** 55-64
[4] P.J. Slater 1988 Dominating and reference sets in graphs, J. Math. Phys. Sci. **22** 445-455
[5] C.J. Colbourn, P.J. Slater, L.K. Stewart 1987 Locating-dominating sets in series-parallel networks, Congr. Numer. **56** 135-162
[6] A. Finbow, B.L. Hartnell 1988 On locating dominating sets and well-covered graphs, Congr. Numer. **65** 191-200
[7] Foucaud, F., Henning, M. A., Lwenstein, C. and Sasse, T 2016 Locatingdominating sets in twin-free graphs, Discrete Applied Mathematics. **200** 52-58.
[8] G. A. Waspodo, Slamin, Dafik, I. H. Agustin 2016 Bound of Distance Domination Number of Graph and Edge Comb Product Graph, IOP Conf. Series: Journal of Physics: Conf. Series **855** 012-014.
[9] K. Carlson 2016 Generalized books and C_m - snakes are Prime Graphs, Ars.Combin. **80** 215-221

IOP Conf. Series: Journal of Physics: Conf. Series **943** (2017) 012028 doi:10.1088/1742-6596/943/1/012028

Numbered head together with scientific approach in geometry learning

Dwi Indarti, Mardiyana and Ikrar Pramudya

Universitas Sebelas Maret, Jl. Ir.Sutami No. 36A, Surakarta 57126, Indonesia

E-mail: dwitantowi74@gmail.com

Abstract. The aim of this research was to find out the influence of learning model implementation toward student's achievement in mathematics. This research was using quasi-experimental research. The population of the research was all of 7th grade students in Karanganyar. Sample was taken using stratified cluster random sampling technique. The data collection has been conducted based on students' mathematics achievement test. The results from the data analysis showed that the learning mathematics by using Numbered Head Together (NHT) learning model with scientific approach improved student's achievement in mathematics rather than direct learning model particularly in learning object of quadrilateral. Implementation of NHT learning model with scientific approach could be used by the teachers in teaching and learning, particularly in learning object of quadrilateral.

1. Introduction

Mathematics is a very important since it is a basic science in human life, but ironically the quality of mathematics competence in Indonesia is still quite low [1-3]. The rank of Indonesian students' mathematics test score in PISA 2015 is 64th based on 72 countries. Low achievement of students' mathematics achievement is also showed by the results of average score of national exam in mathematics based on academic year of 2016 [4]. The average score of students' junior high school national exam in mathematics is 50.24, which is the lowest score compared to other subjects.

One of the possible reasons of Indonesian students' low achievement in mathematics is inappropriate learning model due to the mathematics learning process [5]. This fact is inline with conducted interview result to some teachers and students in Karanganyar who state that the learning of mathematics is still using direct learning model. Direct instruction is held by a teacher centered model under five steps namely: establishing set, explanation and/or demonstration, guided practice, feedback, and extended practice [6]. In this model, students have obly a little chance to join and being active in the learning process, so that they are not capable in develop their social skills and their interpersonal ability. Teacher's high role in learning process will influence badly toward student's problem solving skill, independence, and curiousity. Quadrilateral is one of the challenging topics in mathematics, according to student's statement. The challenges in quadrilateral topics are divided in understanding the problem, determining the correct strategy, and the reasoning ability due to its solution. According to Hwang, in traditional education environments, geometry is most commonly taught using text, 2D image and mathematical formulas. For area topics, such teaching methods are not highly effective [7].

One of the learning models that can be used in learning mathematics is a cooperative learning model. Cooperative learning is an approach with an access of groupwork to minimize the occurrence of those unpleasant situations and yet maximizes the learning experience through satisfactied result from working on a high-performance team [8].

NHT is one of cooperative learning types. NHT provides an opportunity for students to share their ideas and determine the right answers [9]. The use of NHT could improve student's activity, social skills, and cooperative skills while reducing disruptive behavior [10]. Learning activities through a scientific approach consists of these following steps namely: observing, questioning, experimenting, associating, and communicating [11]. In this study, the NHT was modified by a scientific approach. It is essential to modify the type of NHT with scientific approach since NHT only provides students with a booklet, including problems and incomplete sentences to fill up. Whereas, scientific approach gives authentic and challenging problems to rise student's curiosity and encourage them to work together to find the solutions. Learning object of quadrilateral is one of learning materials in mathematics that can be presented in authentic and applied problems in daily life. Implementation of NHT with scientific approach is expected to facilitate student's knowledge effectively on the learning object of quadrilateral because these two combination of learning method could give students the opportunity to solve the real and contextual problems scientifically. The teachers are no longer dominate the class, but they only direct and motivate the students to learn independently and develop a sense of responsibility. Besides, teacher could also help to trigger student's activity to solve real problems in their daily life based on groups.

Implementation of NHT with scientific approach is expected to improve student's achievement in mathematics, especially on the learning object of quadrilateral. The previous research shows that NHT could improve student's learning outcomes [12]. The purpose of this research is to know the effectiveness of NHT learning model with scientific approach to improve student's achievement in mathematics. In this study, each group of student in the first class is given a worksheet of quadrilateral topic. Other classes will use direct learning model.

2. Method

This research is a quasi-experimental study. Besides, the design of this study is random static group comparison design [13]. A group of subjects is taken from the population and they are divided into experimental and control groups. Experimental group is using NHT learning model with scientific approach in mathematics learning. In the other hand, control group is using direct learning model in mathematics learning. Then, both groups are given the achievement test. The study design is illustrated in Table 1.

Tabel 1. The randomization static group comparison design

Group	Treatment	Achievement Test
Experiment Group	X	T
Control Group	-	T

The research is conducted in Karanganyar, Central Java, Indonesia using 7th grade students as the population. Sample is taken by using stratified cluster random sampling. Schools in Karanganyar are classified into three categories, which are high level, moderate level, and low level bsed on mathematics score in national exam. Then, one school is picked to represent each category. Two classes are taken randomly from each selected school. One class is assigned as experimental class and the other class is assigned as control class. The experimental group consists of 96 students and the control group consists of 96 students. Both are in the same condition.

In this research, there are one independent variable and one dependent variable. Learning model is the independent variable and student's achievement in mathematics is the dependent variable. Instrument of this research is a mathematics achievement test. The test is used to collect the data about the student's achievement in mathematics and it is given right after the treatment. Content validity of the student's achievement in mathematics test is conducted by 3 experts in mathematics. To be able to measure the reliability of the student's achievement in mathematics test, Kuder-Richardson technique is used in this study. Data analysis techniques used in this research is a one-way analysis of variance with same cells. The level of significance is 5% or it can be said as $\propto = 0.05$. Based on the hypothesis, H0 is rejected if F > Ftable [14].

3. Result and discussion

NHT with scientific approach in learning quadrilateral was being conducted in 8 sessions of mathematics class. At the 9th mathematics class session, students are given an achievement test. One session of mathematics class is held along 80 minutes. This test contains 25 items of multiple choice questions. The results of the student's achievement in mathematics are presented in Table 2.

Table 2. The data of students' mathematics achievement

Group	N	Mean	Std.Dev	Variance	Max Score	Min Score
Experimental	96	47.58	15.93	253.80	88	12
Control	96	40.85	15.28	233.58	72	8

The results from Table 2 shows that the experimental group got 47.58 of mean score and 15.93 of standard deviation bsed on achievement test in mathematic in quadrilateral topic. The control group obtained a mean score of 40.85 and the standard deviation of 15.28 from the mathematical achievement test on quadrilateral material. Table 2 shows that student's mean score of experimental group in mathematics achievement is higher than student's mean score of control group. However, hypothesis testing can be performed if those two data are normally distributed and have the same variance. Table 3 shows the normality of mathematical achievement test results for each group by using Liliefors method with 5% of significant level.

Tabel 3. The result of normally test of mathematics achievement

Group	L_{obs}	L_{table}	Decision Test	Conclusion
Experimental	0.089	0.090	H_0 is not rejected	Normal
Control	0.083	0.090	H_0 is not rejected	Normal

Based on Table 3, it can be seen that experimental group has $L_{obs} = 0.089$ less than $L_{table} = 0.090$ and control group has $L_{obs} = 0.08$ less than $L_{table} = 0.09$. It means that H_0 is not rejected in each group. So, it can be concluded that the experimental group and the control group are normally distributed. Homogeneity test is conducted to find out whether the student's achievement in mathematics between two groups has the same variance or not. Table 4 shows homogeneity test results between two groups with a significant level of 5% by using Bartlett method.

Tabel 4. The result of homogeneity test of mathematics achievement

Test	k	$\chi^2{}_{obs}$	$\chi^2{}_{table}$	Decision Test	Conclusion
Mathematics Achievement	2	0.16	3.841	H_0 is not rejected	Homogen

Based on table 4, it can be seen that $\chi^2{}_{obs} = 0.16$ is less than$\chi^2{}_{table} = 3.841$. This means H_0 is not rejected. So, it can be concluded that the experimental group and the control group have the same variance. The summary of the results of variance analysis is presented in Table 5.

Tabel 5. Summary of one-way ANOVA with same cell

Source	S.S	D.F	M.S.S	F_{obs}	$F_{0.05;1,190}$	Decision Test
Learning Model	2173.52	1	2173.52	8.92	3.84	H_{0A} rejected
Error	46301.29	190	243.69	-	-	-
Total	48474.81	191	-	-	-	-

Based on table 5, it can be seen that F_{obs} is less than F_{table}. So, H_0 is rejected. It means that NHT learning model with scientific approach gives different influence on mathematics achievement. The gap of mean score between those two groups are 6.73. To be able to know which one of learning model that gives better influence for student's achievement in mathematics can be seen from the mean score. Based on table 2, the mean score of experimental group is 47.58 and the mean score of control group is 40.85. So, it can be concluded that NHT learning model with scientific approach can improve student's achievement in mathematics rather than direct learning model.

The use of the NHT model with a scientific approach is more effective than the direct learning model. According to Hunter [15], NHT requires all students to respond actively in class and by sharing responses before responding, all students gain access to credible information. This is due in part to the use of mixed-ability teams [16]. The implementation of NHT can increase active student engagement, enhance interpersonal relationship, and improve student learning [17].

4. Conclusion

Based on the results of data analysis and discussion, it can be concluded that the implementation of NHT learning model with scientific approach will effectively improve student's achievement in mathematics rather than direct learning model, especially in teaching and learning about quadrilateral topic. Hence, NHT learning model with scientific approach can be used as an alternative learning model by teacher in the process of learning mathematics toward students. For further research, it can examine the effect of NHT learning model with scientific approach on other materials and can modify NHT learning model with other learning approaches.

5. Acknowledgments

The author thank Dr. Mardiyana, M. Si. and Dr. Ikrar Pramudya, M. Si. in Universitas Sebelas Maret for advice in writing of this journal.

6. References

[1] Rafiepour A 2017 The gap between theory and action trough a model for classroom inquiry *International Journal on Emerging Mathematics Education* **1** 185

[2] Syamsuri, Purwanto, Subanji, and Irawati S 2017 Using APOS theory framework: Why did students unable to construct a formal proof? *International Journal on Emerging Mathematics Education* **1** 135

[3] Aziz T A, Pramudiani P, and Purnomo Y W 2017 How do college students solve logarithm questions? *International Journal on Emerging Mathematics Education* **1** 25

[4] BSNP 2016 *Laporan Hasil Ujian Nasional Tahun Pelajaran 2015/2016* (Jakarta: Kemdikbud)

[5] Prahmana R C I, Kusumah Y S, and Darhim 2017 Didactic trajectory of research in mathematics education using research-based learning *J. Phys.: Conf. Ser.* **893** 012001

[6] Arends R I 2001 *Learning to Teach* (New York: Mc Graw Hill Companies)

[7] Hwang W, Su J and Dong J 2009 A Study of Multi-Representation of Geometry Problem Solving with Virtual manipulatives and Whiteboard System *Educational Technology & Society* **12** 229

[8] Felder M R and Brent R 2007 *Cooperative Learning Active Learning: Models from the Analytical Sciences* ed P A Mabrouk (Washington, DC: American Chemical Society)

[9] Lie A 2005 *Cooperative Learning* (Jakarta: Gramedia Widiasarana Indonesia)

IOP Conf. Series: Journal of Physics: Conf. Series **943** (2017) 012028 doi:10.1088/1742-6596/943/1/012028

[10] Haydon T, Maheady L & Hunter W 2010 Effects of Numbered Heads Together on The Daily Quiz Scores and on-Task Behavior of Students with Disabilities *Journal of Behavioral Education* **19** 222
[11] Hosnan M 2014 Pendekatan Saintifik dan Kontekstual dalam Pembelajaran Abad 21 (Bogor: Ghalia Indonesia)
[12] Nasrun 2016 The Use Cooperative Learning with Numbered Head Together Model Improve The Students' Mathematics Subject *IOSR Journal of Mathematics* **12** 113
[13] Budiyono 2003 *Metode Penelitian Pendidikan* (Surakarta: UNS Press)
[14] Budiyono 2016 *Statistika untuk Penelitian Edisi Ke-2* (Surakarta: UNS Press)
[15] Hunter W C 2015 *Education and Treatment of Children* **38** 355
[16] Sundayana R, Herman T, Dahlan J A and Prahmana R C I 2017 Using ASSURE learning design to develop students' mathematical communication ability *World Transactions on Engineering and Technology Education* **15** 245
[17] Kagan S and Kagan M 2009 *Kagan's Cooperative Learning* (San Clemente, CA: Kagan Publishing)

IOP Conf. Series: Journal of Physics: Conf. Series **943** (2017) 012029 doi:10.1088/1742-6596/943/1/012029

Photography activities for developing students' spatial orientation and spatial visualization

Aan Hendroanto[1], Frans van Galen[2], D van Eerde[2], R C I Prahmana[1], F Setyawan[1] and A Istiandaru[1]

[1]Universitas Ahmad Dahlan, Jl. Prof. Dr. Soepomo, S.H. Yogyakarta 55164, Indonesia
[2]Utrecht University, Heidelberglaan 8, 3584 CS Utrecht, the Netherlands

E-mail: aan.hendroanto@pmat.uad.ac.id

Abstract. Spatial orientation and spatial visualization are the foundation of students' spatial ability. They assist students' performance in learning mathematics, especially geometry. Considering its importance, the present study aims to design activities to help young learners developing their spatial orientation and spatial visualization ability. Photography activity was chosen as the context of the activity to guide and support the students. This is a design research study consisting of three phases: 1) preparation and designing 2) teaching experiment, and 3) retrospective analysis. The data is collected by tests and interview and qualitatively analyzed. We developed two photography activities to be tested. In the teaching experiments, 30 students of SD Laboratorium UNESA, Surabaya were involved. The results showed that the activities supported the development of students' spatial orientation and spatial visualization indicated by students' learning progresses, answers, and strategies when they solved the problems in the activities.

1. Introduction

Spatial ability is widely known to be positively related to student's performance in learning mathematics [1, 2, 3, 4]. Many also have confirmed this positive correlation between spatial ability and achievements in mathematics for both young learner and also higher level students [3, 5]. Considering its importance, we must pay more attention to it so that students can develop their spatial ability maximally.

According to Hendroanto et al., spatial ability is related to student's mental ability in manipulating, understanding, or interpreting information related to the use of space [6]. Spatial ability consists of two major aspects that determine student's spatial ability [7]. These two aspects are spatial orientation and spatial visualization. Spatial orientation is about how good student's to not get confused if the different perspective of certain objects presented [8]. Based on this definition, it suggests that spatial orientation do not need mental movements or imagery manipulations. On the other hand, spatial visualization is more about mental movements and manipulations of mental images [7, 9]. For instance, students create mental images of a cube and the rotate them or even construct its net in mind. If students have good score on these two aspects of spatial ability than it can be concluded that the students have a good spatial ability. Therefore, all efforts in developing student's spatial ability shall focus on the development of student's spatial orientation and spatial visualization.

Some studies have tried to develop student's spatial ability through creative activities and learning such as the study of Risma et al. and Revina et al. [10, 11]. The result showed signs of development of student's spatial ability. In the study, they used physical objects and asked students to observe them by looking and drawing its views from many different perspectives. Therefore, to support student's spatial ability, activities involving physical objects are better than having theoretical and passive activities. Furthermore, the use of many different perspectives of an object also will trigger student's development of spatial orientation and spatial visualization.

In the explanation above, developing spatial orientation and spatial visualization requires a good activity in which it involves the use of physical objects and different perspectives or views. One of the activities that have all these requirements is photography activity. Photography is an activity of capturing moments, pictures of objects, or sceneries by using tools as a camera [12]. This activity always involves physical objects to capture and perspective in order to get a good result. In addition, photography nowadays becomes the most popular activity in the world as the technological advances provide people with gadget like compact cameras, smartphones, or tablets. These gadgets have a camera as their main feature to take pictures. Almost every people take selfish every day and post them on their social media. Hence, students nowadays are really familiar with photography activities. This familiarity can be an advantage for designing a learning activity.

The current study aims to design learning activities involving spatial orientation and spatial visualization based on the context of photography activity to support the development of student's spatial ability. In addition, the result is also meant to contribute to the local instruction theory of developing student's spatial ability. The design is based on the theory of Realistic Mathematics Education (abbreviated as RME) which has five tenets. They are the use of context, the use of model, the use of students' contribution, students' interactivity, and intertwinement [13, 14, 15].

2. Method

In order to achieve the aim, this study uses design research as the research approach. There are three advantages of using design research to design learning activities. First, design research enables the researcher to develop both activities and instruction theory at the same time [16]. Second, design research has a direct link between theory and its application [17]. Third, design research gives the researcher a possibility to have a deep analysis of student thinking [18]. In the other words, design research makes researcher's mind and student's mind into one. Therefore, it enables the researcher to understand further students thinking, learning, and their development.

There are three phases of the study: preparation and designing, teaching experiment, and retrospective analysis [16, 18, 19]. A series of activities is developed in the preparation and designing phase. In addition, an initial Hypothetical Learning Trajectory is developed. In the teaching experiment, two cycles of implementation are conducted involving 30 second-grade elementary students of SD Laboratorium UNESA, Surabaya. The first cycle involves 6 students as a pilot experiment and the second involves 24 students and 1 homeroom teacher. The first and the second cycle were conducted in March and April 2014. Data collection uses three sources from recorded classroom observations, student's written works, unstructured interviews, and the result of students' pretest and posttest. These data are analyzed qualitatively to support one another. To ensure validity and reliability of data collection, all materials are consulted with 5 experts of RME.

3. Result and discussion

There are three phases in this study from designing process until teaching experiment phase. The discussion of each phases are discussed in detail below.

3.1 Preparation and designing

To begin the process of developing the activities, an initial Hypothetical learning trajectory (HLT) and initial tasks were designed as can be seen in Table 1. Afterwards, three activities were design based on the photography activity and the tenets of RME. The first lesson, playing with the camera part 1, aims to give students experiences of the spatial orientation. This is also a preliminary activity to do mental movements and creating mental images. The second activity is called playing with the camera part 2. Students were asked to investigate photos of object from many different sides to assist student's ability to create mental images. The last activity is a drawing activity based on the context of reporting on newly discovered temples. This activity is to develop spatial visualization ability and to help students moving from using physical objects into using its distant representations.

Table 1. Initial design of the activities

Task	Activity	Hypothesized Results
Spatial orientation task	Observing physical objects	- Students will develop initial concept of spatial orientation and visualization - Students will be familiar with 3D objects.
	Finding the position of given standard views of 3D objects	- Students will develop spatial orientation ability by investigating the position of standard views. - Students will be familiar with distant representations.
Spatial visualization task	Drawing standard views of 3D objects	- Students will develop spatial visualization by drawing the standard views from given object.

3.2 Teaching experiment

Two teaching experiments were conducted in this phase. First, involving 6 students in two small groups in order to test the initial HLT and the activities whether it works or not. The findings of the first cycle are published in the SEA DR Conference, Palembang in 2015. In first and second activity, students experienced and developed spatial orientation indicated by their developing strategies to solve the problems in the activity. Moreover, students were also able to differentiate correct photos from wrong photos in the second activity. Students also slowly worked without the use of camera to view the object. This shows a progress in their mental movements and images. In lesson 3, students' strategies to identify the view slowly change into work without building block indicating that they performed mental movements based on distant representation of the objects. However, the students failed to perform this ability in the posttest. Therefore, we conclude that more bridging activities are needed to help students moving from concrete objects to representations.

Based on the result in cycle 1, the activities were revised and redesign to further help students developing their spatial ability. The initial HLT was also revised to include students' progress in cycle 1. Afterwards, the activities were implemented in a mathematics class consisting of 24 second-grade students and 1 homeroom teacher. All the activities were done in small groups of 4 or 5 students. The result of cycle 2 is discussed in details below.

Activity 1: Playing with the camera part 1

There are three problems presented in this activity. First, students have to determine visible buildings viewed from many different perspectives by using a model of camera. Second, students describe in words the position of building in the photo. Third, students have to find positions where they could capture two, three, or all the buildings. During the discussion, students used both a model of camera and digital camera to help them looking the buildings. How the buildings set on the table can be seen in Figure 1. To investigate them, students moved around the table and used the camera model to see the building as if they took pictures of them from each position. On the other hand, some groups preferred to use the digital camera to capture the view of the buildings. But, most groups agreed that using the camera model was easier. To describe the position of each building in the second activity, students used their body orientation such as "on my right", "on my left", or "in the back since I cannot see". They did not use cardinal directions like "on the north", "west", or "east". At this level, students really need the camera to help them visualizing the images. Without the camera, they confused to differentiate the view from the bird eye angle from standard view.

Figure 1. Model of buildings and students' investigations

Activity 2: Playing with the camera part 2

In the second activity, students were given a problem of finding the original position where some certain photo being taken. There were 12 photographs to investigate, but among these 12 photos, there were 4 misfit photos. Therefore, students have to be careful in order to pick the correct photos and find their position. At the beginning, the teacher gave each group objects in the photo, camera models and a digital camera. However, during the investigation, digital camera was not used as it was for discussion session to confirm their answers. Students' strategy to find the position of the photos, mostly were moving around the objects while holding the photo. They looked for a match or similarity between what they saw on the photo and the view of the real object in front of them. At first, all students used the camera model to see the object. After several photos, they began to stop using the camera model and directly see the object from bird eye view. To distinguish the wrong photos, students firstly looked for their positions by moving around. If they did not find a similarity between the photos and the view, they would consider the photos wrong. On the other hand, some of them found the wrong photos by identifying the difference between the photos and the object from a certain perspective. One of the examples is in Figure 2 where students points a part of the object on the photo and wrote "*terbalik*" or in English it means "reversed".

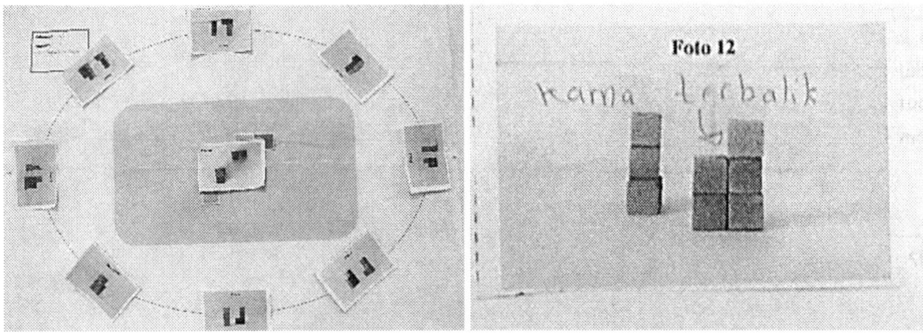

Figure 2. Fourth group's answer for the activity in lesson 2

Activity 3: Reporting new temples

There were two problems about drawing standard views in the third activity. The first problem is about drawing standard views of a temple constructed by students using exactly 8 cubes. In the second problem, the students were asked to determine and draw the standard views of an object in a photo as can be seen in Figure 3. In the investigation, some students used the camera model to guide their drawing while the others did not use it. Some students drew the standard views of their temple like the view from bird eye perspective. They confused to imagine the views of the object to draw. To help them, the teacher asked the students to use camera model and compare their drawing with the view from the camera. By doing this, they realized their mistakes and began to perceive standard view of the object from a correct angle or stand point. One of the examples for the first problem in activity 3 is in Figure 3.

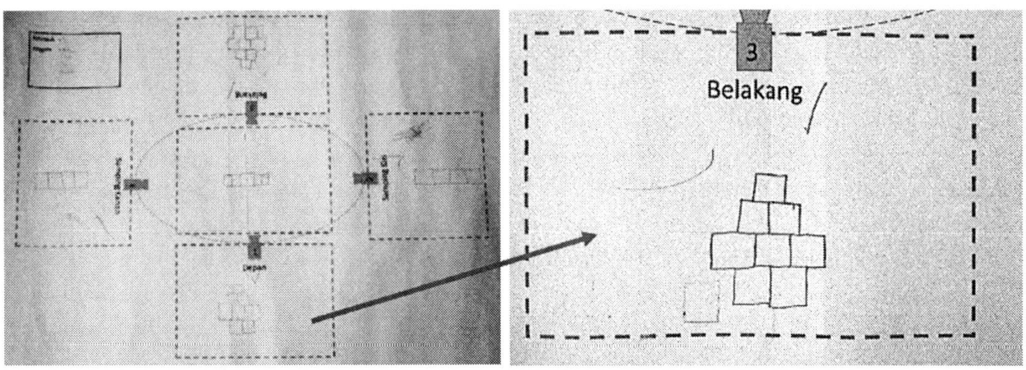

Figure 3. Fourth group's drawings for the first problem in lesson 3

To do drawing in the second problem, students built the object by using building blocks so that they can see directly from each position. However, some confused to count the number of the blocks needed to build the object because the underneath parts were not visible in the photo. Based on the observation, students drew the object's standard views by seeing the model they built from desired positions and then they compared the view to the view of the object in the photo. After getting the image of the standard views, they drew the views slowly from one cube to another until the views were complete.

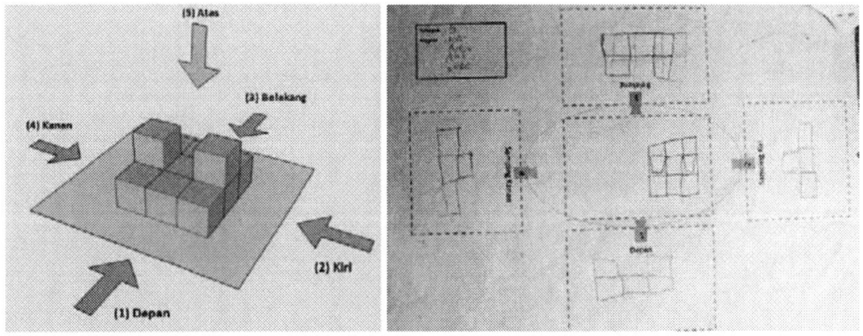

Figure 4. Fourth group's drawings for the first problem in lesson 3

3.3 Retrospective analysis

Based on the description of the teaching experiment, several points about student's spatial ability can be identified. Before describing student's development during the activities, we have to come back to what spatial ability is. As mentioned in the introduction, spatial ability is determined by two major aspects. They are spatial orientation and spatial visualization [7]. Thus, to determine student's development of their spatial ability, one must shows the progress of student's spatial orientation and spatial visualization. Therefore, in this retrospective analysis, the discussion is focused only on student's spatial orientation and spatial visualization.

In activity 1, students moved around the table and used the camera model to see the building as if they took pictures of them from each position and to describe position, they used terms like "on my right", "on my left", or "in the back since I cannot see". Students really need the camera to help them visualizing the images. Without the camera, they confused to differentiate the view from the bird eye angle from standard view. This activity helps students experience early concept of spatial orientation and spatial visualization. The camera model indeed guides them to get the view of the object. Without the camera, they lost to bird eye images. In the second activity, they develop further by beginning to work without the camera and they can directly identify the views to solve the problem. This indicates that students had been able to visualize the images just by looking from the bird eye perspective. Thus, students develop basic spatial visualization ability because they manipulate the images of the object in their mind and visualize the standard view. Misfit Photos also helps students to develop their spatial orientation as they identify the views of the objects from different perspectives in the activity. Some students demonstrate that they make relation they got from photos and the object which suggest a development of spatial orientation. According McGee, spatial orientation ability means the ability to understand and operate relationships of different information in space with respect to the observer's position [7, 8].

In the last activity, the model bridge students to get pass their problem of visualizing the views of an object in a representation or a photo. This means concrete objects became a shortcut for the students to imagine the object in the photo. By looking the concrete object, they got the presence of the objects, explain and reason their drawing just by looking the photos. Slowly, students develop mental movements by imagining their strategy of moving around in the first activity. This can be proved by comparing the result of their pretest and posttest.

IOP Conf. Series: Journal of Physics: Conf. Series **943** (2017) 012029 doi:10.1088/1742-6596/943/1/012029

Prior to the teaching experiments, students attended pretest to have an insight of their initial spatial ability. Related to the students' spatial orientation, they were asked to find the position where given photos were captured. The result showed that only 7 out of 24 answered correctly. These 7 students were known to be high achievers in the class. This result suggested that students have low spatial orientation ability because they got confuse when the perspective was changed [8]. On the other hand, regarding students' spatial visualization, they were asked to draw standard views of a given object in a photo. One of the most common answers of this problem is in Figure 5. The figure shows that students could not visualize the view of the objects which suggest they have poor ability to do mental manipulation of the objects.

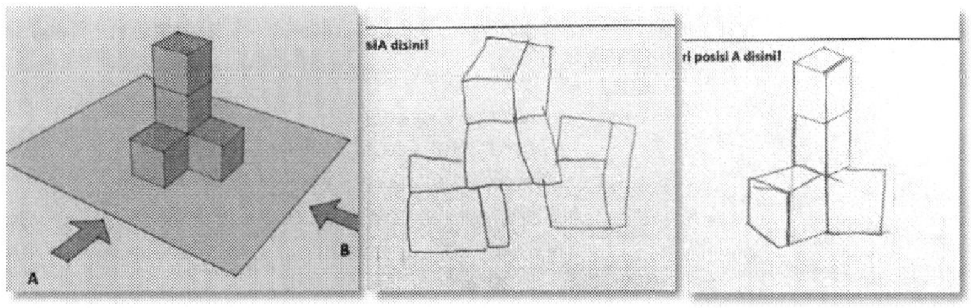

Figure 5. Students' answers for one of the problem in the pretest

The results in the pretest are very different from the posttest as there were 15 students answered correctly the first problem. Meanwhile, there were 19 students drawing correctly the second problem. We must emphasize that during the investigation in the third activity, students were able to draw standard views by using concrete model the built as guidance. On contrary, in the posttest, they were not provided with such a thing. This means students in the posttest shows an ability to determine the object's views without the use of model. This finding suggests that they performed mental movement by imagining the model in their mind. Table 2 illustrates the summary version of the findings of student's development in each activity.

Table 2. Designed activities and learning result

Task	Activity	Results
Spatial orientation task	Observing physical objects	- Preliminary experiences of spatial orientation and spatial visualization - Preliminary experiences of mental movement and creating mental images
	Finding the position of given standard views of 3D objects	- Supporting students spatial orientation - Preliminary support for spatial visualization - Preliminary experiences of mental movement and creating mental images
Spatial visualization task	Drawing standard views of 3D objects	- Supporting students spatial visualization - Supporting students' ability to do mental movement and create mental images

IOP Conf. Series: Journal of Physics: Conf. Series **943** (2017) 012029 doi:10.1088/1742-6596/943/1/012029

4. Conclusion

Based on the description of the result and the retrospective analysis, photography activities indeed support the development of students' spatial ability. During the investigation in the activities, students shows a progress of their developing ability in spatial orientation and spatial visualization. This progress can be looked in their performance during the activities of when they solved problems of investigating photographs.

For instance, observing objects from different positions in activity 1 and finding the positions of photos in activity 2 greatly contribute to develop student's spatial orientation ability. Moreover, the activities gave students preliminary experiences about mental images and mental movements. Drawing standard views in activity 3 develop student's spatial visualization ability. The activities helped them visualizing the views and working with the objects' representations.

References

[1] Newcombe N S 2010 Picture This: Increasing Math and Science Learning by Improving Spatial Thinking *American Educator* **34** 29

[2] Pittalis M and Christou C 2010 Types of reasoning in 3D geometry thinking and their relation with spatial ability *Educational Studies in mathematics* **75** 191

[3] Cheng Y L and Mix K S 2014 Spatial training improves children's mathematics ability *Journal of Cognition and Development* **15** 2

[4] Höffler T N 2005 Individual differences in spatial abilities in *The Cambridge handbook of visuospatial thinking* 121

[5] Guay R B and McDaniel E D 1977 The relationship between mathematics achievement and spatial abilities among elementary school children *Journal for Research in Mathematics Education* **8** 211

[6] Hendroanto A, Budayasa I K, Abadi A, van Galen F and van Eerde H A A 2015 Supporting Students' Spatial Ability in Understanding Three-Dimensional Representations *Proc. in the Third South East Asia Design/Development Research (SEA-DR) Conf.* (Palembang: Universitas Sriwijaya)

[7] McGee M 1979 Human spatial abilities: psychometric studies and environmental, genetic, hormonal, and neurological influences *Psychological Bulletin* **86** 889

[8] Tartre L 1990 Spatial orientation skill and mathematical problem solving *Journal for Research in Mathematics Education* **21** 216

[9] D Clements 2003 Geometry and spatial thinking in young children (New Jersey: Lawrence Erlbaum Associates)

[10] Risma D A, Putri R I I and Hartono Y 2013 On developing students' spatial visualization ability *International Education Studies* **6** 1

[11] Revina S, Zulkardi, Darmowijoyo and van Galen F 2011 Spatial visualization task to support students' spatial structuring in learning volume measurement *Indonesian Mathematical Society Journal on Mathematics Education* **2** 127

[12] Adams D and Kurniawan S 2014 A blind-friendly photography application for smartphones *ACM SIGACCESS Accessibility and Computing* **108** 12

[13] A Treffers 1887 Three dimensions: A model of goal and theory description in mathematics instruction-The Wiskobas Project (Dordrecht: Reidel)

[14] Zulkardi 2002 Developing a learning environment on realistic mathematics education for Indonesian student teachers (Twente: University of Twente)

[15] Gravemeijer K P E, Bowers J and Stephan M 2003 Chapter 4: A Hypothetical Learning Trajectory on Measurement and Flexible Arithmetic *Journal for Research in Mathematics Education. Monograph* **12** 51

[16] Gravemeijer K P E and Cobb P 2006 Design research from a learning design perspective *Educational design research* **1** 17

[17] van den Akker J, Gravemeijer K P E, McKenney S and Nieveen N 2006 Introducing educational design research *Educational design research* **1** 3

[18] van Eerde H A A 2013 Design research: Looking in to the heart of mathematics education *Proc. in the First South-East Asian Design Research (SEA-DR) Conf.* (Palembang: Universitas Sriwijaya)

[19] Cobb P, Confrey J, Lehrer R and Schauble L 2003 Design experiments in educational research *Educational researcher* **32** 9

Interictal Epileptiform Discharges (IEDs) classification in EEG data of epilepsy patients

J W Puspita[1], G Soemarno[1], A I Jaya[1] and E Soewono[2]

[1]Mathematics Study Program of Science Faculty, Tadulako University, Palu, Sulawesi Tengah, Indonesia.
[2]Department of Mathematics, Institut Teknologi Bandung, Jawa Barat, Indonesia.

juni.wpuspita@yahoo.com

Abstract. Interictal Epileptiform Dischargers (IEDs), which consists of spike waves and sharp waves, in human electroencephalogram (EEG) are characteristic signatures of epilepsy. Spike waves are characterized by a pointed peak with a duration of 20-70 ms, while sharp waves has a duration of 70-200 ms. The purpose of the study was to classify spike wave and sharp wave of EEG data of epilepsy patients using Backpropagation Neural Network. The proposed method consists of two main stages: feature extraction stage and classification stage. In the feature extraction stage, we use frequency, amplitude and statistical feature, such as mean, standard deviation, and median, of each wave. The frequency values of the IEDs are very sensitive to the selection of the wave baseline. The selected baseline must contain all data of rising and falling slopes of the IEDs. Thus, we have a feature that is able to represent the type of IEDs, appropriately. The results show that the proposed method achieves the best classification results with the recognition rate of 93.75 % for binary sigmoid activation function and learning rate of 0.1.

1. Introduction

Electroencephalography (EEG) is supporting tools for establishing epilepsy diagnosis when the clinical history indicates that the patient has recurrent seizures. Patients with epilepsy will show Interictal Epileptiform Discharges (IEDs) in the EEG recording [1]. Generally, IEDs consist of spike waves or complexes polyspike waves, and sharp waves. Spike waves has duration of 20-70 ms, while sharp waves has duration of 70-200 ms [2]. The two waves have amplitude bigger than $30\mu V$ [2]. However, this difference in duration is related to the size of the epileptic zone which would certainly have an impact on clinically significant differences in epilepsy patients [3].

Seizure detection and classification using EEG signal have been widely published [4-10], as well as IEDs. Among such studies, Valenti [11] presented automatic detection of interictal spikes using data mining models. Sharma [12] studied a novel scheme for the validation of an automated classification method for epileptic spikes by comparison with multiple observers. Automated Event Detection of Epileptic Spikes Using Neural Network is proposed by Khanwani [13], while Zacharaki [14] studied a machine learning approach for automated detection of epileptiform discharges in low dimensional embedding space. However, all methods only proposed the classification or detection of IEDs and non-IEDs. Not many studies have discussed the classification of IEDs types. Automatic identification of two types of IEDs, repeated sharp waves and runs of sharp-and-slow-waves, using support vector machine is proposed by Chang [15]. In 2017, Puspita [16] proposed Bayesian approach to classify spike waves and sharp waves based on the feature resulting from Walsh transformation. Moreover, a simpler method in classifying spike wave, sharp wave and Wicket spike, is also presented by Puspita [17] and Jaya [18]. In contrast to the previous study, in this study, we used frequency feature with correction baseline to improve classification accuracy. The selected baseline must contain all data of rising and falling slopes of the IEDs. We also compute the statistical feature, such as mean, standard deviation, and median, of each wave. Furthermore, Backpropagation Neural Network is applied to classify spike waves and sharp waves based on frequency and statistical feature.

The organization of this paper is organized as follows. Section 2 describes the methods for extraction feature and Backpropagation Neural Network as a classifier. The results and discussion are presented in section 3. Finally, the conclusion is offered in section 4.

2. Methods

In this study, we construct IED profile based on frequency and statistical feature for each wave. Furthermore, the Backpropagation Neural Network is used to classify the IED profile into one of two classes, namely spike wave class or sharp wave class.

2.1. Feature Extraction

A total of 109 EEG recordings used in this study were taken from Hasan Sadikin Bandung hospital. Furthermore, the EEG recordings in European data format (edf), that is localized in time interval containing IEDs based on the observation of an EEGer, are converted into ASCII (American Standard Code for Information Interchange) data using Polyman software.

In this study, we choose two baselines that must contain all data of rising and falling slopes of the IEDs, as illustrated in Figure 1. We compute the amplitude A_j, duration Δt_j and statistical feature, such as mean R_j, standard deviation S_j and median M_j, where $j=1,2$, based on each baseline of the IEDs, by using the following formula:

$$A_1 = v_n;\quad A_2 = v_n + v_m;\quad \Delta t_1 = t_n - t_1;\quad \Delta t_2 = t_m - t_n;\quad R_1 = \sum_{h=1}^{n} \frac{v_h}{n};\quad R_2 = \sum_{h=n}^{m} \frac{v_h}{m-n};\quad S_1 = \sqrt{\sum_{h=1}^{n} \frac{(v_h - R_1)^2}{n-1}};$$

$$S_2 = \sqrt{\sum_{h=n}^{m} \frac{(v_h - R_2)^2}{m-n-1}};\quad M_1 = \frac{1}{2}\left(v_{\frac{n}{2}} + v_{\frac{n}{2}+1}\right)$$ and $$M_2 = \frac{1}{2}\left(v_{\frac{m-n}{2}} + v_{\frac{m-n}{2}+1}\right)$$ if n and $m-n$ are even numbers, while $M_1 = v_{\frac{n+1}{2}}$ and $M_2 = v_{\frac{m-n+1}{2}}$ if n and $m-n$ are odd numbers, respectively.

Furthermore, frequency of the waveforms can be extracted by using the following formula [17]:

$$f = \frac{\text{the number of data in 1 second}}{\text{time duration of 1 waveform}} = \frac{500}{\Delta t}$$

with $\Delta t = \Delta t_1 + \Delta t_2$. The features are represented as a 9×1 vector in the following form:

$$x_i = \begin{pmatrix} f_i \\ A_{1_i} \\ A_{2_i} \\ R_{1_i} \\ R_{2_i} \\ S_{1_i} \\ S_{2_i} \\ M_{1_i} \\ M_{2_i} \end{pmatrix}, i = 1, \ldots, k,$$

where k is the number of data. Before the Backpropagation Neural Network is applied to distinguish spike waves and sharp waves, we re-scale the profile data in scale of [0,1] to improve the accuracy by using the following formula:

$$\hat{X} = \frac{X - \min(X)}{\max(X) - \min(X)}$$

where $X = (x_1, x_2, \ldots, x_k)^T$ [11].

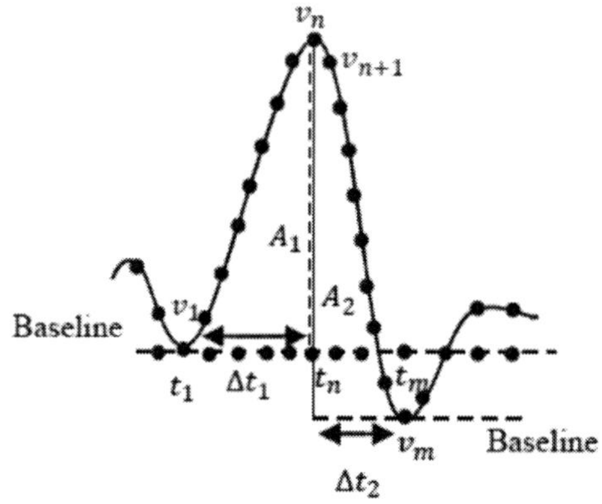

Figure 1. Simulation of feature extraction.

2.2. Backpropagation Neural Network as a Classifier

The algorithm of the Backpropagation Neural Network (BNN) is discussed in detail in [17]. In this study, the architecture of BNN consists of 1 input layer with nine neurons, 1 hidden layer with 2 and 3 neurons, and 1 output layer with 1 neuron. The IEDs data, that has been extracted in the previous subsection, are divided into training and testing data. Furthermore, spike wave is represented as number 0, while sharp wave is represented as number 1.

Let $\hat{x}^*$ be the testing data. $\hat{x}^*$ is classified into spike wave or sharp wave by following classification rules:

- if the output value ≤ 0.5 then $\hat{x}^*$ is spike wave, and
- if the output value > 0.5 then $\hat{x}^*$ is sharp wave.

3. Results and Discussion

In this study, we use 60 data of Spike waves and Sharp waves, respectively, as training data, and 80 data as testing data consisting of 35 data of spike waves and 45 data of sharp waves. Then the classification results using BNN are compared by the EEGer's expertise.

Table 1. Classification Results for Binary Sigmoid activation function with 2 neurons in the hidden layer.

No.	Learning Rate (α)	MSE	Correct classification of Testing Data		Recognition Rate (%)
			Spike	Sharp	
1.	0.05	0.0141	33	34	83.75
2.	0.1	0.0072	33	42	93.75
3.	0.3	0.0089	34	39	91.25
4.	0.5	0.0089	34	39	91.25
5.	0.7	0.0118	34	38	90
6.	0.9	0.0145	32	36	85

Table 2. Classification Results for Binary Sigmoid activation function with 3 neurons in the hidden layer.

No.	Learning Rate (α)	MSE	Correct classification of Testing Data		Recognition Rate (%)
			Spike	Sharp	
1.	0.05	0.0075	32	41	91.25
2.	0.1	0.0064	32	41	91.25
3.	0.3	0.0043	31	36	83.75
4.	0.5	0.0048	31	34	81.25
5.	0.7	0.0042	34	37	88.75
6.	0.9	0.0049	32	34	82.5

Table 1 and table 2 show the classification results using Binary sigmoid activation function in hidden and output layers for some variations of learning rate. The best classification results is achieved by the BNN model using 2 neurons in the hidden layer with a learning rate of 0.1. The recognition rate of this result, i.e. 93.75%, is better than the previous result in [18] which only achieved 88%. Figure 2(a) and 2(b) show the testing results for learning rate of 0.1 using the optimum weight values that minimize the network performance function, namely Mean Square Error (MSE), in the training process.

In this study, we also use others activation function, such as Hyperbolic tangent and Bipolar sigmoid, but the operation in the training process produces undefined numerical results. Moreover, the MSE values for every epoch tend to infinity and sometimes unstable, as shown in figure 3(a) and 3(b).

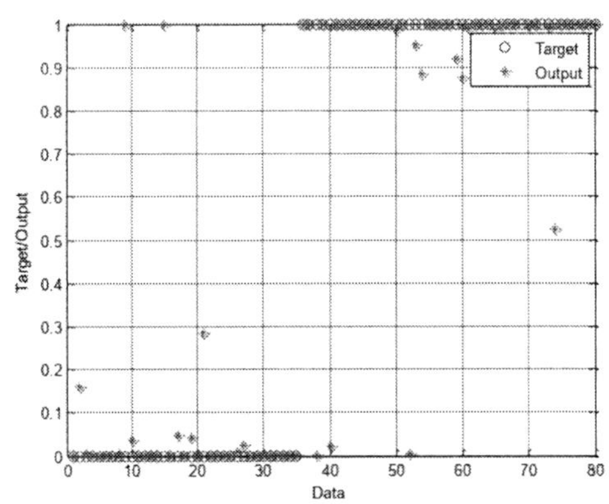

(a)

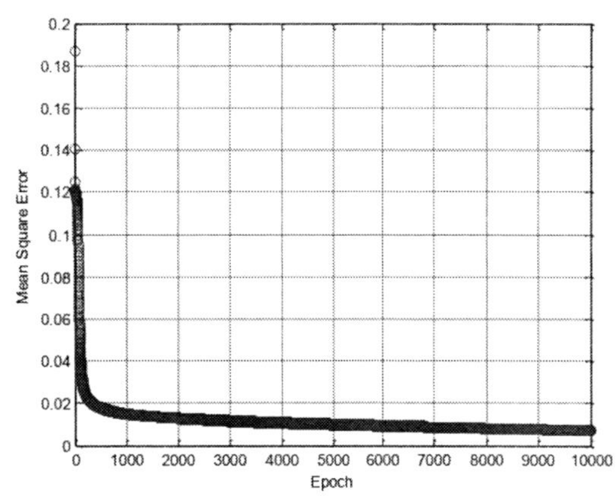

(b)

Figure 2. (a) Testing results for binary sigmoid activation function with $\alpha = 0.1$, (b) The Mean Square Error every epoch.

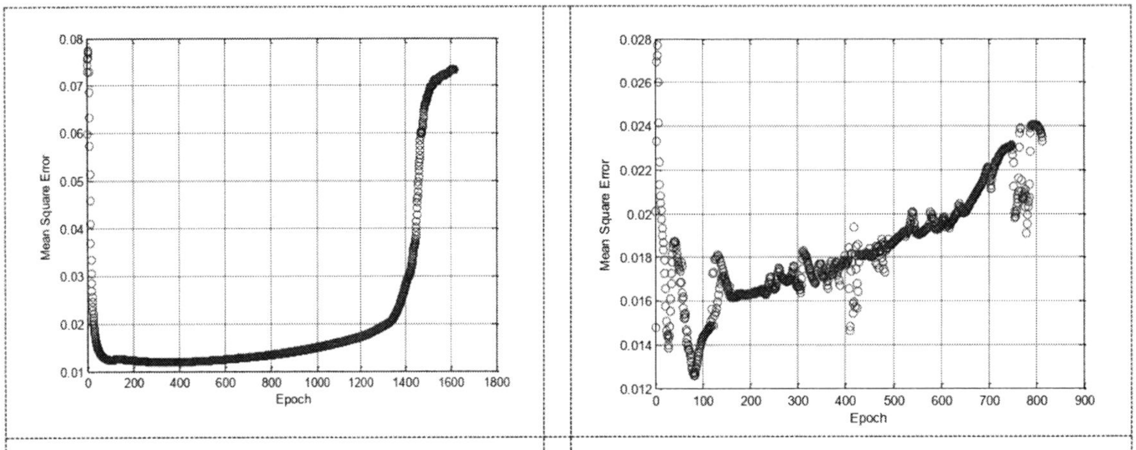

Figure 3. The Mean Square Error every epoch: (a) using bipolar sigmoid activation function with $\alpha = 0.5$, (b) using hyperbolic tangent activation function with $\alpha = 0.5$.

4. Conclusion

In this study, we proposed a simpler model to classify the type of IEDs using Backpropagation Neural Network (BNN) based on frequency and statistical feature, such as mean, standard deviation, and median. The selection of two different baselines for the raising and falling slope of the IEDs affects the frequency feature. The selection features may improve the accuracy. The results show that the hidden layer with 2 neurons for learning rate of 0.1 and binary sigmoid activation function gives the best classification result, where the recognition rate is 93.75%. According to the recognition rate, this result is better than the result of the previous study [18].

Acknowledgments

The authors would like to acknowledge to Pusat Riset dan Kreativitas Matematika (PRKM) Tadulako University and Pusat Pemodelan Matematika dan Simulasi (P2MS) ITB for help in preparing this study.

References

[1] Selvitelli M F, Walker L M, Schomer D L and Chang B S 2010 The relationship of interictal epileptiform discharges to clinical epilepsy severity: A study of routine EEGs and review of the literature *J. Clin. Neurophysiol.* **27** 87-92

[2] Fisch B K 1999 EEG Primer. Basic Principles of Digital and Analog EEG, *Elsevier.*

[3] Jaseja H, Jaseja B 2012 EEG spike versus EEG sharp wave: differential clinical significance in epilepsy *Epilepsy Behav.* **25** 137

[4] Gajic D, Djurovic Z, Di Gennaro S and Gustafsson F 2014 Classification of EEG Signals for Detection of epileptic Seizures Based on Wavelets and Statistical Pattern Recognition *Biomedical Engineering: Applications, Basis and Communications* **26**

[5] Kumar Y, Dewal M and Anand R 2014 Epileptic Seizure Detection Using DWT based Fuzzy Approximate Entropy and Support Vector Machine *Neurocomputing* **133** 271-79

[6] Bandarabi M, Teixeira C A, Rasekhi J and Dourado A 2015 Epileptic Seizure prediction Using Relative Power Features *Clin. Neurophysiol* **126** 237-48

[7] Song J L, Hu W and Zhang R 2016 Automated Detection of Epileptic EEGs Using a novel Fusion Feature and Extreme Learning Machine *Neurocomputing* **175** 382-291

[8] Mursalin M D, Zhang Y, Chen Y and Chawla N V 2017 Automated epileptic seizure detection using improved correlation-based feature selection with random forest classifier, *Neurocomputing* **241** 204-14

[9] Wang L, Xue W, Li Y, Luo M, Huang J, Cui W and Huang C 2017 Automatic Epileptic Seizure Detection in EEG Signals using Multi-Domain Feature Extraction and Nonlinear Analysis *Entropy* **19** 222
[10] Wang Y, Li Z, Feng L, Zheng C and Zhang W 2017 Automatic Detection of Epilepsy and Seizure Using Multiclass Sparse Extreme Learning Machine Classification *Computational and Mathematical Methods in Medicine*
[11] Valenti P, Cazamajou E, Scarpettini M, Aizemberg A, Silva W and Kochen S 2006 Automatic detection of interictal spikes using data mining models *Journal of Neuroscience Methods* **150** 105-10
[12] Sharma N K, Pedreira C, Centeno M, Chaudhary U J, Wehner T, Franca L G S, Yadee T, Murta T, Leite M, Vos S B, Ourselin S, Diehl B and Lemieux L 2017 A novel scheme for the validation of an automated classification method for epileptic spikes by comparison with multiple observers *Clinical Neurophysiology* **128** 1246-54
[13] Khanwani P, Sridhar S and Vijaylakshmi K 2010 Automated Event Detection of Epileptic Spikes Using Neural Network, *Int. J. Comput. Appl.* **2** 14-7
[14] Zacharaki E I, Mporas I, Garganis K and Megalooikonomou V 2016 Spike pattern recognition by supervised classification in low dimensional embedding space *Brain Informatics* **3** 73-83
[15] Chang W D, Cha H S, Lee C, Kang H C and Im C H 2016 Automatic Identification of Interictal Epileptiform Discarges in Secondary Generalized Epilepsy *Computational and Mathematical Methods in Medicine*
[16] Puspita J W, Gunadharma S, Indratno S W and Soewono E 2017 Bayesian approach to identify spike and sharp waves in EEG data of epilepsy patients *Biomedical Signal Processing and Control* **35** 63-9
[17] Puspita J W, Jaya A I and Gunadharma S 2017 Classification of Epileptiform and Wicket Spike of EEG Pattern Using Backpropagation Neural Network *AIP Conference Proceedings* **1825** 020018.
[18] Jaya A I, Puspita J W and Soewono E 2016 Identifikasi Gelombang Spike dan Sharp pada Data EEG Pasien Epilepsi Menggunakan Backpropagation Neural Network *Prosiding Konverensi Nasional Mathematika XVIII, IndoMS.*

The utilization of mind map painting on 3D shapes with curved faces

Ayuk Nur Sholikhah[1, 2], Budi Usodo[1] and Ikrar Pramudya[1]

[1]Universitas Sebelas Maret, Jl. Ir. Sutami No. 36 A Kentingan, Surakarta, Indonesia
[2]SMP Islam Diponegoro, Jl. Kapten Mulyadi No. 221D, Surakarta, Indonesia

E-mail: ayuknursholikhah@gmail.com

Abstract. This paper aims to study on the use of mind map painting media on material with 3D shapes with curved faces and its effect on student's interest. Observation and literature studies were applied as the research method with the sake design of utilization of mind map painting. The result of this research is the design of mind map painting media can improve students' ability to solve problems, improve the ability to think, and maximize brain power. In relation, mind map painting in learning activities is considered to improve student interest.

1. Introduction

Mathematics is a compulsory subject that is given to all the students from elementary schools to universities to equip students with the ability to think logically, mathematical, analytical, systematic, critical and creative, as well as the ability to work together. Besides mathematics also has an important role in the development of modern technologies and disciplines to develop the power of human thought.

TIMSS (Trends in International Mathematics and Science Study) showed that in 2011, Indonesia is ranked 41 out of 45 countries. As for the results of the PISA (program for International Student Assessment) in 2015, showed that Indonesia has increase in the ranks, from rank 63 to 69 for all the countries evaluated. Although there is an increase, but the score of Indonesia is still below the average set by PISA. With the improvement of our education rankings, we should be able to be encouraged to improve education issues that have been unresolved. A lot of breakthroughs, creativity and cooperation of various parties to improve the quality of education in Indonesia are still needed.

Within a week the number of subjects studied by students with limited time was considered too much. Students are required to study all subjects without having enough time to explore the material. It also applies in the subjects of Mathematics, which is incidentally an abstract subject, and requires a lot of time in depth. If students are not optimal in learning the material, it will impact on the result of learning Math.

In addition, the implementation of Mathematics learning becoming less attractive and tend to be boring is also caused by factors such as: from the teacher, such as: a) the role of the teacher is still dominant in the learning process, b) the lack of use of multimedia learning, c) the teacher is less able using multimedia, d) less than the maximum of classroom management. Factors of the students; of course there are several factors to consider as well, including: a) a low student interest, b) lack of attention to the students' learning materials were submitted by teachers, and c) lazy to finish the homework assigned by the teacher. Both of these statements can be concluded that learning is less optimal and low student interest due to factors of teachers and students.

3D shapes with curved faces is one of the materials in class 9 which is a matter of National Examination. The results of the national exam mathematics in Surakarta in 2015/2016 showed the material with curved faces is still below 60%. It is triggered by several factors contained in the study, which is the low interest of students, learning difficulties and less optimal learning process.

Because time is less than optimal and there is low student interest, one of the solutions that require media teachers can support students in optimizing their learning activities. This media is called mind map painting. Mind map painting is a development mind map, which is equal in helping students and teachers in the learning process in the classroom. Through mind map painting, all the key information and the importance of each of lessons can be organized using radians structure that corresponds to the natural mechanisms of the brain, making them easier to understand and remember. Learning media mind map painting that is the audio-visual is also able to evoke the emotions of students so as to increase student interest in learning the material.

Reviewed from the concept of expediency media that the concept of acquisition of one's experience through the media used, the more concrete the media used, the higher the value of the experience gained. According to the experience cone (Cone of Experience) Edgar Dale in educational media Sadiman [1], suggests that a person's learning experience: 7.5% is obtained with a sense view (eye); 13% sense of hearing; and the rest by other senses. Direct experience a bigger part in the analysis of Dale. With the experience high gained provision, the learning material to be absorbed by the pupils more meaningful, and eventually clarify or advance the objectives of a lesson.

2. Literature Review

The study uses literature study. Literature is a way of solving the problem stracing the ancient sources that have been made previously. The kind of literature used in this paper is the literature of primary and secondary literature. Primary literature are works of original writing which contains a study on the new theory, or explanation of an idea in various fields. While the secondary literature is literature that contains information about the primary literature. Secondary literature offers primary literature by way of summarizing index, so the secondary literature does not contain new knowledge, but merely repeating an existing organize knowledge.

Holland [2] said that established of mind mapping to be a valuable teacher with more effectively for helping someone plan, structure projects and assignments more. Testimony that a mind map is an invaluable tool for planning and organizing your thinking for any projects. Boyson [3] said a mind map is an excellent tool for collaborating with others to develop plans or implement key project. It allows you to harness the input of all members of group in a dynamic and creative way. When used for group brainstorming sessions, min mapping was seen to enhance critical thinking and co-operation as well as providing a solid basis for collaborative problem solving. The student enjoyed expressing their opinions in a participative and open climate Groups involved in the sessions [4] [5]. It is hoped that all description all of management job description can be conducted by executing program that model has been made appropriate.

Mind mapping, which uses a multi-sensory learning approach, can support a student's ability to explore associations amongst information because it is a free-form learning technique in which creative thinking is fostered [6]. For the learner, exploring these relationships engages a deep approach to learning rather than a superficial dive [7]. The interconnection created by the branches in the mind map also allows for dual coding [8] of information which supports association and links to be made.

Mind map must be explore and then inform others of diverse teaching and learning strategies that may support the academy's role to enlighten students not only with knowledge but the ability to act upon the knowledge as scholars of teaching and learning [9].

3. Result and discussion

Mind map painting can transform verbal techniques in image visualization in learning activities. Image is one way to help convey the message concretely making it easier for students to strengthen the understanding of the concept of mathematical learning material as a whole. Mind map painting is developed with different colors, pictures/paintings and stories in it. Jensen [10] argues that each color has a wavelength of color ranging from ultraviolet to infrared waves can affect the brain and body are different depending on the person's personality and state of mind. Therefore, the use of color will create a mind map painting more alive, visually stimulating, and adding energy to the creative thinking of students.

Mind map this painting was introduced in Indonesia by Erika [11]. This painting of Mind map is basically a mind mapping presented in the form of paintings. In contrast to the mind map in general the use of colors, images, or photos a simple, mind map painting is more expressive and varied according to the artist's viewpoint. Therefore, the mind map painting does not only improve the ability to think, but also enhances creativity. Erika and Abdul Gafur [11] states that the use of mind mapping painting through Madjoe Jaya Productions has enhanced the learning outcomes as much as 44.36% of the gain score. Here are the differences between mind map and mind map painting (Figure 1 and 2)

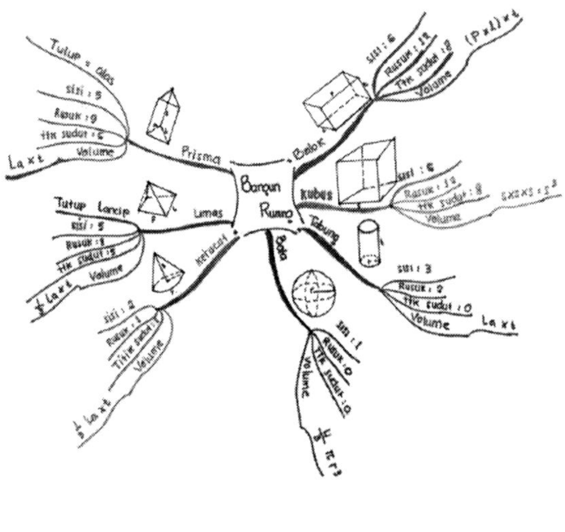

Figure 1. Mind map 3d shapes with curved faces

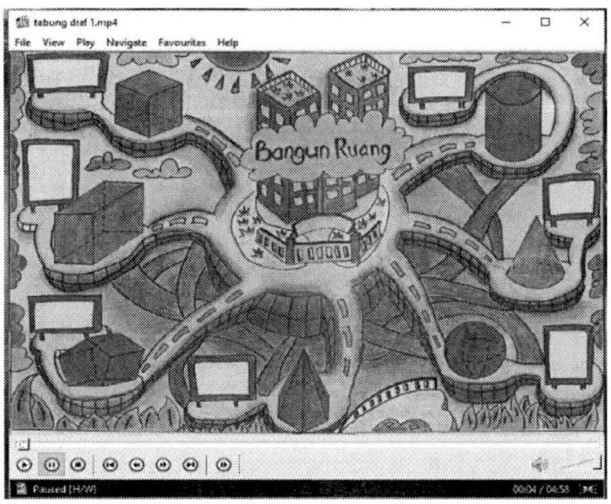

Figure 2. First view mind map painting on 3d shapes with curved faces

Video of mind map painting is divided into three sections corresponding to the basic competencies on the material of 3D shapes with curved faces in order to help students understand the material elements every waking space.

IOP Conf. Series: Journal of Physics: Conf. Series **943** (2017) 012031 doi:10.1088/1742-6596/943/1/012031

Figure 3. Views of cylinder

Figure 4. Views of cone

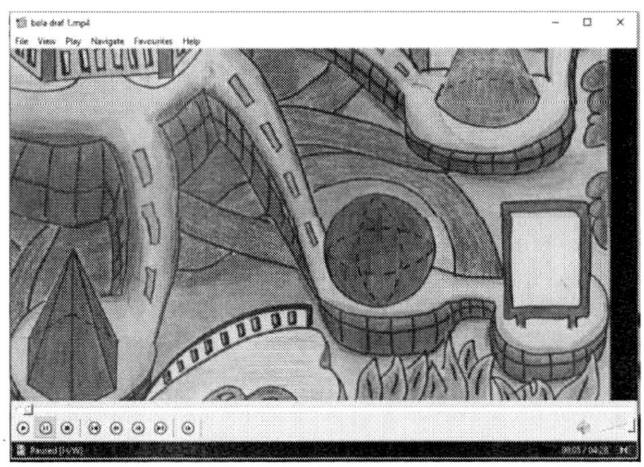

Figure 5. Views of sphere

The presence of mind map painting media in the form of video provides more free time for teachers in performing strategic roles in the classroom, as a facilitator and concept discovery process. Mind map painting media is made with Wacom Intuos tools and Camtasia Studio applications. Mind map painting media will be packed in a compact disc so that it can be used anytime and anywhere according to the needs of learners or students. Mind map painting video is divided into three sections corresponding to the basic competencies on the material 3D shapes with curved faces.

In a video of making this mind map painting, grating instrument in validating a video media refers to the theory [12-15], among others.

Table 1. Lattice instrument

No	Variable	sub-variables	Aspect
1	Instructional	Content	Up to date, clarity, usefulness, and a bias-free
		Design	Stating the purpose, presentation content, application for students, reflection for students, student interaction, integration with the learning environment, student motivation, student learning outcomes and reusability
2	Media	Technical production	design characteristics in accordance the formula visual quality, audio quality, the relationship between the audio and visual equipment, additional material
		Feature Specialty of Video	Paintings of mind map painting, a compelling story and elements and formula geometrical curved faces

4. Conclusion

Mind map mind is presented in the form of paintings and stories are packaged in an instructional video media. Mind map painting media is divided into three sections that are adjusted to the material basic competence. Media will be set forth in a compact disk (CD). The benefits that can be gained in the use of mind mapping painting on the material 3D shapes with curved faces: a) easy to use anytime and anywhere, b) students have to draw to increase students interest, c) to provide more time for teachers to freely undertake strategic roles in the classroom, d) the material can be completed on time, e) more optimal learning process, f) improve students' skills in problem solving, g) improve the ability to think, h) maximize the brain works, i) increased creativity

5. Acknowledgments

The author Ayuk N S is grateful to Allah and blessings on the Rasulullah SAW. Thank you to the parents for the support and the dua, as well as other material. Thanks also to my friends and team who have provided help and support in the process, as well as lecturers, especially Mr. Budi Usodo and Mr Ikrar Pramudya who have provided guidance during this final project.

Reference

[1] Sadiman, et al 2011 *Media Pembelajaran: Pengertian, Pengembangan, dan Pemanfaatannya* (Jakarta: King Grafindo)

[2] Holland B, Holland L and Davies J 2003/2004 An investigation into the concept of mind mapping and the use of mind mapping software to support and improve student academic performance. *Learning and Teaching Projects* 2003/2004, p 89

[3] Boyson G 2009 *The Use of Mind Mapping in Teaching and Learning* (The Learning Institute, Assignment) p 3

[4] Paykoc F, Mengi B, Kamay P O, Ozgur B, Pilli O and Yildirim H 2004 What are the Major Curriculum issues?: The Use of Mind Mapping as a Brainstorming Exercise. *Paper presented at the first International Conference on Concept Mapping* (Spain)

[5] Ismail Z and Rahman S N A 2017 Learning 2-dimensional and 3-dimensional geometry with Geogebra: Which would students do better? *International Journal on Emerging Mathematics Education* **1** 121

[6] Davis B, Sumara D and Luce-Kaper R 2000 *Engaging minds: Learning and teaching in a complex world* (Mahwah NJ: Lawrence Erlbaum Associates)

[7] Biggs J 1987 *Student approaches to learning and studying* (Hawthorn, Vic: Australian Council for Educational Research (ACER))

[8] Kullay R W, Lee J B and Caterino L C 1985 Conjoint retention of maps and related discourse. *Contemporary Educational Psychology* **10** 683

[9] Genevieve Zipp and Chatterine Maher 2013 Prevalence of Mind Mapping as the teaching and learning strategy in physical therapy curricula *Journal of the Scholarship of Teaching and Learning* **12** 21

[10] Jensen E and Karen M 2002 *Brain Million Gygabyte: Buku Pintar Membangun Ingatan Super* (Bandung: Kaifa)

[11] Oktarini, Erika and Abdul Gafur 2013 Formative Evaluation on MaJoe Djaya Instructional Video Produced by Eduartion *Jurnal Inovasi Teknologi Pendidikan* **1** 40-48

[12] Beaudin BP 1993 *Intructional Video Evaluation* (Colorado: Colorado State University, Fort Collins)

[13] Dick, W., Carey, L. & Carey, JO 2009 *The Systematic Design for Instruction (7thed)* (Boston: Pearson Education Inc)

[14] Smaldino S E, Lowther D L, & Russell J D 2008 *Instructional Technology and Media for Learning 9th ed* (New Jers)

[15] Gagne, R M, Briggs, L J, and Wager, WW 1992 *Principles of Instructional design (4thed)* (Orlando: Harcourt Brace Jovanovich Publishers)

IOP Conf. Series: Journal of Physics: Conf. Series **943** (2017) 012032 doi:10.1088/1742-6596/943/1/012032

Ethnomathematics: Exploration in Javanese culture

Irma Risdiyanti and Rully Charitas Indra Prahmana

Universitas Ahmad Dahlan, Jl.Soepomo 55164,Yogyakarta,Indonesia

E-mail: irma.risdiyanti28@gmail.com

Abstract. This research is the exploration study to indicate the correlation between mathematics and Javanese culture. These studies have a purpose of exploring Javanese culture in Yogyakarta that contains mathematics concept namely Batik. The activity of society in making Batik in all regency at Yogyakarta is the focus of this study. The research use ethnography method. The technic to collection data uses principles in ethnography such as observation, interview, documentation, and field note making with the original ethnography description. The result is exploration ethnomathematics in the several motifs of Yogyakarta batik that contains philosophy, deep cultural value, and mathematics concept, especially geometry transform subject.

1. Introduction

Developing and applying mathematical concepts based on daily-life problem situations is a part of student learning process [1]. The activity is also explained by Freudenthal [2] if mathematics is a human activity and must be related to real life. However, nowadays, mathematics is still viewed as something that far from the human life [3]. Karnilah [4] explains that the rational reason if society considers that mathematics is not related to culture, it starts with the behavior of students who not understand to use mathematics to solve the daily life's problem. So this makes society even less the benefit of mathematics. Prabawati [5] also explain that mathematics regarded as a perfect science, the truth is objective and far away from the reality of daily life. Similarly, explain by Suyitno [6] one form of misunderstanding of mathematics is the assumption that mathematics is an exact whose truth is absolute.

Mathematics always taught in school as a cultural subject that involves learning supposedly from universally accepted facts, concept, and content [7-9]. The learning of mathematics that far from daily life and detaches itself from the culture has an impact on the ability of the student in solving mathematical problems related to the real life [9]. The learning process evidenced in the results of PISA (Programme for International Student Assessment) that was analyzed by Stacey [10] has shown that Indonesian student was not able to use mathematical concepts to solve the real-life problems. They are difficult to solve the math problem in the form of the mathematical project because during this time students tend to be taught practical formulas that will be used to answers the exam questions.

Arisetyawan [11] explain that the one reason student not able to use mathematics concepts to solve the real-life problem is because the teacher in the school does not relate the cultural and daily activities in the learning process. So, learning mathematics, students need a bridge between mathematics and culture that is ethnomathematics [12]. It means that ethnomathematics is a cultural-influenced mathematical form. D'Ambrosio [13] explain that ethnomathematics has a purpose to recognize that to doing mathematics there are have different ways with considering academic knowledge developed by a different sector of society and by considering different modes in different cultures.

Indonesia is one country that has much cultural diversity from Sabang to Merauke, including on the Java Island, especially in Yogyakarta known as a cultural city. There are some cultural activities in Yogyakarta. They have implemented some mathematics concept namely the batik home industries activity. The activity is almost doing by society in all regency of Yogyakarta. Danoebroto [14] explain that in the art form of batik is produced the transformation of points, lines or data fields through translation, rotation, reflection or dilation.

Yogyakarta has a typical batik or known as batik *keraton* (palace) which has a philosophy and contain full deep of meaning. Ethnomathematics not only math but also explore the cultural values that existed inside. Researchers are interested in exploring how the exploration of ethnomathematics in Javanese culture, especially in the activities of batik and digging the cultural values in this activity. It expected that the result of this research could bridge math with culture and also cultural values can be used as learning and applied in daily life or classroom learning.

2. Method

Implementation of the research procedures doing from April until June 2017 done with three main steps that is an analysis of the data in pre-field, analysis of the data during of field and analysis of the data overall [12]. The research used qualitative descriptive as the research's type to explain and obtain information overall, extensively and deeply [15].

In the other than, this research uses the ethnographic approach as a research approach. Three research boundaries used to be the base of research which is consisting of the unity society who is just using single language dialect, the unity of society bundled with the boundaries of political administration and the unity of society who have same historical experience, so can get the data from the native source. In this research to selected the boundaries determine based on the boundary set of ethnographic research [12]. The technic of collection data uses principles in ethnography such as observation, interview, documentation, and field note making with the original ethnography description [15].

3. Results and discussion

The research output shows that culture inheritance such as batik still be interested and be kept by Javanese people. Didik [17] explain that Indonesia batik is the Indonesian kingdoms inheritance, while in Yogyakarta batik was heritage by the Mataram Kingdom that became the basic design motif of *keraton* Yogyakarta and *keraton* Surakarta batik. He also said if batik of Yogyakarta does not only have a beautiful motif, yet containing philosophy, full of deep meaning and there are prayer and hope inside.

Batik as one of the cultural heritage needs to be positioned strategically, mainly through education as a means of resistance so that local identity can incise and strengthen local character [18, 19]. In another fact, apparently in batik motif of Yogyakarta making, it is used transformation geometry concepts such as reflection, translation, rotation, and dilatation.

3.1. *The philosophy, cultural values and the application of geometry transformation in motif of Semen Rama batik*

Semen Rama batik is a batik that made in Paku Buwono IV government who had the throne in 1788-1820 M [17]. This motif gives a lesson to his son who has chosen as *Putra mahkota* (princes). In history, the motif of *Semen Rama* batik is taken from Prabu Wijaya's advice to Raden Gunawan Wibisono when he would replace king of Alengka Kingdom after Prabu Dasamuka had passed away. The advice of Prabu Wijaya is called *Hasta Brata* that has eight advice that must be done by the leader or king of a Kingdom. The eight messages are as follows *Indrabrata* symbolized by plants. This symbol has meaning, if it is a leader must to give the prosperity and save the earth. *Yamabrata* symbolized as a mountain or something high; this symbol has meaning if a leader be must give justice to other people. *Sasibrata* symbolized as a star, the meaning of this symbol that if be a leader must give brightness for they are in the darkness. *Suryabrata* symbolized as a *garuda*; the meaning is if be a

IOP Conf. Series: Journal of Physics: Conf. Series **943** (2017) 012032 doi:10.1088/1742-6596/943/1/012032

leader must have a strong heart and no halfway in taking a decision. *Bayubrata* symbolized as flying animal or bird; this is advising the goodness or high position that does not show off the authorization. *Barunabrata* symbolized by a dragon or something that have related to water; the is advising *wellas asih* or the attitude to easy, forgiving fault. *Agnibrata* symbolized by fire as a meaning of power to grab the *angkara murka* (natural appetite or desire) and the power to safeguard other people. *Danabrata* symbolized by heirloom; the meaning is if be a leader must giving reward or blessing to other people. Furthermore, to making motif of *Semen Rama* batik, students can apply geometry transformation seen in Figure 1.

Figure 1. The application of geometry transformation in motif of *Semen Rama* batik

Figure 1 describes the application of geometry transformation in the motif of *Semen Rama* batik, to making this motif can the applied reflection of several ornaments using y-axis in Cartesian coordinate system.

3.2. The philosophy, cultural values and the application of geometry transformation in motif of Kawung batik

The motif of *Kawung* batik has a form like four circles or ellipse that is similar to the coconut or *kolang-kaling* (the fruit of sugar palm) that arranged neatly and geometrically [20]. The four of circles represent the four directions of the wind that pivot on the force of the east as the source of life, north as the dwelling place of God, west as the source of luck and south as the center of everything. The circle's center that pivot the center of power is symbolized as king, in this case, the king as a center surrounded by his people as a center of science, art, culture, religion, government, and economy. The people must obey the king, but a king is also always giving protect to his people. Also, the motif of *Kawung* batik also interpreted as a symbol of the simplicity of a king who is always giving justice and prosperity to his people. Furthermore, to making motif of *Kawung* batik, students can apply geometry transformation that describes in Figure 2.

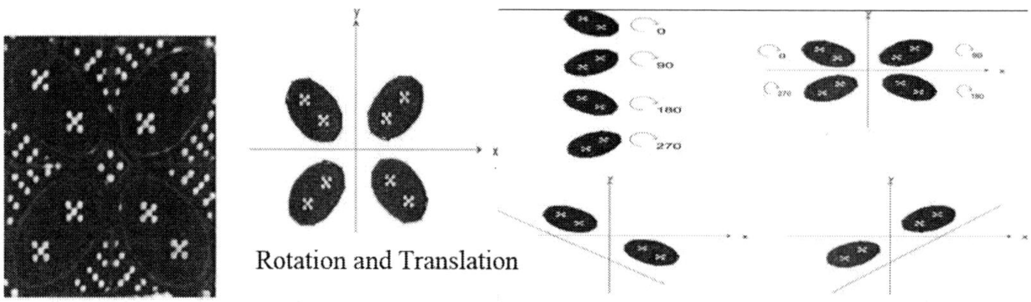

Figure 2. The application of geometry transformation in motif of *Kawung* batik

Figure 2 explains the application of geometry transformation in the motif of *Kawung*. To making this motif can applied reflection of an ornament by x and y-axis and also can using rotation with 0°, 90°, 180° and 270° toward x or y-axis or can also use the translation by the axis in the Cartesian coordinate.

3.3. The philosophy, cultural values and the application geometry transformation in motif of Sidoasih batik

The word of *Sidoasih* came from the word of *sido* that have to mean making something come true and the word of *Asih* that have mean affection [20]. The motif of *Sidoasih* batik describes if to build a family should build with full affection. So, the gotten value from the motif of *Sidoasih* batik is when through the family life it should use affection as the base to the built family. Furthermore, to making motif of *Sidoasih* batik, students can apply geometry transformations that explain in Figure 3.

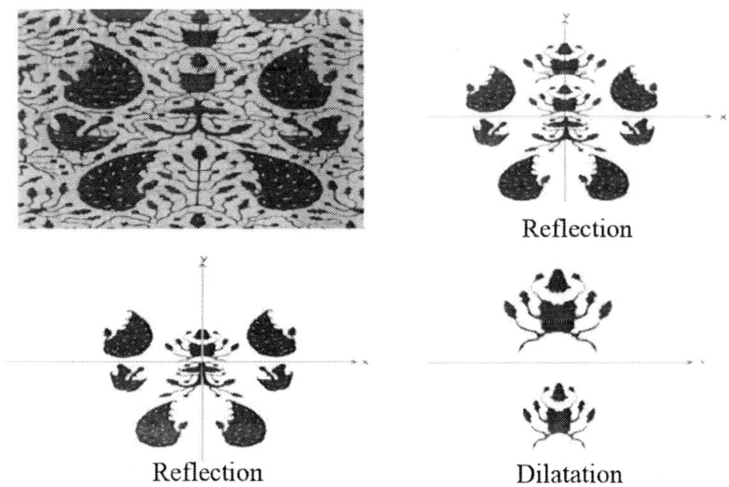

Figure 3. The application of geometry transformation in motif *Sido Asih* batik

Figure 3 describes the application of geometry transformation in the motif of *Sidoasih* batik, to making this motif can apply reflection of several ornaments in the motif using y-axis in Cartesian coordinate system, and also can to applied dilatation of an ornament, so can get the motif that has the same form but have different size.

3.4. The philosophy, cultural values and the application of geometry transformation in motif of Wahyu Temurun batik

Didik [17] explain that the word of *Wahyu Temurun* is mean blessing that gave to people, on this motif, there are have several elements, one of them are semen's motif that have characteristic as plants form. The element in this motif represents blessing that symbolized by *lung-lungan* (the branch of a vine), the word of *lung* in the Javanese language have mean *tetulung* (helping other people). So, this motif has meaning if it wants to get the blessing must help many people and also pray for hopefully the blessing of God. Furthermore, to making motif of *Wahyu Temurun* batik, students can apply geometry transformation that seen in Figure 4.

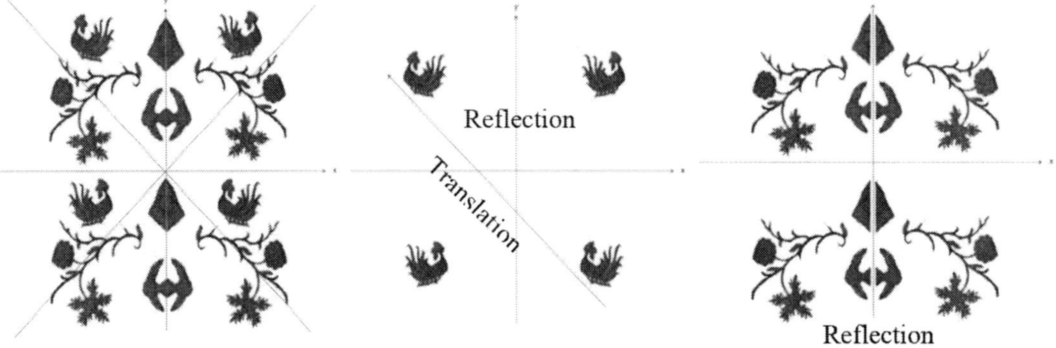

Figure 4. The application of geometry transformation in Wahyu Temurun batik motif

Figure 4 explains the application of geometry transformation in the motif of Wahyu Temurun batik, To making this motif can the applied translation of several ornaments by an axis and also can applied reflection of several ornaments in this motif by an axis in Cartesian coordinate system.

3.5. The philosophy, cultural values and the application of geometry transformation in motif of Parang Pamor batik

Parang Pamor batik is a batik appeared in the Mataram Kuthagedhe Kingdom in XVI centuries, came from the word *Pamor* means the light or light [20]. This batik motif is like knife blade digger radiating light of beautifulness that gives *daya perbawa* or authority. The prosperity, the people who wear the batik motifs *Parang Pamor*, the aura will be appeared or lighting. Furthermore, to make the motif of *Parang Pamor* batik, students can apply geometry transformation that seen in Figure 5

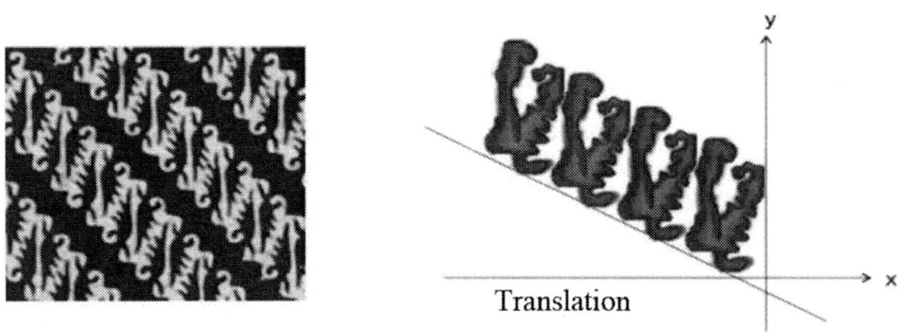

Figure 5. The application of geometry transformation in Parang Pamor batik motif

Figure 5 describes the application of geometry transformation in the motif of *Parang Pamor* batik, to making this motif can the applied translation of an ornament in the motif using an axis in the Cartesian coordinate system.

4. Conclusion

The research shows that culture inheritance such as batik still be interested and be kept by Javanese people. Batik of Yogyakarta does not only have a beautiful motif, yet containing philosophy, full of deep meaning and there are prayer and hope inside. Moreover, also in the making of this motif applied the geometry transformation concepts such as reflection, translation, rotation, and dilatation. From the motif of batik, design student can easily to learning the concept of geometry transformation especially to a student in junior high school because the motif is familiar in the daily life of a student. To make student easily to view the connections of mathematics concept in batik design can be done by making the design of learning using batik design to be a starting point and using activities drawing the motif of batik and introduce the concept of transformation in the design learning.

References

[1] Tanujaya B, Prahmana R C I and Mumu J 2017 Mathematics instruction, problems, challenges, and opportunities: A case study in Manokwari regency, Indonesia *World Transactions on Engineering and Technology Education* **15** 287
[2] Freudenthal H 1991 *Revisiting mathematics education China lectures* (Dordrecht: Kluwer)
[3] Prahmana R C I and Kusumah Y S 2016 The hypothetical learning trajectory on research in mathematics education using research-based learning *Pedagogika* **123** 42
[4] Karnilah N and Juandi D 2013 Study Ethnomathematics: Pengungkapan Sistem Bilangan Masyarakat Baduy *Jurnal Online Pendidikan Matematika Kontemporer* **1** 8
[5] Prabawati M N 2016 Etnomatematika Masyarakat Pengrajin Anyaman Rajapolah Kabupaten Tasikmalaya *Infinity Journal* **5** 1
[6] Suyitno H 2016 *Pengantar Filsafat Matematika* (Yogyakarta: Magnum Pustaka Utama)
[7] Rosa M and Orey D C 2011 Ethnomathematics: the cultural aspects of mathematics *Revista Latioanamericana de Etnomatematica* **4** 32
[8] Sundayana R, Herman T, Dahlan J A and Prahmana R C I 2017 Using ASSURE learning design to develop students' mathematical communication ability *World Transactions on Engineering and Technology Education* **15** 245
[9] Muhtadi D, Sukirwan, Warsito and Prahmana R C I 2017 Sundanese ethnomathematics: Mathematical activities in estimating, measuring, and making patterns *J. on Math. Educ.* **8** 185
[10] Stacey K 2011 The PISA view of mathematical literacy in Indonesia *J. on Math. Educ.* **2** 95
[11] Arisetyawan A, Suryadi D, Herman T, Rahmat C and No J D S 2014 Study of Ethnomathematics: A lesson from the Baduy Culture *International Journal of Education and Research* **2** 10
[12] Abdullah A S 2016 Ethnomathematics in perspective of sundanese culture *J. on Math. Educ.* **8** 1
[13] D'Ambrosio U 1985 Ethnomathematics and its place in the history and pedagogy of mathematics *For the learning of Mathematics* **5** 44
[14] Danoebroto S W 2012 Model pembelajaran matematika berbasis pendidikan multikultural *Jurnal Pembangunan Pendidikan: Fondasi dan Aplikasi* **1** 94
[15] Prahmana R C I, Kusumah Y S and Darhim 2017 Didactic trajectory of research in mathematics education using research-based learning *J. Phys.: Conf. Ser.* **893** 012001
[16] McKenney S and Reeves T C 2013 *Conducting educational design research* (Abingdon: Routledge)
[17] Didik 2017 *Filosofi dan Makna Batik Khas Yogyakarta* (Personal Communication)
[18] Efianingrum A 2011 *Proc. Seminar Batik* (Yogyakarta: Universitas Negeri Yogyakarta) p 17
[19] Koentjaradiningrat 2015 *Pengantar Ilmu Antropologi* (Yogyakarta: Rineka Cipta)
[20] Kusrianto A 2013 *Batik: Filosof, motif dan kegunaan* (Yogyakarta: Andi Offset)

IOP Conf. Series: Journal of Physics: Conf. Series **943** (2017) 012033 doi:10.1088/1742-6596/943/1/012033

Measuring the Indonesian provinces competitiveness by using PCA technique

Ditha Runita[1] and Rohmatul Fajriyah[1]

[1]Department Statistics, Universitas Islam Indonesia, Yogyakarta, Indonesia

E-mail: rfajriyah@uii.ac.id

Abstract. Indonesia is a country which has vast teritoty. It has 34 provinces. Building local competitiveness is critical to enhance the long-term national competitiveness especially for a country as diverse as Indonesia. A competitive local government can attract and maintain successful firms and increase living standards for its inhabitants, because investment and skilled workers gravitate from uncompetitive regions to more competitive ones. Altough there are other methods to measuring competitiveness, but here we have demonstrated a simple method using principal component analysis (PCA). It can directly be applied to correlated, multivariate data. The analysis on Indonesian provinces provides 3 clusters based on the competitiveness measurement and the clusters are Bad, Good and Best perform provinces.

1. Introduction

Indonesia is an archipelagic island country in South East Asia. There are many provinces under Indonesia's auspices. In 2017, Indonesia has 34 provinces. Every province in Indonesia has a unique and distinctive charm. Morever Indonesia has a vast teritory, it's area is 1.905 million km^2. Therefore it will be difficult for each province has similar development. To solve this problem, the central government made a regional autonomy (Otonomi Daerah) policy.

The new policy of decentralization and regional autonomy is outlined in Law No. 22, 1999 concerning "Local Government"1 and Law No. 25, 1999 concerning "The Fiscal Balance Between the Central Government and the Regions." Both these laws are based on five principles: 1) democracy, 2) community participation and empowerment, 3) equity and justice, 4) recognition of the potential and diversity within regions and 5) the need to strengthen local legislatures. These five principles support Indonesia's push for reformasi, which continues to aim to eradicate the practices of corruption, collusion, and nepotism (known as KKN), within the government bureaucracy. One of the specific reasons behind the policy of decentralization and regional autonomy is that a centralized government system cannot possibly administer Indonesia's large population of over 203 million (BPS, 2001) and its diverse socio-cultural and religious background. Strong, competent regional governments and greater autonomy are fundamental requirement for a country as diverse as Indonesia. The main aim of decentralization and regional autonomy is to bring the governments closer to their constituents so that government services can be delivered more effectively and efficiently. This is based on the assumption that district and municipal governments have a better understanding of the needs and aspirations of their communities than the central government [1].

We can rank each province by using PCA and try to find the cluster of each provinces. Therefore, it will build the local competitiveness. Building local competitiveness is critical to enhance the long-term national competitiveness. A competitive local government can attract and maintain successful firms and increase living standards for its inhabitants, because investment and skilled workers gravitate from uncompetitive regions to more competitive ones.

Asia Competitiveness Institute (ACI) defines competitiveness through four different environments, each with their three sub-components, namely [2] :

1) Macroeconomic stability
   a. Regional Economic Vibrancy
   b. Openness to Trade and Services
   c. Attractiveness to Foreign Investors

2) Government and institutional setting
   a. Govenment Policies and Fiscal Sustainability
   b. Instutions, Govemance and Leadership
   c. Competition, Regulatory Standards and Rule of Law
3) Financial, business and manpower conditions
   a. Financial Deepening and Business Efficiency
   b. Labour Market Flexibility
   c. Productivity Performance
4) Quality of life and infrastructure development
   a. Physical Infrastructure
   b. Technological Infrasturcture
   c. Standard of Living, Education and Social Stability

Those environment give us an idea about the condition of provinces. Two environments are specifically related to the economy and two other environment are more political, instituitional and social in character. Macroeconomic stability encompassing aggregate economic conditions. It will be easy to see how strong the potential of a province in attracting foreign investors from this environment. Financial, business and manpower conditions represent condition related to the micro-economy. They include an analysis on the performance of firms as well as challenges that they face in running their companies, also from this environment we will be able to describe the state of the population in that area, the availability of labor in order to see the potential of human resources in the area. Government institutional setting will provide an overview of the rules, policies of local governments in addressing the problems in the region. And how responsive the local government in solving a case. Thus the stability of the government will be seen in the area. As for the quality of life and infrastructure development can be used as a reference in view of the security of the area, the power level of life in society, and the development of technology or infrastructure.

2. Literature Review/Related Works

Malesky, in 2008 [3] with his research tim, make the research under the title The Vietnam Provincial Competitiveness Index: Measuring Economic Governance for Private Sector Development. 2008 Final Report indicated in last year's report, there is evidence that increasing inequality is beginning to appear across the country. Top performing provinces excel at all aspects of economic development, as shown in this report in terms of economic governance, infrastructure, and ICT capability. This group is pulling away from the rest of the country. At the same time, another group that must struggle with weak initial conditions and poor infrastructure has not been able to develop the good governance practices to compensate for their handicap.These are steadily falling behind the performance of their peers. The Provincial Competitiveness Index in Vietnam measured by 10 sub indices they are: Privat Sector Development services, Transparancy, Labor training, Proactivity, Time cost of regulatory compliance, Legal institutions, SOE bias (competition environment), informal charges, Land access and security, and Entry costs. In This research, Malesky diveded the provinces in Vietnam using the provincial competitivness index into 6 classes. The classes are Excellent Province, High Province, Mid-High Province, Average Province, Mid-Low Province and las is Low Province.

In 2013, Tan and Amri [4] in their research under the title Subnational Competitiveness and National Performance: Analysis and Simulation for Indonesia said that the stable growth of Indonesia's economy over the past eight years has provided momentum for investment in the country. One of the approaches taken by the central government is to allow healthy competition between its provinces. The Asia Competitiveness Institute (ACI) responds positively to that policy by ranking the competitiveness of Indonesia's 33 provinces and providing simulation studies on how to improve each province's competitiveness. ACI takes a comprehensive approach to the notion of provincial competitiveness, dissecting it from four major environments: macroeconomics, microeconomics, governance, and quality of life. Drawing on 91 indicators from formal sources as well as ACI's own surveys and interviews, the study aggregates the indicators into 12 sub-environments, reaggregates them into four environments, and finally reaggregates them again into an overall competitiveness index. The conclusion highlights the high level of competitiveness in provinces where the country's major urban regions are situated, as well as those closest to Singapore as the regional trading hub. Provinces endowed with natural resources also have the opportunity to be competitive, but not if they are wanting in

IOP Conf. Series: Journal of Physics: Conf. Series **943** (2017) 012033 doi:10.1088/1742-6596/943/1/012033

aspects such as governance and quality of life. The study's findings invite further research on more specific topics such as labor market flexibility and regional cooperation.

3. Method

The data set that we use for this research is downloaded from Indonesian statistics website, www.bps.go.id, for the years 2014 and 2015. Non-Oil exports data is downloaded from www.kemendag.go.id. From both sources we have 29 variables which are belong to the 4 environments at ACI, namely:

1) Macroeconomic stability
    1. Realization Of Investment In Domestic Investment (Project) Year 2015
    2. Realization Of Investment In Domestic Investment By Province (Billion Rupiah) Year 2015
    3. Gdp Per Capita By Province Year 2014
    4. Non-Oil Export By Province Year 2015
    5. Number Of Foreign Tourist By Province Year 2015
2) Government And Institutional Setting
    6. Total Crime According To The Regional Police, 2000-2015
    7. Crime Rate Per 100.000 Population By Regional Police Year 2000-2015
    8. Percentage Of Completion Crime According To The Regional Police Year 2000-2015
3) Financial, Business And Manpower Conditions
    9. Unemployement Rate Year 2015
    10. Labor Force Particaption Rate Year 2015
    11. Registered Job Seekers Years 2015
    12. Job Vacancies List Year 2015
    13. Placement/Fulfillment Of Labor Year 2015
4) Quality Of Life And Infrastructure Development
    14. Population By Province Year 2015
    15. Human Development Index (HDI) Year 2015
    16. Food Poverty Line Year 2015
    17. Gini Ratio Year 2015
    18. Number Of Poor Year 2015
    19. School Enrollment Rate (7-12 Years Old) Year 2015
    20. School Enrollment Rate (13-15 Years Old) Year 2015
    21. School Enrollment Rate (16-18 Years Old) Year 2015
    22. School Enrollment Rate (19-24 Years Old) Year 2015
    23. Number Of Universities Year 2014
    24. Number Of Students Year 2014
    25. Number Of Educational Personnel Year 2014
    26. Percentage Of Households By Source Of Improved Drinking Water Year 2015
    27. Percentage Of Households By Lighting Source Of Electricity Year 2015
    28. Percentage Of Households By Improved Sanitation Year 2015
    29. Percentage Of Population Having Health Complaint Year 2015

Principal Component Analysis (PCA) is one of famous techniqeus for dimension reduction, feature extraction, and data visualization. In general, PCA is defined by a transformation of a high dimensional vector space into a low dimensional space [5]. Manage and Scariano [6] describe PCA as a nonparametric variable reduction technique well suited for correlated data and one objective of principal component analysis is to reduce a set of correlated variables into fewer uncorrelated variables as linear combinations of the original ones.

In general, the PCA procedures can be explained [7-9] explain as follows.
Suppose we have a random vector X

$$X = \begin{pmatrix} X_1 \\ X_2 \\ \vdots \\ X_p \end{pmatrix}$$

with population variance-covariance matrix

$$var\ (X) = \sum = \begin{pmatrix} \sigma_1^2 & \sigma_{12} & \cdots & \sigma_{1p} \\ \sigma_{21} & \sigma_2^2 & \cdots & \sigma_{2p} \\ \vdots & \vdots & \ddots & \vdots \\ \sigma_{p1} & \sigma_{p2} & \cdots & \sigma_p^2 \end{pmatrix}$$

Consider the linear combinations:

$$\begin{aligned} Y_1 &= e_{11}X_1 + e_{12}X_2 + \cdots + e_{1p}X_p \\ Y_2 &= e_{21}X_1 + e_{22}X_2 + \cdots + e_{2p}X_p \\ &\vdots \\ Y_n &= e_{p1}X_1 + e_{p2}X_2 + \cdots + e_{pp}X_p \end{aligned}$$

Each of these can be thought of as a linear regression, predicting Y_i from X_1, X_2, ..., X_p. There is no intercept, but e_{i1}, e_{i2},...,e_{ip} can be viewed as regression coefficients.

The Y_i is a function of random data and it has a population variance

$$var(Y_i) = \sum_{k=1}^{p}\sum_{l=1}^{p} e_{ik}e_{il}\sigma_{kl} = e_i' \sum e_i$$

$e_i' \sum e_j$ is population covariance of Y_i and Y_j

$$cov(Y_i, Y_j) = \sum_{k=1}^{p}\sum_{l=1}^{p} e_{ik}e_{jl}\sigma_{kl} = e_i' \sum e_j$$

The coefficients of e_{ij} are

$$e_i = \begin{pmatrix} e_{i1} \\ e_{i2} \\ \vdots \\ e_{ip} \end{pmatrix}$$

The ith principal component is built by selecting e_{i1}, e_{i2},...,e_{ip} such that maximizes the variance of the new component.

$$var(Y_i) = \sum_{k=1}^{p}\sum_{l=1}^{p} e_{ik}e_{il}\sigma_{kl} = e_i' \sum e_i$$

The maximization considers two things: the sums of squared coefficients add up to one and the new component will be uncorrelated with all the previously defined components.

$$e_1'e_1 = \sum_{j=1}^{p} e_{1j}^2 = 1$$

$$cov(Y_1, Y_i) = \sum_{k=1}^{p}\sum_{l=1}^{p} e_{1k}e_{il}\sigma_{kl} = e_1' \sum e_i = 0$$

$$cov(Y_2, Y_i) = \sum_{k=1}^{p}\sum_{l=1}^{p} e_{2k}e_{il}\sigma_{kl} = e_2' \sum e_i = 0$$

$$\vdots$$

$$cov(Y_{i-1}, Y_i) = \sum_{k=1}^{p}\sum_{l=1}^{p} e_{i-1,k}e_{il}\sigma_{kl} = e_{i-1}' \sum e_i = 0$$

4. Results and Discussion

The calculations for the average of each environment are shown as following:

Table 1. Average of each ACI environments

Infromations	Macroeconomic Stability(MS)	Government and Institutional Setting (GIS)	Financial, Business and Manpower Conditions (FBI)	Quality of Life and Infrastructur Development (QL)
Averages	8.48994E-17	-6.75522E-17	1.18206E-15	2.35106E-16
Conclusions	Lower Is The Best	Higher Is Best	Lower Is The Best	Lower Is The Best

Table 1 shows the average of each environments from ACI and the the scores for each provinces. It is based on the standardized of original values of each variable for every environments. If the value of variable tends as higher is better then we multiply the standardization with minus, and otherwise.

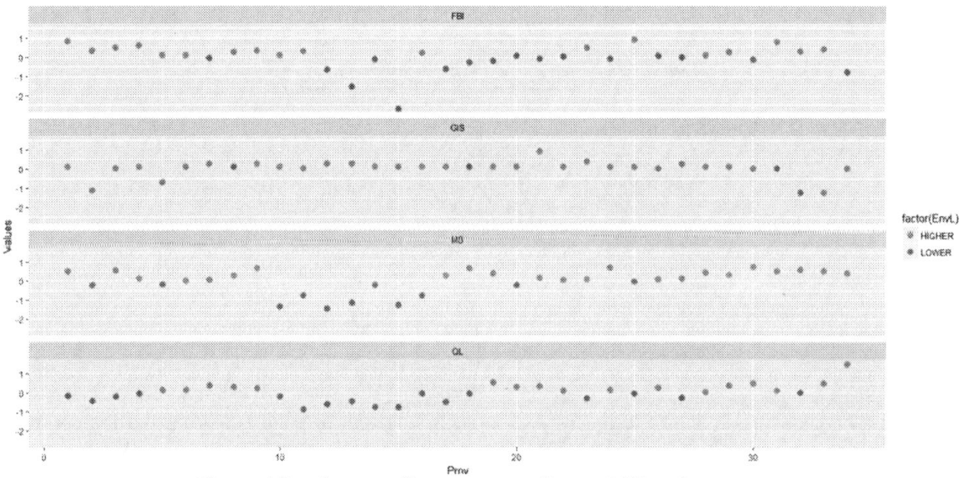

Figure 1 Provinces performance according to ACI environments

Before applying PCA, the standarization is needed to be applied cause the variables have different scales. The processes of finding the principal components by using the standardized variables are equivalent to finding the principal components by using the correlation matrix instead of the covariance matrix. Furthermore, it is used to compute the eigen values and vectors.

```
Importance of components:
                          PC1    PC2     PC3    PC4     PC5     PC6     PC7
Standard deviation     3.0022 2.2405 1.58625 1.5165 1.31291 1.25224 1.09671
Proportion of Variance 0.3108 0.1731 0.08676 0.0793 0.05944 0.05407 0.04148
Cumulative Proportion  0.3108 0.4839 0.57066 0.6500 0.70940 0.76347 0.80494
                           PC8     PC9   PC10    PC11    PC12    PC13    PC14
Standard deviation     1.00250 0.96360 0.8599 0.79844 0.69249 0.65489 0.59768
Proportion of Variance 0.03466 0.03202 0.0255 0.02198 0.01654 0.01479 0.01232
Cumulative Proportion  0.83960 0.87162 0.8971 0.91910 0.93563 0.95042 0.96274
                          PC15    PC16    PC17   PC18    PC19    PC20    PC21
Standard deviation     0.48354 0.43963 0.42625 0.3489 0.30707 0.26177 0.24318
Proportion of Variance 0.00806 0.00666 0.00627 0.0042 0.00325 0.00236 0.00204
Cumulative Proportion  0.97080 0.97747 0.98373 0.9879 0.99118 0.99354 0.99558
                         PC22    PC23    PC24    PC25    PC26    PC27    PC28
Standard deviation     0.2151 0.20626 0.15033 0.09514 0.06984 0.04690 0.01963
Proportion of Variance 0.0016 0.00147 0.00078 0.00031 0.00017 0.00008 0.00001
Cumulative Proportion  0.9972 0.99865 0.99943 0.99974 0.99991 0.99998 0.99999
                          PC29
Standard deviation     0.01218
Proportion of Variance 0.00001
Cumulative Proportion  1.00000
```

IOP Conf. Series: Journal of Physics: Conf. Series **943** (2017) 012033 doi:10.1088/1742-6596/943/1/012033

Figure 2 PCA's Output

The output of PCA, Figure 2, shows that the PC1 gives the highest proportion of variance, 31.08%, then it is used to rank the provinces. Furthermore, its eigen vectors are in the Table 2.

Table 2 Eigen Vectors

No	Variables	Eigen Vectors	No	Variables	Eigen Vectors
1	HDI	-0.166	16	Number Of Universities	-0.306
2	Food Poverty Line	0.105	17	Number Of Students	-0.290
3	Gini Ratio	-0.135	18	Number Of Educational Personnel	-0.309
4	Number Of Poor	0.099	19	Household by Drinking Water Sources	-0.150
5	School Enrollment Rate (7-12 yo)	-0.113	20	Households by Electric Power Sources	-0.170
6	School Enrollment Rate (13-15 yo)	-0.111	21	Households by Improved Sanitation	-0.157
7	School Enrollment Rate (16-18 yo)	0.037	22	Total Crime	0.018
8	School Enrollment Rate (19-24 yo)	0.041	23	Crime Rate	0.046
9	Unemployement Rate	-0.051	24	Percentage of Completion Crime	0.129
10	Labor Force Participant Rate	0.104	25	Health Complaint	-0.121
11	Registered Job Seeker	-0.273	26	Population (Thousands)	-0.288
12	Job Vacancies List	-0.269	27	Non Oil Exprots	-0.169
13	Labor Fulfillment	-0.271	28	GDP	-0.011
14	Investment Realization (Project)	-0.295	29	Foreign Tourist	-0.072
15	Investment Realization (billion)	-0.291			

Table 2 gives the first principal component as follows:

$$Y_1 = e_{1,1}X_1 + e_{1,2}X_2 + \cdots + e_{1,29}X_{29}$$

IOP Conf. Series: Journal of Physics: Conf. Series **943** (2017) 012033 doi:10.1088/1742-6596/943/1/012033

$$
\begin{aligned}
Y\_1 = & -0.166(HDI) + 0.105(Poverty\ Line) - 0.135(Gini\ Ratio) + 0.099\ (Number\ of\ Poor) \\
& - 0.113\ (School\ Enrollment\ Rate\ (7-12\ yo)) \\
& - 0.111\ (School\ Enrollment\ Rate\ (13-15\ yo)) \\
& + 0.037\ (School\ Enrollment\ Rate\ (16-18\ yo)) \\
& + 0.041\ (School\ Enrollment\ Rate\ (19-24\ yo)) \\
& - 0.051(Open\ Unemployement\ Rate) + 0.104(Labor\ Force\ Participant\ Rate) \\
& - 0.273(Registered\ Job\ Seeker) - 0.269(Job\ Vacancies\ List) \\
& - 0.271(Labor\ Fulfillment) - 0.295(Realization\ Investment\ (Project)) \\
& - 0.291(Investment\ Realization\ (billion)) - 0.306(Number\ Of\ Universities) \\
& - 0.290(Number\ Of\ Students) - 0.309(Number\ Of\ Educational\ Personnel) \\
& - 0.150(Household\ by\ Drinking\ Water\ Sources) \\
& - 0.170(Households\ by\ Electric\ Power\ Sources) \\
& - 0.157(Households\ by\ Improved\ Sanitation) + 0.018(Total\ Crime) \\
& + 0.046(Crime\ rate) + 0.129(Percentage\ of\ Completion\ Crime) \\
& + 0.121(Health\ Complaint) - 0.288(Population\ (Thousands)) \\
& - 0.169(Non\ Oil\ Exprots) - 0.011(GDP) - 0.072(Foreign\ Tourist)
\end{aligned}
$$

Therefore, the scores for each province are as in Table 3:

Table 3 The scores and ranking based on PCA

Provinces	PC 1 score	Standardized PC 1 score	Ranking
Aceh	0.437172	0.143654	15
Sumatera Utara	-0.88412	-0.29052	8
Sumatera Barat	0.875282	0.287616	18
Riau	0.02172	0.007137	11
Jambi	0.96015	0.315504	19
Sumatera Selatan	0.245784	0.080764	14
Bengkulu	2.051826	0.674226	28
Lampung	1.125308	0.369774	21
Kep. Bangka Belitung	1.407645	0.462549	23
Kep.Riau	0.184928	0.060767	13
DKI Jakarta	-5.78365	-1.9005	4
Jawa Barat	-7.62536	-2.50568	2
Jawa Tengah	-5.86413	-1.92694	3
DI Yogyakarta	-1.08801	-0.35752	7
Jawa Timur	-9.47617	-3.11385	1
Banten	-2.07725	-0.68258	5
Bali	-0.5356	-0.176	9
Nusa Tenggara Barat	0.082932	0.027251	12
Nusa Tenggara Timur	2.425648	0.797063	29
Kalimantan Barat	1.430153	0.469946	24
Kalimantan Tengah	1.893612	0.622237	25
Kalimantan Selatan	0.645778	0.212202	16
Kalimantan Timur	-0.26374	-0.08666	10
Kalimantan Utara	2.540385	0.834766	32

Provinces	PC 1 score	Standardized PC 1 score	Ranking
Sulawesi Utara	0.743866	0.244433	17
Sulawesi Tengah	1.93902	0.637158	26
Sulawesi Selatan	-1.91624	-0.62967	6
Sulawesi Tenggara	1.064083	0.349656	20
Gorontalo	1.261665	0.414581	22
Sulawesi Barat	2.428965	0.798153	30
Maluku	2.033066	0.668061	27
Maluku Utara	2.542381	0.835422	33
Papua Barat	2.480462	0.815075	31
Papua	4.692438	1.541926	34

Table 3 shows that Jawa Timur hass the first ranking and Papua is the last one. In 2012, Tan and Amri [4] has ranked the competitivenss provinces in Indonesia by using 2010 data. The results are the first ranking is DKI Jakarta and the second is Jawa Timur. It means since 2010, Jawa Timur has improved itself and becomes the best province. The same condition is applied to other provinces. For examples, in 2010 Nusa Tenggara Barat has a ranking 29 and in 2015 Nusa Tenggara Barat has a ranking 12. In Tan and Amri [4] results, DKI Jakarta have become the first ranking in 2000 and 2010. But in 2015 competitiveness, DKI Jakarta has dropped to number 4.

The ranking of provinces based on PCA is determined by the following procedures. The scores that we got from PCA needs to be normalized. The PCA scores are divided into three areas, as in Figure 3

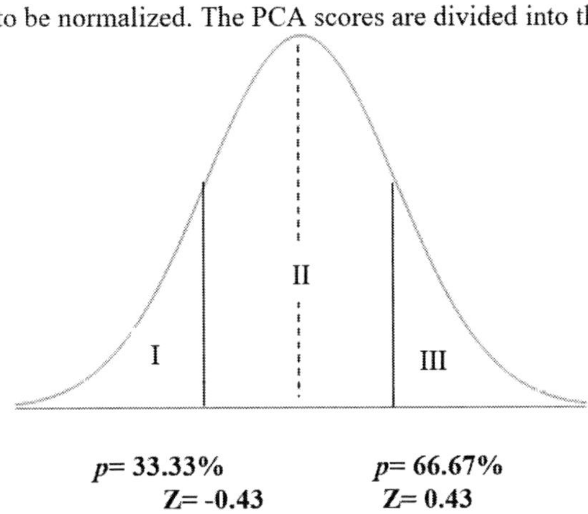

Figure 3 Three clusters for the provinces scores

In Figure 3, the first area represents the best perform provinces, the second one represents the good perform provinces and the third one represents the bad perform provinces. The scores for best perform provinces are less than and or equal to -0.43, for good perform provinces are between -0.43 until 0.43 half right inclusive, and for bad perform provinces are greater than 0.43. The best, good and bad perform provinces based on their scores are shown at Table 4.

Table 4 Members of each PCA clusters

BEST PERFORM PROVINCES	GOOD PERFORM PROVINCES		BAD PERFORM PROVINCES
Jawa Timur	DI Yogyakarta	Aceh	Kep. Bangka Belitung
Jawa Barat	Sumatera Utara	Kalimantan Selatan	Kalimantan Barat
Jawa Tengah	Bali	Sulawesi Utara	Kalimantan Tengah
DKI Jakarta	Kalimantan Timur	Sumatera Barat	Sulawesi Tengah
Banten	Riau	Jambi	Maluku
Sulawesi Selatan	Nusa Tenggara Barat	Sulawesi Tenggara	Bengkulu
	Kep. Riau	Lampung	Nusa Tenggara Timur
	Sumatera Selatan	Gorontalo	Sulawesi Barat
			Papua Barat
			Kalimantan Utara
			Maluku Utara
			Papua

Table 4 shows the members of every cluster using PCA. The best perform provinces has 7 members, good perform provinces contains 16 members and for bad perform provinces has 12 members. If we compare the average of each ACI environment in Table 2 and the scores of each provinces, then

a. from the best perform provinces, there are some provinces that worst than the average score. For example, in Financial, Bussiness and Manpower environment, there are two provinces have lower scores than its average, DKI Jakarta and Banten. For Macroeconomy stability environment Sulawesi Selatan has lower score than its average. In two others environments, each member of the Best perform provinces have better scores than the average.
b. from the good perform provinces, with 16 members. In Macroeconomic stability environment there are 5 provinces have scores better than the average, namely DI Yogyakarta, Sumatera Utara, Kep. Riau, Sulawesi Utara and Jambi. In Govenment and Institutional Setting environment, there are two provinces have scores lower than the average, namely Nusa Tenggara Barat and Lampung. In Financial, Bussiness and Manpower Condition environment, there are 4 provinces have better scores than its average, namely DI Yogyakarta, Bali, Nusa Tenggara Barat and Kalimantan Selatan. For Quality of Life and Infrastructure development envronment, there are six provinces have lower scores than its average. They are Sumatera Selatan, Kalimantan Selatan, Jambi, Sulawesi Tenggara, Lampung and Gorontalo
c. from the last cluster is Bad perform Provinces which has 12 members, in Macroeconomic stability envronment, it is only Kalimantan Barat has a core better than its average. In Government and institutional setting environment, there are two Provinces which have lower scores than its average. They are Maluku Utara and Maluku. In Financial, bussiness and manpower condition environment, four Provinces, Kep. Bangka Belitung, Papua Barat, Maluku Utara and Maluku, have lower scores than its average. And for the last environment, Quality of life and inftrastructure development, only Maluku Utara has a better score than its average.

5. Conclusion and Remarks

Measuring the competitiveness of every provinces in Indonesia is a challenging task. The result from competitiveness can be used to make some government's rules to make the provinces in Indonesia better. This results can also be used by some investors to decide where they can invest their money acording to its competitiveness. In this paper we show that the principal component analysis can be used to ranking the provinces competitiveness based on 29 variables, where Jawa Timur is the best perform provinces (the most competitive) and Papua is the least competitive. Further investigation could be applied to measure the provinces competitiveness simultaneously and thoroughly by adding more information through some additional variables as in the ACI ([2] and [4]).

References

[1] Usman, Syaiku 2002 *Regional Autonomy in Indonesia: Field Experiences and Emerging Challenges* (Jakarta: The Semeru Research Institute)

[2] Tan, Khee G 2014 Assessing competitiveness of ASEAN-10 economies *Int.J. Economics and Business Research* **8**

[3] Malesky, Edmund 2008 The Vietnam Provincial Competitiveness Index: Measuring Economic Governance for Private Sector Development *Final Report Vietnam Competitiveness Initiative Policy Paper* **13**

[4] Tan, Khee G and Mulya Amri 2013 Subnational Competitivenss and National Performance: Analysis and Simulation for Indonesia *JCC: The Business and Economics Research Journal* **6**

[5] Jang, Sujin 2014 Basics and Examples of Principal Component Analysis (PCA) https://www.projectrhea.org/rhea/index.php/PCA_Theory_Examples.

[6] Manage, Ananda B and Scariano, Stephen M 2013 An Introduction Application of Principal Components to Cricket Data *Journal of Statitics Education* **21**

[7] PennState Eberly College of Science. *Lesson 11: Principal Component Analysis Procedure.* https://onlinecourses.science.psu.edu/stat505/node/51.

[8] Sharma, Subhash 1996 Applied Multivariate Techniques (Canada: John Wiley & Sons)

[9] Timm, Neil H 2002 Applied Multivariate Analysis (New York: Springer)

IOP Conf. Series: Journal of Physics: Conf. Series **943** (2017) 012034 doi:10.1088/1742-6596/943/1/012034

The teacher understanding of probability using scientific approach

Taufik Fajar Gumilang[1,2], Budi Usodo[2] and Ikrar Pramudya[2]

[1]SMA Al Islam 1 Surakarta, Jl. Honggowongso No 94, Laweyan, Surakarta, Indonesia
[2]Universitas Sebelas Maret, Jl. Ir. Sutarmi No. 36 A Kentingan, Surakarta, Indonesia

E-mail: gilangboyolali.03081992@gmail.com

Abstract. The purpose of curriculum 2013 preparing Indonesian people to have the ability to live as individuals and citizens who are faithful, productive, creative, innovative and affective is the objective. The subject of this research is the teacher of mathematics class XI SMA Al Islam 1 Surakarta with the criteria have received training on curriculum 2013. How teachers can integrate between knowledge, skills, and attitudes related to character. The results of this study are as follows asking students to hear the explanation given by the teacher is observing, asking questions to students about the material being studied is questioning activity, students read books is activities collect information, teachers instill an attitude of cooperation and responsibility to the students is association activities, asking students to write down the results of group discussions in front of the class is communicating activities.

1. Introduction

In the curriculum education system is one component that has an important role, not only formulated about the goals to be achieved, but also provides an understanding of learning experiences that must be owned by students [1-4]. According to Ornstein and Hunkins [5] states that as a guideline in guarding the course of education, the ideal curriculum within 7-10 years should be evaluated or reviewed to conform to the times, science and technology, and the dynamics of society.

The curriculum and learning are inseparable although both have different positions. Dawson and Schibeci [6] state that school science curriculum has to prepare students for their future roles as citizens among technologies which will have a significant impact on their lives, while learning is a process that occurs in the interaction of learning and teaching between teachers and learners.

Ma and Kishor [7] mathematics educators have considered the connection between students' attitudes toward mathematics, and their achievement in the subject as one of the major concerns. In line with that opinion Handler [8] teachers must have comprehensive understanding of their content areas and methods for communicating knowledge to students. So that in learning mathematics, interaction between teacher and student very important role in forming student attitude.

According to the 2015 study of PISA (Program for International Student Assessment), a study focusing on literacy, math, and science, shows that the new Indonesia ranking can occupy 69 of 76 countries. While the results of the study TIMSS (Trends in International Mathematics and Science Study) shows Indonesian students are ranked in 36 of 49 countries. From these results are low in the ability to understand complex information, theory, analysis and problem solving, use of tools, procedures, and problem solving and investigate [9]. This is due to the fact that the number of test materials questioned in TIMSS and PISA is not contained in the Indonesian curriculum.

The city of Surakarta is one of the areas that has implemented the curriculum 2013. The curriculum is being implemented in senior high school in the academic year 2013. For schools that have implemented the curriculum 2013 should teachers have gained understanding and training on the curriculum 2013. In the implementation there are still teachers who complained about his unpreparedness in implementing the curriculum 2013. This is in accordance with the results of interviews conducted to teachers in MAN 2 Surakarta who have implemented the curriculum 2013 that says teachers are still confused in implementing learning and assessing the results of learning based on the curriculum 2013. In the process of learning teachers only prioritize assessment on the aspect of knowledge without any integration of knowledge, skills, and attitudes.

The results of pre-research, about the implementation of curriculum 2013 in SMA Al Islam 1 Surakarta by interviewing mathematics teachers, this is because SMA Al Islam 1 Surakarta also has been using the curriculum 2013. Show that the integration of knowledge, skills, and attitudes that teachers do to students in the learning process of mathematics is to set an example to comply with the prevailing rules, always check the presence, give the same attention, give questions that can provoke curiosity, give an opportunity to express opinions, always try to appreciate opinions, and remind or punish those who do not carry out its duties and obligations.

It is also found that the teacher's frequent assessment only measures cognitive achievement without measuring the development of student attitudes, so that in its development the students only pursue the value. The conclusion of the above explanation is the need for a correlation between knowledge, skills, and attitudes to bring up scientific learning. The purpose of this study is to describe the curriculum 2013; integration of knowledge, skills and attitude with a scientific approach; learning mathematics with a scientific approach; as well as mathematical learning and assessment based on the curriculum 2013.

2. Method

2.1 Forms and strategies research

Moleong [10] states that qualitative research is a research procedure that produces descriptive data in the form of words written or spoken of people and observed behavior. In accordance with the characteristics of the qualitative data this research strategy using qualitative descriptive method. Capturing data using observation and interview methods. Data sought in this study is data about the learning process of mathematics based curriculum 2013, the integration of knowledge, skills and attitude of learning mathematics with a scientific approach, and The process of assessing the knowledge, skills and attitudes based on the curriculum of 2013.

2.2 Data and data source research

In this study, the data used to obtain information about teachers' understanding of the learning and assessment process based on the curriculum in 2013 in the form of words. This data was obtained by researchers through interviews with a mathematics teacher at SMA Al Islam 1 Surakarta. The data used to obtain information about the process of learning and curriculum based assessment of 2013 is the record of learning activities that take place in class. This data was obtained through observation of the learning process in the classroom.

Source of data to be utilized in this study is that the informant or informants, activities, and documents. Informant or informants is a high school math teacher Al Islam 1 Surakarta. Informants were selected based on criteria have the capacity, long teaching and have received training on the curriculum of 2013. Vice Principal Curriculum SMA Al Islam 1 Surakarta is one of the sources of data. Mathematics teaching and learning activities at the time of presenting the material. Document the learning of mathematics is mathematics lesson plan. The data source is needed as a complement to the data implementing the learning.

Subjects in this study were Sugiharti, S.Pd and Ruqoiyah, S. Pd as high school math teacher Al Islam 1 Surakarta. Subjects were selected based on criteria already received training in the national curriculum in 2013 from the Department of Education. Based on the information obtained by investigators of the research subject, the subject has been taught mathematics at SMA Al Islam 1 Surakarta since 1998.

2.3 Data collection techniques

Data collection techniques used in this study is the observation and interviews. Budiyono [11] says that the methods of the interview or interviews are data collected through conversations between researchers and research objects or respondent. In this case, the interviewer hold a conversation so that interviewees are willing to open issued his opinion. Usually requested is not ability but information about something.

Interview in this research is an unstructured interview. Interviews were conducted aimed at obtaining information on teachers' understanding of the learning and assessment process based on the curriculum of 2013. In addition, interviews were used to obtain additional information related to the process of learning and assessment in the classroom [12, 13].

Observations were applied in this study is the observation of teachers in implementing the learning process in class. Observations were conducted aimed at obtaining information mathematics learning process of imparting knowledge, skills, and attitudes to scientific approach as well as the process of assessing the knowledge, skills and attitudes based on the curriculum of 2013.

2.4 Research instruments

The main instrument in this study is the researchers themselves who aims to search for and collect data directly from the data source. Instrument first aid in this research is interview guide was created by researchers as a tool in the field of data retrieval. Guidelines for the interview were made as a reference in the interview to the subject of study to obtain information in the form of teachers' understanding of the learning and assessment process based on the curriculum 2013. Second auxiliary instrument in this study was the observation and video cameras. This observation guide contains activities that teachers and students during the learning process and an assessment based on the curriculum 2013. While the video camera is used to capture the learning process in the classroom, so that later researchers were able to play back the video lessons to deepen her study of the learning and assessment process based on curriculum 2013.

Validation data used in this study includes several activities. In this study, researchers observed carefully and thoroughly to mathematics learning activities in the classroom. Researchers carefully and meticulously transcribe the recording and analyzing the data, in this way, the data can be described in a systematic and comprehensive.

In this study, the researchers checked the data by clarifying the findings of researchers with the research subjects. After the research subjects to check the findings of researchers and investigators reported data in accordance with the actual situation, then the data is valid. But if, according to the subject of research findings researchers did not correspond to the real situation, the researchers conducted a discussion with research subjects and when the discussion of research subjects give the correct data, then the data the researchers used to supplement existing data

In this, researchers perform data validation by using triangulation of time, i.e. with the data match the learning process of mathematics at the first observation and the second observation. Each execution data the same learning process later declared invalid. Data that is not the same as it was said other findings.

Data analysis was performed on two recordings outcome data retrieval chosen and supported four interviews with teachers and 6 interviews with three students. Based on the results of two recordings of the data collection, the data obtained from each of the data collection will complement each other (snowball). Two results of the data collection in the form of a recording transcribed mathematics learning activities carefully and in detail from the beginning to the end of the learning activities. Thus, obtained a transcript of a collection of statements of teachers and students at two selected data are retrieval results.

Data analysis techniques used in this study follows the model of interactive analysis. Sutopo [14] state that there are three components of the data analysis: data reduction, data presentation and conclusion drawing and verification.

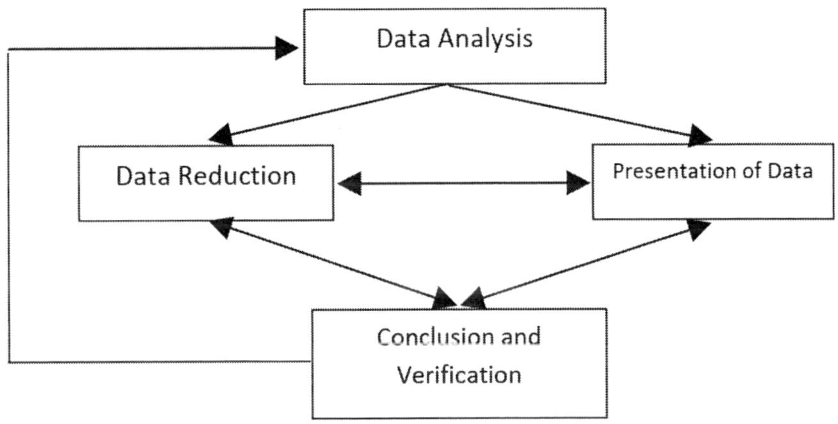

Figure 1. Figure Sutopo Interactive Analysis [14]

2.5 Research procedure

The procedure for collecting each data is described. Information about knowledge of learning process and appraisal based on curriculum 2013 is obtained by interview method to class XI mathematics teacher by using recorder. This information gathering was conducted once on 7 February 2017, before the researchers made observations about the learning and assessment process.

Information about the mathematics learning process that imparts knowledge, skill, and attitude with scientific approach is obtained by observation method on the subject matter of opportunity by using handy cam recorder. The data collection is done in order to know how the activity of observing, asking, gathering information, associating, and communicating happened in learning process of opportunity material, and also want to know how to plant the knowledge, skill, and attitude that happened.

This information gathering is done at the time of delivery of opportunity material that is happened as many as three times meeting. In this study, the data obtained from three observations is selected observation data at all meetings. The data is chosen because the data is the most complete data, this is because at every learning process that takes place apply scientific approach. The selected observational data is on the first data with the purpose of learning to determine how to calculate the probability of an event, the second data with the purpose of learning to determine the permutation formula and on the third data collection with learning objectives to determine the formula of the combination.

Information on the process of assessing students' knowledge, skills and attitudes based on the curriculum 2013 is obtained by observation method on the subject matter of opportunity by using handy cam recorder. This information gathering was conducted during the delivery of opportunity material which was held three times.

This information gathering was conducted during the delivery of opportunity material which was held three times. In this study, the data obtained from three observations were selected observational data at the first, second and third observations of data. The data is selected because the data is the most complete data, this is because at the first, second, and third meeting of teachers to assess the aspects of knowledge, skills, and attitudes. The selected observational data is on the first data with the purpose of learning to determine how to calculate the probability of an event, the second data with the purpose of learning to determine the permutation formula and on the third data collection with learning objectives to determine the formula of the combination.

3. Result and discussion

Data on Teacher Understanding in Learning and Assessment Process Based of Rule Curriculum 2013. Information: "R" is the researcher and "T" is the teacher

R: "what is the learning process in the 2013 curriculum?"

T: "there is a 2013 curriculum learning process centered on students, so students are required to be more active in building knowledge while the teacher only as a facilitator only. So suppose in material A, the teacher only convey the purpose of learning course and the teacher facilitates the students to achieve the learning objectives by providing a problem and then students complete the discussion, read or summarize themselves. So the students try to find their own, suppose a formula, the student can find their own formula."

Data of interview result, obtained by teacher understanding about learning process based on curriculum 2013 that is student-centered learning process. In the learning process students are required to be more active in terms of knowledge and teachers only as a facilitator. So, the teacher simply tells the learning objectives by giving a problem, and then the student finishes by discussion, questioning, reading, or summarizing himself.

Competence to be developed through observation activities can train students actively, questioning activities can make students able to formulate a question, information gathering activities can make students appreciate the opinions of others, information processing activities can foster mutual cooperation, communicating activities, can make students brave express their opinions [4].

R: "What competencies want to develop through these activities?"

T: "If observing activities that competence to be developed can train activc students."

R: "Active how?"

T: "So by giving a problem to the students it can make students arise curiosity of students so they actively seek information about the given problem"

R: "What about the other activities?"

T: "If the questioning activity that can make students able to formulate a question. So by asking questions to the students, the question arises with a question mark to the students so they can ask questions again."

T: "If the activity of collecting information can make the students appreciate the opinions of others, collecting information can only be through discussions with friends, but it can also make students independent in learning."

R: "So indirectly the attitude aspect is also embedded?"

T: "Yes, in the activity of processing information also aspects of attitude is also embedded like cooperation. If the activities to communicate can make students dare to express their opinions."

In the curriculum 2013 the assessment process is conducted not only at the end, meaning that the assessment is done during the learning process takes place. So, if the assessment on the curriculum 2013 aims to evaluate whether students have achieved the expected competencies.

R: "The learning process cannot be separated from the assessment, for the assessment process on the 2013 curriculum is like what?"

T: "In the 2013 curriculum the assessment process is done not only at the end, meaning that the assessment is done during the learning process takes place. If the first time we do the assessment only in the end just like a daily test or six month only. So if the assessment on the curriculum 2013 aims to evaluate whether students have achieved the expected competencies. If there are students who have not achieved the competence given remedy like duty."

According to the teacher the learning process on the curriculum 2013 is centered on students, while the teacher only as a facilitator. So the teacher just tells learning objectives only, then teachers facilitate students to achieve goals learning by giving a problem, then completing the student with discussion or question and answer, reading or summarizing you [15]. Other than that, in the learning process is not only the aspect of the knowledge implanted to the students but the learning process in the curriculum 2013 instill aspects of attitude and skills as well

4. Conclusion

The teacher has not understood all the activities on scientific approach, i.e. on observing activities and asking activities. On assessment process, according to teacher assessment process on curriculum 2013 done during the learning process. To assess the knowledge aspect can be done by giving the quiz at the end of the learning and the instrument in the form of questions description, attitude aspects by making observations during the learning process and instrument in the form of observation sheet, skill aspect with practice test or portfolio and instrument in the form of rubric.

Observing activities conducted by teachers by asking students to listening to the teacher's explanation of sample space, sample point, and frequency hope. Questioning activities conducted by teachers by inviting students question and answer related to the material being studied that is about determining opportunities and frequency of expectation of an event, this activity is chosen because teachers can fostering an active attitude of students in the learning process. Activities collecting information that occurs i.e. students reading books and student worksheet for answer questions or questions given by the teacher. On activities collecting this information teachers instill an active attitude to the students. Associating activities that occur are students using the information obtained from books and student worksheet to answer questions given by the teacher.

Reference

[1] Akinoglu O and Tandogan R O 2007 The effects of problem-based learning in science education on students' academic achievement, attitude and concept learning *Eurasia Journal of Mathematics, Science & Technology Educational* **71** 3

[2] Rafiepour A 2017 The gap between theory and action trough a model for classroom inquiry *International Journal on Emerging Mathematics Education* **1** 185

[3] Wilkins J L M 2010 Elementary school teacher's attitudes toward different subject *International Journal of Teacher Educator* **1** 23

[4] Sundayana R, Herman T, Dahlan J A and Prahmana R C I 2017 Using ASSURE learning design to develop students' mathematical communication ability *World Transactions on Engineering and Technology Education* **15** 245

[5] Ornstein A and Hunkins F 2009 *Curriculum design. curriculum: foundations, principles and issues* (Boston: Pearson Allyn and Bacon)

[6] Dawson V and Schibeci R 2003 Western Australian high school students' attitudes towards biotechnology processes *Journal of Biological Education* **38** 7

[7] Ma X and Kishor N 1997 Assessing the relationship between attitude towards mathematics and achievement in mathematics: a meta-analysis *Journal for Research in Mathematics Education* **1** 26

[8] Hendler B 2010 Teachers as curriculum leader: A consideration of the appropriateness of that role assignment to classroom-based practitioners *International Journal of Teacher Leadership* **3** 32

[9] Stacey K 2011 The PISA view of mathematical literacy in Indonesia *Journal on Mathematics Education* **2** 95

[10] Moleong L J 2013 *Metode Penelitian Kualitatif* (Bandung: PT Remaja Rosdakarya)

[11] Budiyono 2003 *Metodologi Penelitian Pendidikan* (Surakarta: UNS Press)

[12] Kusdinar U, Sukestiyarno, Isnarto, and Istiandaru A 2017 Krulik and Rudnik model heuristic strategy in mathematics problem solving *International Journal on Emerging Mathematics Education* **1** 205

[13] Lamichhane B R and Belbase S 2017 Images of mathematics held by undergraduate students *International Journal on Emerging Mathematics Education* **1** 147
[14] Sutopo H B 2002 *Pengantar Penelitian Kualitatif* (Surakarta: Universitas Sebelas Maret Press)
[15] Suhaimi Z, Shahrill M, Abbas N A H, Tengah K A, Roslan R, and Yusof N 2017 Exploring the use of journal writing in mathematics classroom *International Journal on Emerging Mathematics Education* **1** 41

IOP Conf. Series: Journal of Physics: Conf. Series **943** (2017) 012035 doi:10.1088/1742-6596/943/1/012035

Addition of fraction in swimming context

R I I Putri[1], M S Gunawan and Zulkardi[1]

[1]Universitas Sriwijaya, Jl. Srijayanegara, Palembang, Indonesia
[2]Universitas Islam Negeri Raden Fatah, Palembang, Indonesia

E-mail: ratu.ilma@yahoo.com

Abstract. This study aimed to produce learning trajectory that can help students in learning fractions by using swimming context. The study involved 37 fourth grade students with different capabilities in Elementary School IBA, South Sumatra, Indonesia. This study used an instructional theory called Indonesian version of Realistic Mathematics Education (PMRI). This research used design research method with three stages: preliminary design, the design experiment, and retrospective analysis. Several techniques used for collecting data including a video recording of students interaction in the group discussion, students' work, and interviewing the students. To conclude, the swimming context could stimulate students' informal knowledge about the meaning of fractions in which it can be used in the additional learning either the same denominator or different denominator.

1. Introduction

Building the understanding of fractions for elementary school students is not easy to do. Learning fractions in Indonesia tend to focus on procedures only. The problem is a variety of fractions meaning is one of the difficulty causes in the learning fractions [1]. The students should be given the widest possible opportunity to explore the meaning of fractions before students study the relationship between fractions and operations on fractions [2]. Another problem faced by students is they got difficulty in adding the fractions, especially with different denominator fractions. The difficulty in conventional teaching is caused by the hazy understanding of fractions itself. Students do not understand the actual meaning of fractions, but they are forced to continue the learning process by operating fractions. Also, students tend to follow the way given by teacher mechanistically. The teacher shows a method, with some examples, which pupils then apply to similar [3].

Mathematics has always been one of the most significant issues in an individual's life [4]. Children need mathematics when they graduate from school or when they start working [4]. Mathematical competence is increasingly recognized as crucial for study in a wide range of science disciplines [5]. Learning should be more meaningful so that students are more active and understanding of the subjects taught. Understanding how mathematical conceptions are constructed can help in thinking about teaching with the aim of encouraging learning [6]. Therefore, it is necessary a promising approach be applied, that is Realistic Mathematics Education, in Indonesia [7]. PMRI selected because in line with Curriculum 2013 in which its objectives is to provide knowledge to the students completely and not fragmented. The curriculum 2013 emphasizes the involvement of the students to find the lesson concepts through the teachers' role as a facilitator. PMRI provide opportunities for teachers and students to interact between the teacher with students and students with students. With the interaction, the learning atmosphere in the classroom is expected to be more conducive to achieving the learning objectives. PMRI is an approach that starts from the real things for the students, and emphasize the skills of discussing the process with classmates so that at last their discovery that they can use to solve problems either individually or in groups [8]. The situation of students' lives is not only what is the real extent of the students' views but also all imaginable of students, affordable by their imagination [9, 10]. Mathematics must be close to children and be relevant to everyday life situation, therefore the contextual situation that is relevant and familiar to students need to be elaborated within mathematics learning [11]. Learning starts from a real thing so that students can get involved in the learning process significantly. In PMRI, the teaching is built on informal knowledge of students; it is important to give students the opportunity to explore some of the situations of daily lives where the fractions play a role [9].

PMRI requires a context which is close to children to help students in understanding the lesson [8]. Freudenthal stated education should lead students to use a variety of situations and opportunities to reinvent mathematics in their way [12, 13]. Mathematical concepts arise from the process of mathematization-starting from accomplishment which related to the context (context-link solution), students gradually developed it as the device and comprehend the mathematics to a more formal level [13]. Therefore, in this study using real context that can be imagined by the students to stimulate informal knowledge to the formal knowledge. The swimming context was chosen because it can represent fractions using measurements. The measurement concerns the identification of a length which uses the length as a part in determining the length [14]. The shape of the pool is one model that allows representing parts of the whole.

From the above discussion, the aim of this paper is to develop a theory of learning to assist students to understand the concept of the adding fractions using the swimming context.

2. Method

2.1. *The subject of the research*

The study involved 37 fourth grade students of IBA Palembang, South Sumatra, Indonesia. Subjects were selected randomly with heterogeneous capabilities that will be divided into groups with the heterogeneous capabilities. Those students had not studied about fractions yet before. Six students participated in the pilot experiment to make some adjustments to the Hypothetical Learning Trajectory (HLT). Then, thirty-one students participated in the teaching experiment.

2.2. *The design of the research*

This study used research design as the proper approach to achieving the objectives of the research. Research design consisted of three stages, namely the preliminary design, the design of experiment (pilot experiment and teaching experiment), and retrospective analysis [15].

In the perspective of design research, the objective of the first stage was to formulate the local instructional theory (LIT) which can be elaborated and perfected while performing experiments [16, 17]. At the preliminary stage, researcher examined the literature about fractions, the concept of supporting fractions material (meaning of fractions, the addition of fractions with the same denominator, and the addition of fractions with the different denominator), the fractions learning model, the swimming context, the fractions learning in the curriculum, PMRI approach, and design research which is used as research methods. After that, the researcher discussed with the teacher who will teach at the teaching experiment stage to learn more about the condition and capabilities of students. After receiving information from the teacher, researcher designed HLT as an image of fractions learning sequence using swimming context with PMRI approach.

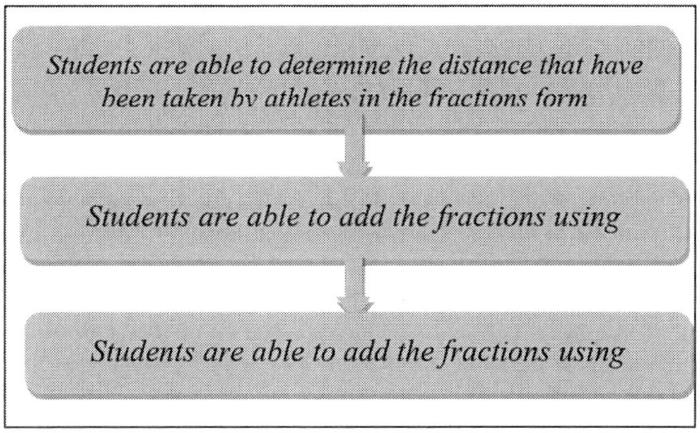

Figure 1. HLT

Swimming was chosen because it can be represented through Athlete's Distance Measurement theory while swimming in the certain time can be determined the parts. In this activity, students also used fractions bar as a model for studying fractions. Through the bar, students used a smaller field to find the names of fractions for the parts that were provided [14].

After that, the prototype which had been designed at the preliminary design stage was tested at the experiment stage. The purpose of teaching experiment is to test and improve LIT which is developed at an early stage and to develop an understanding of how it works [16]. The teaching experiment has purposed to collect data to answer the research questions [17]. The design experiment stage was conducted in two cycles, pilot experiment, and teaching experiment.

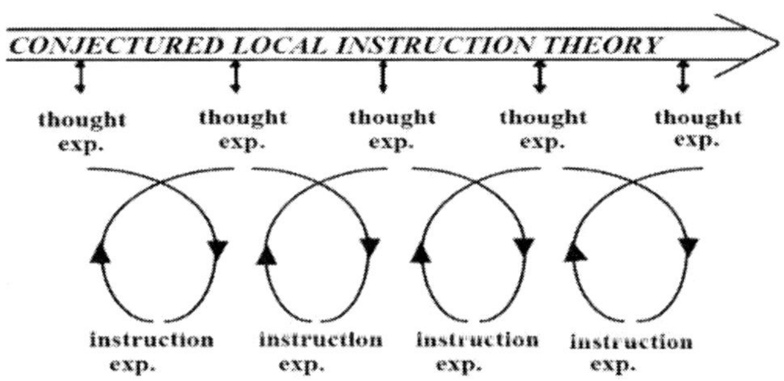

Figure 2. A cumulative cyclic process in design research [16, 17]

The first cycle was conducted as an experiment design; this stage involved six students with each of the two high-ability students, two medium-ability students, and two low-ability students. The purpose of this pilot experiment was to adjust the content and sequence of activities that have been arranged and to increase the learning activities in order to gain a better design in the next cycle [17], the second cycle was conducted as the actual experiment activity, in this cycle, involving 31 students with heterogeneous capabilities, and the activities were divided into groups consisting of students with heterogeneous capabilities. A reflection on the learning process at the end of each lesson concerning the strong and weak points of the subjects was also done [17].

2.3. Data collection and data analysis

Some activities for data collection such as interviews, video observation, and students' answer sheets. Before conducting the learning, researcher and teacher discussed learning which will be carried out. The role of the researcher in this learning activities was to ask the students some additional questions, observe the learning activities, coordinate activities, and to make changes to activities which are required to provide relevant information on the research [18, 19].

During the experiments, learning activities thoroughly and students' discussion activity was recorded. In the process of discussion, the teacher while interviewing the members of the group asked about how far their understanding toward the problems on the activity sheet. After the teaching experiment, we collected the students' answers on the answer sheet. This data was analyzed to obtain insights from the learning process of students and to investigate how far the learning objectives were achieved.

Data analysis was conducted in a retrospective analysis stage. At this stage, HLT which is used as a guideline in analyzing all the data on the experimental stage compared with the actual learning in the classroom. The conclusion from this analysis is used as an answer to the problem formulation [17]. The main result is not a worked design, but the reasons of how it works [15, 16, 17].

3. Results and discussion

3.1. Activity 1: Understanding the meaning of fraction

The problems are given in activity 1 related to elements of fractions. At the beginning of learning, the teacher gave apperception by asking "Have you swum?", "Tell your experience!", "How do you think of a swimming athlete can be declared victorious in the race?". This apperception activity made students enthusiastic in answering the teacher's questions. This activity also aimed to inform the context that will be used during the learning process.

Next, the teacher gave instruction to the students to sit in groups and gave LAS 1 to students, and they started working in groups. Students discussed and did the activity and the teacher as a facilitator and to observe students' activities. After finishing the discussion and working LAS 1, the students presented the results of their answers. Problems presented in LAS 1 were part of athletes that have been taken while swimming. It can be seen in the following figure.

Figure 3. Problem in activity 1

On the first question, it was asked the length of swimming pool. Then, on the second question, it was asked the distance of athletes. After that, the students determined which parts which are reached by athletes. From those answers, students' understanding of the meaning of fractions began to be built. Students were able to know that the part that has been taken by athletes was one of the fractions meanings. At last, the students concluded that the distance which is reached by the athletes was the numerator, while the length of the pool was the denominator of the fraction. The results of the students' answers can be seen in Figure 4.

To clarify the students' answers, here are following transcripts of conversations with students.

Teacher :How far the distance that has taken by the athlete?
Student :Half of the pool
Teacher :Why did you say that?
Student :Let's see. The end of the pool here, and here is the beginning of the pool (pointing 0 and 50)
Teacher :What is the meaning of 25?
Student :25 is the location where he arrives
Teacher :Oh ... what about the 50's?
Student :It is the whole pool length

Teacher :Did you know in the fractions, what is the meaning of the number 25?
Student :Numerator, and here is the denominator (pointing to the number 50)

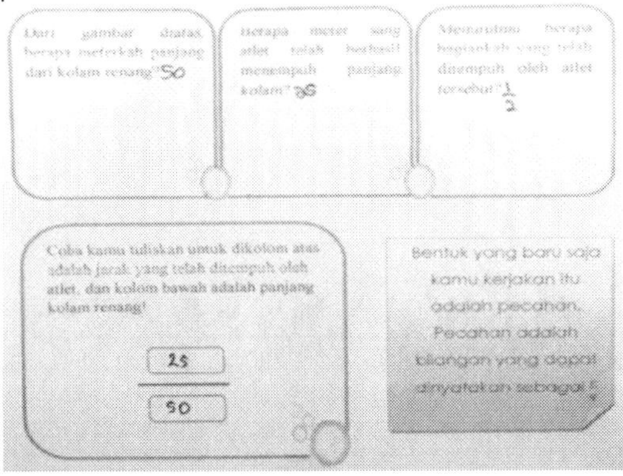

Translation:
1. From the picture above, how long the pool? (in meters)
2. How much distance taken by athlete? (in meters)
3. How much part of the pool has been taken by athlete?
4. Fill the top box is a section taken by athlete and the bottom box is the length of the pool!

Conclusion:
The form that you write is faction. Fractions are numbers that can be expressed by $\frac{a}{b}$.

Figure 4. The student's answer in activity 1

The conversation above describes how the student begins to understand the situation of this problem. The student concluded that the distance of that athlete reach the swimming pool is the part that has been taken, while the length of the pool is the whole should be taken. So that when answering the question "how many parts have been taken by the athlete?" He was able to answer half the pool. The answer was the expected answer; the students knew that the meaning of parts in question was about the form of fractions. Then, from that problem, he was led to know the elements of fractions; those are the numerator and denominator. By answering the last question, he was able to answer that the distance enumerated by the athlete was the numerator and the distance of the whole pool was the denominator. From this, it was concluded that the students' understanding of the real meaning of fractions had shown good results.

3.2. Activity 2: Adding the fractions with the same denominator

In this activity, students were asked to resolve the problem where an athlete reached distance in certain seconds, and then the two distances added together. In this activity, students were asked to use the fraction bar in the process of finding the result of the fractions addition. Students' activity started from shading the bar based on the distance covered for 20 seconds; then students shaded more another fraction bar based on the distance in the 30 seconds. Furthermore, the students cut one fraction bar by its shading, after that the students patched the shading which has been cut to the top of the intact bar. From the bar, it was got the addition results of the parts that have been taken by athletes. Students' answer can be seen in Figure 5.

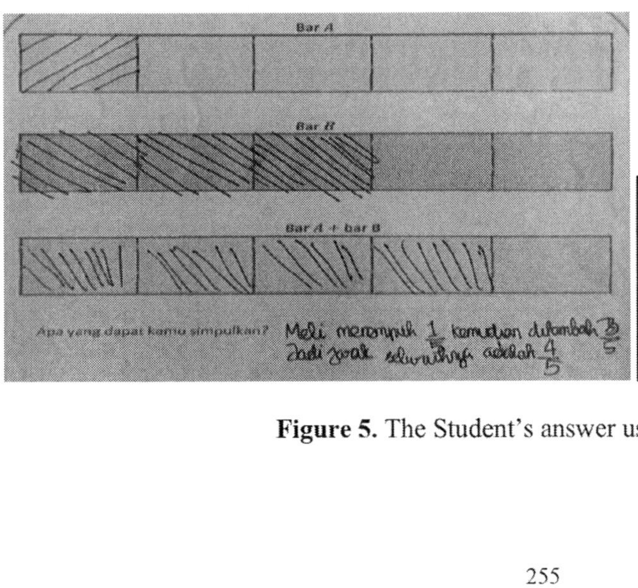

Translation:
What can you conclude?
Meli has been take $\frac{1}{5}$ parts of the pool, then she swim again as far as $\frac{3}{5}$ parts of the pool.
So, the result is $\frac{4}{5}$

Figure 5. The Student's answer using fraction bar

In Figure 5, students can shade properly in the parts which have been determined on the questions. Then, by the bar helping, students added those parts and moved it to the answers' column. He re is the following conversation transcript of the students' answers.

Teacher : *How much does yellow represent?*
Student$_1$: $\frac{1}{5}$
Teacher : *How many regions will be shaded?*
Student$_1$: *1*
Teacher : *Of how many regions?*
Student$_{1,2}$: *5 regions*
Teacher : *And how about the blue one?*
Student$_{1,2,3}$: $\frac{3}{5}$
Teacher : *Then what should we do?*
Student$_2$: *There! (Student attached Bar A and B)*
Teacher : *So how much is that?*
Student$_1$: $\frac{4}{5}$ *(Students shaded the results boxes)*
Teacher : *So how much is the sum?*
Student$_{1,2}$: $\frac{4}{5}$
Teacher : *What can you conclude?*
Student$_1$: $\frac{1}{5}+\frac{3}{5}=\frac{4}{5}$
Teacher : *When the bottoms have the same number, you can directly add them up.*
Student$_1$: *Yes, you can.*

The conversation above illustrates activities in one of the groups in adding the same denominator fractions using a fraction bar media. We can see how students develop their understanding of the concept of fractions. Fraction bar which is used likened to a pool, and then the students shading the first distance reached by athlete in the 20 seconds, then shading the second distance reached by athlete in 30 seconds. From this bar shading activity, understanding of the meaning of fractions is required. Students knew that athletes have to take what parts of the pool, which will both parts should be added together. Seeing the conversations and results of students' answers, fraction bar was very helpful to solve the problem. Fraction bar represented trajectory of swimming pool in the smaller units, so that it can be folded or torn to fractions addition activities. Students were able to answer questions on the problem of the same denominator well.

The next problem in this activity was the students were asked to add fractions with the same denominator by using the number line. The number line had been in the column, the students only determined the exact point of the swim athlete's distance. The distance was determined by the students should be added using the number line. On these problems, students should add up within two athletes who were swimming. Students determined the distance of the first athlete on the number line, and then it is added with the second athlete's distance. So that students can draw the conclusion that the sum of both these athletes can be seen from the number line that has been drawn by the students. Here are the students' answers shown in Figure 6.

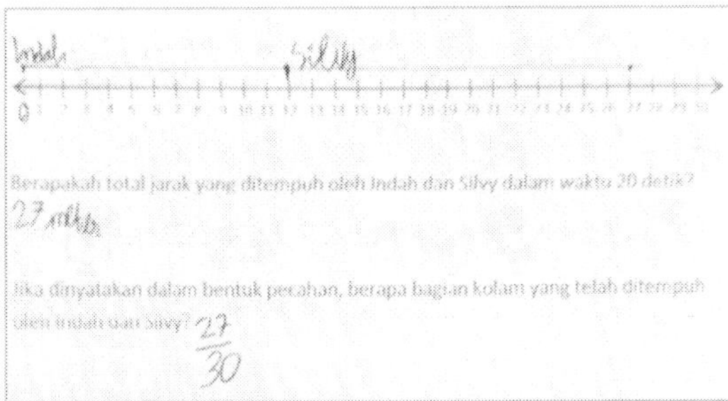

Translation:

1. How the overall distance that has been taken by Indah and Silvy in 20 seconds? (in meters)
2. If expressed as a fraction, how many parts have been taken by Indah and Silvy?

Figure 6. The Student's answer using fraction bar number line

Figure 6, students correctly determined the point that represented the athlete's distance. From the students' answers, it can be seen that students have understood how to add fractions with the same denominator by using the number line.

3.3. Activity 3: Adding the fractions with different denominator

In this activity, students were asked to solve problems where athletes from Japan No. 5 and Korea No. 8 reached distance at the certain second, then the two distances added together. The problem can be seen from Figure 7.

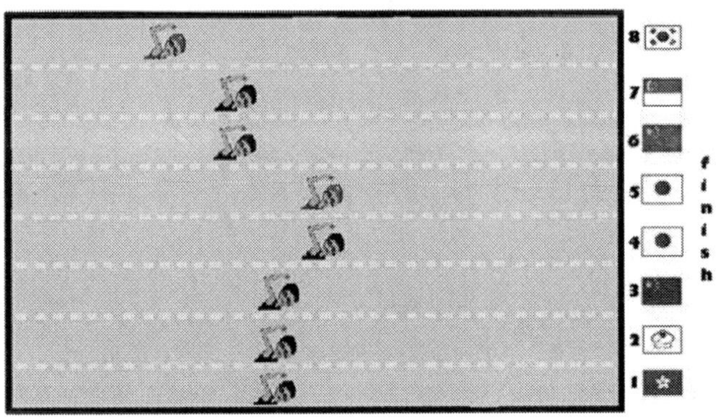

Figure 7. Problem in Activity 3

In this activity, students were asked to use the fraction bar in the process of finding the result of the addition of fractions. The way using fraction bar on this activity was equivalent to the fraction bar on activity 2. Students were given fraction bar with different parts, and each student shading fraction bar by a predetermined part. Then, the students cut one of fraction bar and patched it to the intact fraction bar, after that the students calculated the amount of the existing parts in the fraction bar. Pictures of students' activity can be seen in Figure 8.

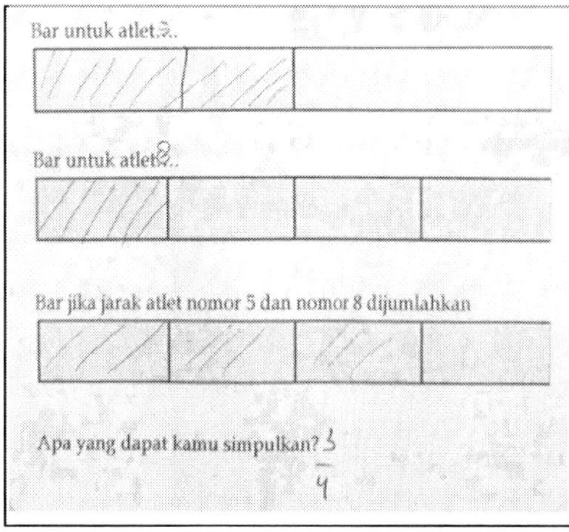

Translation:
1. Bar for athlete number 5
2. Bar for athlete number 8
3. Bar for the addition of athlete number 5 and 8's travelled
4. What can you conclude? $\frac{3}{4}$

Figure 8. The student's answer using fraction bar

In Figure 8, students could add fractions with the same denominators. Students added up the two fractions by using a bar that has been given. Then, they moved the results on the activity sheet that has been provided. This is in accordance with HLT that has been designed by researcher. For more details can be seen in the following transcript.

Teacher: How did you point the athlete position in number 5?
Student: Measured, half of the pool.
Teacher: How do know it's half of the pool?
Student: Because it's in the middle.
Teacher: Is there any other way? How about using the bar?
Student: We fold it.
Teacher: How do you show a half when you fold it?
Student: It was 2, when you fold them, it becomes $\frac{1}{2}$
Teacher: Now for number 8, which bar will you use?
Student: This one, the athlete is still in the $\frac{1}{4}$ *of the box. So you fold it, if it's collapsible, fold it again, and again. Done.*
Teacher: Oh .. so you folded it, then how much is left?
Student: 1
(He cuts the bar)
Teacher: The bar you cut, how much does it show?
Student: $\frac{1}{2}$*, then you attach them together.*
Teacher: Then how much is it?
Student: $\frac{3}{4}$

In the next activity, the students resolved different problems but still used the swimming context. Students were required to add the two athletes' distance. The two athletes' distance was known by using part of the pool that has been taken. However, in this question, the pool part used fractions with different denominator. Then, students solved the problem using a number line. Figure 9 shows the student's work. He correctly added the fractions using a number lines.

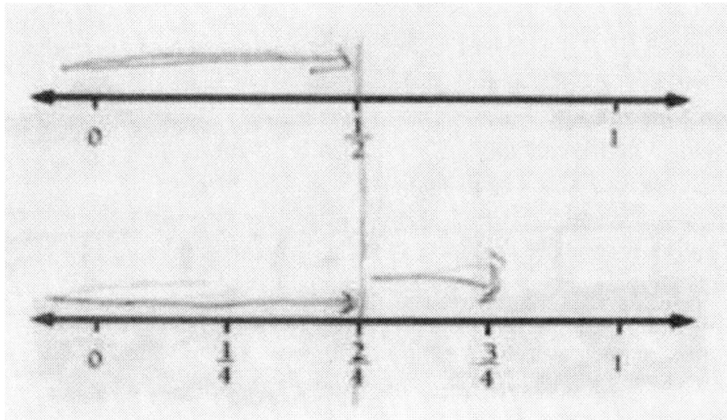

Figure 9. The student's answer using number line

Also, one of the groups have understood that equating the denominator can be done by finding the Least Common Multiple from both denominators. Students have known that the addition of fractions with different denominator cannot be done directly, but it must equalize the denominator first.

This learning was designed to produce the learning trajectory in the learning of addition of fractions material by using swimming context. There were three learning activities of addition of fractions on the learning trajectory design that have been designed and implemented previously. The first activity helped students in understanding the actual meaning of fractions through the swimming context. The second activity helped students in adding fractions with the same denominator through problems of adding distance from the specified swimming athletes. The third activity helped students in adding fractions with different denominator through the problems of adding distance of the two swimming athletes. All of these three activities conducted during the three sessions of learning.

Several previous studies have also been getting maximum results in learning fractions using the PMRI approach, showed that through a series of activities that have been done to help students in learning fractions [21]. Other results indicated by the research that showed that the results of the use of learning media in the form of fractional card can support students' understanding in comprehending the students' fractions from informal stage to formal stage [22].

In develops teaching materials based PMRI approach on summation material fractions also showed the good results, namely learning process of students using teaching materials of the addition of fractions PMRI approach based extremely guide students to develop ideas and foster the creativity in solving problems [1]. From the results of the researches on the fractions material that use PMRI, show that the PMRI approach is good enough to be applied in either the fractions material or other materials [23].

4. Conclusion

Using the results and discussions, it can be concluded that the activities in the student worksheet has been designed to help students understand and resolve problems in an addition of fraction intuitive informally to the formal problem-solving. The things in the learning trajectory in this study to assist students in learning the real meaning of fractions using swimming context and experience in learning meaningfully and fun given that the activity using fraction bars which can help students add fractions.

5. Acknowledgements

Researchers expressed acknowledgement to KEMRISTEKDIKTI-Indonesian Directorate General of Higher Education and Technology has funded this research in the skim of Hibah Pasca.

References

[1] Ullya, Zulkardi and Putri R I I 2010 Desain bahan ajar penjumlahan pecahan berbasis pendidikan matematika realistic (PMRI) untuk siswa kelas IV sekolah dasar negeri 23 Indralaya *Jurnal Pendidikan Matematika* **4** 86

[2] Sari E A P, Juniati D and Patahudin S M 2012 Early fractions learningof 3rd grade students in SD laboratorium Unesa *Journal on Mathematics Education* **3** 17

[3] Tok S, Bahtiyar A and Karalök S 2015 The effect of teaching mathematics creatively on academic achievement, attitudes towards mathematics, and mathematics anxiety *International Journal of Innovation in Science and Mathematics Education* **23** 1

[4] Tok S and Keskin A 2012 The effect of fast draw learning strategy on the academic achievement and attitudes towards mathematics *International Journal of Innovation in Science and Mathematics Education* **20** 1

[5] Jackson D C, Johnson E D and Blanksby T M 2014 A practitioner's guide to implementing cross-disciplinary links in a mathematics support program *International Journal of Innovation in Science and Mathematics Education (formerly CAL-laborate International)* **22** 67

[6] Roig A I, Linaves S and Penalva, M C 2012 Different moments in the participatory stage of the secondary students' abstraction of mathematical conceptions *Bolema, Rio Claro (SP)* **26** 1345

[7] Stacey K, Almuna F, Caraballo RM, Chesne JF, Garfunkel S, Gooya Z, …, Zulkardi, Z 2015 PISA's Influence on Thought and Action in Mathematics Education. In Stacey K, Turner R (eds.) *Assessing Mathematical Literacy* DOI 10.1007/978-3-319-10121-7_15

[8] Zulkardi 2002 *Developing a learning environment on realistic mathematics education for Indonesian student teachers* (Ph.D Thesis University of Twente, Enschede, the Netherlands)

[9] Van Den Heuvel-Panhuizen M 2003 The didactical use of models in realistic mathematics education; an example from a longitudinal trajectory on percentage *Educational Studies in Mathematics* **54** 9

[10] Putri R I I 2014 *Evaluasi program pelatihan Pendidikan Matematika Realistik Indonesia (PMRI) bagi guru matematika Sumatera Selatan* (Presented on Seminar Nasional Implementasi Kurikulum)

[11] Ekawati R and Kohar A W 2016 Innovative professional developtment within PMRI in Indonesia *International Journal of Innovation in Science and Mathematics Education* **24** 1

[12] FrudenthaL H 1991. *Revisiting Mathematics Education-China Lectures.Dordrecht* (Kluwer Academic Publisher)

[13] Hasratuddin and Karnasih I 2017 Improving critical thinking and emotional intelligence capabilities of secondary school students through realistic mathematic education approach *International Journal of Innovation in Science and Mathematics* **5** 1

[14] Van De Walle J A, Karp K S and Bay Williams J M 2013 *Elementary and Middle School mathematics Teaching Developmentally (Eight Ed.)* (United States of America: Pearson Educations)

[15] Gravemeijer K 2004 Local instruction theories as means of support for teachers in reform mathematics education. *Mathematical Thinking and Learning* **6** 105

[16] Gravemeijer K and Cobb P 2006 *Design research from the learning design perspective.* In Van den Akker J, Gravemeijer K, McKenney S and Nieveen, N. (Eds.). Educational Design Research (London: Routledge)

[17] Bustang, Zulkardi Z, Darmawijoyo, Dolk M and Van Eerde D 2013 Developing a local instructional theory for learning the concept of angle through visual field activities and spatial representations. *International Education Studies* **6** 58

[18] Van Nes F and Van Eerde D 2010 Spatial structuring and the development of number sense: A case study of young children working with blocks. *The Journal of Mathematical Behavior* **29** 145

[19] Risma D A, Putri R I I and Hartono Y 2013 On developing students' spatial visualization ability. *International Education Studies* **6** 1

[20] Cobb P, Confrey J, Didessa A A, Lehrer R and Schauble L 2003 Design experiments I educational research. *Educational Researcher* **32** 9

[21] Khuriyati L, Hartono Y and Somakim 2015 Desain pembelajaran operasi pecahan menggunkaan kertas berpetak di kelas IV *Jurnal Paradikma* **8** 62
[22] Zabeta M, Hartono Y and Putri R I I 2015 Desain pembelajaran materi pecahan menggunakan pendekatan PMRI di kelas VII *Beta Jurnal Pendidikan Matematika* **3** 98
[23] Putri R I I, Zulkardi Z, "Fraction in Shot-Put: A Learning Trajectory," in *Research and Education for Developing Scientific Attitude in Sciences and Mathematics*, AIP Conference Proceedings 1868, edited by C. Kusumawardani *et al.* (AIP Publishing, New York, 2017)

IOP Conf. Series: Journal of Physics: Conf. Series **943** (2017) 012036 doi:10.1088/1742-6596/943/1/012036

Optimal control of nutrition restricted dynamics model of Microalgae biomass growth model

R Ratianingsih[1], Azim[1], N Nacong[1], Resnawati[1], Mardlijah[2] and B Widodo[2]
[1]Tadulako University, Jl. Soekarno Hatta Palu, 94111, Indonesia
[2]Sepuluh Nopember Institute of Technology, Keputih Sukolilo, Surabaya, Indonesia

E-mail: ratianingsih@yahoo.com

Abstract. The biomass of the microalgae is very potential to be proposed as an alternative renewable energy resources because it could be extracted into lipid. Afterward, the lipid could be processed to get the biodiesel or bioethanol. The extraction of the biomass on lipid synthesis process is very important to be studied because the process just gives some amount of lipid. A mathematical model of restricted microalgae biomass growth just gives 1/3 proportion of lipid with respect to the biomass in the synthesis process. An optimal control is designed to raise the ratio between the number of lipid formation and the microalgae biomass to be used in synthesis process. The minimum/ Pontryagin maximum principle is used to get the optimal lipid production. The simulation shows that the optimal lipid formation could be reach by simultaneously controlling the carbon dioxide, in the respiration and photosynthesis the process, and intake nutrition rates of liquid waste and urea substrate. The production of controlled microalgae lipid could be increase 6.5 times comparing to the uncontrolled one.

1. Introduction

Microalgae growth were discussed in many studies [1-8]. Its growth models were studied in several papers, i.e Droop model [9], which discusses a simple model that represents its growth restricted by nitrogen. The nitrogen is consider as dependently intracellular growth factor [10]. To optimize the microalgae neutral lipid production, the microalgae growth is controlled by maintaining the dilution rate up to reach steady state condition through the nutrition restriction.

This paper considers two restricted nutrition, i.e liquid waste and urea, as the nutrition of microalgae and optimizes its growth by controlling the ratio of the CO_2 absorption rate due to respiration process and the glucose formation rate due to photosynthesis during the day. The other influenced process are the lipid synthesis process that mobilized the CO_2 and its released because of respiration process in the night. The phenomenon is expressed in a differential equations system that influenced by two added controlled parameters. The first one is due to maintain the ration in order to guarantee the sufficient of CO_2 in glucose formation because of photosynthesis process, while the other is due to nutrition absorption process. Those two parameters control is placed in an appropriate ways and is expected to raise the microalgae lipid production.

The referred model is governed and is analyzed its stability. This paper discuses the optimization of the microalgae lipid production using maximum Pontryagin criteria [11]. The optimum solution is derived for an objectives function that constructed to maximize the number of glucose, biomass and lipid formation and minimize the number of liquid waste and urea nutrition absorption for such Hamiltonian function and dynamics constrain. A numerical simulation is shown to have a better drawing of the optimal growth of microalgae to have the maximum microalgae lipid production.

2. The mathematical model

The mathematical model governed is constructed by consider respectively the growth of CO_2, glucose formation, microalgae biomass, lipid formation and nutrition. The model is considered to support an optimal strategy of growth and lipid accumulation based on experimental design. In this paper the model is constructed based on the lipid synthesis diagram of microalgae in Figure 1.

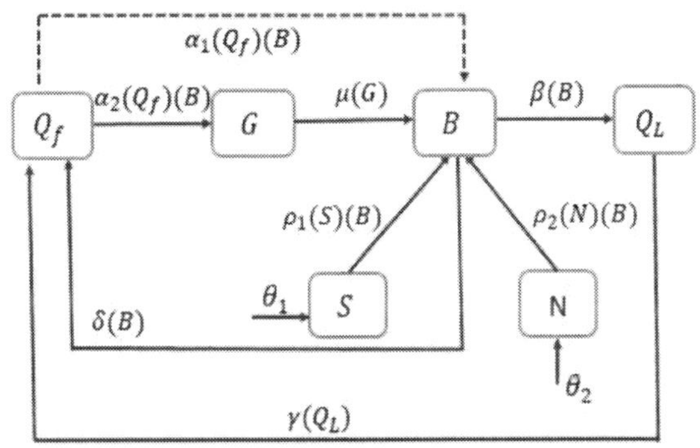

Figure 1. The lipid synthesis diagram

2.1. The dynamical system

The diagram shows that the number of CO_2 functional (Q_f) that decreases because of its absorption in the respiration process by biomass (*B*) and photosynthesis process that forms glucose (*G*), while its increasing is influenced by the CO_2 mobilization due to lipid (*L*) synthesis process and CO_2 released due to the night respiration process. The number of glucose (*G*) increasing is dependently related to the effectiveness of glucose (*G*) formation, while its decreasing is related to the transition of glucose (*G*) that used to growth the microalgae biomass (*G*). The biomass (*G*) growth is supported by two kinds of nutrition. Externally, the nitrogen (*N*) comes from the domestic waste and influenced urea (*S*), while internally, glucose (*G*) is formed from photosynthesis process. The process needs a sufficient of CO_2 that release from the respiration of microalgae at night and mobilized from the lipid (*L*) synthesis by extracting the microalgae biomass (*B*). We also have to remind that some amount of CO_2 will also absorbed by respiration process.

The representation of the model is stated in the following nonlinear dynamical system

$$\frac{dQ}{dt} = \delta B + \gamma L - \alpha_1 QB \quad (1.a)$$

$$\frac{dG}{dt} = \alpha_2 QB - \mu G \quad (1.b)$$

$$\frac{dB}{dt} = \mu G + \rho_1 S + \rho_2 N - \beta_1 B \quad (1.c)$$

$$\frac{dL}{dt} = \beta_2 B - \gamma L \quad (1.d)$$

$$\frac{dS}{dt} = \theta_1 - \rho_1 S \quad (1.e)$$

$$\frac{dN}{dt} = \theta_2 - \rho_2 N \quad (1.f)$$

where parameters α_1, α_2, β_1, β_2, γ, δ, μ, θ_1, ρ_1 and $\rho_2$are respectively stated the CO_2 absorption rate due to respiration process, glucose formation rate due to photosynthesis process, glucose loss and biomass formation rate due to photosynthesis process, biomass growth rate due to the domestic waste nutrition, domestic waste nutrition absorption rate of microalgae, Nitrogen absorption rate of microalgae, lipid formation rate due to biomass synthesis process, CO_2 release rate due to respiration process and CO_2 mobilization due to synthesis process. The last two equations of the model are consider separately such that the solutions could be determined directly for such initial value problem, $S(0) = S_o$ and $N(0) = N_o$ such that the model represents the dynamic interaction between the variables of CO_2 functional (Q), glucose (*G*), biomass (*B*), and lipid (*L*).

2.2. *The critical point*

The critical point of the system stated in (1.a) – (1.f), that represents the timeless or stagnant critical point $T(Q^*, G^*, B^*, L^*)$ where $Q^* = \frac{\delta+\beta_2}{\alpha_1}$, $G^* = -\frac{SN\alpha_2(\delta+\beta_2)}{(-\beta_1\alpha_1+\alpha_2(\delta+\beta_2))\mu}$, $B^* = -\frac{SN\alpha_1}{-\beta_1\alpha_1+\alpha_2(\delta+\beta_2)}$ and $L^* = -\frac{\beta_2 SN\alpha_1}{(-\beta_1\alpha_1+\alpha_2(\delta+\beta_2))\gamma}$. The expression of the critical point T requires $\frac{\alpha_1}{\alpha_2} < \frac{\delta+\beta_2}{\beta_1}$ that interpreted the importance of maintaining the ratio of the CO_2 absorption rate due to respiration process and the glucose formation rate due to photosynthesis during the day.

2.3. *The stability*

Stability appears as the importance of maintaining the nutrition sufficiency. The stability of the critical point is determined from the Jacobian matrix [12] that gives the following fourth order characteristic polynomial of λ

$$P(\lambda) = A(a_o\lambda^4 + a_1\lambda^3 + a_2\lambda^2 + a_3\lambda + a_4)$$
$$\text{where } A = \alpha_1\big(-\beta_1\alpha_1 + \alpha_2(\delta + \beta_2)\big)$$
$$a_o = A$$
$$a_1 = -\alpha_1(\alpha_1{}^2SN + \beta_1(\beta_1 + \mu + \gamma)\alpha_1 - (\delta + \beta_2)(\beta_1 + \mu + \gamma)\alpha_2)$$
$$a_2 = -SN(\beta_1 + \mu + \gamma)\alpha_1{}^3 - \beta_1\big((\beta_1 + \gamma)\mu + \gamma\beta_1\big)\alpha_1{}^2 + (\delta + \beta_2)\big((\gamma + 2\beta_1)\mu + \gamma\beta_1\big)\alpha_2\alpha_1 - \mu\alpha_2{}^2(\delta + \beta_2)^2$$
$$a_3 = -SN\big((\beta_1 + \gamma)\mu + \gamma\beta_1\big)\alpha_1{}^3 + \mu\big(SN\alpha_2\delta - \beta_1{}^2\gamma\big)\alpha_1{}^2 + 2\mu\alpha_2\gamma\beta_1(\delta + \beta_2)\alpha_1 - \gamma\mu\alpha_2{}^2(\delta + \beta_2)^2$$
$$a_4 = \alpha_1\mu A\gamma SN$$
$$SN = \rho_1 S + \rho_2 N$$

The stability of the critical point is identified using Routh Hurwitz criteria [13] such that the identification is determined from the sign changing of the first coulomb of Table 1 that are $a_o, a_1, b_1 = \frac{a_1a_2 - a_oa_3}{a_1}, b_2 = a_4, c_1 = \frac{b_1a_3 - a_1b_2}{b_1}$ and $d_1 = b_2 = a_4$. The positive sign of a_o is guaranteed from the existence requirement, while the positive sign of a_1 gives $\frac{\alpha_1}{\alpha_2} < \frac{\beta_1(\beta_1+\mu+\gamma)(\delta+\beta_2)}{SN\alpha_1+\beta_1(\beta_1+\mu+\gamma)}$. This result comes to a new criteria for $\frac{\alpha_1}{\alpha_2}$, that is $\frac{\alpha_1}{\alpha_2} < \min\left(\frac{\delta+\beta_2}{\beta_1}, \frac{\beta_1(\beta_1+\mu+\gamma)(\delta+\beta_2)}{SN\alpha_1+\beta_1(\beta_1+\mu+\gamma)}\right)$. The identification results of the sign of b_1, c_1 and d_1 is tabulated in Table 2. The table mentions the critical point, the existence and stability requirements.

Table 1. Routh-Hurwitz Table

λ^4	a_0	a_2	a_4
λ^3	a_1	a_3	0
λ^2	b_1	b_2	0
λ^1	c_1	0	0
λ^0	d_1	0	0

Table 2. Parameter stability identification

Critical Point	Existence requirement	Stability Requirement	Description
$T(Q^*,G^*,B^*,L^*)$	$\frac{\alpha_1}{\alpha_2} < \frac{\delta+\beta_2}{\beta_1}$	$\max\left(\frac{(\mu+\beta_1)(\delta+\beta_2)J}{(\beta_1+\gamma)(\mu+\beta_1)(\mu+\gamma)K}, \frac{(\delta+\beta_2)(\mu+\beta_1)J}{L}\right) < \frac{\alpha_1}{\alpha_2} < \min\left(\frac{\delta+\beta_2}{\beta_1}, \frac{\beta_1(\beta_1+\mu+\gamma)(\delta+\beta_2)}{SN\alpha_1+\beta_1(\beta_1+\mu+\gamma)}\right)$	unstable
		$(E-F)(G-H)>0$	stable

where E, F, G, H, J, K, L and P are in a complex following expression of parameters α_1, α_2, β_1, β_2, $\gamma, \delta, \mu, \theta_1, \rho_1$ and ρ_2

$$E = SN^2(\beta_1+\mu+\gamma)\alpha_1{}^5 + SN(\beta_1+\mu+\gamma)^2\alpha_1{}^4 + \left((\mu+\gamma)^2(\delta+\beta_2)\alpha_2 + \beta_1{}^2(\beta_1+\gamma)(\mu+\beta_1)(\mu+\gamma)\right)\alpha_1{}^3 + \alpha_1{}^2 SN + \alpha_2(\delta+\beta_2)(SN\mu\beta_2\alpha_2)\alpha_1{}^2 + \alpha_2{}^2\left((3\mu+\gamma)\beta_1+\gamma(\mu+\gamma)\right)(\delta+\beta_2)^2(\mu+\beta_1)\alpha_1$$

$$F = SN\left((\delta+\beta_2)\beta_1{}^2 + \left((2\delta+3\beta_2)\mu + 2\gamma(\delta+\beta_2)\right)\beta_1\right)\alpha_1{}^3 + 2\beta_1\left(\left(\frac{3}{2}\mu+\gamma\right)\beta_1+\gamma(\mu+\gamma)\right)(\mu+\beta_1)\alpha_1{}^2 + \mu\alpha_2{}^3(\delta+\beta_2)^3(\mu+\beta_1)$$

$$G = \mu(SN\mu\beta_2\alpha_2)\alpha_1{}^2 + 2\mu\alpha_2\gamma\beta_1(\delta+\beta_2)\alpha_1$$

$$H = SN\left((\beta_1+\gamma)\mu+\gamma\beta_1\right)\alpha_1{}^3 + \mu\beta_1{}^2\gamma\alpha_1{}^2 + \alpha_2{}^2\gamma\mu(\mu+\gamma\beta_1)^2$$

$$J = SN\mu\beta_2\alpha_2 - 2\beta_1\left(\left(\frac{3}{2}\mu+\gamma\right)\beta_1+\gamma(\mu+\gamma)\right)$$

$$K = SN\left((\delta+\beta_2)\beta_1{}^2 + \left((2\delta+3\beta_2)\mu+2\gamma(\delta+\beta_2)\right)\beta_1\right) - (\mu+\gamma)^2(\delta+\beta_2)\alpha_2 - \beta_1{}^2$$

3. The optimization model

This paper proposes an optimal control to maintain the internal and external nutrition to have the maximal production of microalgae lipid. Its production is formulated in an objective function such that its optimal value is solved from the *Hamiltonian* function that restricted by the controlled nonlinear dynamical system.

To reach the maximal of microalgae lipid production two programs are proposed. The first one due to maximize the CO_2 consumption needed in photosynthesis and respiration process, while the other one due to minimizing the nutrition by restricting the liquid waste content. The controlled nonlinear dynamical system is designed to optimize the lipid production by two controlled parameters, that are u_1 and u_2. Both parameters respectively represent mixotrophic by providing light with organic carbon [14] and Nitrogen nutrition inlet control [15-16]. The controlled model is stated

$$\frac{dQ}{dt} = \delta B + \gamma L - u_1\alpha_1 QB \quad (2.a)$$

$$\frac{dG}{dt} = u_1\alpha_2 QB - \mu G \quad (2.b)$$

$$\frac{dB}{dt} = \mu G + \rho_1 S + \rho_2 N - \beta_1 B \quad (2.c)$$

$$\frac{dL}{dt} = \beta_2 B - \gamma L \quad (2.d)$$

$$\frac{dS}{dt} = \theta_1 - (1-u_2)\rho_1 S \quad (2.e)$$

$$\frac{dN}{dt} = \theta_2 - (1-u_2)\rho_2 N \quad (2.f)$$

The controlled model in equations (2.a) – (2.f) is solved using *Pontryagin* maximum principle [11] by defining an objective function

$$f(G,B,L,S,N,u_1,u_2,t) = \max \int_0^{tf} \left(G(t) + B(t) + L(t) - S(t) - N(t) + \frac{1}{2}{u_1}^2 + \frac{1}{2}(1-u_2)^2 \right) dt$$

The optimal solution of the controlled model is derived by define the following *Hamiltonian* system

$$H = f(G,B,L,S,N,u_1,u_2,t) + \lambda g(G,B,L,S,N,u_1,u_2,t)$$

Solving the state and co state equations simultaneously

The state equations:

$$\dot{Q} = \frac{\partial H}{\partial \lambda_1} = \delta B + \gamma L - u_1\alpha_1 QB$$

$$\dot{G} = \frac{\partial H}{\partial \lambda_2} = u_1\alpha_2 QB - \mu G$$

$$\dot{B} = \frac{\partial H}{\partial \lambda_3} = \mu G + \rho_1 S + \rho_2 N - \beta_1 B$$

$$\dot{L} = \frac{\partial H}{\partial \lambda_4} = \beta_2 B - \gamma L$$

$$\dot{S} = \frac{\partial H}{\partial \lambda_5} = \theta_1 - (1-u_2)\rho_1 S$$

$$\dot{N} = \frac{\partial H}{\partial \lambda_6} = \theta_2 - (1-u_2)\rho_2 N$$

The co-State equations:

$$\dot{\lambda_1} = \frac{\partial H}{\partial Q} = -u_1\alpha_1 B\lambda_1 + u_1\alpha_2 B\lambda_2$$

$$\dot{\lambda_2} = \frac{\partial H}{\partial G} = 1 - \mu\lambda_2 + \mu\lambda_3$$

$$\dot{\lambda_3} = \frac{\partial H}{\partial B} = 1 - u_1\alpha_1 Q\lambda_1 + u_1\alpha_2 Q\lambda_2 - \beta_1\lambda_3 + \beta_2\lambda_4$$

$$\dot{\lambda_4} = \frac{\partial H}{\partial L} = 1 + \gamma\lambda_1 - \gamma\lambda_4$$

$$\dot{\lambda_5} = \frac{\partial H}{\partial S} = -1 + \rho_1\lambda_3 - (1-u_2)\rho_1\lambda_5$$

$$\dot{\lambda_6} = \frac{\partial H}{\partial N} = -1 + \rho_2\lambda_3 - (1-u_2)\rho_2\lambda_6$$

with the stationer condition

$$\frac{dH}{d\theta} = 0$$

gives the following optimal parameter control

$$u_1 - \alpha_1 QB\lambda_1\alpha_2 + \alpha_2 B\lambda_2 = 0 \Rightarrow u_1 = \alpha_1 QB\lambda_1 - \alpha_2 B\lambda_2$$

$$-(1-u_2) + \lambda_5(u_2\rho_1 S) + \lambda_6(u_2\rho_2 N) = 0 \Rightarrow u_2 = \frac{1}{1+\rho_1 S\lambda_5 + \rho_2 N\lambda_6}$$

4. The numerical solution and discussion

The optimization solution is solved numerically for some given parameter values and initial values of variables. The initial values, i.e $Q(0) = 1, G(0) = 1, B(0) = 0.5$ and $L(0) = 1$, shows that, in the beginning process, the number of CO_2 is assumed in 100% portion as well as the glucose and no lipid formation. The value of parameters refers to [3].

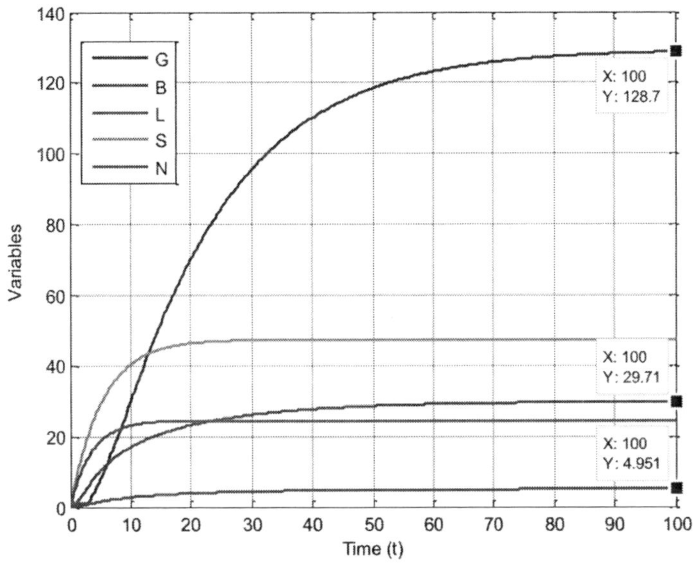

Figure 2. The Dynamic of Glucose, Biomass, Lipid, Urea and Liquid Waste

Figure 2 shows that the portions of glucose, biomass and lipid are successfully to be increased. The successes of the increasing value of the variables with respect to time is also supported by figure 3 that describe the greater value of controlled variables comparing to the related uncontrolled glucose, biomass and Lipid. A similar characteristic is also figured for the CO_2 formation in figure 4 that comparison of it in the beginning and time of intervals. At the beginning the controlled CO_2 still less than the uncontrolled one, but after that it turns to be greater. The production of controlled microalgae lipid could be increase 6.5 times comparing to the uncontrolled one.

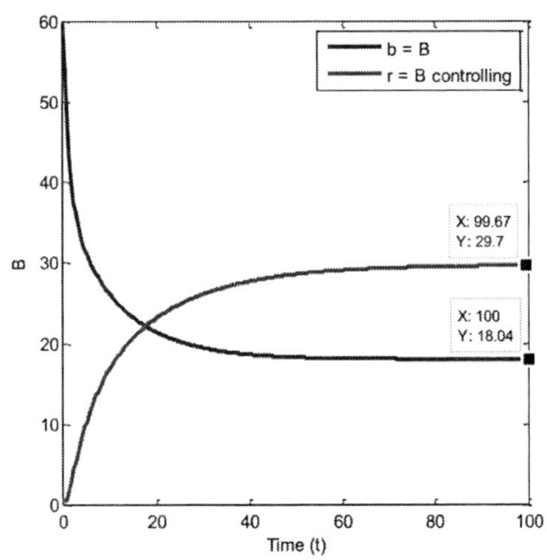

(a)

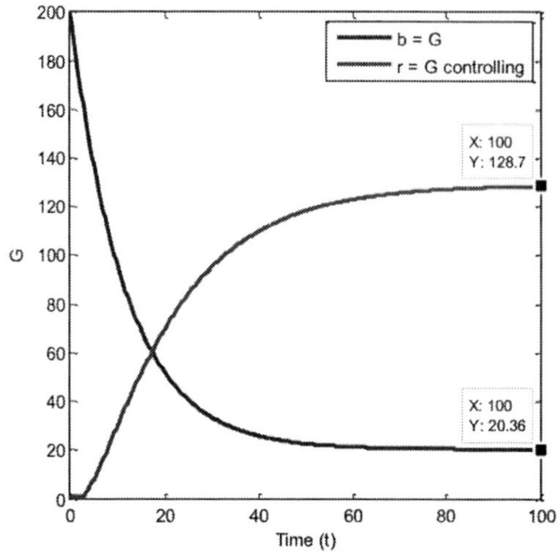

(b)

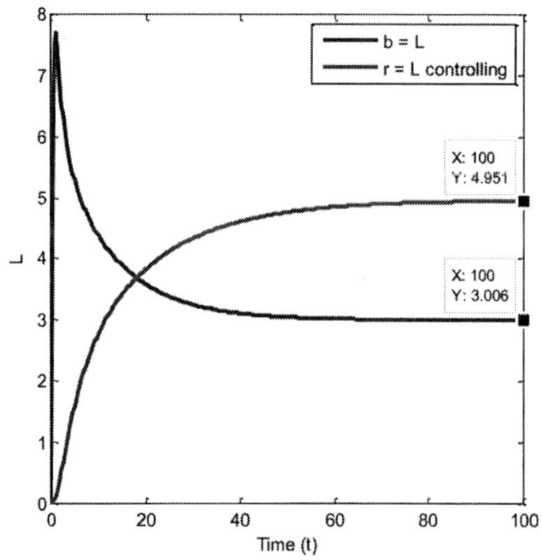

(c)

Figure 3. The Comparison of The Controlled and Uncontrolled of (a) Glucose, (b) Biomass and (c) Lipid

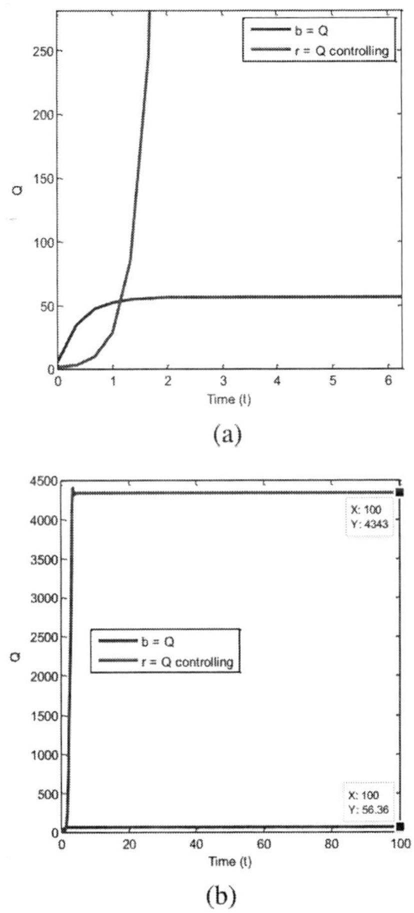

Figure 4. The Comparison of The Controlled and Uncontrolled of CO_2 formation
(a). In the beginning time of interval (b). In the end time of interval

5. The concluding remarks

The mathematical model of microalgae biomass growth that its nutrition is restricted externally and internally had been derived. The optimal solution of the model that solved using maximum *Pontryagin* principle shows that the microalgae lipid production could be maximized increase. This result is potentially to be further studied by consider the cost production of the lipid synthesis process.

Acknowledgement

All authors convince a great acknowledgement to *Pusat Riset dan Kreatifitas Matematika* (PRKM) *Program Studi Matematika FMIPA Universitas Tadulako* and Mathematics Department of Sepuluh Nopember of Institute Technology for many inspiring discussion. Many thanks are also dedicated to Directorate General of Higher Education of Ministry of Research, Technology & Higher Education, for funding this research and Dr. Mardlijah, M.T and Prof. Basuki Widodo for the intensive discussion.

References

[1] Metting F 1996 Biodiversity and application of microalgae *Journal of Industrial Microbiology* **17** 477
[2] Chisti Y 2007 Biodiesel from microalgae *Biotechnology Advances* **25** 294

IOP Conf. Series: Journal of Physics: Conf. Series **943** (2017) 012036 doi:10.1088/1742-6596/943/1/012036

[3] Gerardo M, Van Den Hende S, Venaeren H, Covard T, Skill S C 2016 Harvesting of microalgae within a biorefinery approach a review of the developments and case studies from pilot-plants *AAlgal Research* **11** 248
[4] Li Y, Horsman M, Wang B, Wu N and Lan C Q 2008 Effect of nitrogen sources on cell growth and lipid accumulation of green algae Neochloris oleabundans. *Appl. Microbiol. Bio Biotechnol* **81** 629
[5] Huntley M E., Redalje D G 2006 CO2 mitigation et renewable oil from photosynthetic microbes: a new appraisal *Mitigation and.Adaptation Strategies for Global Change*
[6] Widjaja A 2009 Lipid production from microalgae as a promising candidate for biodiesel production *Makara Teknologi* **13** 47
[7] Harun R, Danquah M K and Forde G M 2009 Microalgal biomass as a fermentation feedstock for bioethanol production *Journal of Chemical Technology and Biotechnology* **85** 199
[8] Droop M R 1968 Vitamin B12 and marine ecology. IV. The kinetics of uptake growth and inhibition in Monochrysis lutheri. *J. Marine Biological Association* **48** 689
[9] Droop M R 1973 Some thoughts in nutrient limitation in algae *J. Phycol* **9** 264
[10] Mairet F, Bernard O, Masci P, Lacour T and Sciandra A 2011 Modeling neutral lipid production by the microalga *Isochrysis* aff. *galbana* under nitrogen limitation *Bioresources Technology* **102** 142
[11] Vinter R 2002 Optimal Control *The Maximum Principle* (New York, CRC PRESS) chapter 6 201
[12] Campbell S L and Haberman R 2008 Introduction to Differensial Equations with Dinamycal System *Nonlinear System of Differential Equations in The Plane* (New Jersey: Princeton University) chapter 6 332
[13] Ogata K 2010 Moderen Control Engineering *Transient and Steady-State Response Analysis* fifth edition (New Jersey: Prentice Hall) chapter 5 159
[14] Wan M, Liu P, Xia J, Rosenberg J, George A, Oyler G, Betenbaugh M and Nie Z. 2011 The effect of mixotrophy on microalgal growth, lipid content, and expression levels of three pathway genes in Chlorella sorokiniana *Appiedl Microbioogyl Biotechnology* **91** 835
[15] Mona Arnold (Ed.). Espoo 2013. Sustainable algal biomass products by cultivation in waste water flows *VTT Technology* **147** 84
[16] Olofsson M, Lamela T, Nilsson E, Bergé J P, del Pino V, Uronen P and Legrand C 2014 Combined Effects of Nitrogen Concentration and Seasonal Changes on the Production of Lipids in Nannochloropsis oculata *Marine Drugs* **12** 1891

Mathematical problem solving ability of sport students in the statistical study

E F P Sari[1], Zulkardi[2] and R I I Putri[2]

[1,2]Universitas PGRI Palembang, Jl. Jend A. Yani, Palembang, Indonesia
[2]Universitas Sriwijaya, Jl. Bukit Besar, Palembang, Indonesia

E-mail: ekafitrips@univpgri-palembang.ac.id

Abstract. This study aims to determine the problem-solving ability of sport students of PGRI Palembang semester V in the statistics course. Subjects in this study were sport students of PGRI Palembang semester V which amounted to 31 people. The research method used is quasi experiment type one case shoot study. Data collection techniques in this study use the test and data analysis used is quantitative descriptive statistics. The conclusion of this study shown that the mathematical problem solving ability of PGRI Palembang sport students of V semester in the statistical course is categorized well with the average of the final test score of 80.3.

1. Introduction

Realized or not in our daily lives we often find information related to statistics, whether presented through electronic media and print media. So it can be said that statistics have an important role in human life. The course of statistics is one of the subjects taught in universities, especially in the university PGRI Palembang. This course is expected to help students in completing the final task of thesis, especially on the research methodology.

Although the course of statistics is very important in college, but generally less interested students to study especially in sports education program. At the University of PGRI Palembang sports education program, the statistical courses are taught in the fifth semester. By studying the statistics course, students are expected to develop analytical, systematic, conclusive, diligent and disciplinary skills in solving problems in mathematics and in other fields. This is reinforced by the opinion of Kholidi and Sahat [1] which states that one of the indicators of good quality learning is the high level of knowledge as well as the interaction of students to the material taught in real life.

The facts on the ground are not in accordance with the existing learning objectives. The difficulties of students in translating statistical problems, complexity in the process of calculation at each stage to make conclusions, coupled with the level of accuracy of students in the calculations are still lacking, and student self-distrust causing students easily desperate in solving statistical problems. Likewise that is recognized by some students of Sports Education Program Education University PGRI Palembang.

Problem solving skills are the mathematical skills that every student should have. Problem solving is said to be the heart of mathematics, this is because of the importance of problem solving skills in learning mathematics. NCTM [2] says that problem solving has two functions in mathematics learning. First problem solving is an important tool to learn math. Both problem solving can equip students with knowledge and tools.

The importance of problem-solving skills is also expressed by Branca, as quoted by Effendi [3] that problem-solving skills are the heart of mathematics. The problem solving ability of students has relevance to the stage of solving mathematical problems. According to Polya [4] the mathematical problem solving stage includes: (1) understanding the problem, (2) making a plan of completion, (3) executing the plan, and (4) looking back. It is intended that students are more skilled in solving mathematical problems, that is skilled in carrying out procedures in solving problems quickly and carefully, especially on the statistical problems that are often faced by students. Statistics is one of the subjects that must be followed by every sem. V sports students of Universitas PGRI Palembang. Therefore this course becomes very important, point to help students in completing the final task (thesis).

Problem is a situation where the individual wants to do something but does not know the way or action needed to get what he wants. Hudojo, as quoted by Yuwono [5] states that something is called a problem for students if: (1) the questions confronted to learners must be understandable to the learners, but the question must be a challenge for him / her to answer, 2) the question can not be answered by routine procedures that learners already know.

The problem for someone is not necessarily a problem for others. This is because of the possibility that the other person has ever encountered and solved such a person's problem. A problem that comes to someone causes the person to at least attempt to solve the problem he is facing. So he has to use various means such as thinking, trying, and asking to solve the problem. Even in this case, the process of solving the problem between one person and another person may differ. According to [6] problem solving is a planned process that needs to be done in order to obtain a certain solution of a problem that may not be immediately available. Polya [4] defines problem solving as an attempt to find a way out of a difficulty. According to Goldstein and Levin as quoted by Rosdiana & Misu [7] problem solving has been defined as a high-level cognitive process that requires modulation and control over routine or basic skills.

At the time of solving mathematical problems, students are faced with several challenges such as difficulty in understanding the problem. This is because the problem faced is not a problem that has been faced by previous students. There are several troubleshooting steps introduced by mathematicians and mathematics teachers such as the problem-solving stage according to Polya, Krulik and Rudnick, and Dewey. Schoenfeld, as quoted by Ellison [8] states that is not a teaching on strategies that can cause differences in solving problems, more than that, practice problem solving then make a difference. According to [9] students need to do things like accept challenges from problems, plan problem-solving strategies, implement strategies, and re-examine the solutions obtained.

According to Polya [4] the four stages of Polya problem solving are detailed as follows.

1. Understanding the problem (understand the problem)

The first stage of problem solving is understanding the problem. Students need to identify what is known, what is there, the number, relationships and related values and what they are looking for. Some suggestions can help students understand complex issues: (1) to ask questions about what is known and sought, (2) explain the problem according to the sentence itself, (3) link it to other similar problems, (4) focus on the part which is important from the problem, (5) developing the model, and (6) drawing the diagram.

2. Make a plan (devise a plan)

Students need to identify the operations involved as well as the strategies needed to solve the given problem. This can be done by students such as: (1) guessing, (2) developing a model, (3) sketching diagrams, (4) simplifying problems, (5) identifying patterns, (6) creating tables, (7) experiments and simulation, (8) working inversely, (9) testing all possibilities, (10) identifying sub-goals, (11) making analogies, and (12) sorting data / information.

3. Implement the plan (carry out the plan)

What is applied clearly depends on what has been planned before and also includes the following: (1) interpreting the information given into a mathematical form; and (2) execute a strategy during the process and ongoing calculations. In general at this stage students need to retain the chosen plan. If such a plan cannot be done, then the student can choose another way or plan.

4. Looking back (looking back)

The following aspects need to be taken into consideration when reviewing the steps previously involved in solving the problem, namely: (1) recheck all important information that has been identified; (2) checking all counts already involved; (3) consider whether the solution is logical; (4) look at other alternative solutions; and (5) read the question again and ask yourself if the question has really been answered.

As for the problem in this research is "How the problem solving skills mathematical students PGRI Palembang semester V students in the statistics course?". This study aims to determine the problem-solving ability of mathematical students of PGRI Palembang sports semester V in the statistics course.

2. Method

The method of this research is one case shoot study where there is only one class of sample that is class which become experiment which implemented without control class (comparison class). Data collection techniques in this study used the test and data analysis used is quantitative descriptive statistics [10].

3. Result and discussion

Data of students' mathematical problem solving ability is seen from the result of the test which is given at the 4th meeting. The test is given in the form of essays with categories easy, medium, medium, easy and moderate. About 4 test questions about mathematical problem solving students.

Table 1. The value of students every meeting

Average		
Meeting 1st	Meeting 2nd	Meeting 3rd
64.40	72.65	83.24

Table 1 shows that at the first meeting it was sufficient with an average score of 64.40 at the second meeting was good with an average score of 72.65 whereas in the third meeting the average was classified as good with an average of 83.24.

At the meeting to four students are given a test that is as much as 4 essay-shaped questions. The test results were followed by 31 students, with the highest score being 100 while the lowest score was 50. From the final test obtained students 'mathematical problem solving ability of the average value of 80.30 This means that students' mathematical problem solving abilities are categorized as well as the results of the mathematical problem solving ability of the student per indicator on the test can be seen in Table 2.

Table 2. Average value of end tests ability resolution mathematical student per-indicator

No	Indicators	Average	Result
1	The ability to identify problems	80.2	Very good
2	Planning a solution with a mathematical model	87.5	Very good
3	Resolving the problem as planned	89.4	Very good
4	Interpret solutions	58.9	Enough
	Average	80.30	

The students' mathematical problem solving abilities on the first meeting test were averaged 64.40. At the second meeting the students began to be active and began to get used to the statistical material so that the impact on the average value of the second meeting is 72.65. At the third meeting, students get good grades, this is because students have started to get used to the statistical material with an average value of 83.24.

From the test results obtained the average value of mathematical problem solving ability of student each indicators and obtain an average of 80.30. And the lowest indicator is on inductor 4 makes conclusions with an average of 58.9 less categories, this is because students forget to write a conclusion or wrong when making conclusions. While the highest indicator lies in indicator 3 solve the problem with an average of 89.4 well categorized this is because students have understood the concept of the problem and have understood the purpose of the problem.

4. Conclusion

Based on the results of research that has been done, it can be concluded that students' mathematical problem solving skills are categorized well with the average of the final test score of 80.3. While the average per indicator students' mathematical problem solving abilities on the indicator 1 reaches an average of 80.2 with very good category, in indicator 2 reaches an average of 87.5 with excellent category, indicator 3 reaches 89.4 average rating with excellent category, and indicator 4 reaches 58.9 average rating with enough category.

References

[1] Kholidi M dan Saragih S 2011 Peningkatan Kemampuan Koneksi dan Pemecahan Masalah Matematika Siswa SMA Melalui Pembelajaran Kooperatif *Jurnal Pendidikan Matematika Paradikma* **5** 166-185

[2] National Council of Teachers of Mathematics 2000 *Principles and Standards for School Mathematics* (Reston,VA: NCTM)

[3] Effendi L A 2012 Pembelajaran Matematika dengan Metode Penemuan Terbimbing untuk Meningkatkan Kemampuan Representasi dan Pemecahan Masalah Matematis Siswa SMP *Jurnal Penelitian Pendidikan Universitas Pendidikan Indonesia* **13** 1-10

[4] Polya G 1973 *How to solve it: A new aspect of mathematical method* (NJ : Princeton university press)

[5] Yuwono A 2010 *profil siswa sma dalam memecahkan masalah matematika di tinjau dari tipe kepribadian* Doctoral dissertation (Surakarta: Universitas Sebelas Maret)

[6] Fauziah A 2010 Peningkatan Kemampuan Pemahaman dan Pemecahan Masalah Matematika Siswa AMP Melalui Strategi REACT (Relating, Experiencing, Applying, Cooperating, Transfering) *Forum Kependidikan* **30**

[7] Misu L and Rosdiana 2013 pengembangan teori pembelajaran perilaku dalam kaitannya dengan kemampuan pemecahan masalah matematik siswa di SMA *Proc Seminar Nasional Matematika dan Pendidikan Matematika* (Yogyakarta: UNY)

[8] Ellison J G 2008 Increasing problem solving skills in fifth grade advanced mathematics students *Journal of Curriculum and Instruction* **3** 15

[9] Minarni A 2012 Pengaruh Pembelajaran Berbasis Masalah Terhadap Kemampuan Pemecahan Masalah Matematis *Proc Seminar Nasional Matematika dan Pendidikan Matematika* (Yogyakarta: UNY)

[10] Creswell 2009 *Research Design Pendekatan Kualitatif, Kuantitatif, dan Mixed Edisi Ketiga.* (Yogyakarta: Pustaka Pelajar)

IOP Conf. Series: Journal of Physics: Conf. Series **943** (2017) 012038 doi:10.1088/1742-6596/943/1/012038

Developing Ill-defined problem-solving for the context of "South Sumatera"

S Arifin[1,2], Zulkardi[2], R I I Putri[2], Y Hartono[2] and ESusanti[2]

[1]Islamic State University of Raden Fatah, Jl. Prof. K.H. Zainal Fikry, Palembang, Indonesia
[2]Sriwijaya University, Jl. Padang Selasa 524, Palembang, Indonesia

E-mail: sujinal@radenfatah.ac.id

Abstract. This study aims to produce a valid and practical ill-defined problem-solving for context South Sumatera. The subject of the research is three students of the first semester of undergraduate students in the mathematics department of Raden Fatah State Islamic University. This study use development studies that consist of preliminary and prototyping. In preliminary stage have been analysis content curricula, indicator, and strategies of problem-solving. Meanwhile, in prototyping stage only consist of self-evaluation, expert review, and one-to-one. The data were collected through a walkthrough, interview, and test. The data were validated using expert review, but in practice, the data were obtained from test and interview to subject of the research. This studies produced two valid and practical problem-solving. The first problem is about "Benteng Kuto Besak", and the second problem is about "Monpera". From the expert review, the conclusion can be drawn that two problems which are developing are ill-defined problem-solving, and valid from content, construct, and its language. Besides that, the problems are practical because all students know and understand what the problems goal, but not the solutions.

1. Introduction

There are five abilities students will have through mathematics learning. These capabilities include communication skills, reasoning, connection, representation, and problem-solving [1]. One of the most important skills is the ability of problem-solving because it is the heart of mathematics and can not be separated from the learning of mathematics. Problem-solving ability is the ability to find and apply ideas that students have in other fields of study in everyday life [2].

Problem-solving is the process of identifying a problem, developing possible solution paths, and taking the appropriate course of action [3]. Problem-solving is very important because good problem-solving skills empower someone not only in their personal life but also in their professional life. In the current fast-changing global economy, everyone often identifies everyday problem-solving as crucial to the success of their organizations, and problem-solving can be used to develop practical and creative solutions and to show independence and initiative to employers.

In Indonesia, students problem-solving ability is still low [4, 5]. Many teachers and students experience difficulties, not only in teaching but also on developing and enhancing problem-solving skills [2]. There are a lot of factors that cause the above problems, for example, still, at least problem-solving in school books. This has an impact on teachers and students. Teachers rarely give problem-solving and students rarely do problem-solving.

Patnani [6] adds that the lack of problem-solving skills not only faced by students but also by undergraduate students. The number of complaints occurs in the world of work because undergraduates seem unprepared facing the world of work and they are still confused to face the problem. Meanwhile, They are expected to be able to solve problems. This is due to the quality of college graduates who are considered still inadequate in divergent thinking. Although there is the problem that teachers or lecturer once gave to their students a matter in the form of mathematical symbols.

The formal mathematics problem makes the excuse that mathematics is dry, difficult, and unattractive, and some students assume that mathematics is not related to human activity. Therefore, it is necessary to find and add context to the math problem [7]. Also, the use of local context can assist students in understanding the mathematical phenomena of their own life experience perspective [8]. This means that context can not only be used in learning, but also in judgments. One context that can be used is the context of South Sumatera.

South Sumatera is one of the provinces in Indonesia. The local context in the province of South Sumatera is very much and diverse, for example from the aspect of tourism. There are several aspects of this context in South Sumatera that belong to the tourism aspect of culinary, historical place, and natural beauty. Previous research has developed many problem-solving problems [8], but no one has developed it in the context of South Sumatera. This study aims to produce a valid and practical ill-defined problem-solving for the context of South Sumatera.

A Problem is a situation which there is no obvious solution [9, 10]. A problem exists when there is a situation you want to resolve, but no solutions are readily apparent. Problem-solving is a complex task [11]. Problem-solving is the process by which the unfamiliar situation. A situation that is considered as a problem to one person may not be a problem to someone else. Problem-solving also can define as the ability to understand what the goal of the problem and what rules could be applied represent the key to solving the problem. Problem solving is also one of the competency in PISA mathematics [12].

Problem-solving is consist of two types: the ill-defined problem and the well-defined problem [13]. The ill-defined problems are those that do not have clear goals, solution paths, or expected solution. The well-defined problems have specific goals, clearly defined solution paths, and clear expected solutions. Problem-solving is the subject of a major portion of research and publishing in mathematics education. A lot of studies are founded on the problem-solving writing of George Polya. In the case of problem-solving, it may be speculated that a solvers' idea about the nature of mathematics may affect the way in which a solution is sought [14].

Capraro and Cifarelli [15] give an example, each student will have opportunities to obtain their unique solutions. They can respond to the problem in some significant way with many different solutions. Students also will have more opportunities to make comprehensive use of their mathematical knowledge and skills to solve the problem. They also will choose their favorite strategies to obtain answers and create their unique solutions.

The well-defined problem is contradictive to the ill-defined problem. The well-defined problem sometimes calls as the formal problem. The formal problem has some characteristics. They are as follows: (1) have a defined given initial situation; (2) have a defined goal; (3) have a defined set of resources; and (4) have some ownership [16]. The term problem-solving sometimes includes posing and solving problems, posing and accomplishing tasks, posing and answering questions, and posing and making decisions. Problem-solving is an integral component of every academic discipline [17]. Polya [16] mentions that there are many reasonable ways to solve problems. The skill at choosing an appropriate strategy is best learned by solving many problems. Students will find choosing a strategy increasingly easy. A partial list of strategies includes: (1) guess and check; (2) make an orderly list; (3) eliminated possibilities; (4) use symmetry; (5) consider special cases; (6) use direct reasoning; (7) solve an equation; (8) look for a pattern; (9) draw a picture; (10) solve a simpler problem; (11) use a model; (12) work backwards; (13) use a formula; and (14) be ingenious.

2. Method

This research uses part of Design Research type development study [18] so that every development stage of questions consist of the preliminary stage and formative evaluation stage [19]. The formative evaluation stage consists of self-evaluation and prototyping, but at prototyping stage, it only occurs at the expert review and one-to-one [20].

The preliminary stage begins by doing curriculum content analysis, determining indicators of problem achievement, and the strategies students use in problem-solving. Next, the results of this preliminary are designed grid, draft questions, and rubric assessment that refers to the strategy used in problem-solving. The developed problem contains the context of South Sumatera, especially the historic place in South Sumatera. Furthermore, self-evaluation stage, the problem device has been developed in its own evaluation by the researcher. This is done with a purpose to know the shortcomings of the draft questions that have been developed so that the resulting draft is feasible to be validated by experts. The next stage is prototyping. The prototyping stage begins with expert validation. Expert validation is defined as testing the quality of the questioning device by the experts viewed from three aspects namely content, constructs, and language.

Along with validation by experts, a one-to-one stage is performed. This is done in order to know the quality and practicality of problem-solving. This stage involves three students of mathematics education of UIN Raden Fatah Palembang. They have various skills (high, medium, low). At the one-to-one stage, students are asked to read and examine the matter, and then ask students to solve the problem. There are several instruments used in this study. They are walkthrough, interview, and test. The walkthrough is used to collect data during expert validation. At the time of one-to-one, there are two instruments are used: interview and test. Interviews are used to collect data of students' understanding of the given problem. While the test is used to see student strategies in solving the problem.

3. Result and discussion

3.1. Preliminary stage

This stage begins with curriculum analysis and collects theories related to the concept of problem-solving. The results of this stage are generated grids, draft questions, and alternative strategies used in solving the problem. All the questions developed contain the context of South Sumatera. At this stage result six problem-solving with the context of South Sumatera.

Table 1. Grid of Problem-solving

Number	Indicator	Context	Content	Strategy
1	Use nonstandard unit sizes to resolve issues related to shape	Gate of Benteng Kuto Besak (BKB)	Space and shape	Consider special cases
2	Use a mathematical model to determine the maximum number of visitors	monument of struggle (Monpera)	Change and relationship	Eliminated possibilities
3	1. Using the concept of scale to determine the distance of two points 2. Using the concept of speed to determine the location of a place	Dempo Mountain	Quantity	Use a formula
4	1. Predict the wide area of Gelora Sriwijaya using satellite imagery scale 2. Predict the capacity if there is a standard size	Sports Area (Gelora Sriwijaya)	Space and shape	Consider special cases Use a formula
5	Determine the broad ratio of the two pensions	monument of struggle (Monpera)	Space and shape	Draw a picture
6	Determine the percentage increase in the number of visitors	Temam Waterfall	Uncertainty and Data	Use a formula

3.2. Expert review

The main validator of this research is three lecturers in Mathematics Education Department. Stage occurs in two cycles, because of the first walkthrough result of many things to be updated, especially indicators, the language on the problem, and strategy of completion.

Table 2. Result of Expert Review

Aspect	Comment from Validator
Content	- There are some issues that are categorized not about problem-solving
	- Situation and context offered still tend to be camouflaged
	- The description of the problem does not help much in solving the problem
Construct	- There is a problem with too many pictures
	- No pictures that do not support problem-solving
	- The information on the picture is less clear
Language	- There is still an elusive term
	- Use good and correct Indonesian (EYD)

3.3. One-to-one

In line with the implementation of expert validation, a one-to-one stage is performed. At this stage, there are three students of Mathematics Education UIN Raden Fatah who became the subject of research. These three students have different abilities. Here are the initials and categories of students' abilities.

Table 3. The Initials and Categories of Students' Abilities

Initial Name	Ability
SrW	High
SiW	Medium
MI	Low

The three students are asked to read and examine the problem on the matter. Furthermore, researchers asked some questions to see the responses, constraints, and understanding of students to solving the problem. Here are some questions asked (Table 4)

Table 4. Questions in Interview

Number	Question
1	Do you know the meaning of the words in the questions?
2	Can you mention the problem in your own sentences?
3	Is the information sufficient to solve the problem?
4	What do you need to get it done?
5	Is there anything you do not know from the problem?
6	Is the condition of the questions sufficient for you to determine the things you do not know?
7	Have you ever encounter a problem like this?
8	Do you know the concepts or theorems that can be used to solve the problem?
9	What strategies do you use in solving the problem?

From the interview result, it can be concluded that MI students with low ability level tend to be confused when interpreting the problem. This happens to almost all of the question numbers. MI not only cannot solve it but also feel confused when reading and declaring problems in his own language. The problem-solving and some excerpts of the interview with MI on problem number 1 can be seen in Figure 1.

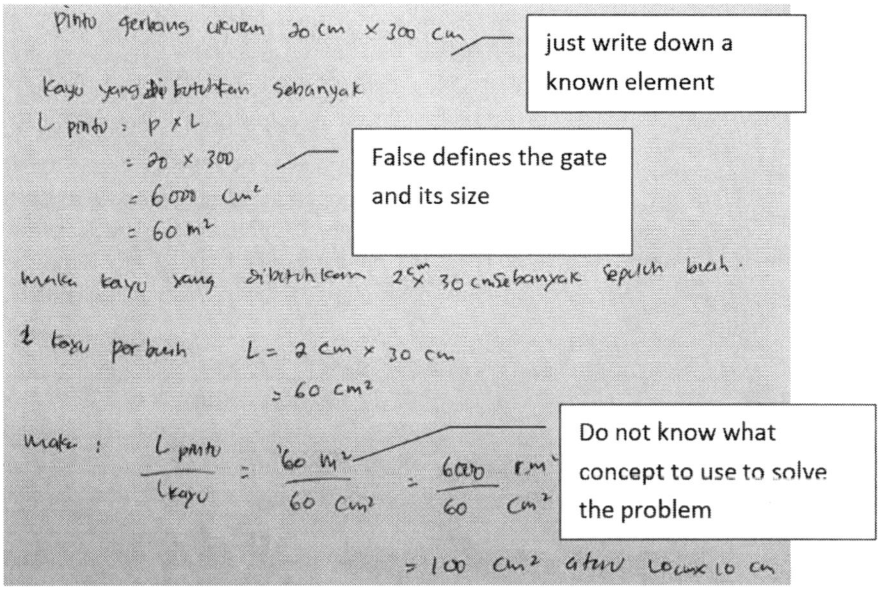

Figure 1. MI's Answer on Problem Number 1

Researcher : "Can you express the problem in your own words?"
MI : "not yet, Sir. I'm confused how to say it"
Researcher : "What is needed to solve the problem?"
MI : "gate size Sir"
Researcher : "Do you know what concepts or theorems can be used to solve the problem"
MI : "I do not know, sir"

The same findings were obtained from highly skilled students (SrW). SrW is not only able to interpret and express problems with its own words when most of the required information does not exist, but SrW has also been able to plan what will be used to solve the problem. Srw has also seen problems like this though in different forms. The problem-solving and some excerpts of the interview with SrW on problem number 5 can be seen in figure 2.

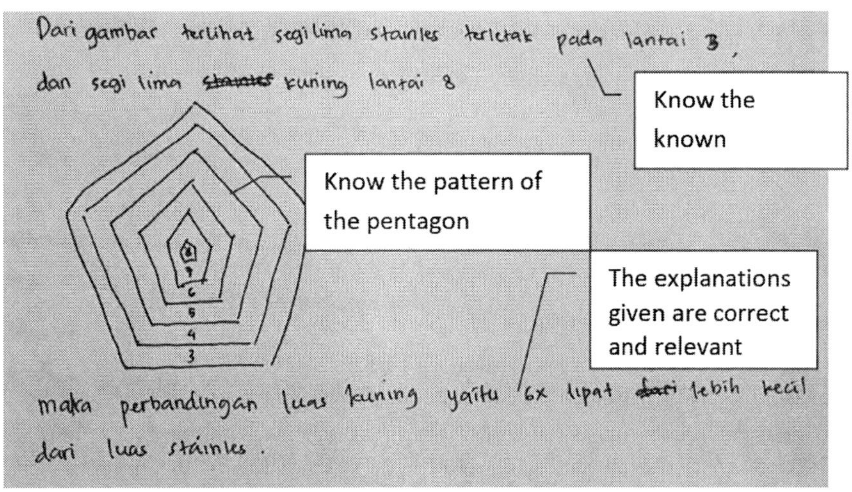

Figure 2. SrW's Anser on Problem Number 5

Researcher : "Have you ever seen a problem like this?
SrW : "Never Sir, but the problem is different. Usually because of the comparison of triangle or square.
Researcher : "Do you know the concepts or theorems that can be used to solve the problem?"
SrW : "know enough, the concept used is the concept of comparison".

The same findings were obtained from medium-ability students (SiW). SiW is capable of interpreting and raising problems with its own words when most of the required information already exists in the description of the problem, but SiW encounters problems if the required information is unclear, for example on the interior of Monpera, as only the images presented on the problem and there is no explanation of the pentagonal measures. The problem-solving and some excerpts of the interview with SiW on problem number 1 can be seen in Figure 3.

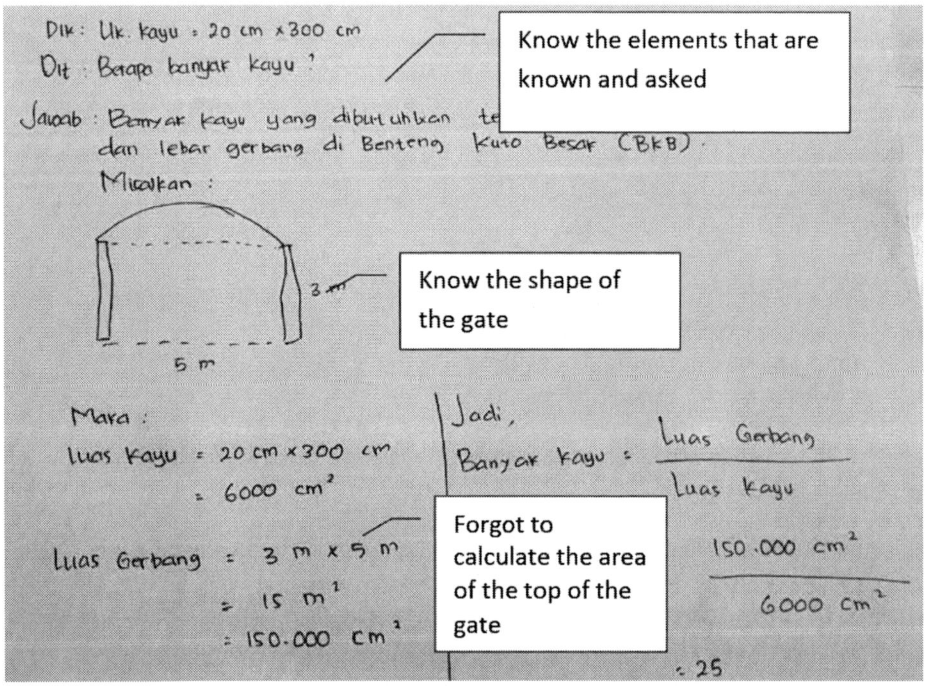

Figure 3. SiW's Answer on Problem Number 1

Researcher: "What is needed to solve the problem?"
SiW : "Specifically for Problems 1 and 5, I need additional information about the size of the gate and the size of a pentagon with a pink yellow". Because if not known, then the problem can not be solved.

Based on the results of expert validation and one-to-one, it can be concluded that only 2 of the 6 questions that have been designed are categorized as a valid problem-solving problem. This is not only reflected in the responses provided by expert validation, but also from the responses, constraints, and understanding of the students.

Although almost all of the problems given can be solved by the students, regarding of responses, obstacles, and understanding and settlement strategies, it can be recommended that all of these problems are categorized as practical. Based on the results of this research, there are only two of six problems that are categorized as valid and practical problems, namely the number 1 problem with the context of "Benteng Kuto Besak" and problem number 5 with the context of "Monpera". The question numbers 1 and 5 have been declared valid and practical can be seen in Figure 4 and Figure 5.

IOP Conf. Series: Journal of Physics: Conf. Series **943** (2017) 012038 doi:10.1088/1742-6596/943/1/012038

Unit 1. Gate Of Benteng Kuto Besak

The construction of Benteng Kuto Besak was initiated by The first Sultan Mahmud Badaruddin who reigned in 1724-1758 and the construction was completed by his successor, Sultan Mahmud Badaruddin II who reigned in 1776-1803. To maintain its sustainability, the government of Palembang plans to create a wooden gate. If the size of wood is 20 cm x 300 cm. Determine how much wood is needed for the gate?

Figure 4. Problem Number 1

Unit 5. Interior Of Monpera

Source: http://www.1001wisata.com/wp-content/uploads/2015/01/lantai-monpera.jpg

Monpera is a struggle monument made to commemorate the hero's service in South Sumatera. Monpera resembles a five-pointed jasmine. The height of the Monpera building reaches 17 meters, has eight floors, and 45 fields/lane. The numbers represent the date of the proclamation of independence of the Republic of Indonesia August 17, 1945. If you are in the Monpera and looked up the ceiling of the building it will look like the picture above. Can you determine the ratio of the area of a pentagon with a yellow area of a pentagonal area (stainless)? Describe the strategy you use in solving the problem?

Figure 5. Problem Number 5

From the result of prototyping has produced two problems which categorized problem-solving. Both questions contain the context of South Sumatera. Problem number 1 contains the context of "Benteng Kuto Besak" with space and shape content and the strategy used for problem-solving is consider special cases. While problem number 5 is still related to content space and shape but its context "Monpera" and strategy draw a picture.

The difference in completion given by the student depends on the difference in the student's level of ability [21, 22]. Students who are low in understanding have low-level students. MI students make many mistakes in solving problems, not only wrong concepts but also wrong in choosing settlement procedures. The error is caused by not understanding the related concepts [23-25]. Different answers are obtained from students who are categorized have ability medium level and high level. Si W (medium-level students) already understands the concept but is still wrong in choosing the settlement procedure, while SrW (high-level students) is not only able to identify and understand the problem, but also able to complete and provide supporting arguments from the solution.

4. Conclusion

This studies produced 2 valid and practical problem-solving. The first problem is about "Benteng Kuto Besak", and the second problem is about "Monpera". From the expert review, the conclusion can be drawn that two problems which are developing are ill-defined problem-solving, and valid from content, construct, and its language. Besides that, the problems are practical because all students know and understand what the goal of the problems, sometimes they don't know the solutions.

References

[1] Paridjo and Waluya St B 2017 Analysis mathematical communication skills students in the matter algebra based NCTM *Journal of Mathematics* IOSR-JM **13** 60

[2] Mursiadi R 2013 Model pembelajaran matematika tipe group investigation untuk meningkatkan kemampuan pemecahan masalah matematis dan self-concept siswa MTs *Tesis* (Bandung: Universitas Pendidikan Indonesia)

[3] Al-Tarawneh H A 2012 The main factors beyond decision making *Journal of Management Research* **4** 1-22

[4] Kusaeri 2006 Profil kemampuan dasar matematika mahasiswa jurusan tadri pendidikan matematika fakultas tarbiyah *Qualita Ahsana* **8** 2

[5] Sari S Eliati S and Fauzan A 2014 Pengaruh pendekatan pembelajaran berbasis masalah terhadap kemampuan pemecahan masalah matematika siswa kelas VIII SMP Negeri 1 Padang *Jurnal Pendidikan Matematika* **3** 54

[6] Patnani M 2013 Upaya meningkatkan kemampuan problem-solving pada mahasiswa *Jurnal Psikogenesis* **1** 130

[7] The Cornerstone Tech Prep 1999 *Teaching Mathematics Contextually* (Texas: CORD)

[8] Charmila N, Zulkardi and Darmawijoyo 2016 Pengembangan Soal Matematika Model Pisa Menggunakan Konteks Jambi *Jurnal Penelitian dan Evaluasi Pendidikan* **20** 198-207

[9] Ong R 2000 The Role Of Reflection In Student Learning: A Study Of Its Effectiveness In Complementing Problem-Based Learning Environments Centre for Educational Development *Centre for educational development* (Singapore)

[10] Carson J 2007 A problem with problem-solving: teaching thinking without teaching knowledge *The Mathematics Educator* **17** 7

[11] NCTM 2000 *Principles and Standards for School Mathematics* (Reston VA: NCTM)

[12] Stacey K et al 2015 PISA's Influence on Thought and Action in Mathematics Education, in *Assessing Mathematics Literacy*, edited by K. Stacey and R. Turner (Springer, Switzerland) pp 275–306

[13] Holyoak K J 1990 Problem-solving *Thinking: An Invitation to Cognitive Science* **3** 117-146

[14] De Hoyos M, Gray E and Simpson A 2004 students' assumptions during problem-solving *The 28th International Conference of the International Group for the Psychology of Mathematics Education* (Bergen: University of Norway) p 2
[15] Capraro M M, Capraro R M and Cifarelli V V 2007 What are students thinking as they solve open-ended mathematics problems? *Proc of the ninth international conference of Mathematics Education in a Global Community* (Charlotte, NC)
[16] Moursund D 2016 *Learning Problem-solving Strategies by using games: A guide for Educator and Parents* (Eugene. OR: Information Age Education)
[17] Polya G 1985 *How to Solve it: A New Aspect of Mathematical Method* (NJ: Princeton University Press)
[18] Putri R I I and Zulkardi 2017 Fraction in shot-put: A learning trajectory *AIP Conference Proceedings*
[19] Zulkardi Z 2002 Developing a learning on realistic mathematics education for Indonesia students teachers *Dissertation* (Nederland: University of Twente)
[20] Tessmer M 1998 *Planning And Conducting Formative Evaluations: Improving The Quality Of Education And Training* (London: Kogan Page)
[21] Felder R M and Brent R 2005 Understanding Student Differences *Journal of Engineering Education* **94** 57
[22] Dunlosky J R et al. 2013 improving students' learning with effective learning techniques: Promising directions from cognitive and educational psychology *Psychological Science in the Public Interest* **14** 4
[23] Nurhasanah F, Kusumah Y S and Sabandar J 2017 Concept of triangle: Examples of mathematical abstraction in two different contexts *International Journal on Emerging Mathematics Education* **1** 53
[24] Tanujaya B, Prahmana R C I, and Mumu J 2017 Mathematics instruction, problems, challenges, and opportunities: A case study in Manokwari regency, Indonesia *World Transactions on Engineering and Technology Education* **15** 287
[25] Syamsuri, Purwanto, Subanji, and Irawati S 2017 Using APOS theory framework: Why did students unable to construct a formal proof? *International Journal on Emerging Mathematics Education* **1** 135

IOP Conf. Series: Journal of Physics: Conf. Series **943** (2017) 012039 doi:10.1088/1742-6596/943/1/012039

The effectiveness of robust RMCD control chart as outliers' detector

Darmanto and Suci Astutik

Study Program of Statistics, Department of Mathematics, Brawijaya University, IDN

E-mail: darman_stat@ub.ac.id

Abstract. A well-known control chart to monitor a multivariate process is Hotelling's T^2 which its parameters are estimated classically, very sensitive and also marred by masking and swamping of outliers data effect. To overcome these situation, robust estimators are strongly recommended. One of robust estimators is re-weighted minimum covariance determinant (RMCD) which has robust characteristics as same as MCD. In this paper, the effectiveness term is accuracy of the RMCD control chart in detecting outliers as real outliers. In other word, how effectively this control chart can identify and remove masking and swamping effects of outliers. We assessed the effectiveness the robust control chart based on simulation by considering different scenarios: n sample sizes, proportion of outliers, number of p quality characteristics. We found that in some scenarios, this RMCD robust control chart works effectively.

1. Introduction

Application of statistics as a science and analysis data method is used widely in almost all fields. By statistics, following data collecting, the researchers do their data analysis and further sometimes the result is used in decision making. To get valid and correct result of data analysis, it requires in choosing and using statistical method correctly. In addition, the researchers also have to follow right step of data analysis. One of the steps towards obtaining an appropriate data analysis is testing and detecting an outlier data. The outliers may be often assumed as a mistake or error, but in some condition they bring important information about whole data. If this step is ignore, it may lead to wrong model construction, got biased parameter estimation and incorrect result of data analysis. Then, identification of outlaying observations is an important part to modelling and data analysis [1].

Hawkins [2] defines an outlier *as an observation that deviates so much from other observations as to arouse suspicion that it was generated by a different mechanism.* Analogue with Hawkins, Johnson [3] defines an outlier *as an observation in a data set which appears to be inconsistent with the remainder of that set of data.* An outlier is divided into univariate and multivariate. Almost all of the univariate methods in outlier detecting based on the assumption of prior distribution of the data, which is assumed to be identically and independently distributed (i.i.d). In addition, there are inappropriate tests for detecting univariate outliers further assume that the distribution parameters and the kind of expected outliers are have to known [4].

In univariate outlier is detected separately to each by each variable. But, in multivariate cases, outlier should be detected to all variables simultaneously. For many cases multivariate data can not be detected as outliers when variables do not have correlation. Multivariate outlier detection is could be worked when multivariate analysis is performed, and the interaction among different variables are compared within the class of data [5]. Figure 1 is a simple example which presents data point having two variables, x and y. From the figure can be seen that the lower left observation is clearly a multivariate outlier. If two these variables are measured separately with respect to the spread of values along the x and y axes, observations fall close to the centre of the univariate distribution. In a brief, the test for outliers should take into reason the correlation between the two variables, which in this case appear abnormal.

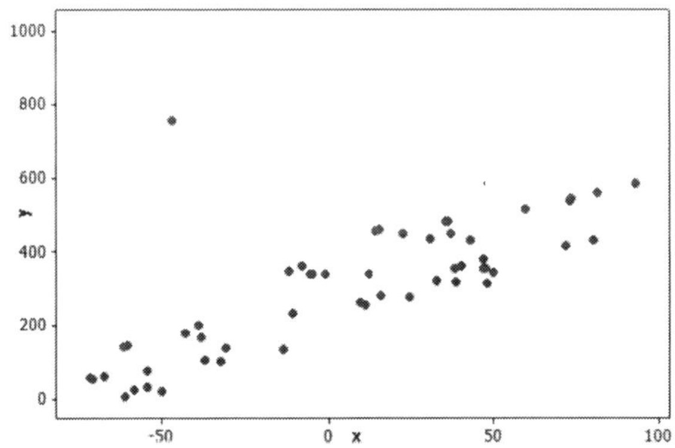

Figure 1. Multivariate Outlier

In many multivariable observations cases, there are data sets with multiple outliers. The multiple outliers are subject to *masking* and *swamping* effects. Even though not mathematically exact, Acuna and Rodriguez [6] give an illustration understanding for these two effects as follows: ***Masking effect***, it is said that one outlier masks a second outlier. If the second outlier can be detected as an outlier, but not in the presence of the first outlier. Thus, if the first outlier is eliminated from data the second instance is appeared as an outlier. ***Swamping effect,*** it is said that one outlier swamps a second observation, if the next can be detected as an outlier because of the presence of the first one. It means that, after the first outlier is deleted from data sets, the following observation be a normal.

In industrial manufacturing, especially for multivariate process controlling, Hotelling-T^2 control chart is widely used [7]. The control chart is closely-related to multivariate outlier detection method. It assumes the event where the multivariate stream of measures represents a stochastic process, and the require procedure in detecting outlier is online. The Hotelling-T^2 control chart uses Mahalanobis distance method to calculate the statistic of T^2. The formula of Mahalanobis distance given in Eq. (1) below

$$MD_i^2 = \left(\mathbf{X}_i - T(\mathbf{X})\right)' C(\mathbf{X})^{-1} \left(\mathbf{X}_i - T(\mathbf{X})\right) \tag{1}$$

Where $\mathbf{X}_i$ is an m observations random vector, $T(\mathbf{X})$ is a p-dimensional estimated location parameter and $C(\mathbf{X})$ is a $p \times p$ estimated scale parameter. The observation of $\mathbf{X}_i$ follows p-variates Normal distribution with population mean vector $\boldsymbol{\mu}$ and variance-covariance matrix $\boldsymbol{\Sigma}$, MD_i^2 has χ^2 distribution by p degree of freedom [8]. Analogue with this formula of *mahalanobis* distance, statistic T^2 in the Hotelling-T^2 control chart we calculate

$$T_i^2 = \left(\mathbf{X}_i - \bar{\mathbf{X}}\right)' \mathbf{S}_X^{-1} \left(\mathbf{X}_i - \bar{\mathbf{X}}\right) \tag{2}$$

Where $\mathbf{X}_i = \left(X_{i1},\ldots,X_{ip}\right)', i = 1,\ldots,m,$ are the m p-variate observations with the sample mean $\bar{\mathbf{X}} = \sum_{i=1}^{m} \mathbf{X}_i$ and variance-covariance matrix $\mathbf{S} = \frac{\sum_{i=1}^{m}\left(\mathbf{X}_i - \bar{\mathbf{X}}\right)\left(\mathbf{X}_i - \bar{\mathbf{X}}\right)'}{m-1}$. A large value of T_i^2 (compare to specific distribution: *Beta* or *F* distribution) indicates that the process has shifted which is statistically sentenced by the identification of usual patterns in the control chart [9].

As shown in (2), the Hotelling-T^2 uses the classical sample vector mean and sample variance-covariance matrix to estimate the population vector mean and variance-covariance matrix. However, mean vector and covariance matrix which are estimated classically and used in T^2 are very sensitive to the presence of outliers data. In addition, these estimators also are marred by the effects of masking and swamping of the outliers. In 1998, Sullivan and Woodall [10] showed that the classical estimators are not effective in detecting shifts in the process mean and they proposed a control chart which uses the successive difference of observations in estimating a variance-covariance matrix, but the control chart is not effective in detecting process shift. To overcome the problem, robust estimators are strongly recommended. There are several robust estimators are previously had been proposed. Some of the robust estimators are minimum variance ellipsoid (MVE) and minimum covariance determinant (MCD) which are proposed by Roesseeuw [11], [12]. By developing MCD, Roesseeuw and Van Zomeren [13], Lopuhaa and Rousseew [14], and Willems et.al. [15] proposed re-weighted MCD (RMCD). The RMCD estimators have robustness characteristics as same as MCD estimators such as breakdown point, affine equivariance and asymptotic normality.

The following years, in 2003, Vargas [16] used robust estimators whether MVE or MCD in application of control chart for identifying the outliers. These robust control charts did not use the exact distribution of the classical T^2 control chart, so he estimated the control limits based on simulation. Jensen et al. (2007) also did simulation to estimated the control limits and tabulated these estimation for sample sizes m = 10,…, 100, dimensions $p = 2,\ldots,$ 10 and confidence level $1-\alpha = 0.95$. Chenouri, et.al [9] proposed robust control chart by RMCD in detecting the outliers. They did like Jensen et. al. to estimated the robust estimators by generating 10,000 observations in Phase I and then used it to calculate statistic T_i^2 in Phase II. They also tabulated these estimation for dimensions p = 2,…, 10 and confidence level $1-\alpha = 0.95;0.99;0.999$.

Our approach in detecting the outliers and monitoring the multivariate observations differ in three ways of Vargas [16], Jensen et.al. [17] and Chenouri et.al. [9]. We propose to use robust estimators of RMCD in application of the robust control chart. We assessed the effectiveness the robust control chart based on simulation by considering different scenarios: n sample sizes, proportion of outlier, number of p quality characteristics. The effectiveness term is accuracy of the robust control chart in detecting outlier as an outlier. In other word, how effectively this control chart can identify and also remove the masking and swamping effects of outliers.

IOP Conf. Series: Journal of Physics: Conf. Series **943** (2017) 012039 doi:10.1088/1742-6596/943/1/012039

2. Robust Estimation and Robust Control Chart of RMCD

2.1. Robust Estimation of RMCD

The robustness can be identified based on two properties are affine equivariance and breakdown point. An estimator that has affine equivariance characteristic means the estimator has a stable value even though the data has been transformed and rotated.

Definition: ***Affine Equivariance*** (Olive [18])

Assume that $\mathbf{X}$ is $m \times p$ observations data matrix. Let again assume that $\mathbf{B} = \mathbf{1b}'$ where $\mathbf{1}$ is an $m \times 1$ vector of ones and $\mathbf{b}$ is $p \times 1$ constant vector then *i*th row of $\mathbf{B}$ is $\mathbf{b}_i' = \mathbf{b}'$ for i = 1, 2, ..., *m*. For such a matrix $\mathbf{B}$, consider the affine transformation $\mathbf{Z} = \mathbf{XA} + \mathbf{B}$ where $\mathbf{A}$ is any non-singular $p \times p$, then the multivariate location and dispersion estimators of $\mathbf{t}$ and $\mathbf{S}$ would be said *affine equivariance* if:

$\mathbf{t(Z) = t(XA + B) = A't(X) + b}$, and

$\mathbf{S(Z) = S(XA + B) = A'S(X)A}$.

Proposition: If $(\mathbf{t}, \mathbf{S})$ is *affine equivariance*, then:

$$d_i^2(\mathbf{X}) \equiv d_i^2[\mathbf{t}(\mathbf{X}), \mathbf{S}(\mathbf{X})] = d_i^2[\mathbf{t}(\mathbf{Z}), \mathbf{S}(\mathbf{Z})] \equiv d_i^2(\mathbf{Z}).$$

Based on the definition and the proposition, will be showed that the *mahalanobis* distance estimated are invariant under affine equivariance. Since $\mathbf{Z} = \mathbf{XA} + \mathbf{B}$ has *i*-th row, $\mathbf{z}_i' = \mathbf{x}_i'\mathbf{A} + \mathbf{b}'$, then

$$\begin{aligned}
d_i^2(\mathbf{Z}) &= [\mathbf{z}_i' - \mathbf{t}(\mathbf{Z})]' \mathbf{S}^{-1}(\mathbf{Z})[\mathbf{z}_i' - \mathbf{t}(\mathbf{Z})] \\
&= [(\mathbf{x}_i\mathbf{A}' + \mathbf{b}) - \mathbf{t}(\mathbf{XA} + \mathbf{B})]' \mathbf{S}^{-1}(\mathbf{XA} + \mathbf{B})[(\mathbf{x}_i\mathbf{A}' + \mathbf{b}) - \mathbf{t}(\mathbf{XA} + \mathbf{B})] \\
&= [\mathbf{A}'\mathbf{x}_i + \mathbf{b} - \mathbf{A}'\mathbf{t}(\mathbf{X}) - \mathbf{b}]' [\mathbf{A}'\mathbf{S}(\mathbf{X})\mathbf{A}]^{-1} [\mathbf{A}'\mathbf{x}_i + \mathbf{b} - \mathbf{A}'\mathbf{t}(\mathbf{X}) - \mathbf{b}] \\
&= \{\mathbf{A}'[\mathbf{x}_i - \mathbf{t}(\mathbf{X})]\}' \mathbf{S}^{-1}(\mathbf{X}) \{\mathbf{A}'[\mathbf{x}_i - \mathbf{t}(\mathbf{X})]\} \\
&= \mathbf{A}[\mathbf{x}_i - \mathbf{t}(\mathbf{X})]' \mathbf{S}^{-1}(\mathbf{X})\mathbf{A}'[\mathbf{x}_i - \mathbf{t}(\mathbf{X})] \\
&= [\mathbf{x}_i - \mathbf{t}(\mathbf{X})]' \mathbf{S}^{-1}(\mathbf{X})[\mathbf{x}_i - \mathbf{t}(\mathbf{X})] \\
&= d_i^2(\mathbf{X}). \qquad (\textit{proofed}).
\end{aligned}$$

The breakdown point (BP) is a popular global measure of robustness which is introduced by Donoho and Huber [19]. The value of BP of estimators lays on the closed interval [0,1] [20]. The value of BP means proportion of outlier data can be covered before the estimators of data differ from the good one (no outliers). A higher BP leads to a more robust estimators and the highest attainable BP is 0.5 in the case of median in the univariate case.

Theorem: ***Breakdown Point*** **(BP) MCD** (Lopuhaa and Rosseeuw [13])

Assume that $\mathbf{X} = [\mathbf{X}_1, \mathbf{X}_2, \ldots, \mathbf{X}_m]'$ is a set of m observation data with dimension p-variate and $m \geq p+1$. The estimators $\bar{\mathbf{X}}_{MCD}$ and $\mathbf{S}_{MCD}$ are MCD estimators for mean vector and variance-covariance matrix of **X**. If $p = 1$, then $\text{BP}\left[\overline{\mathbf{X}}_{MCD}\right] = \frac{\frac{(m+1)}{2}}{m}$ and $\text{BP}\left[\mathbf{S}_{MCD}\right] = \frac{\frac{m}{2}}{m}$. When $p \geq 2$,then $\text{BP}\left[\overline{\mathbf{X}}_{MCD}\right] = \text{BP}\left[\mathbf{S}_{MCD}\right] = \frac{\frac{(m-p+1)}{2}}{m}$.

The RMCD is extended of MCD with goal is efficiency and speed of computation. To get the RMCD estimators, C-Step algorithm is applied. There are two stages, the first stage is to sort the observations with the minimum distance, then the second stage is to get the result of a subset that has a determinant of the minimum variance-covariance matrix. Assume again that $\mathbf{X}_1, \mathbf{X}_2, \ldots, \mathbf{X}_m$ is a random sample Normal p-variate distribution with population mean vector $\boldsymbol{\mu}$ and variance-covariance matrix $\boldsymbol{\Sigma}$, then the RMCD algorithm can be described as follows:

Stage I:

1. Determine subset H_1, with data sizes $h = \frac{m+p+1}{2}$ from the previous m sample size.
2. Based on all observations in H_1, calculate mean vector $\overline{\mathbf{x}}_{H_1}$ and variance-covariance matrix $\mathbf{S}_{H_1}$. Then using formula $RD_i^2 = \left(\mathbf{X}_i - \overline{\mathbf{X}}_{RD}\right)' \mathbf{S}_{RD}^{-1} \left(\mathbf{X}_i - \overline{\mathbf{X}}_{RD}\right)$, calculate distance for all $i = 1, 2, \ldots, m$ where $d_{H_1}^2(i) = d_{RD}^2(i)$ with $\overline{\mathbf{x}}_{RD} = \overline{\mathbf{x}}_{H_1}$ and $\mathbf{S}_{RD} = \mathbf{S}_{H_1}$.
3. Ascending sort of $d_{H_1}^2(i)$ to get $d_{H_1}^2(\pi_1) \leq d_{H_1}^2(\pi_2) \leq \ldots \leq d_{H_1}^2(\pi_m)$ where π is a combination in $\{1, 2, \ldots, m\}$.
4. Define $H_2 = \left\{X_{(\pi_1)}, X_{(\pi_2)}, \ldots, X_{(\pi_h)}\right\}$, then calculate mean vector $\overline{\mathbf{x}}_{H_2}$, variance-covariance matrix $\mathbf{S}_{H_2}$, and $d_{H_2}^2(i)$.
5. If $\det\left(\mathbf{S}_{H_2}\right) = 0$, then subset $H_2 = H_1$.
6. Repeat step 1-5 to get different members of subset H (example: 500 times).
7. From 500 subset of H that are calculated the determinant from step 1-5, choose 10 subset which have the smallest determinants.
8. Continue to iteration process for the 10 smallest determinant subset until convergent (kth iteration). The convergent condition is attained for all subset if $\det(\mathbf{S}_{H_1}) \geq \det(\mathbf{S}_{H_2}) \geq \ldots \geq \det(\mathbf{S}_{H_k}) = \det(\mathbf{S}_{H_{k+1}})$.
9. Compare the 10 convergent smallest determinant and choose one the smallest determinant. After get $\overline{\mathbf{x}}_{H_k}$ and $\mathbf{S}_{H_k}$, then calculate the distance by formula below

$$d_{H_k}^2(i) = \left(\mathbf{X}_i - \overline{\mathbf{X}}_{H_k}\right)' \mathbf{S}_{H_k}^{-1} \left(\mathbf{X}_i - \overline{\mathbf{X}}_{H_k}\right) \quad (3)$$

10. To get data distribution consistence which is generated from multivariate normal distribution, then define:

$$\overline{\mathbf{x}}_0 = \overline{\mathbf{x}}_{H_k} \text{ and } \mathbf{S}_0 = \frac{med_{\left(d_{H_k}^2(i)\right)}}{\chi_{p,0.5}^2} \mathbf{S}_{H_k} \quad (4)$$

Stage II:

In the RMCD algorithm, we re-weighted the estimators of location and scale which is calculated from the first stage. The score 1 is given to $d_0^2(i)$ that is not unusual observations (not outlier) statistically, and score 0 for outliers.

1. The stage I estimators $\bar{\mathbf{x}}_0$ and $\mathbf{S}_0$ are used to determined the weight in calculating robust distance

$$w_i = \begin{cases} 1, \text{if } d_0^2(i) \leq \chi^2_{p,0.975} \\ 0, \quad \text{others} \end{cases} \tag{5}$$

2. Based on Eq. (5) then we get [21]

$$\bar{\mathbf{x}}_{RMCD} = \frac{\sum_{i=1}^{m} w_i \mathbf{x}_i}{\sum_{i=1}^{m} w_i}$$
$$\mathbf{S}_{RMCD} = \frac{\sum_{i=1}^{m} w_i (\mathbf{x}_i - \bar{\mathbf{x}}_{RMCD})(\mathbf{x}_i - \bar{\mathbf{x}}_{RMCD})'}{\sum_{i=1}^{m} w_i - 1} \tag{6}$$

From Eq. (6), then

$$d^2_{RMCD}(i) = \left(\mathbf{X}_i - \bar{\mathbf{X}}_{RMCD}\right)' \mathbf{S}^{-1}_{RMCD} \left(\mathbf{X}_i - \bar{\mathbf{X}}_{RMCD}\right), i = 1, 2, \ldots, m. \tag{7}$$

2.2. *RMCD Robust Control Chart*

We propose to use the Hotelling-T^2 control chart with RMCD robust estimators of location and dispersion parameters for detecting and monitoring the outliers that appear in multivariable process. The RMCD estimators shown in Eq. (6) and Eq. (7) substitute location and dispersion parameters in Eq. (2) then we can rewrite below

$$T^2_{RMCD}(i) = \left(\mathbf{X}_i - \bar{\mathbf{X}}_{RMCD}\right)' \mathbf{S}^{-1}_{RMCD} \left(\mathbf{X}_i - \bar{\mathbf{X}}_{RMCD}\right) \tag{8}$$

Where $\bar{\mathbf{X}}_{RMCD}$ and $\mathbf{S}_{RMCD}$ are mean vector and variance-covariance matrix under the RMCD method based on m multivariate observations.

3. Computation and main result

3.1. *Method*

To evaluate robustness performance of the RMCD estimators and to know how effectively the T^2_{RMCD} control chart in detecting the outliers data, the first step is generate simulation data randomly by hierarchy model $f(\cdot) = (1-\varepsilon) f_0(\cdot) + \varepsilon f_1(\cdot)$, where $f_0(\cdot)$ is a base distribution of normal multivariate distribution $N_p(\boldsymbol{\mu}_0, \boldsymbol{\Sigma}_0)$, for instance with p = 2, $\boldsymbol{\mu}_0 = [0,0]'$, $\boldsymbol{\Sigma}_0 = \begin{bmatrix} 1 & 0 \\ 0 & 1 \end{bmatrix}$. The $f_1(\cdot)$ is contaminant distribution which substitute as many as εm in $f_0(\cdot)$; ε is an outlier proportion data. The effect of outliers data using shift outlier that shift mean vector so the centre of data change with hierarchy model below (Hubert and Van Dreissen [22]) :

$$f(\cdot) = (1-\varepsilon) N_p(\boldsymbol{\mu}_0, \boldsymbol{\Sigma}_0) + \varepsilon N_p(\boldsymbol{\mu}^*, \boldsymbol{\Sigma}_0); \boldsymbol{\mu}^* = \boldsymbol{\mu}_0 + \delta \tag{9}$$

The assessment to know how effectively the RMCD robust control chart by simulation considering sample sizes m = 50, 100, 250 and 500; p characteristics quality = 2, 5, 8, 10, 12 and 15; percentage of outliers data = 0%, 5%, 10%, 15%, 20%, 25% and 30% and all scenarios repeated 5.000 times. Then calculate statistic of T^2 classically by Eq. (2) and T^2_{RMCD} by Eq. (8) for each scenario. The last step is count outliers data for each simulation scenario, compare between T^2 and T^2_{RMCD}.

3.2. Manual Algorithm of RMCD

To simplify understanding of RMCD algorithm, in this sub-chapter will be given an example of estimator calculations generated based on RMCD algorithm. The first step is to generate multivariate random data with m = 15 and p = 3 by $\boldsymbol{\mu} = \begin{bmatrix} 3 \\ 3 \\ 3 \end{bmatrix}, \boldsymbol{\Sigma} = \begin{bmatrix} 1 & 0 & 0 \\ 0 & 1 & 0 \\ 0 & 0 & 1 \end{bmatrix}$. The result data is as follows:

$$\mathbf{X}' = \begin{bmatrix} 3.0578 & 3.4929 & 5.1135 & 2.3233 & 1.2720 & 1.4274 & 3.4649 & 3.3828 \\ 2.9334 & 2.6593 & 4.1785 & 2.7495 & 4.1988 & 3.0727 & 1.4649 & 2.7601 \\ 3.1134 & 4.0699 & 2.3903 & 2.7771 & 3.9704 & 3.8960 & 2.9456 & 3.4176 \end{bmatrix.$$
$$\begin{matrix} 1.4765 & 3.2052 & 4.4202 & 3.0558 & 2.5506 & 3.3772 & 2.1032 \\ 2.2610 & 4.7330 & 4.0700 & 1.5055 & 3.5060 & 3.0582 & 1.8471 \\ 0.6338 & 3.3305 & 3.2270 & 4.3926 & 2.1865 & 3.9412 & 2.2278 \end{matrix}\Bigg]$$

Step-:

1. Determine the subset of $H_{1\text{-}1}$ with as many members $h = \frac{m+p+1}{2} = \frac{15+3+1}{2} = 8.5 \cong 9$ as selected randomly from observation m. Suppose the selected observation to be a subset of $H_{1\text{-}1}$ are observations 1, 11, 14, 6, 5, 10, 13, 2, and 9 then the element $H_{1\text{-}1}$ is

$$H_{1\text{-}1} = \begin{bmatrix} 3.0578 & 2.9334 & 3.1134 \\ 4.4202 & 4.0700 & 3.2270 \\ 3.3772 & 3.0582 & 3.9412 \\ 1.4274 & 3.0727 & 3.8960 \\ 1.2720 & 4.1988 & 3.9704 \\ 3.2052 & 4.7330 & 3.3305 \\ 2.5506 & 3.5060 & 2.1865 \\ 3.4929 & 2.6593 & 4.0699 \\ 1.4765 & 2.2610 & 0.6338 \end{bmatrix}$$

and from $H_{1\text{-}1}$ the mean vector estimation and the variance-covariance matrix is obtained

$$\left(\bar{\mathbf{x}}_{H_{1\text{-}1}}, \mathbf{S}_{H_{1\text{-}1}}\right) = \left(\begin{bmatrix} 2.6978 \\ 3.3880 \\ 3.1521 \end{bmatrix}, \begin{bmatrix} 1.2009 & 0.1867 & 0.3222 \\ 0.1867 & 0.6452 & 0.3166 \\ 0.3222 & 0.3166 & 1.2502 \end{bmatrix} \right)$$

With $\det(\mathbf{S}_{H_{1-1}}) = 0.7758$, so as to obtain distance for each observation

$$d^2_{H_{1-1}}(i) = \begin{Bmatrix} 0.538;\ 2.8042;\ 7.5479;\ 0.6639;\ 3.965;\ 2.5964;\ 7.4721;\ 1.4183; \\ 5.7073;\ 3.0595;\ 3.0545;\ 9.9535;\ 0.9794;\ 1.2952;\ 3.7179 \end{Bmatrix}.$$

2. Create a subset of $H_{1\text{-}2}$ whose element is the observation as much as h which has the smallest of $d^2_{H_{1-1}}(i)$ observations 1, 4, 13, 14, 8, 6, 2, 11, and 10. Thus

$$H_{1-2} = \begin{bmatrix} 3.0578 & 2.9334 & 3.1134 \\ 2.3233 & 2.7495 & 2.7771 \\ 2.5506 & 3.5060 & 2.1865 \\ 3.3772 & 3.0582 & 3.9412 \\ 3.3828 & 2.7601 & 3.4176 \\ 1.4274 & 3.0727 & 3.8960 \\ 3.4929 & 2.6593 & 4.0699 \\ 4.4202 & 4.0700 & 3.2270 \\ 3.2052 & 4.7330 & 3.3305 \end{bmatrix}$$

$$\text{with } \left(\bar{\mathbf{x}}_{H_{1-2}}, \mathbf{S}_{H_{1-2}}\right) = \left(\begin{bmatrix} 3.0264 \\ 3.2825 \\ 3.3288 \end{bmatrix}, \begin{bmatrix} 0.7151 & 0.1743 & 0.0585 \\ 0.1743 & 0.4920 & -0.0910 \\ 0.0585 & -0.0910 & 0.3650 \end{bmatrix} \right).$$

obtain $\det(\mathbf{S}_{H_{1-2}}) = 0.1078$. So, it is known $\det(\mathbf{S}_{H_{1-1}}) > \det(\mathbf{S}_{H_{1-2}})$ that for the first subset the member used is $H_{1\text{-}2}$.

3. Repeat steps 1 and 2 in Phase I to form another H_j subset as much as h with j generally 500 times [21].
4. Select 10 H_j subset that yield the smallest determinant value and continue the iteration process until convergent. Table 1 shows the jth subset which gives the smallest determinant value and the end result after iteration until convergent.

Tabel 1. The jth Subset Generates the Smallest Determinant

j th-	Observation Elements $H_{j\text{-}2}$	$\det(\mathbf{S}_{H_{j-2}})$	Observation Elements $H_{j\text{-}k}$	$\det(\mathbf{S}_{H_{j-k}})$	Lots of iterations
24	1, 4, 8, 14, 2, 15, 13, 12, 9	0.02056	1, 2, 14, 15, 13, 8, 4, 12, 9	0.02056	3
95	8, 1, 2, 15, 14, 4, 13, 12, 9	0.02056	1, 2, 14, 15, 13, 8, 4, 12, 9	0.02056	3
238	1, 4, 8, 14, 2, 13, 15, 12, 9	0.02056	1, 2, 14, 15, 13, 8, 4, 12, 9	0.02056	3
281	8, 1, 2, 15, 14, 4, 13, 12, 9	0.02056	1, 2, 14, 15, 13, 8, 4, 12, 9	0.02056	3
19	1, 8, 14, 2, 13, 4, 15, 10, 12	0.02486	1, 14, 2, 13, 4, 8, 15, 12, 10	0.02486	3
250	1, 8, 14, 2, 13, 4, 15, 10, 12	0.02486	1, 14, 2, 13, 4, 8, 15, 12, 10	0.02486	3

263	4, 1, 14, 8, 13, 2, 15, 12, 10	0.02486	1, 14, 2, 13, 4, 8, 15, 12, 10	0.02486	3
280	4, 1, 14, 8, 13, 2, 15, 12, 10	0.02486	1, 14, 2, 13, 4, 8, 15, 12, 10	0.02486	3
296	4, 1, 14, 8, 13, 2, 15, 12, 10	0.02486	1, 14, 2, 13, 4, 8, 15, 12, 10	0.02486	3
376	4 , 1, 13, 15, 14, 8, 2, 10, 12	0.02486	1, 14, 2, 13, 4, 8, 15, 12, 10	0.02486	3

Based on Table 1 it can be seen that the subset H_j which gives the smallest determinant value is the 24th, 95th, 238th and 28th subset with the mean vector estimator and the variance-covariance matrix is

$$\left(\overline{\mathbf{x}}_{H_{24-3}}, \mathbf{S}_{H_{24-3}}\right)=\left(\begin{bmatrix}2.7578\\2.5867\\2.9733\end{bmatrix},\begin{bmatrix}0.4744 & 0.1095 & 0.7353\\0.1095 & 0.3840 & -0.0394\\0.7353 & -0.0394 & 1.3822\end{bmatrix}\right)$$

With robust distance,

$$d_{24-3}^2(i)=\begin{Bmatrix}0.6249;\ 1.2468;\ 123.3981;\ 3.8800;\ 178.4003;\\103.8074;\ 34.6788;\ 2.9567;\ 4.6033;\ 18.1007;\\32.6061;\ 4.5569;\ 2.7776;\ 1.4391;\ 1.9141\end{Bmatrix} i=1,\ldots,15.$$

5. Define $\overline{\mathbf{x}}_0$ and $\mathbf{S}_0$ use Eq. (4) with $med_{\left(d_{H_k}^2(i)\right)}=4.5569$ and $\chi_{p,0.5}^2=2.3659$, thus

$$\left(\overline{\mathbf{x}}_0, \mathbf{S}_0\right)=\left(\begin{bmatrix}2.7578\\2.5867\\2.9733\end{bmatrix},\begin{bmatrix}0.9137 & 0.2109 & 1.4161\\0.2109 & 0.7395 & -0.0758\\1.4161 & -0.0758 & 2.6622\end{bmatrix}\right)$$

With robust distance:

$$d_0^2(i)=\begin{Bmatrix}0.3244;\ 0.6473;\ 64.0677;\ 2.0145;\ 92.6247;\\53.8963;\ 18.0051;\ 1.5351;\ 2.3900;\ 9.3978;\\16.9289;\ 2.3659;\ 1.4421;\ 0.7472;\ 0.9937\end{Bmatrix} i=1,\ldots,15.$$

6. Giving weight to $\overline{\mathbf{x}}_0$ and $\mathbf{S}_0$ that $d_0^2(i)$ obtained in step 5 is used as a weighting basis. Based on the Eq. (5) $\left(\chi_{(3;0.975)}^2=9.3484\right)$, weights are obtained:

$$w_i=\{\ 1,1,\ 0,\ 1,\ 0,\ 0,\ 0,\ 1,\ 1,\ 0,\ 0,\ 1,\ 1,\ 1,\ 1\}, i=1,\ldots,15 \text{ and } \sum_{i=1}^{15} w_i=9.$$

7. Calculate the RMCD robust estimator by Eq. (6) to obtain:

$$\left(\overline{\mathbf{x}}_{RMCD}, \mathbf{S}_{RMCD}\right)=\left(\begin{bmatrix}2.7578\\2.5867\\2.9733\end{bmatrix},\begin{bmatrix}0.4744 & 0.1095 & 0.7353\\0.1095 & 0.3840 & -0.0394\\0.7353 & -0.0394 & 1.3822\end{bmatrix}\right)$$

With robust distance of RMCD as follow:

$$d^2_{RMCD}(i) = \begin{Bmatrix} 0.6249;\ 1.2468;\ 123.3981;\ 3.8800;\ 178.4003; \\ 103.8074;\ 34.6788;\ 2.9567;\ 4.6033;\ 18.1007; \\ 32.6061;\ 4.5569;\ 2.7776;\ 1.4391;\ 1.9141 \end{Bmatrix} i = 1,\ldots,15.$$

3.3. Robustness of RMCD

The robustness of the RMCD estimator can be seen from the consistency of statistical value calculated based on the RMCD method. The mean vector and variance-covariance matrix are said to be robust against the likelihood of emission if the value does not fluctuate greatly (relatively constant) than when there is no outliers. The robustness of the estimator can be determined based on the breakdown point (BP). Estimator RMCD has a maximum BP value of 0.5 which means RMCD estimator can accommodate the existence of up to 50% of total data. If the percentage of outliers data is more than 50%, then the mean vector and variance-covariance matrix will shift from the actual value. As the controls are mean, variance and covariance are calculated from the generated data $N_p(\boldsymbol{\mu}_0, \boldsymbol{\Sigma}_0)$, with $p = 2$, $\boldsymbol{\mu}_0 = [0,0]'$, $\boldsymbol{\Sigma}_0 = \begin{bmatrix} 1 & 0 \\ 0 & 1 \end{bmatrix}$, $m = 20$, percentage of outliers = 0%, 5%, 10%, …, 50% and shifted (δ) = 0; 0,5; 1; 1,5; 2; 2,5; 3.

Figures 2 and 3 show the mean statistical values for the X_1 and X_2 variables respectively calculated by the classical method and the RMCD method. The mean parameters are based on the generated data for each variable X_1 and the variable X_2 is 0. The mean statistic X_1 computed by the classical method estimates the value of the parameter around 0.794 before the shift, while the mean X_1 is calculated with RMCD method is much lower that is around 0.605 value. Similarly, for the mean of X_2, there is no shift of 0.809 discharged value if calculated by the classical method and located around 0.619 by RMCD method.

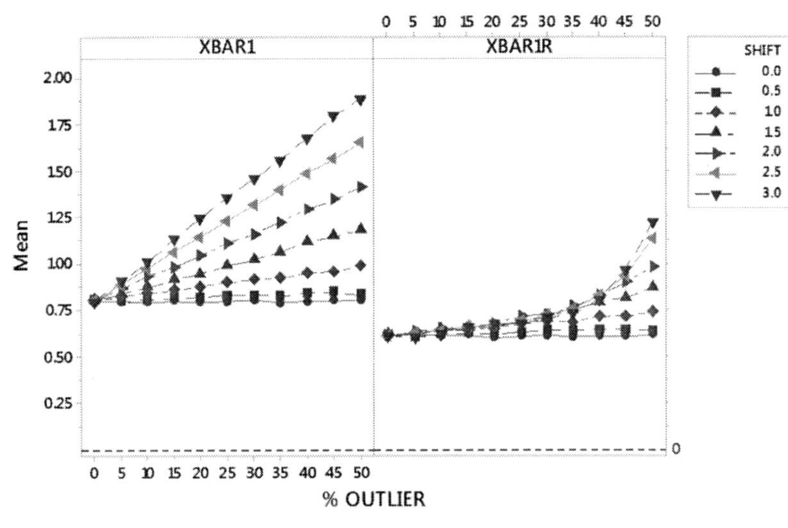

Figure 2. Comparison of Mean Values X_1 Classical Calculations and RMCD with Different Shifts and Percentage of Outliers

When the mean is shifted at a value of 1.5 using the hierarchical function in Eq. (9), the mean X_1 on the 0% percentage is at 0.801 if calculated by the classical method and 0.613 by the RMCD method. This value increases with increasing percentage of data involved in data, eg in percentage of 30% statistically the mean classical X_1 is 1.017 and X_1 RMCD estimates much smaller that is 0.710. Similarly, on the X_2 variable, the mean statistic of classic X_2 estimates the parameters larger than the real, while the mean RMCD X_2 estimates are much smaller than the classical X_2. For example, when the mean is shifted to the number 2 with a 35% percentile percentage, the mean produced using the classical method is 1.221, while the mean calculated using the RMCD method is 0.769.

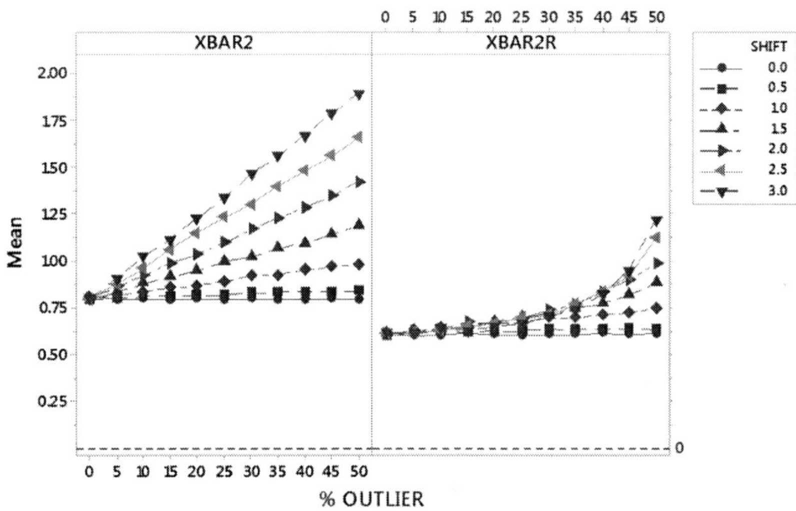

Figure 3. Comparison of Mean Values X_2 Classical Calculations and RMCD with Different Shifts and Percentage of Outliers

The estimated variance generated from the classical method also gives different results than the estimated variance resulting from the RMCD method. As a control, the variance of the generated data for variables X_1 and X_2 is 1 with covariance 0.

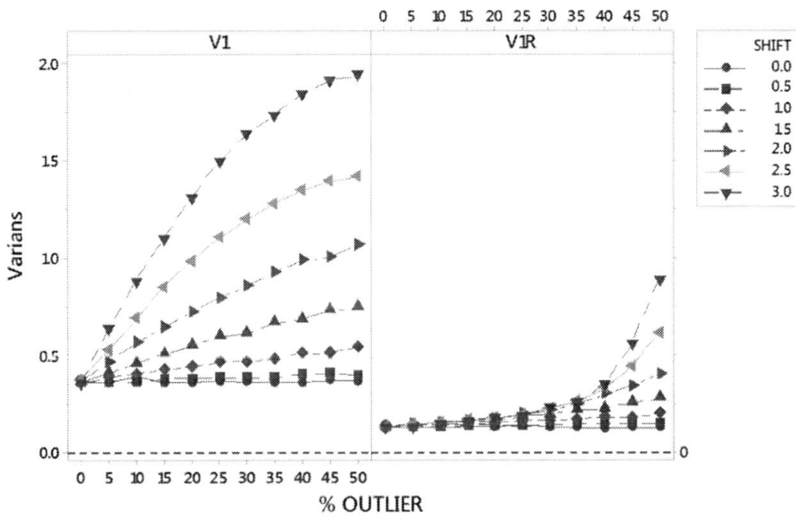

Figure 4. Comparison of Variance Values X_1 Classical Calculations and RMCD with Different Shifts and Percentage of Outliers

Based on the simulation results (Fig. 4 and Fig. 5), the variance is estimated to be less than the initial value, 1. In Figure 4 and Figure 5 it is known that the larger the mean is shifted and the greater the percentage of the outliers involved in the data, the variance estimate also the greater it is. However, the estimation of variance X_1 and X_2 produced by the classical method has a much greater value than that of the RMCD method.

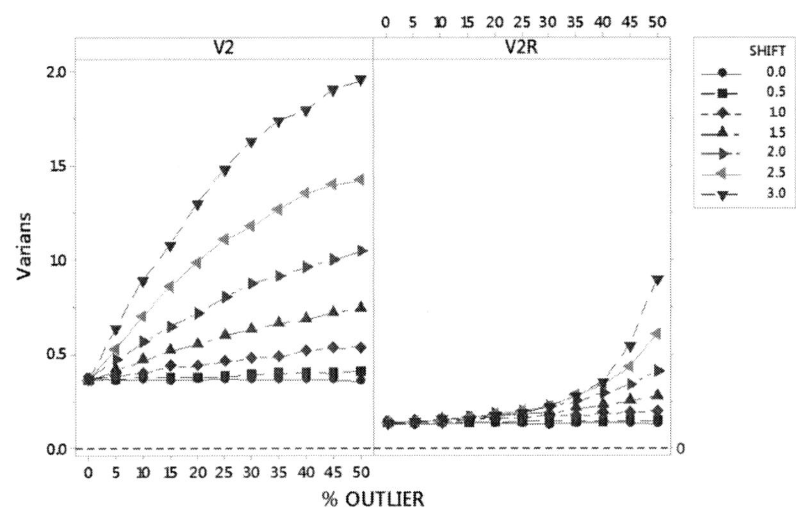

Figure 5. Comparison of Variance Values X_2 Classical Calculations and RMCD with Different Shifts and Percentage of Outliers

3.4. *The Effectiveness of T^2_{RMCD} as An Outlier Detector*

The effectiveness of the statistics T^2_{RMCD} in detecting the presence of a call can be determined based on statistical accuracy in defining the data of the channel as the data of the channel. For example, if the data is included in 3 (three) outliers, then it is said to have 100% effective if the statistic is capable of detecting all three of them as data.

Figures 5, 6, 7 and 8 show the ratio of sensitivity in various p to m = 50, m = 100, m = 250 and m = 500 respectively. Based on Figure 4.5 with m = 50, it is generally known that the percentage of statistical sensitivity in detecting the presence of sine is much greater than the statistics. For example at p = 2, with a 20% percentage of 50 data or in other words there are 10 out of 50 data kept away from the mean, statistics T^2_{RMCD} detect almost exact (103.26%) of the real, while statistics T^2 are only capable of detecting around 7.5% or in other words, statistics T^2 are only capable of detecting 1 of 10 data including the data of the channel. The sensitivity of the T^2_{RMCD} constant control chart to accurately detect the presence of the data is up to p = 5, while at p = 8, 10, 12 and 15 the T^2_{RMCD} statistics detect in excess the existence of outliers up to 3.5 times the actual number of outliers. As for statistics T^2, as the proportion of output and quality characteristics grow, this chart is incapable of detecting altogether presence.

It can also be seen from Figure 5 that the degree of statistic T^2 sensitivity in detecting the presence of outlier decreases as the variables are involved. At a 10% percentage point, the statistic T^2 sensitivity level decreased from 103.48% with p = 2 to 35.36% with p = 5. This decrease continued to 20.76% with the increase of the variable to p = 8, 14,32 % at p = 10, declined again to 9.32% at p = 12 and statistic T^2 only detected 5.68% when the variable involved was 15. Figure 6 also shows that there is over sensitive statistics T^2_{RMCD} in detecting the presence of the outliers at p is equal to 8 for all the proportion of output from the data, whereas at p less than 8 statistics T^2_{RMCD} it is almost appropriate to detect the number of outliers in the data.

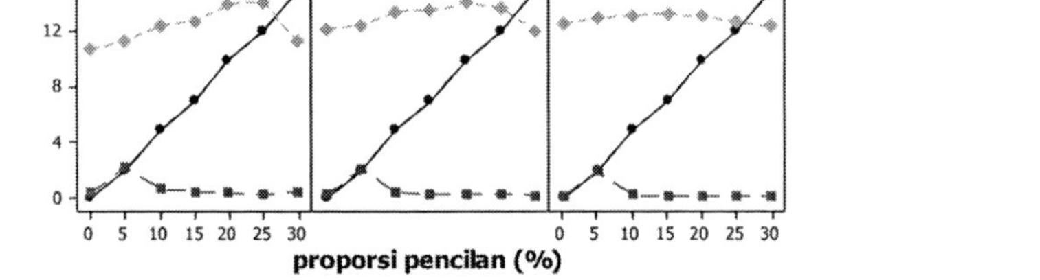

Figure 6. The Effectiveness Comparison Between Statistic T^2 and T^2_{RMCD} for m = 50

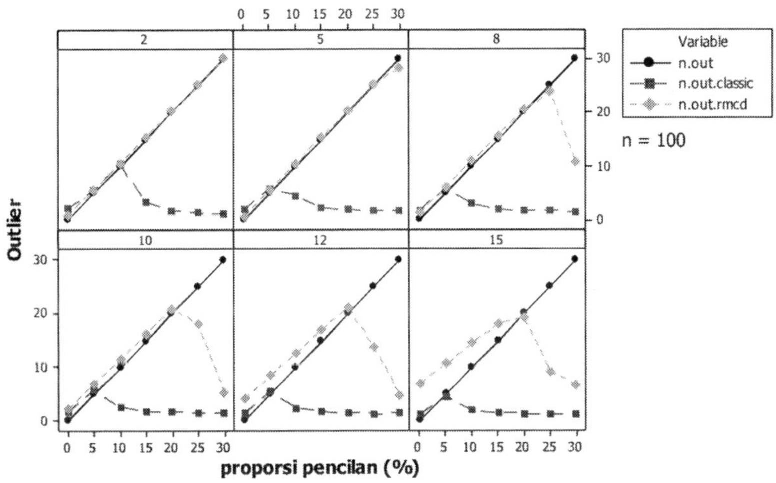

Figure 7. The Effectiveness Comparison Between Statistic T^2 and T^2_{RMCD} for $m = 100$

As shown in Figure 5, Figure 6, Figure 7, Figure 8, and Figure 9 also show a relatively similar pattern of comparison rates of sensitivity. At a proportion of up to 30%, for p less than 8, the statistic T^2_{RMCD} almost accurately detect the presence of the data. As for p more than equal to 8, in the proportion of 0% up to between 20% -25%, the T^2_{RMCD} control chart is almost capable of precisely detecting the presence of outliers, but this sensitivity decreases as the proportion of sine increases by more than 25% (say 30%).

In contrast to statistics T^2_{RMCD}, statistics T^2 are less sensitive in detecting the presence of a call even when the percentage of earnings is 10%. The greater the percentage of enlarged audiences in the data, the lower the degree of statistical sensitivity in detecting its presence. Decrease in the level of statistic T^2 sensitivity also occurs along with the increase in the number of variables at the same percentage of the same. In addition, based on the results of this simulation, it can be seen that the larger the size of the data, then the consistency of the T^2_{RMCD} control chart in detecting the existence of the data with the right way is better.

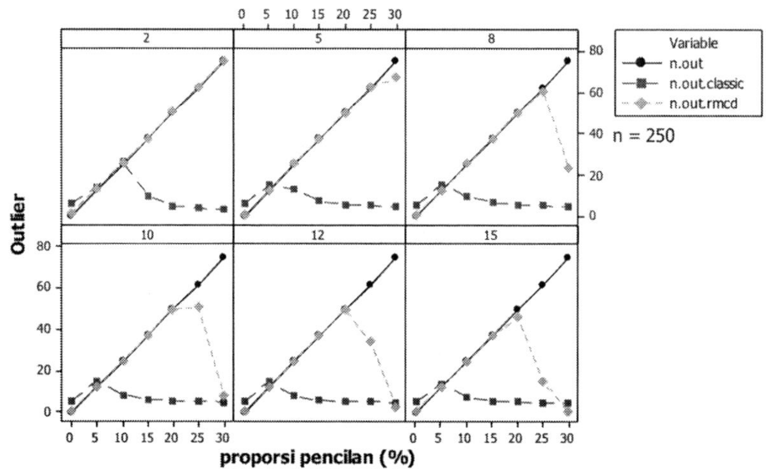

Figure 8. The Effectiveness Comparison Between Statistic T^2 and T^2_{RMCD} for m = 250

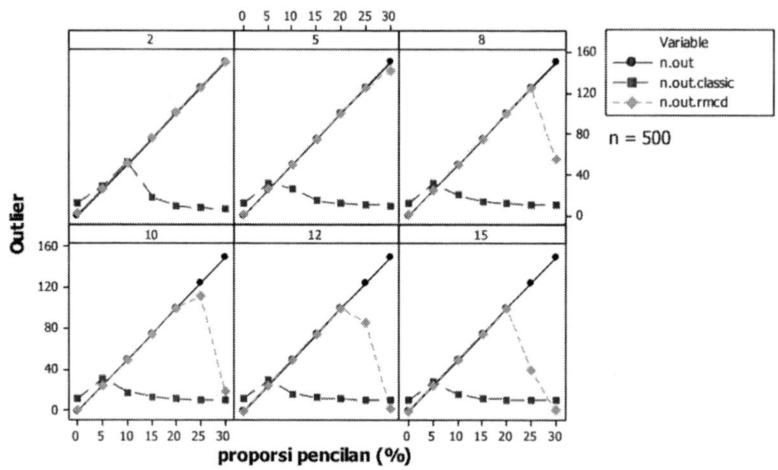

Figure 9. The Effectiveness Comparison Between Statistic T^2 and T^2_{RMCD} for m = 500

4. Conclusion

Based on the results, there are two conclusions that can be taken. Firstly, on the size of the data m = 50 with p less than equal to 5 for all proportions of outliers (less than equal to 30%), the control chart almost precisely detects the presence of the data of the call, whereas at p more than equal to 8 occurs over sensitive. Secondly, In the data size m = 100, 250 and 500 or greater, for all p with a proportion of less than 25%, the control chart almost accurately detects the presence of the data. As for the proportion of outliers more than 25%, occurs less sensitive.

Acknowledgement
We thank Faculty of Mathematics and Natural Sciences, Brawijaya University for funding our research by BOPTN 2017 and also deeply thank Ari Prasetyo, S.Si for many helping in programs and many discussing.

References

[1] Liu H, Shah S, and Jiang W 2004 On-line outlier detection and data cleaning *Computers and Chemical Engineering* **28** 1635
[2] Hawkins D 1980 *Identification of Outliers* (London: Chapman and Hall)
[3] Johnson R 1992 *Applied Multivariate Statistical Analysis* (New Jersey: Prentice Hall)
[4] Barnett V, and Lewis T 1994 *Outliers in Statistical Data* (New York: John Wiley)
[5] Ben-Gal I 2005 *Outlier Detection in: Data Mining and Knowledge Discovery Handbook: A Complete Guide for Practitioners and Researchers* (Tel Aviv: Kluwe Academic Publishers)
[6] Acuna E, and Rodriguez C A 2004 Meta analysis study of outlier detection methods in classification *IPSI* (Venice: University of Poerto Rico)
[7] Montgomery, D C 2009 *Introduction to Statistical Quality Control* (New York: John Wiley & Sons)
[8] Midi H, and Shabbak 2011 A Robust Multivariate Control Chart to Detect Small Shift in Mean *Mathematical Problems in Engineering* **2011** 1
[9] Chenouri S, Stefan H S, and Asokan M V 2009 A Multivariate Robust Control Chart For Individual Observation *Journal of Quality Technology* **43** 259
[10] Sullivan J H, and Woodall W H 1998 Adapting Control Charts for the Preliminary Analysis of Multivariate Observations *Communications in Statistics, Simulation and Computation* **27** 953
[11] Rousseeuw P J 1985 Multivariate Estimation with High Breakdown Point. In *Mathematical Statistics and Applications, Section B* **1** 283
[12] Rousseeuw P J, and Yohai V 1984 Robust Regression by Means of S-Estimators *Robust and Nonlinear Time Series Analysis. Lecture Notes in Statistics* **26** 256
[13] Rousseeuw P J, and Van Zomeren B C 1990 Unmasking Multivariate Outliers and Leverage points *Journal of American Statistical Association* **85** 633
[14] Lopuhaa H P, and Rousseeuw P J 1991 Breakdown Points of Affine Equivariant Estimators of Multivariate Location and Covariance Matrix *The Annals of Statistics* **19** 229
[15] Willems G, Pison G, Rousseeuw P J, and Van Alest S 2002 A Robust Hotelling Test *Metrika* **55** 125
[16] Vargas J A 2003 Robust Estimation in Multivariate Control Charts for Individual Observations *Journal of Quality Technology* **35** 367
[17] Jensen W A, Birch J B, and Woodall W H 2007 High Breakdown Estimation Methods for Phase I Multivariate Control Chart *Quality and Reliability Engineering International* **23** 615
[18] Olive D J 2005 *Applied Robust Statistics* (Illinois: Southern Illinois University)
[19] Donoho D L, and Huber P J 1983 The Notion of Breakdown Point In *A Festschrift for Erich L. Lehmann in Honor of His Sixty-Fifth Birthday* (Belmont: Wadsworth)
[20] Davies P L, and Gather U 2007 The Breakdown Point – Examples and Counter Examples *REVSTAT-Statistics Journal* **5** 1
[21] Rousseeuw P J, and Van Driessen K 1999 Fast Algorithm for Minimum Covariance Determinant Estimator *Technometrics* **41** 212
[22] Hubert M, and Van Driessen K 2004 Fast and Robust Discriminant Analysis *Computational Statistics and Data Analysis* **45** 301

IOP Conf. Series: Journal of Physics: Conf. Series **943** (2017) 012040 doi:10.1088/1742-6596/943/1/012040

Mathematics understanding and anxiety in collaborative teaching

B I Ansari [1] **and N Wahyu** [2]

[1]Mathematics Education Department, Serambi Mekkah University, Banda Aceh, Indonesia
[2]Mathematics Education Department, Syiah Kuala University, Banda Aceh, Indonesia

E-mail: bansuansari58@gmail.com

Abstract. This study aims to examine students' mathematical understanding and anxiety using collaborative teaching. The sample consists of 51 students in the 7th-grade of MTs N Jeureula, one of the Islamic public junior high schools in Jeureula, Aceh, Indonesia. A test of mathematics understanding was administered to the students twice during the period of two months. The result suggests that there is a significant increase in mathematical understanding in the pre-test and post-test. We categorized the students into the high, intermediate, and low level of prior mathematics knowledge. In the high-level prior knowledge, there is no difference of mathematical understanding between the experiment and control group. Meanwhile, in the intermediate and low level of prior knowledge, there is a significant difference of mathematical understanding between the experiment and control group. The mathematics anxiety is at an intermediate level in the experiment class and at a high level in the control group. There is no interaction between the learning model and the students' prior knowledge towards the mathematical understanding, but there are interactions towards the mathematics anxiety. It indicates that the collaborative teaching model and the students' prior knowledge do not simultaneously impacts on the mathematics understanding but the mathematics anxiety.

1. Introduction

Mathematical understanding is one of the cognitive aspects that need to be developed in mathematics classrooms. This is in line with the regulation of the Indonesian Minister of Education and Culture number 22 comment 2006 that one of the important abilities in mathematics learning is the ability to understand concepts [1]. The ability of understanding is the prior knowledge required for all students to achieve other cognitive abilities. The National Council of Teachers of Mathematics (NCTM) also implicitly mentioned the importance of mathematical understanding that the vision in school mathematics learning is that students understand what they learn [2]. Therefore, a student is considered as having a high mathematical proficiency if the student masters mathematical concepts.

Affective aspects are also necessary for students. One of the important affective aspects of mathematics learning is mathematics anxiety. Mathematics anxiety is a students' emotional reaction based on previous unpleasant experiences which impact the future learning [3]. Students who achieve good results are likely to avoid anxiety. On the other hand, students who achieve poor results are likely to hold high anxiety. Therefore, mathematics anxiety is one's attitudes showing fear or nervousness in solving mathematics problems or in mathematics learning. Anxiety has a negative effect on learning. Students with high anxiety tend to work harder but their understandings tend to be poorer resulting in the increase of the anxiety. Students with medium mathematics anxiety are able to handle the problems in learning mathematics [4]. It is important to maintain the anxiety to be at the ideal level. Collaborative learning is one of the strategies that can be applied to manage anxiety in order to improve understanding.

2. Literature review

Review of some literature regarding mathematics understanding and collaborative teaching is presented in this section.

2.1. Mathematics understanding and anxiety

Mathematics understanding is one of the aims of each lesson as teachers are students' facilitator to understand the concepts being presented. Mathematics understanding can be measured using five indicators: (1) the ability to present the concepts learned, (2) the ability to classify objects based on the requirements to form the concepts, (3) the ability to apply concepts algorithm, (4) the ability to present concepts in various mathematical representations and (5) the ability to relate various concepts [5].

Students are usually confused and anxious when they face word problems causing the loss of mathematical concepts learned previously. Students need to manage their level of anxiety so that they will not lose the concepts learned. Students require to control the anxiety in learning mathematics as the anxiety is the disturbing tense and nervousness when one faces number manipulation and solves mathematics problems; this leads to one's being forgetful and loses her/his confidence [6]. Therefore, teaching model resolving students' anxiety in solving mathematics problems is needed.

2.2. Collaborative teaching

Collaborative teaching is one of the teaching model used to deal with anxiety as well as to improve students' understanding. Collaborative teaching is the form of teaching in which two or more teachers involved in managing the classrooms, starting from the planning stage to the students' evaluation; thus, teachers can improve their quality of learning [7]. One of the applications of the collaborative teaching is one teach-one assist. This type consists of two teachers involving one teacher explains in the classroom and the other assists the learning by supervising students who have difficulties in understanding the learning materials. Students are given many opportunities to be guided when they have materials they have not fully understood [8]. The guidance during the exercises process results in students having more time and chances to consult the teacher. Therefore, students can avoid any excessive anxiety or fear in proposing questions for any confusing materials.

The methods enable the improvement of students' understanding of concepts. Students do not only provide positive responses to the collaborative teachings but also are actively involved in the teaching and learning processes which in turn has a positive impact on their academic achievement [9].

3. Method

The method of this research is presented in a form of sampling technique, test administration, research procedure, and treatment.

3.1. Sample

The sample of this study is obtained from one of the Islamic public junior high schools in Jeureula. It was Madrasah Tsanawiyah Negeri (MTsN) Jeureula, located in one of the suburb areas near Banda Aceh. The sample consists of 51 students of 7th-grade around the age of 13 to 14. Based on their prior knowledge, the sample was classified into high, intermediate and low before being divided into the experiment group (26 students) and the control group (25 students). Pre-test, treatment, and post-test were given to the sample. All classes involved were heterogeneous in terms of the ability (Table 1).

Table 1. Sample based on prior knowledge

Prior Knowledge	Group		Total
	Experiment	Control	
High	6	4	10
Medium	12	16	28
Low	8	5	13
Total	26	25	51

3.2. Test

The test consists of five word problems related to the linear equations and linear inequalities in one variable. Some of the final problems are presented as follow.

- Kadir has x candies, Manda has twice as much as Kadir's, and Iwan has 10 candies more than Kadir's. Given that the sum of Kadir's, Manda's and Iwan's candies are 30, write the mathematical model of the problem and find out the number of candies owned by each person?
- Mr. Rifki sells a television for 1,650,000 rupiahs. He earns 10% profit from the sales. How much is the purchase price of the television?
- It takes six months to build a building by employing 100 workers, how long it takes to build the building when 50 people are employed?

3.3. Procedure

The sample of two groups is obtained randomly for the experiment and control group. The experiment group is treated using the collaborative teaching model in groups of five students and the control group is treated using the non-collaborative teaching model. Both groups are classified into the group of high, intermediate and low prior knowledge. The data collection was from early March 2017 to April 2017.

3.4. Treatment

The topics of linear equations and inequalities in one variable and both qualitative and quantitative ways as well as creating a mathematical model to solve problems were taught to both sample groups. The collaborative teaching used in the learning is the one teach-one assist type. The model is modified with two or more teachers in the classroom. When one teacher explains in the classroom, the other teacher assists the teaching and learning by monitoring the students who have difficulties in understanding the materials presented and solving problems. The core of this model is that the teachers work together in managing the classroom including planning the lesson and evaluating the students. After the treatment, post-test (containing five word problems) were administered to the students to examine students' understanding of the concepts. Furthermore, a student questionnaire consisting of 25 items were also conducted to measure students' mathematics anxiety level.

4. Results and discussion

The mathematics understanding and anxiety based on the research variables and prior knowledge is presented in Table 2.

Table 2. Mathematical understanding and anxiety based on students' prior knowledge

Research Variable	Mathematics Understanding				Mathematical Anxiety				Conclusion
	Experimental		Control		Experimental		Control		
Prior Knowledge	Mean	SD	Mean	SD	Mean	SD	Mean	SD	
High	83.2	13.3	80.5	8.5					Ho is
Intermediate	77.4	8.6	68.7	8.3	52.15	5,5	69.08	5.03	rejected
Low	66.3	10.7	59.3	9.4					
Mean	75.6		69.5						

Table 2 indicates that based on the Mann Whitney test (1-tailed), at the significant level of 95% ($\alpha = 0.05$), there is no significant difference of students' mathematical understanding between the experiment and control group at the level of high prior knowledge. However, there is a significant difference at the level of intermediate and low prior knowledge. The level of students' mathematical anxiety in the class taught using collaborative teaching is ideal ($46 \leq x \leq 69$) compared to the class taught without collaborative teaching [10]. This means that the majority of students have a high mathematical anxiety when they solve mathematics problems in the non-collaborative teaching classroom. This is supported by the correlation between the mathematical anxiety and mathematical understanding that is high ($r = 0.679$), indicating that students with higher mathematical anxiety have lower mathematical understanding of concepts.

The interaction between learning model and prior knowledge is presented in Table 3.

Table 3. Interaction between learning model and prior knowledge

Source	Type I Sum of Squares	*df*	F	Sig.	Conclusion
Model (combination)	1983.962[a]	5	4.016	.004	H_0 rejected
Learning Model	124.254	1	1.257	.268	H_0 accepted
Prior Knowledge	1148.412	2	5.811	.006	H_0 rejected
Learning Model*Prior Knowledge	172.488	2	0.873	.425	H_0 accepted
Error	4446.665	45			
Total	6430.627	51			

Table 3 indicates that there is no significant interaction between learning factor and student grouping on students' mathematical understanding ($\alpha > 0.05$). The students' ability is not influenced by the learning model and prior knowledge simultaneously. The influence is only found for the students with high prior knowledge but it has no impact for the students with medium and low prior knowledge. On average, the post-test results show that students from collaborative teaching have a higher score compared to the non-collaborative teaching classroom.

The interaction between learning model and prior knowledge and mathematical anxiety is presented in Table 4.

Table 4. Interaction between learning model and prior knowledge

Source	Type I Sum of Squares	*df*	F	Sig.	Conclusion
Model (combination)	268.115[a]	5	1.863	.002	H_0 rejected
Learning Model	236.052	1	8.199	.006	H_0 rejected
Prior Knowledge	10.470	2	.182	.834	H_0 accepted
Learning Model*Prior Knowledge	66.640	2	1.157	.024	H_0 rejected
Error	1295.571	45			
Total	1563.686	50			

The results presented in Table 4 indicates that the interaction between learning factor and student grouping is significant ($\alpha = 0.024$). This means that there is an interaction between learning factor and student grouping.

Furthermore, we can make a qualitative comparison between students' mathematical understanding of the experiment and control group, as presented in Table 5.

Table 5. Correlations between mathematical anxiety and mathematical understanding of experiment group (*) and control group (**)

Anxiety level	Mathematics understanding (*)			N	Mathematics understanding (**)			N
	H	I	L		H	I	L	
High (H)	-	--	--	--	--	--	12	12
Intermediate (I)	10	14	--	24	8	2	--	10
Low (L)	--	--	2	2	--	--	3	3
Total	10	14	2	26	8	2	15	25

Table 5 indicates that mathematics anxiety in the control group is high while the mathematical anxiety in the experiment group is ideal for students' mathematical understanding being higher. The results of this study show that the mathematical anxiety of students in the collaborative teaching classroom is ideal (medium). This indicates that there is a difference of anxiety between students who are taught by the team of teachers and those who are taught by one teacher.

A study found that one teacher class can have a negative impact on students' performance due to the lack of time, energy and ideas in addressing individual student's case [11]. This results of this study are in line with the study reported that students provide positive responses towards the collaborative teaching-learning and are actively involved in the teaching and learning processes which in turn positively influence their academic performance [9].

When students' mathematical anxiety is under control, it will be a beneficial stimulus in both teaching and learning processes as well as in the examination [12]. The collaborative learning enables students to avoid high mathematical anxiety. It allows students to freely discuss with teachers regarding each material that has not been fully understood. The two teachers are the facilitators, moving around the classroom to guide students when the lesson start, enabling students to avoid the nervousness and fear of asking questions. This setting overcomes the possible problems arise better than the class that is having one teacher only. This means that the collaborative teaching strategies are successful in controlling students' mathematical anxiety so that it does not have a negative influence towards students' ability in learning.

Furthermore, the impact of collaborative teaching on the mathematical anxiety was also implied in the findings mentioning that of the major cause of anxiety is teacher's authority [13, 14]. This indicates that an authoritarian teacher can kill students' creativity. Therefore, teachers have to be able to eliminate prejudice of mathematics.

5. Conclusion

This study shows that the students who are taught using collaborative learning have a higher mathematical understanding and on average have an ideal mathematical anxiety (intermediate level). Students in the non-collaborative class have a lower mathematical understanding and higher mathematical anxiety (high level). In general, there is no interaction between learning factor and student grouping (high, intermediate, and low) on students' mathematical understanding, however, there is an interaction for students with medium and low prior knowledge.

References

[1] MOEC 2006 *Regulation of minister of education and culture number 22 comment 2006* (Jakarta: BSNP)
[2] NCTM 2000 *Principles and standards for school mathematics* (Reston: National Council of Teacher of Mathematics) p 127
[3] Martinez J G R 1987 Preventing math anxiety: A prescription *Academic Therapy* **23** 117

[4] Peker M 2009 Pre-service teachers' teaching anxiety about mathematics and their learning styles *Eurasia Journal of Mathematics, Science, & Technology Education* **5** 335
[5] Kilpatrick J 2001 *Adding it up: Helping children learn mathematics* (Washington, DC: National Academy Press)
[6] Mahmood S and Khatoon T 2011 Development and validation of the mathematics anxiety scale for secondary and senior secondary school students *British Journal of Art and Social Sciences* **2** 169
[7] Sandholtz J H 2000 Interdisciplinary team teaching as a form of professional development *Teacher Education Quarterly* **27** 39
[8] Khashan K H 2014 Conceptual and procedural knowledge of rational numbers for Riyadh elementary school teachers *Journal of Education and Human Development* **4** 18
[9] Marilyn 2008 Co-teaching: A simple solution that isn't simple after all *Journal of Curriculum and Instruction (JoCI)* **2** 9
[10] Puteh M 2016 Mathematics anxiety and its relationship with the achievement of secondary students in Malaysia *International Journal of Social Science and Humanity* **2** 6
[11] Mirza M S and Zafar I M 2014 Impact of collaborative teaching (CT) on mathematics students' achievement in Pakistan *Journal of Research & Reflections in Education (JRRE)* **8** 13
[12] Daneshamooz S, Alamolhodaei H and Darvishian S 2012 Experimental research about effect of mathematics anxiety, working memory capacity on students' mathematical performance with three different types of learning methods *ARPN Journal of Science and Technology* **2** 4
[13] Zakaria E, Nordin N M, Ahmad N A and Erlina A 2012 Mathematics anxiety and achievement among secondary school students *American Journal of Applied Sciences* **9** 1828
[14] Syamsuri, Purwanto, Subanji, and Irawati S 2017 Using APOS theory framework: Why did students unable to construct a formal proof? *International Journal on Emerging Mathematics Education* **1** 135

IOP Conf. Series: Journal of Physics: Conf. Series **943** (2017) 012041 doi:10.1088/1742-6596/943/1/012041

Teaching strategies in the learning of highest common factor and lowest common multiple

N L A Halim[1], H-C Li[1], M Shahrill[1] and R C I Prahmana[2]

[1] Sultan Hassanal Bolkiah Institute of Education, Universiti Brunei Darussalam, Bandar Seri Begawan, Brunei Darussalam
[2] Universitas Ahmad Dahlan, Jl. Pramuka Kav. 5, Pandeyan, Yogyakarta, Indonesia

E-mail: masitah.shahrill@ubd.edu.bn

Abstract. The main purpose of this study is to improve students' understanding of Highest Common Factor (HCF) and Lowest Common Multiple (LCM). Several alternative teaching strategies were integrated in the Year 7 lessons involving 20 students from one secondary school in Brunei Darussalam. The categories identified in the teaching strategy were the application of group work, embedding real-life problems, using presentations and the traditional drilling practice method. An open-ended survey was disseminated to collate the students' feedback, and among the questions posed was in relation to the different types of lesson activities utilised within each of the teaching strategies. Although 55% of the students preferred the traditional drilling practice on the calculation of HCF and LCM, mainly due to the ease of the task as opposed to activities involving real-life problems, the findings also indicated that most students favour sharing their ideas in a healthy competition learning style between each other or between the groups.

1. Introduction

Learning Highest Common Factor (HCF) and Lowest Common Multiple (LCM) have several points of importance, such as calculating when two events will occur at the same time given different recurring periods, and to group two or more sets of items into their largest possible amount. One of the difficulties that students may have when solving real-life problems involving HCF and LCM, is the implication of grasping the key concepts and understanding the problem and the solution [1, 2]. The difficulties in understanding the concepts of LCM are based on the teaching practices at the formal level that are often regarded as procedural and manipulative [2]. Previous studies have reported that the meaning of HCF and LCM often confuses many students [2, 3]. To begin with, students need to have an understanding of the terms 'a factor' and 'a multiple'. For example, 4 x 8 = 32, 4 and 8 are factors of 32, and 32 is a multiple of 4 and 8. According to the Year 7 mathematics textbook currently being used in Brunei, the students are instructed to solve HCF and LCM by using the prime factorisation method or the continuous division method [4]. Heng [4] defined the HCF as "the Highest Common Factor of two or more numbers is the largest common factor of these numbers" (p 8) while "a common factor for two or more numbers is a factor that is common to all numbers" (p 8). Meanwhile, the Lowest Common Multiple of two or more numbers is defined as "the smallest common multiple of these numbers" [4, p 9]. The LCM can be found by listing out the multiples of numbers and by choosing the smallest numbers among the common multiples.

1.1. Perspectives involving the teaching and learning of HCF and LCM

In a previous study from Brunei, a number of students were asked to find the HCF of 12 and 24, and very few could obtain the solution answer of 12 through deduction, having realised that 12 is a factor of 24 [5]. Suffolk [5] added that a lot of schools did not even teach the basic meaning of HCF and LCM. When the question "*What is the highest common factor of 12 and 16?*" was posed to the students, the teachers themselves deemed the question unreasonable as they only taught the students how to solve it and not what it meant. Suffolk [5] concluded that this method of doing without thinking could have severe consequences towards the students' learning capabilities, not just within the topic of HCF and LCM, but towards learning other topics in mathematics as well. In fact, there is a shortcut when calculating HCF and LCM with the use of either continuous division or factor tree of prime factorisation method, which is by using a larger divisor, instead of using prime number as the divisor, for the two or more numbers [5]. However, he expressed that students were using the shortcut method without prior learning to the reason why the lowest prime factor needed to be used first as the divisor for the numbers. He also argued that while students appeared to know how to use these methods of prime, there was no afterthought towards the answer or any discussions regarding it, and it was unclear whether the students really understood the concept that underlies the operation.

In a different study involving 248 students from Pakistan, Mohyuddin and Khalil [3] identified potential misconceptions experienced by the primary school students in the learning of mathematics involving HCF and LCM. One of the questions posed was to calculate the LCM of 12 and 18. They found that one-third (77 students) answered correctly (the answer is 36), and almost two-thirds responded using the HCF method, which indicated that the students might be confused between HCF and LCM. Furthermore, they found that 42 students thought that the number 18 was the LCM while the rest could not answer the question. These results suggest that the concept of HCF and LCM has yet to be fully understood [3]. The other question that they gave was about real-life problem question that is, "*A class was divided into two groups of children. In one group there were 16 children whereas 12 were in the other group. Minimum how many apples are needed so that if divided in the first or the second group, the apples are divided completely?*" [3, p 142]. This question required the students to apply their understanding of LCM. The results obtained were 33% of the students gave the correct answer, while 32% picked the answer that would have been the answer to the HCF method, and the rest were unable to answer this question. Statistically, disregarding the students who did not provide an answer, there was a 50-50 distribution of students using methods to solve either using HCF or LCM. This suggests that when attempting a question, the students were only able to relate the question to HCF or LCM if the problem question was stated clearly, as portrayed in the first question. However, when faced with a worded problem without clearly stating the use of HCF or LCM, the students had an inability to relate the question to the term LCM due to a lack of understanding [3].

1.2. The strategies used in teaching

Real-life applications are essential to help students acquire logical thinking and to show them the relevance of mathematical concepts throughout their learning [6]. In addition to the explanations, along with the aid of diagrams, pictures or graphs may help provide the students with clear conceptual understanding. A significant increase in students' understanding and achievements could occur with the application of real-life examples in problem solving [7-9]. Allowing students to portray their own ideas on real-life application may help enhance their understanding and level of interest on the subject [7]. Meanwhile, Triyani et al. [2] used the story telling approach of the Legend *Putri Dayang Merindu* (LPDM) as their teaching strategy. They identified the LPDM's role in facilitating students' capacity to improve conceptual understanding in LCM by solving situational problems based on the LPDM and eventually moving to formal solution of LCM. Accordingly, previous studies have shown positive effects by using story telling in order to achieve meaningful and enjoyable experiences in the learning of mathematics [2, 10-12].

Khatoon and Akhter [13] designed a teaching strategy using innovative collaboration along with group learning to aid the learning of slow learners. The results suggested that through the use of this teaching strategy, there was significant improvement in the performances of slow learners when comparisons were made of their midterm and final term results. In addition, the students in their study showed increased involvement when given the opportunity to work with their peers. Having students working in groups allows students to initiate discussions, deepen their interactions with the subject and develop critical thought processes [13-15]. Johnsen [16] examined the effects of group work on students when establishing mathematical understanding and achievements. The study involved 13 students subdivided into two groups. One group applied the teacher-based learning while the other group centred on student-based learning. The overall results were that the students improved relations between themselves and worked better as a group rather than individually. Johnsen [16] concluded that the group performed much better if they worked collectively for a longer period of time.

Having students to give presentations of their work has proven to be significant as it enhances their communication and public speaking skills [17-19]. Sharing mathematical ideas through presentations, and the application of grouped or paired work provides students the opportunity to discuss and share ideas with one another, to improve on their task, and to provide and receive constructive criticisms on the existing ideas [19]. Consequently, partaking in discussions with the entire class also promotes the opportunities to listen and challenge other peers' ideas [19].

2. Methodology

The main purpose of this study is to improve students' understanding of HCF and LCM. It also aims to investigate the effects of using HCF and LCM questions based on real-life situations to further facilitate the students' understanding of the topic. Integration of real-life situations into the prospect of HCF and LCM may allow the students to relate to the question on a personal level, instead of undergoing a step-by-step process of solving the equations. This present study involves teaching HCF and LCM by implementing several alternative teaching strategies through a series of intervention lessons using an action research framework.

2.1. Participants and intervention design

The participants in this present study consisted of 20 students (5 males and 15 females) from a Year 7 class in one of the secondary schools in the Brunei-Muara District. All the relevant permissions from the school, teachers, parents and the students were obtained before initialising this study. The first author took on the role as the teacher and conducted a total of five lessons during the course of the intervention. Each lesson lasted about 50 minutes and the lessons consisted of revision extensions, drilling practices, learning activities, pair work, group work, problem-based learning and presentations. Descriptions of each of the intervention lessons are described as follows.

The first lesson of the intervention focused upon the methods and calculation to solve HCF and LCM. The students were provided with a list of direct calculations of the HCF and LCM problems, and they were initially instructed to work out the solutions on their own. Subsequently, the students were paired to let them check each other's answers while having discussions between them. Finally, the solutions were discussed together on the whiteboard with the opportunity for other students in the class to ask questions. The type of work involved is pair work and the activities involved recalling the definitions of, and the drilling practice on the calculations of HCF and LCM. Figure 1(a) represents the exercise questions and Figure 1(b) shows a sample of a student's work in the first lesson.

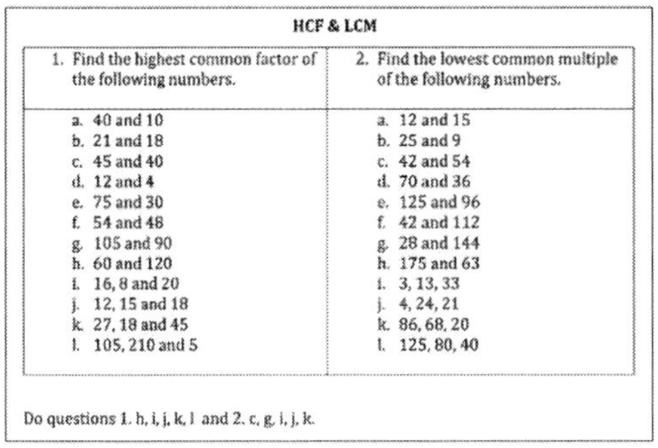

HCF & LCM

1. Find the highest common factor of the following numbers.	2. Find the lowest common multiple of the following numbers.
a. 40 and 10	a. 12 and 15
b. 21 and 18	b. 25 and 9
c. 45 and 40	c. 42 and 54
d. 12 and 4	d. 70 and 36
e. 75 and 30	e. 125 and 96
f. 54 and 48	f. 42 and 112
g. 105 and 90	g. 28 and 144
h. 60 and 120	h. 175 and 63
i. 16, 8 and 20	i. 3, 13, 33
j. 12, 15 and 18	j. 4, 24, 21
k. 27, 18 and 45	k. 86, 68, 20
l. 105, 210 and 5	l. 125, 80, 40

Do questions 1. h, i, j, k, l and 2. c, g, i, j, k.

(a)

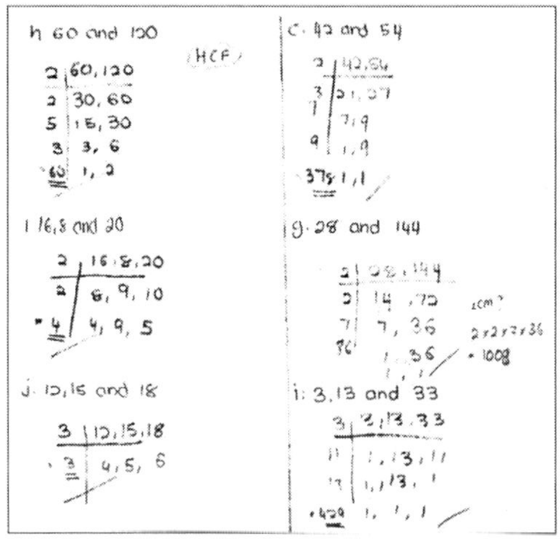

(b)

Figure 1. Samples of questions and a student's work from Lesson 1

The second lesson integrated the application of real-life problems in HCF and LCM. The students were also taught to identify hints within the worded problems so as to assess the nature of the question on whether it was HCF or LCM. The students were provided with a few practice questions (Figure 2(a)) to understand more on the topic with the help of teaching aids, such as accurate measurements of paper mats and number of sweets to help them visualise the problem and practically experience the problem. The students were given the chance to try solving the questions by working in groups. Figure 2(b) represents a sample of a student's work in the second lesson.

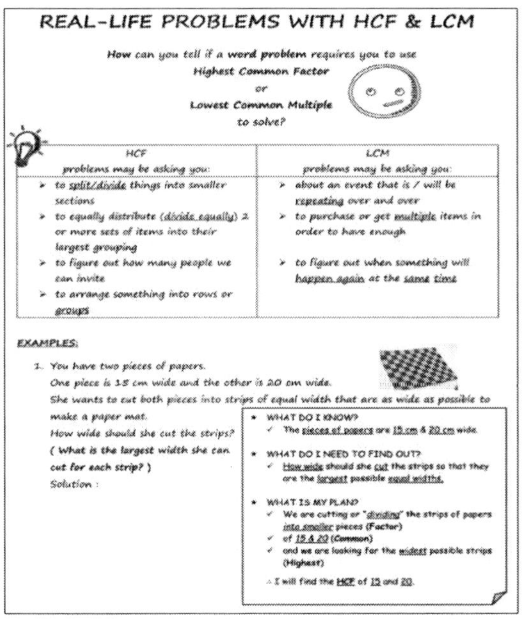

(a)

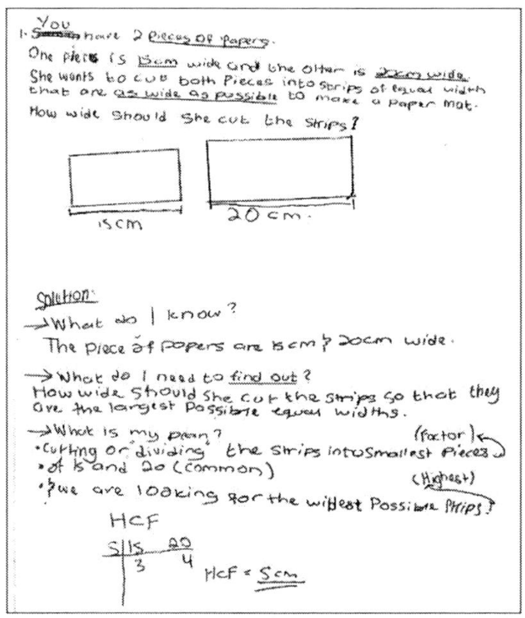

(b)

Figure 2. Samples of questions and a student's work from Lesson 2

In the third lesson, the students continued solving the questions from the previous lesson (shown in Figures 3(a) and 3(b)). Additionally, each group had to discuss and create a question relating to real-life application of HCF and LCM for the other groups to answer in the following lesson.

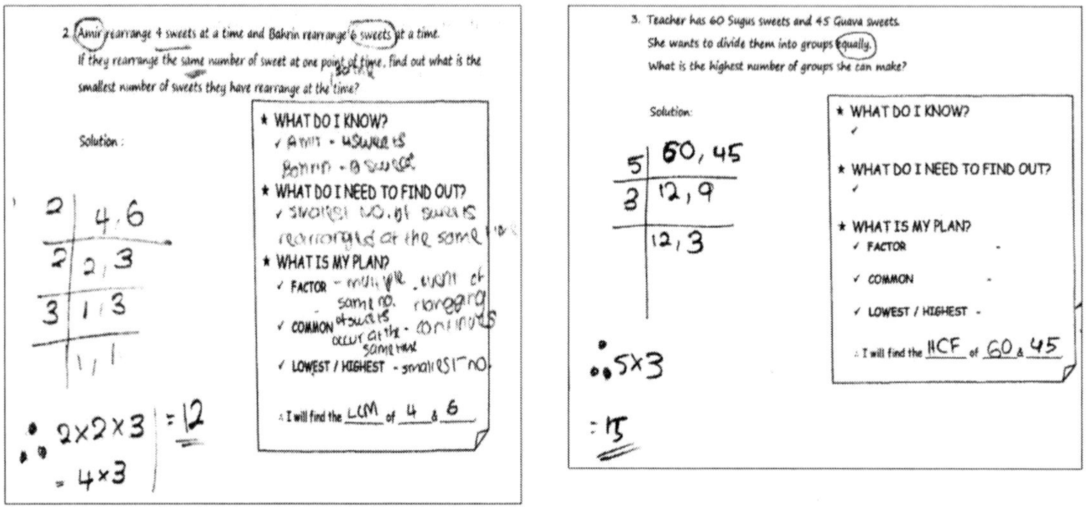

(a) (b)

Figure 3. Samples a student's work from Lesson 3

The fourth lesson involved distributing the questions created from the third lesson to the other groups other than their own. The students were allocated time to discuss and answer the question. Each group had to prepare their responses on a blank piece of A3 sized paper in order to present them in the next lesson. Selected samples of the group work are provided in Figures 4(a) and (b).

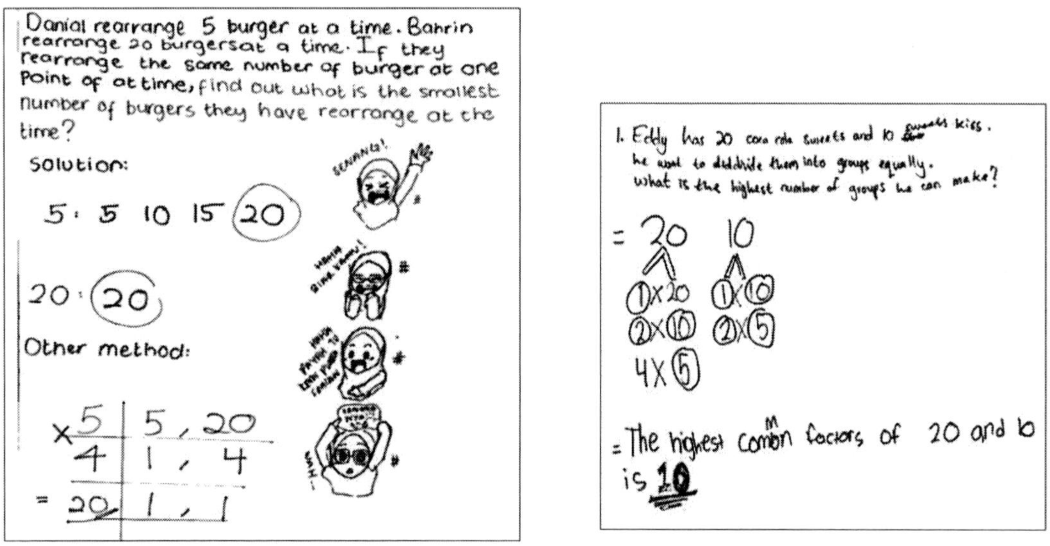

(a) (b)

Figure 4. Selected samples of the group work from Lesson 4

In the final lesson, each of the 6 groups presented their work with the solutions explained by one of their nominated group members (refer to sampled works in Figure 4). Each group that presented were also enquired if the correct methods were used to solve the problems given to them. They also had the opportunity to check with the respective creators of the questions for verification of the worked solutions. This activity allowed all the students to gain and share ideas within the allocated discussion time. The final task given to the class was a question that involved problem-based learning on solving real-life problems using HCF and LCM. This question required higher-order thinking skills in order to solve it. Once the group task was completed, discussions were done together with the teacher's guidance. Displayed in Figures 5(a) and (b) are the selected samples of the group work from Lesson 5.

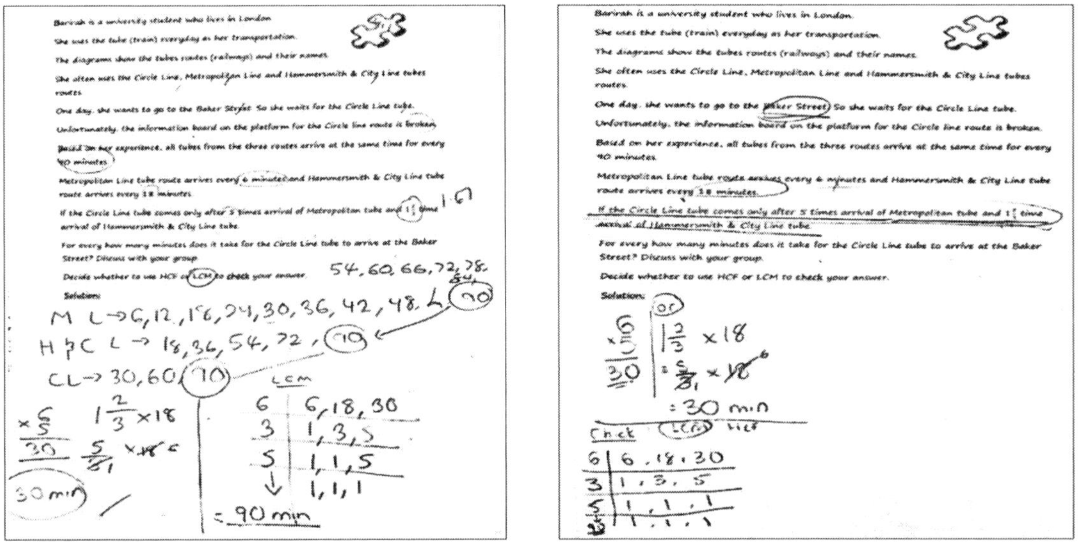

(a) (b)

Figure 5. Selected samples of the group work from Lesson 5

2.2. Data collection and analysis

An open-ended survey was distributed to gather the students' feedback. The survey assessed the students' attitudes towards the intervention and their level of understanding to real-life problems in HCF and LCM. The students were given 30 minutes to complete the survey. The collected data were analysed quantitatively and qualitatively through the students' responses from the survey. Consequently, the results reported in this paper will only focus on the selected questions presented in the survey that were, firstly, related to the different categories of lesson activities employed within each of the teaching strategies during the intervention lessons; secondly, recalling the keywords for HCF and LCM; and thirdly, extracting suggestions for improvements in solving real-life problems involving HCF and LCM.

3. Results

First to be presented here are the students' feedbacks regarding the five lesson activities summarised in Table 1. For this survey question, the students were required to rate from 1 (the most favourite) to 5 (the least favourite) from the series of lesson activities and also to state the reasoning for their choices. The results of the students' rating for each of the five activities are shown in Figure 5.

Table 1. The summarised lesson activities in the five lessons

Lesson	Lesson Activity
1	A – Practice on calculation of HCF and LCM
2	B – Real-life problems (paper mats and sweets)
3	C – Design a question and answer other group's question
4	D – Draw, present and show your work to the class
5	E – Solving problem-based learning question

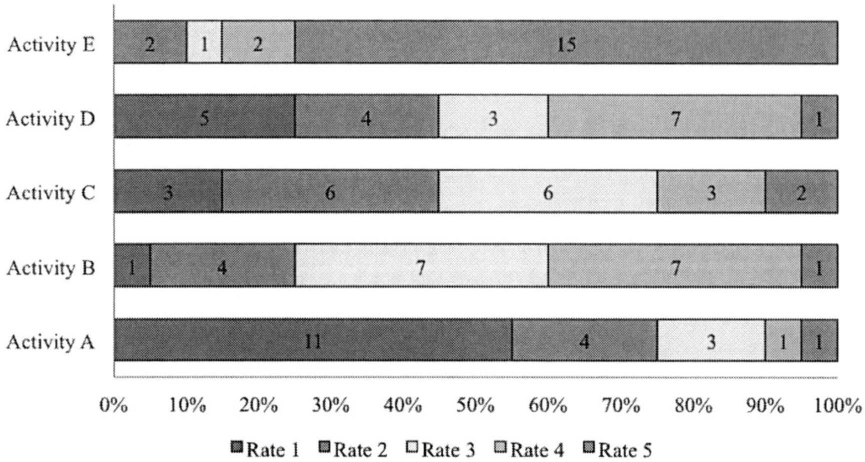

Figure 6. The students' rating for the five activities

The first activity (Activity A) from the first intervention lesson consisted of a calculation exercise on finding the HCF and LCM. As indicated in Figure 6, in total, 11 students (55%) chose this activity as their most favourite activity. From the reasons given, there were 6 students who stated that this activity was 'easy' and they were able to solve the problems without support from the teacher. Furthermore, they claimed that the questions given were understandable. The students who were not in favour of this activity were mainly confused between HCF and LCM.

The second activity (Activity B) involved practice on solving real-life problems with HCF and LCM. This activity posed quite a challenge on the students and it was clearly shown by their reactions to this lesson. Based on the students' responses, 2 students claimed that the activity was easy while 5 students had difficulties during this activity, 4 students had trouble understanding and 3 students were confused. One possibility of this occurrence may be due to the lack of exposure for these students to real-life situations involving HCF and LCM. Hence, this kind of activity, especially the questions given were new to them. Nevertheless, Students #15 and #17 enjoyed the activity as they stated that the activity was fun and easy. In addition, Student #15 stated that she could learn more skills, which meant that she had acquired new knowledge and new solving strategies during the intervention.

Activity C involved expressing their planning and creativity to produce their own question on HCF and LCM, with respect to real life situations in a group work setting, and subsequently, using this knowledge to solve other group's created question. For this activity, half of the class responded with positive feedback where they enjoyed creating the question and also answering other group's question. The same group of students also felt that the questions were easy. Student #4 expressed confidence in his individual capabilities and Student #15 believed that she could improve greatly in this topic. Similar to previous activities, there were still some students who had difficulties understanding and were confused regarding this topic.

The fourth activity (Activity D) also involved their creativity to re-image the received question from other groups and their solved answer on an A3 size poster for their presentation to the class. There were 13 students (65%) who gave favourable feedback, claiming that the activity was exciting, easy, interesting and increased their confidence. Student #8 thought that the purpose in conducting this particular activity was to show that they were able to solve the real-life problems. Meanwhile, Student #9 expressed that the reason they did the activity was so that they could share their knowledge, strategies and ideas to the class especially to those who lacked understanding in this topic. In contrast, there were only 2 students who claimed that this activity was difficult to understand and 1 student did not understand at all. There may be several reasons as to why this student did not understand, such as lacking of understanding towards the purpose of this activity, and another may be from stage fright in standing in front of the class and present a possible incorrect answer. Otherwise this specific topic may be too advanced for the student at that time. It may also be due to lack of planning and creativity in answering and creating the questions.

The final activity (Activity E) of the intervention lesson involved solving the problem-based learning question. This activity was administered to promote higher-order-thinking. According to Figure 6, more than half the class (75%) rated this activity as their least favourite among the five activities. During the lesson, the teacher gave additional support in order to guide the students to the correct solution. Even though 11 students found this question difficult and they required additional time to grasp the question, yet some of the remaining students found this activity to be fun and easy.

In a subsequent question of the survey, the students were asked to recall the keywords that were related to both HCF and LCM. Almost all the students were correct and this showed that the students were able to recollect the keywords within the real-life worded problems. These could be attributed to the effectiveness of the intervention lessons, which enabled the students to not only gain some knowledge but to also understand and recall upon newly acquired knowledge. One student however wrote the term 'maximum' under the LCM part highlighting the confusion trends that were previously observed, or the student might have forgotten the term 'minimum' since she understood that the LCM involved the term 'smallest'. Another possible reason might be that the student did not fully understand the topic [20, 21] especially involving the real-life word problems. Despite that, the student was able to answer perfectly the keywords under HCF.

From the survey, the students were then asked to make suggestions on how to improve their solving capabilities with respect to real-life problems involving HCF and LCM. From the collated comments, there were some suggestions to improve the lessons by having frequent discussions. There are benefits from having frequent discussions in class such that the students could share more of their ideas and the teacher could provide better guidance to the students [17-19, 22]. There is also a possibility that more discussions could boost the students' performances when they solve real-life problems [8, 9, 17-19, 22-25], especially involving HCF and LCM.

4. Conclusions

Conducting the intervention lessons incited the students to strengthen their prior knowledge on HCF and LCM through discovering and experiencing real-life situation examples. These results were also supported by the procured survey questions in which the findings suggested that the students had difficulties in understanding the questions initially but after the intervention lessons, some students found the real-life problem questions were easy to solve. However, there were still a few students who needed more time to grasp the knowledge. Furthermore, some students also found the intervention lessons with the addition of activities to be enjoyable, fun and exciting in comparison to a normal class routine. Sufficient lesson plan preparation by the teacher was essential in providing suitable working exercises for the students. In addition to these exercises, the students were able to experience practical exercises during the intervention lessons and real-life problems in HCF and LCM. Referring back to Figure 6, more than half of the sampled students chose the first lesson with Activity A as their most and second most favourite lesson. This finding suggests that their preference in choosing the traditional drilling practice on the calculation of HCF and LCM, rather than choosing the activities involving real-life problems, was mainly due to the ease of completing the task activity. Importantly, the findings also indicated that most students favoured the group work approach because they were able to eventually share their ideas to the class with healthy competition learning style between each other or between groups. Grouped discussions helped the students to pose different strategies and applying group work helped them to plan and decide the best approach to solve the real-life problems. With the teacher's guidance, the students associated their strategies with their prior knowledge and understanding. Thus, these actions facilitated the students to be more aware of their real-life surrounding which they might relate the problem to HCF and LCM, and foster their metacognitive process to think creatively.

References

[1] Kolitsch S and Kolitsch L 2011 Greatest common factors and least common multiples with venn diagrams *Louisiana Association of Teachers of Mathematics Journal* **5** 1–7
[2] Triyani S, Indra Putri R I and Darmawijoyo 2012 Supporting student's ability in understanding least

common multiple (LCM) concept using storytelling *Journal on Mathematics Education* **3** 151–64
[3] Mohyuddin R G and Khalil U 2016 Misconceptions of students in learning mathematics at primary level *Bulletin of Education and Research* **38** 133–62
[4] Heng H H 2013 *SPN21 Mathematics Year 7* (Singapore: Marshall Cavdish Education)
[5] Suffolk J 2007 Making the teaching of mathematics more effective *Proceedings of the Redesigning Pedagogy: Culture, Knowledge and Understanding Conference Singapore*
[6] Khalid M 2006 Mathematical thinking in Brunei curriculum: Implementation issues and challenges *Progress Report for Collaborative Studies on Innovations for Teaching and Learning Mathematics in Different Cultures (II) - Lesson Study Focusing on Mathematical Thinking* (Japan: University of Tsukuba)
[7] Bomar M 2009 Real life problem solving in eighth grade mathematics *Action Research Projects* (Nebraska: Math in the Middle Institute Partnership)
[8] Chong M S F and Shahrill M 2016 The use of an emerging framework to explore students' cognitive competency *Indian Journal of Science and Technology* **9** 78812
[9] Abdullah N, Shahrill M, Tan A, Chong M S F, Suhaili A S and Adnan N In-Press The common problem solving strategies used by junior high school students in solving non-routine problems *Advanced Science Letters* **23**
[10] Gradanidis G and Hoogland H 2002 *Mathematics as Story* (Ontario: University of Western Ontario)
[11] Goral M B and Gnadinger C M 2006 Using storytelling to teach mathematics concepts *Australian Primary Mathematics Classroom* **11** 4–8
[12] Price R R and Lennon C 2009 *Using Children's Literature to Teach Mathematics* (North Carolina: Quantile)
[13] Khatoon S and Akhter M 2010 An innovative collaborative group learning strategy for improving learning achievement of slow learners *Journal of Research and Reflections in Education* **4** 142–60
[14] Arends I R 2000 *Learning to Teach* (London: McGraw-Hill Companies)
[15] Totten S, Sills T, Digby A and Russ P 1991 *Cooperative Learning: A Guide to Research* (New York: Garland)
[16] Johnsen S 2009 Improving achievement and attitude through cooperative learning in math class *Action Research Projects* (Nebraska: Math in the Middle Institute Partnership)
[17] Alshare K and Hindi N M 2004 The importance of presentation skills in the classroom: Students and instructors perspectives *Journal of Computing Sciences in Colleges* **19** 6–15
[18] Nor H N H M and Shahrill M 2014 Incorporating the use of poster and oral presentations as an alternative assessment in the teaching of secondary mathematics *Proceedings of the 2nd International Conference on Social Sciences Research* (Kota Kinabalu: ICSSR) pp 369–78
[19] Protheroe N 2007 What does good math instruction look like? *Principal* **87** 51–54
[20] Alam S K S 2009 Identifying and sequencing of the elementary minor concepts of the highest common factor (H.C.F.) and the lowest common multiple (L.C.M.) of numbers *Journal of the Korea Society of Mathematical Education Series D: Research in Mathematical Education* **13** 49–61
[21] Alam S K S 2016 Mistakes in H.C.F. and L.C.M. of algebraic expression **7** 44–48
[22] Manjanai S N N P and Shahrill M 2016 Introducing the flipped classroom strategy in the learning of year nine factorization *International Journal of Interdisciplinary Educational Studies* **11** 35–55
[23] Tanujaya B, Prahmana R C I, and Mumu J 2017 Mathematics instruction, problems, challenges, and opportunities: A case study in Manokwari regency, Indonesia *World Transactions on Engineering and Technology Education* **15** 287
[24] Prahmana R C I, Kusumah Y S and Darhim 2017 Didactic trajectory of research in mathematics education using research-based learning *J. Phys.: Conf. Ser.* **893** 012001
[25] Wahyu K, Amin S M and Lukito A 2017 Motivation cards to support students' understanding on fraction divisions *International Journal on Emerging Mathematics Education* **1** 99

IOP Conf. Series: Journal of Physics: Conf. Series **943** (2017) 012042 doi:10.1088/1742-6596/943/1/012042

tscvh R Package: Computational of the two samples test on microarray-sequencing data

Rohmatul Fajriyah[1], **Dedi Rosadi**[2]

[1]Department of Statistics, Universitas Islam Indonesia, Yogyakarta, Indonesia
[2]Department of Mathematics, Universitas Gadjah Mada, Yogyakarta, Indonesia

E-mail: rfajriyah@uii.ac.id

Abstract. We present a new R package, a tscvh (two samples cross-variance homogeneity), as we called it. This package is a software of the cross-variance statistical test which has been proposed and introduced by Fajriyah ([3] and [4]), based on the cross-variance concept. The test can be used as an alternative test for the significance difference between two means when sample size is small, the situation which is usually appeared in the bioinformatics research. Based on its statistical distribution, the p-value can be also provided. The package is built under a homogeneity of variance between samples.

1. Introduction

The two sample tests problem in bioinformatics study actually based on the sample size in microarray experiment which is usually small. Therefore, to decide that there is a significant difference between two means by implementing the parametric t test will create a problem. As Lönnsted and Speed [7] pointed out the large mean and large t are caused by the outlier and small denominator SE respectively.

Tusher [13], Efron [2], Smyth [12] have modified the two samples t-statistic. If the samples come from normal distribution, in case $\sigma_x^2 = \sigma_y^2$, the test statistic has t_{m+n-2} distribution under $\mathbf{H}_0$ (see e.g. [10]). In case $\sigma_x^2 \neq \sigma_y^2$ then we use the approximation distribution as in [15] and [14].

Other the `t.test` function in R, those concepts have been implemented in the software packages, for instance `limma, SpatialPack, geoR, SScore, and Samr.`

Fajriyah [4] introduces the cross-variance concept and present a new approach to test the mean difference between two samples based on it. It is stated that the cross-variance is different from the cross-covariance, where cross-covariance is a function that gives the covariance of one process with the other at pairs of time points [16] and cross-variance is defined as using the second sample average to compute the first sample variance vice versa.

The purpose of the paper is to provide a general overview of the `tscvh` package (version 1.0), which is based on the homogeneity of the cross-variance test. The paper is organized as follows. In Section 2 the cross-covariance concept and test are introduced. Section 3 introduces the package main function and the implementation by providing some examples on how to use the package and the data come from genomics study. We end by comparing `tscvh` to other `R` packages for two samples test and describe future plans for extending the package for the heterogeneity cross-variances test. Finally, section 4 is the concluding remarks.

2. Cross variance concept and the proposed test

2.1. The Cross-Variance Concept

Definition 2.1 *Suppose we have two independent samples, X_i and Y_j; $i = 1, 2, ..., m$ and $j = 1, 2, ..., n$. Their sample mean and variance are denoted by $\overline{X}, \overline{Y}$ and V_x, V_y. Let*

$$V_x^* = \frac{\sum_{i=1}^{m}(X_i - \overline{Y})^2}{m-1} \quad \textit{and } V_y^* = \frac{\sum_{i=1}^{n}(Y_i - \overline{X})^2}{n-1}$$

*be the cross-variance for each sample X and Y respectively. The cross-variance between samples **X** and **Y** is defined as*

$$\boldsymbol{T} = \frac{V_x^a + V_y^a}{2}, \tag{1}$$

where $V_x^a = \frac{V_x}{V_x^*}$, $V_y^a = \frac{V_y}{V_y^*}$.

The T can also be written as

$$T = \frac{V_x}{2V_x + 2\frac{m(\overline{Y}-\overline{X})^2}{m-1}} + \frac{V_y}{2V_y + 2\frac{n(\overline{Y}-\overline{X})^2}{n-1}} = Z_1 + Z_2 \tag{2}$$

In what follows, we assume that

(i) the sample sizes are equal

(ii) X_i and Y_i are i.i.d. normally distributed with unknown means and known variances σ_x^2, σ_y^2.

Based on Rao [10] the Equation (2) can be written as

$$T = \frac{U}{2U + 2abV} + \frac{S}{2S + 2bcV} \tag{3}$$

with

$$U = \frac{(n-1)V_x}{\sigma_x^2}, \quad S = \frac{(n-1)V_y}{\sigma_y^2}, \quad V = \frac{n(\overline{Y} - \overline{X})^2}{(\sigma_x^2 + \sigma_y^2)}$$

and

$$a = \frac{1}{\sigma_x^2}, \quad b = (\sigma_x^2 + \sigma_y^2), \quad c = \frac{1}{\sigma_y^2}.$$

To compute the distribution of T in Equation (3), it can be done by considering the fact that (see e.g. [10]) U, V and S are independent.

2.2. The Cross-variance Test

Under normality asumption of X and Y then U, V and S are independent, where V is $\chi^2_{(1)}$ distributed and U, S are $\chi^2_{(n-1)}$ distributed.

In the case of $\sigma_x^2 = \sigma_y^2$, we estimate the V_x and V_y by the least square estimator of the pooled variance $S_p^2 = \frac{V_x+V_y}{2}$, therefore the Equations (2) becomes

$$T^* = \frac{U^*}{U^* + 4V^*} \tag{4}$$

where $U^* = \frac{(n-1)(V_x+V_y)}{\sigma_x^2}$ $V^* = \frac{n(\bar{Y}-\bar{X})^2}{2\sigma_x^2}$, and U^* and V^* are independents.

The pdf of T^* is derived from the ratio of linear combination of chi-square random variables [9]. First, let $Y = 1 + 4\frac{V^*}{U^*}$, where V^* is distributed $\chi^2_{(1)}$ and U^* is distributed $\chi^2_{2(n-1)}$. Second, the pdf of T^* is computed by taking $T^* = \frac{1}{Y}$. In the folowing computation, the chi-square distribution will be represented as the Gamma distribution.

Suppose $G = \frac{V^*}{U^*}$ and if we take $U^* = H$, then we get $V^* = GH$. Further we got the Jacobian of this transformation variable random is h. Because V^* and U^* are independents, then the joint probability function of G and H is

$$f_{G,H}(g,h) = f_{V^*,U^*}(gh,h).h \tag{5}$$

where

$f_{V^*,U^*}(gh,h) = f_{V^*}(gh).f_{U^*}(h)$, $f_{V^*}(gh) = \frac{(gh)^{\alpha_1-1}e^{-\frac{gh}{\beta_1}}}{\beta_1^{\alpha_1}\Gamma(\alpha_1)}$, and $f_{U^*}(h) = \frac{(h)^{\alpha_2-1}e^{-\frac{h}{\beta_2}}}{\beta_2^{\alpha_2}\Gamma(\alpha_2)}$. Therefore

$$f_{G,H}(g,h) = \frac{g^{\alpha_1-1}h^{\alpha_1+\alpha_2-1}e^{(-\frac{(1+g)}{\beta}h)}}{\beta^{\alpha_1+\alpha_2}\Gamma(\alpha_1)\Gamma(\alpha_2)}, \quad \beta_1 = \beta_2 = \beta \tag{6}$$

The pdf of g then

$$f_G(g) = \int_0^\infty f_{G,H}(g,h)dh = \frac{g^{\frac{1}{2}-1}}{B\left(\frac{1}{2}, n-1\right)(1+g)^{n-\frac{1}{2}}} \tag{7}$$

G is distributed beta of second kind.

The next step is determining the distribution of Y. We define $Y= 1 + 4G$ and by using the transformation random variable, where $G = \frac{Y-1}{4}$, the pdf of Y is computed as follow

$$f_Y = \frac{4^{n-1}(y-1)^{\frac{1}{2}-1}}{B\left(\frac{1}{2}, n-1\right)(y+3)^{n-\frac{1}{2}}}, \quad 1 \leq y \leq \infty \tag{8}$$

The pdf of T^* is

$$\begin{aligned} f_{T^*}(t^*) =& \frac{4^{n-1}(\frac{1}{t^*}-1)^{\frac{1}{2}-1}}{B\left(\frac{1}{2}, n-1\right)(3+\frac{1}{t^*})^{n-\frac{1}{2}}}\frac{1}{t^{*2}} \\ =& \frac{4^{n-1}t^{n-2}(1-t^*)^{\frac{1}{2}-1}}{B\left(\frac{1}{2}, n-1\right)(1+3t^*)^{n-\frac{1}{2}}}, \quad 0 \leq t^* \leq 1 \end{aligned} \tag{9}$$

Furthermore the cdf of T^* analytically is computed as

$$F_{T^*}(t_0^*) = \int_0^{t_0^*} f_{T^*}(t^*)dt^* = \frac{4^{n-1}}{B\left(\frac{1}{2}, n-1\right)} \times \left[\sum_{k=0}^{\infty}(-3)^k \binom{n+k-\frac{3}{2}}{k} B\left(x, n+k-1, \frac{1}{2}\right)\right] \tag{10}$$

where $B(x,\alpha,\beta)=\int_0^x t^{\alpha-1}(1-t)^{\beta-1}dt$ is the incomplete beta function see e.g. [5].

We reject the null hypothesis of the equality of mean of two independent samples if $T^* < t^*_{0,\alpha}$ or $P(t^* < T^*_0) = F_{T^*}(T^*_0) = \text{p-value} \le P(t^* < t^*_{0,\alpha}) = F_{T^*}(t^*_{0,\alpha}) = \alpha$.

Theorem 2.2 *Let $\sigma^2_x = \sigma^2_y$ and the V_x and V_y is estimated by the pooled variance $S^2_p = \frac{V_x+V_y}{2}$, then $T*$ in equation (4) has a pdf*

$$f_{T^*}(t^*) = \frac{4^{n-1}t^{n-2}(1-t^*)^{\frac{1}{2}-1}}{B\left(\frac{1}{2},n-1\right)(1+3t^*)^{n-\frac{1}{2}}}, \quad 0 \le t^* \le 1 \tag{11}$$

Equation (11) is the probability function for the cross-variance test in the **tscvh** package.

3. The package

3.1. Main functions of the package

There are 4 main functions in the *tscvh* package, namely

(i) `cumr` This function is a cumulative probability density function for the Tstar random variable. It provides a cumulative probability density function value of the Tstar

(ii) `fr`: This function is a density function for the Tstar random variable. It provides a probability value of the Tstar

(iii) `quantr`: This function provides a value for the qth quantile of the Tstar

(iv) `Tstf`: This function is a main function for doing the first cross variance test. A test for testing the difference between two samples

3.2. The implementation

3.2.1. Installing the package

- `tscvh_1.0.tar.gz` is the tarball package
- `install.packages('tscvh')` or
- `R CMD install ~/Downloads/tscvh_1.0.tar.gz`

3.2.2. The data The data set we use in this paper is the data used in Corrada Bravo et al. [1]. We take the second probeset, which is the 238493*_at* as a microarray data example for the package. The data comes from the `antiProfilesData` package in Bioconductor. In this data set, there are 5339 probesets (genes) and 68 samples. It contains a colon normal tissue and cancer samples, from different tissues: colon, rectum mucosa and sigmoid colon mucosa. This data set is from the Affymetrix platform.

3.2.3. How to use the package

- **Step 1**. After installing the package then its library has to be called before using it.
  `library(tscvh)`
- **Step 2**. Write or load the data
  Data set 1

```
u=(0.962552767,  0.849068175, -0.006143108, -1.797205349,  1.885255606,
 0.914242952, -1.171969005, -0.700293945, -0.611782849,-1.042330679,
  -0.461310875,  4.722690836,  1.597091975,  0.640279134,  0.684615816 )
v=(1.86832946,  0.14578081,  0.49505374,  1.75326037,  0.91047225,
 1.94170092 , 3.84792213,  0.32982953, -0.06117995, -0.72727366,
 1.07775479,  1.42560956,  2.21901185,  0.96661355 , 2.17089092)
```

Data set 2

```
x=(2.8538601, 0.5906237, 0.6176468, 0.8219640, 2.0970868, 0.1187993,
1.5110154, -0.1979436)}
y=(1.86832946, 0.14578081, 0.49505374, 1.75326037, 0.91047225, 1.94170092,
3.84792213, 0.32982953, -0.06117995, -0.72727366,1.07775479, 1.42560956,
2.21901185, 0.96661355 2.17089092 )
```

- **Step 3**. Called the `Tstf` function

```
Tstf(d1=x,d2=y,alpha=0.05)
Tstf(d1=u,d2=v,alpha=0.05)
```

The results can be seen at Table 1 below.

Table 1. P-values and decision from the **tscvh** and t tests

Sample	t test		proposed test	
	p-value	Decision	p-value	Decision
Data set 1 (u,v)	0.1299	Accept	0.129	Accept
Data set 2 (x,y), max	0.7173	Accept	0.715	Accept
Data set 2 (x,y), ave	0.7173	Accept	0.716	Accept

3.2.4. Comparison to the ***t.test*** Based on Table 1, we can see that the decision from the `tscvh` package is equal to the `t.test`. The slightly different is in the p-value only. In the `tscvh` package we use 3 digits after comma, therefore for the `Data 1`, the package provides the p-value as 0.129. For the `Data 2`, where the sample size is different, then based on Fajriyah [4] the package gives you two best options: `maximum` and `average` and from both of them, the decisions are equals.

The two examples give the insight on how to use the package and show that the decision are equal to the `t.test`. Therefore, this package can be considered as an alternative package to compare means of two samples, in case the variance of the two samples is homogeneous. In the example above, we only show you the two-sided hypothesis testing. The one-sided right and left can be implemented in the package as well

4. CONCLUDING REMARKS

We introduce the concept of cross-variance and derive a method to test the statistically significance of the means between two samples where the variance is homogeneous. The tscvh package is built with the R program language under the S3 class. It is intended for the boinformatics data, where the replication is usually not many, but it can be used for data from other fields as well. In this paper we compare the result based on the available method, `t.test`. In the future the package will be developed under the S4 class and included the heterogeneity case.

Acknowledgement. The authors would like to thanks to the Indonesian Government for its financial support of the Research Grant No. 012/DirDPPM/70/DPPM/ PascaDoktor-KEMENRISTEKDIKTI/IV/2017, under the Ministry of Research, Technology and Higher Education, Republic of Indonesia and Universitas Islam Indonesia, Jogjakarta, Indonesia.

IOP Conf. Series: Journal of Physics: Conf. Series **943** (2017) 012042 doi:10.1088/1742-6596/943/1/012042

References

[1] Bravo HC, McCall M and Irizarry RA 2017 antiProfilesData: Normal colon and cancer preprocessed affy data for `antiProfile building` R package version 1.14.0

[2] Efron B, TIbshirani R, Storey J D, and Tusher V 2001 "Empirical Bayes Analysis of a Microarray Experiment" *Journal of the American Statistical Association*, **96** 1151-60

[3] Fajriyah R 2014 *Micorarray Data Analysis: Background Correction and Diferentially Expressed Genes*, Dissertation. Graz, Austria: Graz University of Technology

[4] Fajriyah R 2016 *Introducing a cross variance concept and its application*, 12th ICMSA Proceeding, IEEE CFP 16 H58 ART

[5] Gradshteyn I S and Ryzhik I M 2007 *Table of integrals, series, and products. 7th edition* (New York: Academic Press)

[6] Kennedy Richard E, Kerns Robnet T, Kong X, Archer, K J, and Miles, M F 2006 "SScore: an R package for detecting differential gene expression without gene expression summaries", *Bioinformatics* **22** 127274

[7] Lönnstedt I and Speed T P 2002 Replicated Microarray Data, *Statistica Sinica*, **12** 31-46

[8] Phipson B, Lee S, Majewski I J, Alexander W S, and Smyth, GK 2016 Robust hyperparameter estimation protects against hypervariable genes and improves power to detect differential expression"", *Annals of Applied Statistics* **10** 94663

[9] Provost S and Rudiuk E M 1994 The exact density function of the ratio of two dependent linear combinations of chi-square variables *Annals of the Institute of Statistical Mathematics*, 46 55771

[10] Rao C R 1971 *Linear statistical inference and its applications. Second edition*(New York: Wiley)

[11] Ritchie ME, Phipson B, Wu D, Hu Y, Law CW, Shi W, and Smyth GK 2015 " limma powers differential expression analyses for RNA-sequencing and microarray studies", *Nucleic Acids Research* **43** e47

[12] Smyth G K 2004 "Linear models and empirical Bayes methods for assessing differential expression in microarray experiments", *Statistical Applications in Genetics and Molecular Biology* **3** article 3

[13] Tusher V, Tibshirani R and Chu G 2001 "Significance analysis of microarrays applied to the ionizing radiation response", *PNAS* **98** 5116-21

[14] Welch B L 1938 The significance of the difference between two means when the Population variances are unequal *Biometrika* **29** 350-362

[15] Welch B L 1947 The generalization of students problem when several different population variances are involved *Biometrika* **34** 2835

[16] `https://en.wikipedia.org/wiki/Cross-covariance`, last access November 8, 2017. At 12.55 PM.

IOP Conf. Series: Journal of Physics: Conf. Series **943** (2017) 012043 doi:10.1088/1742-6596/943/1/012043

Performance assessment in algebra learning process

Ida Lestariani, Imam Sujadi and Ikrar Pramudya

Universitas Sebelas Maret, Jl. Ir. Sutami 36A, Surakarta, Indonesia

E-mail: idha.laney@gmail.com

Abstract. The purpose of research to describe the implementation of performance assessment on algebra learning process. The subject in this research is math educator of SMAN 1 Ngawi class X. This research includes descriptive qualitative research type. Techniques of data collecting are done by observation method, interview, and documentation. Data analysis technique is done by data reduction, data presentation, and conclusion. The results showed any indication that the steps taken by the educator in applying the performance assessment are 1) preparing individual worksheets and group worksheets, 2) preparing rubric assessments for independent worksheets and groups and 3) making performance assessments rubric to learners' performance results with individual or groups task.

1. Introduction

Assessment is the process of gathering information and processing to measure the achievement of learning outcomes of learners. Assessment is an important part of education [1]. For educators, gather information about learners' learning and information about the learning process. For learners, the assessment is the process of their information about learning progress. Assessment has become an important key to improving the quality of education [2]. The quality of education can be achieved by improving the quality of the learning system and assessment [3].

Assessment by educators is an assessment done by educators/teachers of a subject that aims to measure learners' understanding. Assessment can be done by educators by making assessment instruments including test format and assessment. *Permendikbud No 23 Tahun 2016* has set provisions made by educators in assessing the learning outcomes of learners. Provision of educators in conducting an assessment of learners about the knowledge and skills aspects of and the usefulness of the assessment, assessment mechanism, assessment procedures, and assessment instruments undertaken [4].

Assessment in the Curriculum 2013 requires educators to conduct authentic assessments that include an assessment of attitudes, knowledge, and skills [4]. In mathematics subjects, the authentic assessment used is the assessment of knowledge and skills [5]. Assessment as a form of implementation of *KD* 3 and *KD* 4 in the curriculum. In this study focused on performance assessment. Performance assessment is one of the techniques in performing the skill assessment contained in authentic assessment [6].

Performance assessment is an assessment that requires students to perform performance, not answering or selecting answers from a range of possible answers [7]. Performance assessment refers to mathematical tasks and situations that enable learners to demonstrate their understanding and thinking

IOP Conf. Series: Journal of Physics: Conf. Series **943** (2017) 012043 doi:10.1088/1742-6596/943/1/012043

to apply their knowledge and skills in a variety of contexts [8]. Performance assessment is also a procedure for the use of tasks aimed at knowing how well students have learned [9]. Assessment by educators in accordance with the assessment standards set by the government. This is contained in the 2013 curriculum assessment standard. The skills aspect assessment procedure is carried out through the steps of a) assessment planning; b) develop assessment instruments; c) carry out the assessment; d) utilize the assessment results, and e) reporting the results of the assessment of numbers with a scale of 0-100 and description [4].

Performance assessment that refers to mathematical tasks is given by educators in the form of problem-solving questions. The problem-solving stage created by educator refers to the solving stage by Polya. The problem-solving stages by Polya consist of 4 stages: (1) understanding the problem, (2) planning the problem solving, (3) solving the problem according to the second step plan, and (4) re-checking the result [10].

The purpose of this study is to describe the implementation of performance assessment at the problem-solving stages according to by Polya in the mathematics learning process. Performance assessment is one of the authentic assessments that not only to measure what the learners know but to emphasize what the learners do [6]. While in school, educators rarely make an assessment with non-tests [4]. In relation to the valuation function that is very important for educators and learners, it has become the educator's skill to be able to do a good assessment according to the standards of assessment that exist in the Curriculum 2013.

2. Method

Data obtained by observation, interview, and documentation. Researchers make observations on the mathematics learning process in class X, XI, and XII with different educators. Based on the observations obtained subjects who often perform performance assessment of educator class X. So selected class X teachers as research subjects. Researchers make observations on the implementation of assessments made by the educators. With the help of observation guidelines, researchers recorded information needed to support research data. Information obtained by researchers in the form of research implementation process on the mathematics learning process. Based on the observation, the subjects were interviewed for performance assessment conducted at the time of learning. The researcher asked the educator about the assessment process done as a form of validation of observational data obtained by the researcher. From the results of the interview, then performed the method of documentation of performance assessment results conducted on student learning outcomes and instruments made by educators in assessing the performance of learners. This documentation is an assessment instrument that has been created by educators. Then the data analysis technique is done by data reduction, data presentation and conclusion [11].

3. Results and discussion

During the learning process, digital cameras are used to record all the learning activities in the classroom. Learning activities undertaken by educators in the classroom is from the beginning of giving materials to obtain material conclusions and assessment of learners learning outcomes. The stage observed by researchers in the learning process is an observation at the stage of implementation of the assessment. Overall research results, assessment stages conducted by educators are 1) plan assessment; 2) develop assessment instruments; 3) carry out the assessment; 4) reported scoring results in the form of numbers with a scale of 0-100 [4].

The results of field observations indicate that, in the planning and development stage of the educator's assessment instrument, the stages of assessment are commonly undertaken by the educator such as determining the objectives of the study, preparing the grid, making questions/problems that will be used for performance assessment and rubric assessment. However, there are several stages of performance appraisal in accordance with *Permendikbud No 23 Tahun 2016* on assessment standards that are not fully implemented by educators.

There are several stages of performance assessment that are not applied by educators. Based on interview results, the review stage is not always done by educators because the problem used for performance assessment has been used before. Time not included on the assignment/question sheet as it was submitted at the time of the award. Educators do not also write down the steps that need to be done to show the performance because, in accordance with the task given, learners are considered capable of performing stages of learners of the problems done.

The individual task and problem-solving stage using Polya's stage made by educator [10]. The educators are given the individual task in the mathematics learning process to know the performance of the individual learners. Educators also create descriptive scoring rubric problem-solving according to Polya's steps to support a learners' performance assessment.

The educator made the tabulation of performance assessment scores. Based on the learning outcomes of learners who have scored a scale of 1-4 then educators enter the value of learners on the value sheet-like. The score is the total score of the 1- 4 range that has been summed. The final grade obtained by learners is obtained from the provisions made by the educator.

In addition to self-directed tasks, educators create group assignments for performance appraisals. Educators create questions for group task, scoring rubric, and rubric assessments. The group tasks created by educators aim to gain an assessment of the learner's activities that occurred during the discussion [13]. Assessment is done on learner's performance, behavior or interaction [12]. The group tasks are created to test learners making answers or outcomes that show their knowledge and expertise [13].

During group discussions, educators doing observation activities learners in group work. Educators conduct an assessment of the performance of learners with group assessment rubric through the behavior of the song during a discussion. Assessment of the performance of the process of applying the learner's knowledge in completing the task. Educators assess the performance of learners ranging from how to discuss, exchange opinions with the group to get the results of the discussion. To help educators perform group performance assessment, educators make the assessment rubric.

After the learner has finished working on the group task within the time period determined by the educator then collects the worksheet's learners. Next, the educator asks to present the results of his discussion [13]. When learners present the learning outcomes, the educator provides an assessment of the learner's performance with the help of a performance assessment rubric [14, 15]. When other groups respond, educators observe and then assess the learners. This makes it easier for educators to know the ability of learners in performing group tasks. Educators also perform performance assessments to each individual. This is done educators so that learners can show the performance and understanding of the discussion.

4. Conclusion

The results of the research indicate that the steps taken by the educator in applying the performance assessment are preparing individual worksheets and group worksheets, rubric assessments for independent worksheets and groups, and making performance assessments rubric to learners' performance results with individual or groups task. In the implementation of performance assessment on the learning process, educators ask learners to present the results of the discussion to the front of the class. Educators assess the ability of learners to express ideas in solving problems using the rubric that has been prepared.

References

[1] Dimas A, Cari, Suparmi dan Handhika, J 2016 Pengaruh bentuk penilaian tp kemampuan siswa untuk menjawab permasalahan dengan benar *Jurnal Int Ilmu Pengetahuan dan Ilmu Pengetahuan* **1** 40

[2] Hariadi 2017 Instrument Development of Authentic Assessment for Manipulative Fundamental Motor Skill at Elemntary School *Conf Seies: Material Science and Engineering* **180**

[3] Mardapi, D 2012 Pengukuran, Penilaian dan Evaluasi Pendidikan (Yogyakarta: Nuha Litera)

[4] Peraturan Menteri Pendidikan dan Kebudayaan No 23 2016 Tentang Standar Penilaian Indonesia

IOP Conf. Series: Journal of Physics: Conf. Series **943** (2017) 012043 doi:10.1088/1742-6596/943/1/012043

[5] Septiani A dan Rustaman 2017 Implementation of Performance Assessment in STEM (Scien, Technology, Engineering, Mathematics) Education to Detec Science Process Skill Conf Series *Journal of Physics* **812**

[6] Alimudin 2014 Penilaian dalam Kurikulum 2013 *Proc Seminar Nasional* **01** 23

[7] Asmawi Z 2005 *Tes dan Asesment di SD* (Jakarta: Universitas Terbuka)

[8] Marzono, R J D Pickering and McTinghe J 1993 Assessment Student Outcomes Performance Assessment Using the Dimensions of Learning Model (Alexandria: *ASCD Publication*)

[9] Enger and Yanger 2001 *Assessing Student Understanding in Science* (California: Corwin Press, Inc)

[10] In'am A 2014 The Implementation of the Polya Method in Solving Euclidean Geometry Problems *Int Education Studies* 7 149

[11] Salmayzury, Ruslan dan Pristiwaluyo 2015 Evaluasi Program Pembelajaran Matematika di SMA Negeri Watansoppeng *Jurnal Penelitian dan Evaluasi Pendidikan* **1** 1

[12] Direktorat Tenaga Kependidikan Depdiknas 2004 Standar Kompetensi Guru (Jakarta: Depdiknas)

[13] Sa'dijah 2009 Asesmen Kinerja dalam Pembelajaran Matematika *Jurnal Pendidikan Inovatif* **2** 92

[14] Putra H D, Herman T and Sumarmo U 2017 Development of student worksheets to improve the ability of mathematical problem posing *International Journal on Emerging Mathematics Education* **1** 1

[15] Tanujaya B, Prahmana R C I and Mumu J 2017 Mathematics instruction, problems, challenges, and opportunities: A case study in Manokwari regency, Indonesia *World Transactions on Engineering and Technology Education* **15** 287

IOP Conf. Series: Journal of Physics: Conf. Series **943** (2017) 012044 doi:10.1088/1742-6596/943/1/012044

Primary school student teachers' perception to *Pendidikan Matematika Realistik Indonesia* (PMRI) instruction

A Fauziah[1,2], R I I Putri[2], Zulkardi[2] and Somakim[2]

[1]STKIP PGRI, Jl Mayor Toha, Lubuklinggau Sumsel, Indonesia
[2]Sriwijaya University, Bukit Besar, Palembang, Indonesia

E-mail: annafauziah21@yahoo.com

Abstract. This article aimed to report the perceptions of the students of primary school education to PMRI. PMRI or Realistic Mathematics Education (RME) in Indonesian version is one of the promising mathematics learning innovations developed in Indonesia. The research method consisted of three steps, namely preliminary, teaching experiment, and retrospective Analysis. The participants were six students of the primary school teacher education. In the second phase, the participants took an PMRI lesson. Then, they filled in the perception questionnaire (open and closed). The results of the study showed that the participants agreed that learning by realistic mathematics education principles helped them understand the topic.

1. Introduction

PMRI (*Pendidikan Matematika Realistik Indonesia*) is one of the innovations in learning mathematics in Indonesia. It has been developed in Indonesia since 2001 [1]. RME is a Dutch-derived mathematical approach introduced by Freudenthal stating that mathematics is a human activity and math must be related to the real world [2]. According to Freudental, students should have the opportunity to reinvent mathematics, using well-chosen tasks, with the help of teachers.The learning process will occur if the knowledge learned by the students are meaningful to them [3].

Besides as a learning innovation, PMRI is also a reform movement of mathematics education. It is not only a method of learning mathematics, but also it encourages the transformation of social culture [4]. Sembiring [5] adds that PMRI is an approach oriented to the technical ability of mathematics education based on problem-solving. He points out that the reform of mathematics education is based on two pillars, namely: (1) the ability of teachers to create a problem-oriented classroom culture and invites students in interactive learning and (2) designing learning activities that can encourage re-invention of mathematics together with teachers' ability to help the rediscovery process. Thus, the PMRI requires a change in the attitude of teachers in teaching and treating students.

One of the challenges faced in the dissemination of PMRI in Indonesia is the education of teachers, especially the Primary School Teacher Education [5]. Sembiring [5] further stated that primary school teacher education is at the forefront of PMRI development. If candidates for primary school teacher education have been given teaching on PMRI during their study and then after the graduation, they are expected to become teachers, these primary school teacher education graduates are ready to teach using PMRI in their school. Also, learning PMRI in primary school teacher education can also be one way to overcome the problem of Indonesian mathematics education, which is still encountered by high school even college students who cannot solve the problem of how many basic concepts in mathematics. Fauzan [6] argues that it is because of something wrong in mathematics education in primary schools. The learning of the basic concepts is given as something that has been made, not through the process of rediscovering so that the concept becomes less meaningful for the students. One concept that is often not understood by students is about angular topics. Several studies have shown that students are often mistaken about the definition of angles, angular representation, angular size and right angles in different orientations [7-9].

Based on some of the above facts, the researcher then conducted PMRI lesson in primary school teacher education class and looked at how the participant reaction after the learning was given. The researchers provided materials of angular topics including right angles, acute angles and obtuse angles. Therefore, this study aimed to see how the perception of primary school teacher education students on PMRI after they were given a lesson on angular topics.

2. *Pendidikan matematika realistik Indonesia* (PMRI)

In PMRI, mathematics learning must be related to the real world and mathematics is a human activity [2]. Based on the mathematical view as a human activity, there are three principles in realistic mathematics learning by Gravemeijer [10], namely (1) Guided reinvention and progressive mathematizing (guided discovery and progressive mathematics); (2) Didactical phenomenology (didactic phenomena); (3) Self-developed models. According to these principles, mathematics learning should give students the opportunity to understand and process the rediscovery of mathematics itself. Mathematics is not given as a finished thing but through a mathematical activity called mathematization. The situation used in learning mathematics should start from the real situation prior to formal mathematics. Then, the principle of building an independent model as a bridge for students from the real to the abstract or from informal to formal mathcmatics. Thc first model is connected to the real world and then gradually solves the problem. The model emerges from the student's own activities and serves as a catalyst for the process toward formal mathematics. Sometimes the model can also be given to the students but in this case, the model helps move from the thinking process of students to more formal mathematics

Based on the three principles of realistic mathematical approach, there are five characteristics according to Gravemeijer [10] as follows:

1. Use of contextual problems (contextual problems as an application and a starting point of the intended mathematics).
2. Use of models or bridging by vertical instruments (the focus is on building models, schemes, and symbols rather than giving rules or formal maths).
3. Use of students' contribution (a major contribution to learning is coming from self-constructing students, who deliver students from informal to formal math).
4. Interactivity (explicit negotiation, intervention, discussion, cooperation and evaluation among students and teachers) is an important element in the learning process where an informal strategy is used as a lifter to achieve formal mathematics.
5. An intertwining of learning strands (a holistic approach causes the learning strands to be inseparable, even the linkages of the learning sequences are used in solving problems).

The PMRI study is designed and based on the six principles of learning RME by Van den Heuvel-Panhuizen [11] as follows:

1. Activity principle
   This principle is an interpretation of mathematics as a human activity [12]. Students participate actively in the learning process. Students are faced with a problem situation so that students are able to develop a way of finding mathematical concepts based on the information provided [11].
2. Reality Principle
   These principles include two things. First, the importance to reach the learning objectives of mathematics, including the student's ability to apply mathematics in the problem of "real-life", and

second, mathematics learning must start from the problem situation which means giving students the opportunity to build the concept during problem-solving [13].

3. Level principle
   Students must pass a wide variety of levels of understanding, ranging from an informal context to how concepts and strategies are related [13]. In other words, students begin learning with the informal level situation related to the issue of context. then, using a mathematical model concretely representing the mathematical object. This level is called a level model of depicting a concrete model close to the situation described in the problem [14]. Next level is the level of transition, known as a model for, which focuses on the completion of the procedures used to solve a new problem situation. The latter level is a purely formal.
4. Intertwinement principle
   This principle means that the mathematics content domain does not stand alone but isintegrated with other contents [13]. For example, the content is not only related to the angle but also associated with other contents in mathematics.
5. Interactivity principle
   Learning mathematics is not only an activity of its own but also a social activity [13]. Learning provides students the opportunity to share their strategies and discoveries with others.
6. Guidance principle
   This principle refers to the principle of guided reinvention proposed by Freudenthal. Students are given guided opportunities until rediscovering mathematical concepts by balancing between the power of instruction and freedom of learning [12].

3. Method

This study using design research method consisted of three stages: preliminary analysis, teaching experiment, and retrospective analysis [15, 16]. This article only discussed the related perceptions of the students after being given the lesson. To obtain students' perceptions, a perception questionnaire (closed and open) was distributed to the participants who had been given PMRI on the topic of angles (right angle, acute angle and obtuse angle). The study involved six students of primary school teacher education (a small group) as part of the pilot experiment in the teaching experiment phase.

The questionnaire used a likert scale excluding the hesitant options. It asked the participants to explain each choice answer statement. The questionnaire was based on the six indicators of the principle of learning RME (reality principle, activity principle, level principle, interactivity principle, intertwinement principle, and guidance principle). Each indicator consisted of one positive statement and one negative statement.

4. Result and discussion

Table 1 shows the results of the perception questionnaire as indicated by the six principles of PMRI learning. Overall, the average participant agreed (score more than 3) that the learning occurred with the PMRI principles. Although there were three indicators (reality principle, principle level, interaction principle), there was a participant who disagreed (score 2), but after being confirmed by asking for an explanation of the intent of the statement, the participant agreed.

Table 1. Participant perception's about learning

Indicator	N	Score	Mean	Max	Min
Reality principle	6	20	3.33	4	2
Activity principle	6	22	3.67	4	3
Lecel principle	6	18	3	4	2
Intertwinement principle	6	20	3.33	4	3
Interaction principle	6	19	3.17	4	2
Guidance principle	6	20	3.33	4	3

Table 2 is the result of the perception of each statement indicating that most participants agreed that learning taking place in the context of daily life (score 2.8). The context helped in understanding the topic (score 3.8). The average participant also agreed that the learning made them perform mathematical activities (score 3.8), participate actively in learning (score 3.5), and the provided problems help the participants understand the topic gradually (score 3.3). As for the statement that the problem was easy and did not make the participant think hard, there were 3 participants disagreeing (score 2.7). Furthermore, most participants strongly agreed that learning provide an opportunity for them to share resolution strategies with others (score 3.5), active participation (score 3.2), problems given in relation to other topics (score 3.3), learning did not only discuss angular topics, but also others (score 3), activity sheet guided participants to understand the topic (score 3.3), the teachers only functioned as facilitators (score 3.3).

Table 2. Participant perceptions for each item

Indicator	P1		P2		P3		P4		P5		P6	
Statement	1	2	3	4	5	6	7	8	9	10	11	12
Score total	17	22	23	21	20	16	21	19	20	18	20	20
Mean	2.8	3.7	3.8	3.5	3.3	2.7	3.5	3.2	3.3	3	3.3	3.3
Max	4	4	4	4	4	3	4	4	4	4	4	4
Min	2	3	3	3	2	2	3	3	3	3	3	3

In a perceptual questionnaire, in addition to a closed statetement, the questionnaire asked the participants to clarify the choice of selected answers.

First, on the reality principle indicator, there were two statements : (1) Learning about angular topics using real context in everyday life; (2) The context used does not help me in understanding the topic being taught. Some participants provided explanations regarding the choice of answers as follows

Participant 1: I agree because I think using the real context makes me better understand the learning of the angle rather than I use an unreal context, which only confuses me.

Participant 4: Because using the real context, frankly speaking it makes me more passionate about completing the task and looking for answers and also adding to the general knowledge. Context helps me more easily in understanding the topics that are on the subject being taught

Participant 6: Disagree because the images in the real-life context are sometimes less symmetry and less similar to the original form of the building so as to make it confusing and to examine what type requires accuracy and conspiracy in calculating the number of angles as in the image roster that makes me confused where the acute angle and right angle because the picture is too complicated to see.

Then, the second principle consisted of two statements as follows : (1) Ongoing learning does not give me the opportunity to do math activities; (2) I actively participate in learning. Most participants agreed with the statements by providing the following explanation:

Participant 2: Even in the learning that goes on I always do math activities like counting angle count, drawing corner etc. I actively participated in the group and conveyed the results of the discussion

Participant 3: The learning I do has a mathematical element in which I can go directly into the world of mathematics and further understand the lesson. Given a problem-solving context, I co-administered and responded and provided ideas, ideas or more on the topic being discussed.

Furthermore, the statements for the principle level indicator were as follows (1) The provided problem helps me understand the topic gradually; (2) The given problem is easy and does not make me think hard. Concerning the second statement, there was a participant agreeing that the given problem was easy because they just forgot the material and thought a little hard because they had to recall.

Participant 1: Problems in learning angles are arranged gradually and make me better understand the material than my previous learning lessons. Observing the roster image for the angle made me think harder and more thoroughly, as well as on conblock images and the Pentagon building. I did not think there was a picture of a pentagon building, and that made me think hard.

Participant 5: The given problems helped me a lot in understanding the gradual topic of learning from the things I thought that had no relevance to the mathematics and certain topics, now that I have understood it. I do not agree if it was aneasy problem and did not make me think hard. Although it looked trivial, it turned out to be the problem that could not be done carelessly and it made me think hard.

Participant 3: Agreed if the given problem was easy and did not make me think hard because of the topic I learned before, but maybe I forgot a little, and as result I rather thought hard and tried to recall.

On the interactivity principle, two statements were given as follows: (1) Learning gives me the opportunity to share the strategies I use in solving problems for others; (2) I solve problems in individual learning. Participant agreed with these two statements by giving the following explanation.

Participant 4: By sharing the strategy, it made us open with friends and express our opinions. If friends had different opinions, we could search together, so we knew the true midpoint/answer. I worked with friends, sharing information we knew.

Forward, on the intertwinement principle, the first statement was a positive statement, while the second one was a negative statement as follows : (1) Problems related to other topics in learning; (2) Learning only about angle topics. All participants agreed that the learning not only provided with the topic of right angles, acute angles, and obtuse angles, but also linked them to topics of other geometries.

Participant 2: Problems were given regarding flat and angular issues, as well as other knowledge such as starfish, rooster and conblock on the floor.

Participate 6: Problems assigned to / related to other topics, namely the image problem in the form of a flat wake of pentagon and hexagon and look for large angles on each wake flat.

Then, on guidance principle, two statements were given concerning guiding principle as follows: (1) The activity sheet provided does not lead me to understand the topic being studied; (2) The teacher functions as a facilitator when the learning process takes place. The explanations given by participant are as follows :

Participant 4: Precisely with the activity sheet guiding step by step made me understand the topic being studied. The teacher helped us in many ways not only with the material but also helped me think to be more careful in working on a problem and she could bring us into the given context.

Participant 5: The activity sheet I received helped me understand the given topic even though I understood it gradually. The teacher functioned as a facilitator because she facilitated and guided me so that I had a better understanding and could understand the given topic.

The results showed that the participants felt that the learning occurred according to the PMRI principles. Learning took place under the principle of reality because the problems were given in relation to the daily life. The given contexts were related to roster, starfish, *conblock* and pentagon building. Figure 1 is rooster image that given on the first activity sheet. The Paricipants felt that the given contexts allowed them to understand the topics. Based on these results, Bustang et al [17], Putri and Zulkardi [18] have also stated that the context can make learning meaningful.

Figure 1. Roster, the context given on the first activity sheet

Following with the principle of activity, the participants were asked to perform mathematical activities such as counting the number of right angles, acute angles and obtuse angles contained in the given roster images. By doing these mathematical activities, they felt actively involved in the learning process. One of the interesting things was when they were asked to create a different form of roster containing right angles, acute angles and obtuse angles. Two participant groups drew the following shape (Figure 2).

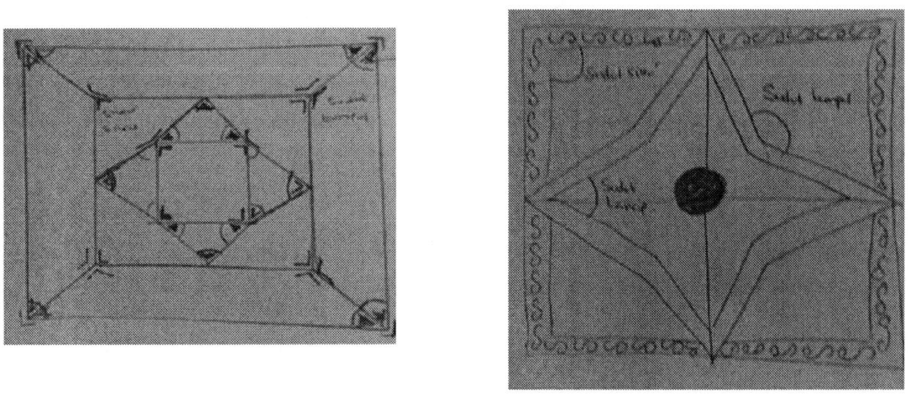

Figure 2. Students' answers

Then the learning began with the participant predicting the sizes of acute angle and obtuse angle. However, to be able to predict the size of an angle, the participant had to have an understanding of the number of inner angles contained in the polygon. This suggests that the change from one level to another in learning requires the scaffolding of others and the sequence of learning activities that are structured for new knowledge to be gained on the basis of prior knowledge [19]. This is in line with the preliminary level in PMRI learning. The learning also used the interactivity principle because participants were actively involved in small group learning, completed the activity sheet, and had the opportunity to share the strategies they used to solve the problems. Similarly, in the intertwinement principle, the topics given in the learning were related not only to angles but also to polygon topics or facets. This enabled the participants to remember recalling the related topics to solve the given problem. The learning was conducted with the guidance principle, because the participants felts that the teacher was indeed a facilitator directing and guiding them to recall the related topics such as the definition of angles, right angles, acute and obtuse angles. She also helped them when they found difficult to solve the problems by providing guidance clues that could make them find the solutions by themselves. This is the importance of the guiding principles of learning that teachers can predict and anticipate the student's understanding and new skills [14, 15, 18]

The results of this study are similar to those conducted by Zulkardi [2]. He found out the positive results on the given RME course. They were satisfied with the lectures given because there are were different topics from different contexts. The RME course participants provided by him also gained a better understanding of RME's principle. Several researches also showed that RME materials had a positive influence on student and student teachers in Indonesia [20-24]. So as Ndlovu [19], which states that teachers after being given RME lessons provide positive outputs and will adopt RME learning activities in their own classrooms.

5. Conclusion

The quantitative results show the results of the students'perceptions of the given PMRI learning, and the qualitative results show the explanation of the answers given by the participant. On the average, most participants agreed that the given learning was in following with the PMRI principles. They felt it was new in their learning because they rarely began the learning by using the context. They did not just listen; they got more active in math activities in the classroom. It was easy for them to understand the topis and they felt there was something new in the learning process even though the topic had been taught previously. This indicates that PMRI learning in primary school teacher education can provide benefits for prospective primary school teachers, judging from their positive reactions and perceptions on the learning that has been given.

References

[1] Putri R I I, Dolk M and Zulkardi 2015 Professional development of PMRI teacher for introducing social norms *Journal on Mathematics Education* **1** 11

[2] Zulkardi Z 2002 Developing a learning environment on realistic mathematics education for Indonesian Students Teacher *Dissertation* University of Twente

[3] Gravemeijer K 2010 Realistic mathematics education theory as a guideline for problem- centered, inter-active mathematics education. In R K Sembiring, K Hoogland and M Dolk (Eds) *A decade of PMRI in Indonesia* (Utrecht: APS International)

[4] Sembiring R K, Hadi S, and Dolk M 2008 Reforming mathematics learning in Indonesia classrooms through RME *ZDM Mathematics education* **40** 927

[5] Sembiring R K 2010 Pendidikan Matematika Realistik Indonesia (PMRI): Perkembangan dan tantangannya *IndoMS-JME* **1** 11

[6] Fauzan A 2002 Applying Realistic Mathematics Education (RME) in teaching geometry in Indonesian primary schools *Dissertation* (Enschede : University of Twente)

[7] Mitchelmore M and White P 1998 Development of angle concepts: a framework for research *Mathematic Education Research Journal* **10** 4

[8] Clements D Hand Burns B A 2000 Students' development of strategies for turn and angle measure *Edu. Stud.in Math* **4** 31

[9] Keiser J M 2004 Struggles with developing the concept of angle: comparing sixth-grade students' discourse to the history of the angle concept *Matematics Thinking and Learning* **6** 285

[10] Gravemeijer K 1994 Educational development and developmental research in mathematics education *Research in mathematics education* **25** 443

[11] Van den Heuvel-Panhuizen M 2000 Mathematics education in Netherland: a guided tour *Freudenthal institute cd-rom for ICME9* (Utrecht : Utrecht University)
[12] Freudenthal 1991 *Revisiting mathematics education China lectures* (Dordrecht: Kluwer Academic Publishers)
[13] Van den Heuvel-Panhuizen M and Drijvers P 2014 Realistic Mathematics Education *Encyclopedia of mathematics education* Dordrecht, Heidelberg, New York, (London: Springer)
[14] Van den Heuvel-Panheuzen M 2003 The didactical use of models in realistic mathematics education: an example from longitudinal trajectory on percentage *Educational studies in mathematics* **54** 9
[15] Prahmana R C I and Kusumah Y S 2016 The hypothetical learning trajectory on research in mathematics education using research-based learning *Pedagogika* **123** 42-54
[16] Prahmana R C I, Kusumah Y S and Darhim 2017 Didactic trajectory of research in mathematics education using research-based learning *J. Phys.: Conf. Ser.* **893** 012001
[17] Bustang, Zulkardi, Darmawijoyo, Dolk M and Van Erde D 2013 Developing a local instruction theory for learning the concept of angle through visual field activities and spatial representations *International Education Studies* **6** 58
[18] Putri R I I and Zulkardi 2017 Fraction in shot-put: a learning trajectory *Proc. AIP Conf.* 4th (Yogyakarta: Universitas Negeri Yogyakarta)
[19] Ndlovu M 2014 The Effectiveness of a teacher professional learning programme : the perceptions and performance of mathematics teachers *Phytagoras* **35** 1
[20] Sundayana R, Herman T, Dahlan J A and Prahmana R C I 2017 Using ASSURE learning design to develop students' mathematical communication ability *World Transactions on Engineering and Technology Education* **15** 245
[21] Hadi S 2002 Effective teacher professional development for the implementation of realistic mathematics education in indonesia *Dissertation* (Enschede: University of Twente)
[22] Armanto D 2002 Teaching multiplication and division realistically in Indonesian primary school: a prototype of local instruction theory *Dissertation* (Enschede: University of Twente)
[23] Tanujaya B, Prahmana R C I and Mumu J 2017 Mathematics instruction, problems, challenges, and opportunities: A case study in Manokwari regency, Indonesia *World Transactions on Engineering and Technology Education* **15** 287
[24] Wahyu K, Amin S M, and Lukito A 2017 Motivation cards to support students' understanding on fraction divisions *International Journal on Emerging Mathematics Education* **1** 99

IOP Conf. Series: Journal of Physics: Conf. Series **943** (2017) 012045 doi:10.1088/1742-6596/943/1/012045

The estimation of time-varying risks in asset pricing modelling using B-Spline method

Nurjannah, Solimun and Adji Rinaldo

Department of Statistics, Brawijaya University, Malang, Indonesia

E-mail: nj_anna@ub.ac.id

Abstract. Asset pricing modelling has been extensively studied in the past few decades to explore the risk-return relationship. The asset pricing literature typically assumed a static risk-return relationship. However, several studies found few anomalies in the asset pricing modelling which captured the presence of the risk instability. The dynamic model is proposed to offer a better model. The main problem highlighted in the dynamic model literature is that the set of conditioning information is unobservable and therefore some assumptions have to be made. Hence, the estimation requires additional assumptions about the dynamics of risk. To overcome this problem, the nonparametric estimators can also be used as an alternative for estimating risk. The flexibility of the nonparametric setting avoids the problem of misspecification derived from selecting a functional form. This paper investigates the estimation of time-varying asset pricing model using B-Spline, as one of nonparametric approach. The advantages of spline method is its computational speed and simplicity, as well as the clarity of controlling curvature directly. The three popular asset pricing models will be investigated namely CAPM (Capital Asset Pricing Model), Fama-French 3-factors model and Carhart 4-factors model. The results suggest that the estimated risks are time-varying and not stable overtime which confirms the risk instability anomaly. The results is more pronounced in Carhart's 4-factors model.

1. Introduction

The main problem in the dynamic asset pricing model literature is that the set of conditioning information is unobservable therefore some assumptions have to be made. Hence, the estimation requires additional assumptions about the dynamics of risks. [1] suggests the use of a rolling window ordinary least squares (OLS) estimation of the market model. This data-driven approach has the advantage of no parameterization but requires the prior selection of the window length. The other estimators in the family of recursive least squares have been considered. Time-varying conditional betas have been nonparametrically estimated by [2], [3] and [4] assuming that betas vary smoothly over time and possibly nonlinearly.

In this paper, we try to estimate time-varying risks using nonparametric or pure data driven methodologies. We use the B-spline method which has not been widely used in finance. Since their introduction by [5], varying coefficient models have become an increasingly popular option for dimension reduction in nonparametric regression with multiple predictors.

Some of the previous studies have applied the varying-coefficient model such as functional regression analysis shown in [6] or [7], and the analysis of longitudinal data, e.g., [8]. The kernel estimation is proposed by [9], two-step estimate by [10], and smoothing spline by [11].

The different methods have different advantages. Smoothing spline is basically similar to kernel regression. The advantages of splines are their computational speed and simplicity, as well as the clarity of controlling curvature directly. Kernels however are easier to program, easier to analyze mathematically and extend more straightforwardly to multiple variables however it takes more time to run. It is also well known that the spline regression has a lot of advantages. The optimal convergence rate, in the nonparametric condition quartile function, partially linear model and partially linear model with mixed effect [12], can be obtained by the B-spline estimators of the nonparametric function.

In addition, asymptotic normality about B-spline estimators of the nonparametric regression is established in [13]. Thus, the B-spline approximation is considered in this paper. One important contribution of this paper is the application in Indonesian stock market. The literature on beta estimation is mainly applied to developed countries data. In contrast, Indonesia is an emerging market that provides the opportunity of analyzing the sensitivity of different beta estimators in this market.

2. The nonparametric time-varying risk estimator

2.1. General model

Let y be the response variable, x and t be the covariant vectors. The most general approach to model y is as a multivariate smooth nonparametric function (x,t). Unfortunately, smoothing estimation of a general multivariate nonparametric regression may require large sample sizes when the dimensionality of the covariant is high, and the smooth results may be difficult to interpret. This phenomenon is called 'the curse of dimensionality'.

A useful alternative is to impose some specific structure between y and (x,t). As a dimensionality-reduction approach, we consider the varying-coefficient model of [5], which has the form

$$y = x_1\beta_1(t_1) + \cdots + x_p\beta_p(t_p) + \varepsilon \tag{1}$$

where $E(\varepsilon) = 0, E(\varepsilon^2) = \sigma^2$ and $\beta_l(\cdot), l = 1, \ldots, p$ are the smoothing functional coefficients.

When $t_i's$ are the same, models (1) can also be written as

$$y = x_1\beta_1(t) + \cdots + x_p\beta_p(t) + \varepsilon \tag{2}$$

A number of methods are proposed to estimate the functional coefficient in the model (2). Among them are the penalized least squares, the local likelihood for generalized varying-coefficient model, kernel estimation, and also smoothing spline. In this paper we will focus on B-spline method due to some advantages such as their computational speed and simplicity, as well as the clarity of controlling curvature directly [12].

Suppose we have observations $y_1, \ldots, y_n$ from the varying-coefficient model (2), and denote the observed values of x_j and t for i^{th} case as x_{ij} and t_i. Then, model (2) has the form

$$y_i = x_{i1}\beta_1(t_i) + \cdots + x_{ip}\beta_p(t_i) + \varepsilon_i \tag{3}$$

The functional coefficients, defined on $[a,b]$ are assumed $\beta_l(t) \in C^m[a,b]$ i.e., $\beta_l(t)$ has m continuous derivatives. The function $\beta_l(t)$ will be approximated by function $s_l(t)$ in the class $S(m,Z)$. The set $S(m,Z)$ is the collection of all polynomial spline of order m (degree m-1) with the knots $Z = \{a = z_0 < z_1 < \cdots < z_k < z_{k+1} = b\}$. By using the least squares criterion, the B-spline estimator of order m for $\beta_l(t)$ is defined to the least squares minimizer $\hat{\beta}_l(t) \in S(m,Z)$ corresponding to

$$\sum_{i=1}^{n}\left(y_i - \left(X_{i1}\hat{\beta}_1(t_i) + \cdots + X_{ip}\hat{\beta}_p(t_i)\right)\right)^2 =$$

$$\min_{s\epsilon S(m,Z)} \sum_{i=1}^{n} \left(y_i - \left(X_{i1}s_1(t_i) + \cdots + X_{ip}s_p(t_i)\right)\right)^2 \tag{4}$$

2.2. *B-Spline estimation of the time-varying model*

Let $Z = \{a = z_0 < z_1 < \cdots < z_k < z_{k+1} = b\}$ be a subdivision of the interval $[a, b]$ by k distinct points. For each (fixed) set of knots Z, the class $S(m, Z)$ is a linear functional space of dimension $N(= k + m)$. Any function $s(t) \in S(m, Z)$ has the form

$$s(t) = \sum_{i=1}^{N} \alpha_i \pi_i(t) = \pi'(t)\alpha \tag{5}$$

where $\pi'(t)$ are *normalized* B-splines and $\pi(t) = (\pi_1(t), \dots, \pi_N(t))'$. Suppose that the *non-decreasing sequence* (v_i) is obtained from $\{z_i\}_{i=1}^{k+1}$ by repeating z_0 and z_{k+1} each exactly m times. For $m \geq 2$, the B-spline basis for the family $S(m, Z)$ is formed by the following $k + m$ *normalized* B-spline [15].

$$\pi_i(x) = (v_{i+m} - v_i)[v_i, \dots, v_{i+m}](v - x)_+^{m-1}$$

$i = 1, \dots, N$ where $[v_i, \dots, v_{i+m}]\phi$ denote the mth divided differences on the $(m+1)$ points $[v_i, \dots, v_{i+m}]$ of the function ϕ and a_+^n is mean a^n if $a > 0$ and 0 otherwise.

The B-spline estimators of functional coefficients of model (2) are defined to be

$$\sum_{i=1}^{n} \left(y_i - \left(x_{i1}\hat{\beta}_1(t_i)\right) + \cdots + \left(x_{ip}\hat{\beta}_p(t_i)\right)\right)^2$$

$$= \min_{\alpha_1 \epsilon R^N l=1,\dots,p} \sum_{i=1}^{n} \left[y_i - \left(x_{i1}\pi'(t_i)\alpha_1 + \cdots + \left(x_{ip}\pi'(t_i)\alpha_p\right]^2\right. \tag{6}$$

Let

$$Y = (y_1, \dots, y_n)'$$

$$D = \begin{bmatrix} x_{11}\pi'(t_1) & \dots & x_{1p}\pi'(t_1) \\ \vdots & \ddots & \vdots \\ x_{n1}\pi'(t_n) & \dots & x_{np}\pi'(t_n) \end{bmatrix}$$

$$\alpha = \left(\alpha_1', \dots, \alpha_p'\right)' \tag{7}$$

The estimator of the lth functional coefficient $\beta_l(t), l = 1, \dots, p$ is

$$\hat{\beta}_l(t) = \pi'(t)\hat{\alpha}_l \tag{8}$$

Hence,

$$\hat{\beta}(t) = I_p \otimes \pi'(t) \cdot \hat{\alpha} = I_p \otimes \pi'(t) \cdot (D'D)^{-1}D'Y \tag{9}$$

where $\otimes$ is Kronecker product

3. Data

The data comprise of the market value, book-to-market value and returns on 407 stocks traded on the Indonesian Stock Exchange over the period of July 1997 - December 2010. From this set of stocks, those with the longest possible time series were chosen. Hence the sample reduced to 177 stocks. The returns are computed assuming continuous compounding of returns. This analysis uses monthly data that spans 144 observations over the period from January 1999 to December 2010. The JSX Composite Index is used as a proxy for the market portfolio while SBI (Interest rate of Bank Indonesia) is used as a proxy for the risk-free rate.

The portfolio formation may conceal the security related information present in individual stocks. This is important especially for emerging market research such as in Indonesia where the market is driven by a few blue chip stocks. Therefore this study will perform tests at the portfolio level as well as at the firm level. This study analyses size-sorted portfolio. When forming size-sorted portfolios, market value is used to sort the stocks into 15 portfolios, from the lowest to the highest based on the trading volume. The first and the last portfolio consists of 17 stocks, while the rest comprises 11 stocks each. The portfolio return is calculated as the equally weighted average return of the stocks in the portfolio.

For each year from July of year $t-1$ to June of year t, we rank the stocks based on their size and book-to-market ratio. We then use these two rankings to calculate a 50% breakpoint for size, and 30% and 70% breakpoints for book-to-market. The stocks are subsequently sorted into two size groups and three book-to-market groups based on these breakpoints. In addition, the stocks above the 50% size breakpoint are designated B (for big) and the remaining 50% are designated S (for small). In addition, the stocks above the 70% book-to-market breakpoint are designated H (for high), the middle 40% are designated M (for medium) and the firms below the 30% book-to-market breakpoint are designated L (for low). We form six value-weighted portfolios, S/L, S/M, S/H, B/L, B/M and B/H as the intersection of size and book-to-market groups. Note that the number of firms in each of the six portfolios varies. This procedure is repeated every year. SMB is the equally-weighted average of the returns on the small portfolios minus the returns on the big stocks portfolios.

Similarly, HML is the equally-weighted average of the returns on the value stock portfolios minus the returns on the growth stock portfolios: For each month from July of year $t-1$ to June of year t, stocks are ranked by their size and prior performance. The size is based on the market value at the end of June in year $t-1$ whereas the prior performance is based on the previous 11-month nominal stock return lagged one month. One month is skipped between portfolio formation and holding periods in order to avoid any potential microstructure biases. Winners (W) are the top 30% of the total stocks with the highest average prior performance. Losers (L) are the bottom 30% of the total stocks with the lowest average prior performance. Neutral (N) are the remaining 40% of the stocks. Six value-weighted portfolios (S/W, S/N, S/L, B/W, B/N, and B/L) are formed at the intersection of size and prior performance. The equally weighted monthly returns on the six portfolios are calculated each month over the 12 months following portfolio formation. WML (winner minus loser) is the equally-weighted average of the returns on the winner stock portfolios minus the returns on the loser stocks portfolios.

4. Results and discussion

4.1. Model estimated

The asset pricing model relates the expected return on an asset to its systematic risk or betas. These betas are the sensitivity of the asset return to changes in the return of the market portfolio and some other factors such as the size effect, value effect and momentum effect. Three well-known asset pricing models are given in Equation

- CAPM

$$R_{it} = \alpha_i + \beta_i^M MR_t + \varepsilon_{it} \tag{10}$$

- Fama-French

$$R_{it} = \alpha_i + \beta_i^M MR_t + \beta_i^S SMB_t + \beta_i^H HML_t + \varepsilon_{it} \tag{11}$$

- Carhart

$$R_{it} = \alpha_i + \beta_i^M MR_t + \beta_i^S SMB_t + \beta_i^H HML_t + \beta_i^W WML_t + \varepsilon_{it} \tag{12}$$

where:

R = firm or portfolio excess return ; β = systematic risk ; MR = market excess return
SMB = Small minus Big; HML = High minus Low; WML = Winner minus Loser

M, S, H = estimator index for market, size and value respectively;
i = cross section index; t = time index

The CAPM specifies a relation between the expected risk premium on an individual asset (portfolio) and the associated systematic risk the expected excess return on a risky asset is equal to the expected excess return on the market portfolio multiplied by the systematic risk (market beta) of that particular asset. Fama and French in [16] expand the CAPM by adding size (SMB) and value factors (HML). It considers the empirical fact that value and small cap stocks outperform the market on regular basis. By including these two additional factors, the model adjusts for the tendency for poor performance, making this model a better model for evaluating manager performance. Fama and French model does well in explaining many existing market anomalies, except for the short-term momentum strategy documented by [17]. In order to capture the one-year momentum anomaly, Carhart in [18] adds a fourth factor into the Fama French, which is nearly orthogonal to the three factors. Carhart's fourth factor is based on longing on winner-stocks as well as shorting on loser-stocks on the basis of their returns over the previous year. To mimic such a momentum factor (WML) or premium, WML is defined as the difference between the return on a portfolio of winner-stocks and the return on a portfolio of loser-stocks.

4.2. Findings

There are 177 firms included in the analysis but we will discuss several selected firms in more detail. For comparison purposes, the selected firms are the same firms as discussed in the previous paper. The firms are PT Bakrieland Development Tbk, PT Buana Finance Tbk, PT Davomas Abadi Tbk, PT Indal Aluminium Industry Tbk, PT Maskapai Reasuransi Indonesia Tbk, and PT Prima Alloy Steel Universal Tbk.

Figure 1 shows the estimated values of the nonparametric time-varying betas together with their confidence intervals. These curves were computed using cubic B-splines with selected knot of 8. As illustrated in Figure 1, the estimated betas are not stable over time and most of them are significantly different from zero. However, there is no clear pattern of how these betas vary over time. Most of the betas have fluctuated around their mean. In some stocks such as Bakrieland Dev, Maskapai Reasi Indo and Prima Alloy Steel, the betas fluctuated less compared to Davomas Abadi and Buana Finance which have fluctuated more. Generally, the market betas and the value-effect betas slightly decrease as time increases while the size-effect betas and momentum-effect betas are mostly stable overtime. In summary, 167 out of 177 stocks (95%) market betas are significantly different from zero, 90% size-effect betas are significantly different from zero, 91% value-effect betas are significantly different from zero and 67% momentum-effect betas are significantly different from zero.

The results reveal that generally the nonparametric time-varying beta estimates differ across the stocks. Figure 2 presents the boxplot of nonparametric time-varying beta for all firms. From Figure 2 we can see that different stocks have different beta characteristics. The beta of some stocks have more spread, large range and interquartile range, few outliers, and evidence of asymmetric distribution, while some firms have more dense beta distributions.

Generally, from 177 firms, we observe structural breaks during 2001-2002 and 2009-2010 where the magnitude of the betas change the direction. These are the periods where the presidential elections were held in Indonesia. Figure 3 presents the boxplot of the four betas for 15 size-sorted portfolios. The market beta in this case show evidence of fluctuation. Moreover, the betas tend to slightly increase as the portfolio size is larger. In contrast, the size-effect betas tend to decrease as the portfolio size increases. On the other hand, the value-effect betas and the momentum-effect betas of the portfolios appear to be less sensitive to portfolio size. There is also evidence of asymmetric in the betas.

IOP Conf. Series: Journal of Physics: Conf. Series **943** (2017) 012045 doi:10.1088/1742-6596/943/1/012045

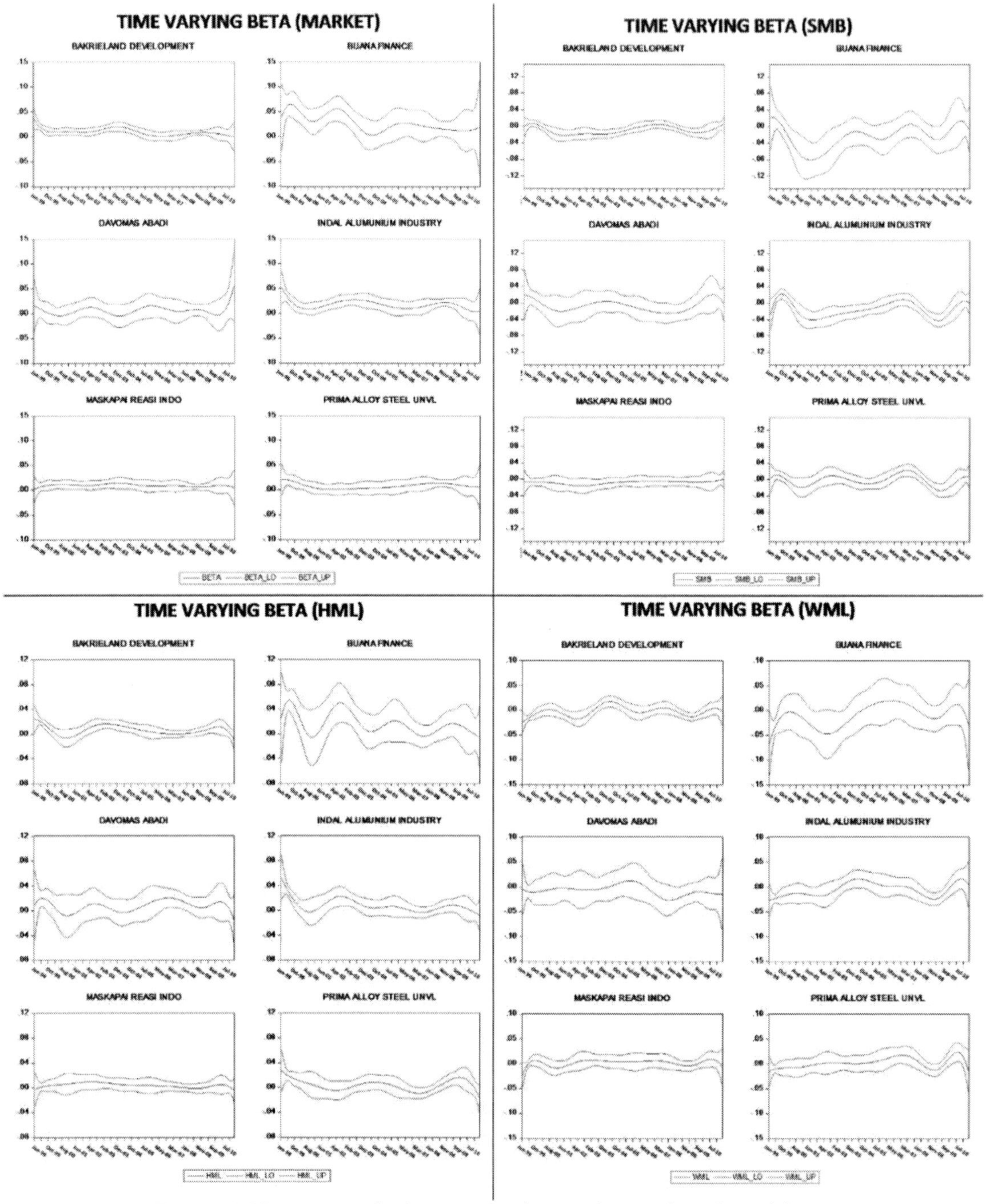

Figure 1. Nonparametric time-varying beta estimates for selected firms

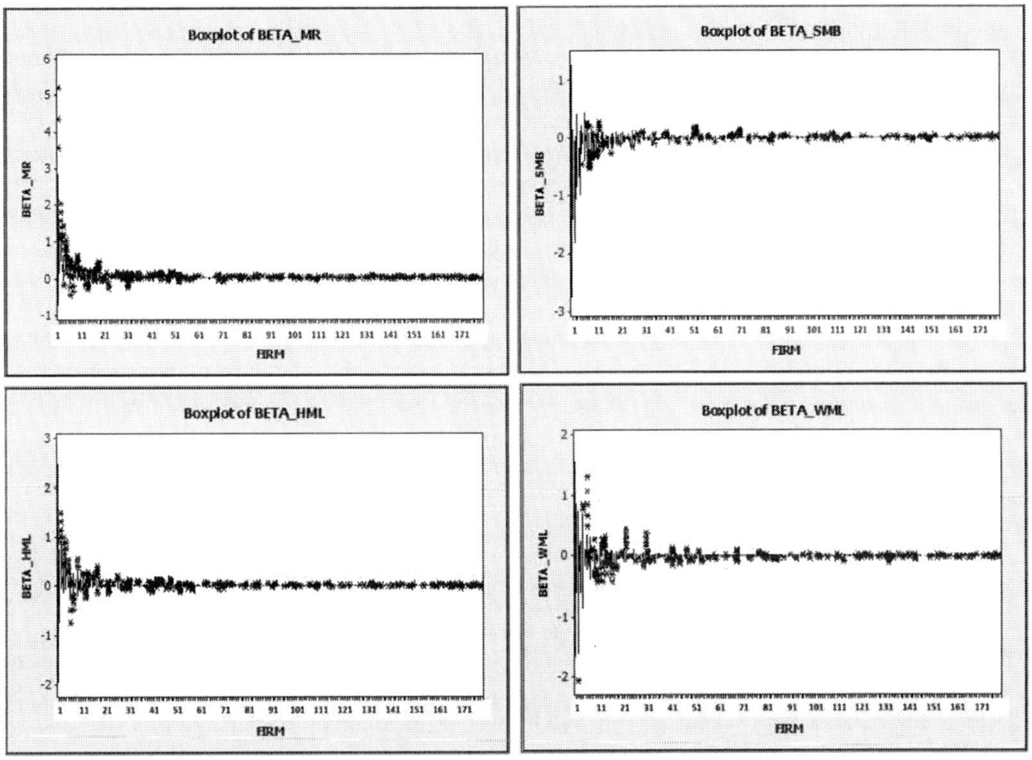

Figure 2. Boxplot of nonparametric time-varying betas for firm-level analysis

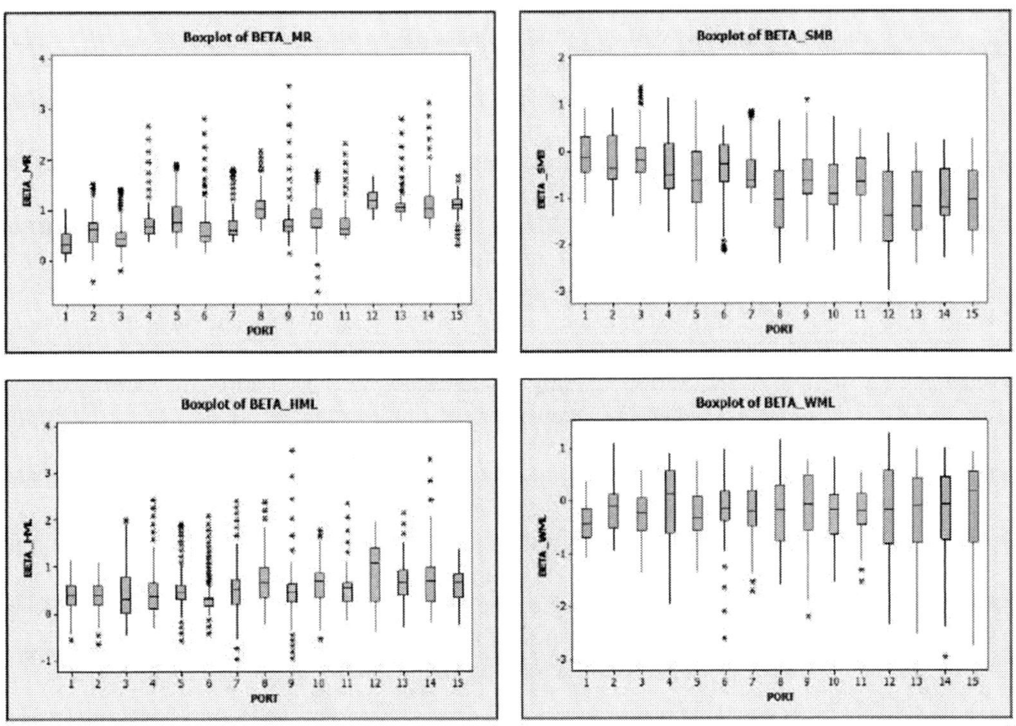

Figure 3. Boxplot of nonparametric time-varying betas for size-sorted portfolio

IOP Conf. Series: Journal of Physics: Conf. Series **943** (2017) 012045 doi:10.1088/1742-6596/943/1/012045

5. Conclusion

This paper investigates the estimation of time-varying asset pricing model using B-Spline method. The flexibility of the nonparametric setting avoids the problem of misspecification derived from selecting a functional form. The results reveal that the estimated betas are not stable over time and most of them are significantly different from zero. Different stocks (portfolios) have different beta characteristics. The betas of some stocks (portfolios) are more spread while the betas of some others have relatively dense distributions. Further, there is evidence that in many cases distributions are asymmetric. There are instances where the betas fluctuate in opposite directions over short time intervals. Generally the nonparametric betas of large market cap portfolios are larger than in the small market cap portfolios. The accuracy of betas, as the measure of systematic risk is crucial in determining investment strategies and pricing individual equities in security analysis and company valuation. The results show that the beta risk changes through time with changing economic environment. Dynamics of time variation of betas also differ across industries. For investors, the time-varying of the risks are important features that may be considered in their asset allocation and risk management decisions.

References

[1] Fama E F and MacBeth J D 1973 Risk return and equilibrium: Empirical tests *Journal of Political Economy* **81** 607
[2] Esteban M V and Orbe-Mandaluniz S 2010 A nonparametric approach for estimating betas: the smoothed rolling estimator *Applied Economics* **42** 1269
[3] Li Y and Yang L 2011 Testing conditional factor models: A nonparametric approach *Journal of Empirical Finance* **18** 972
[4] Ang A and Kristensen D 2012 Testing conditional factor models *Journal of Financial Economics* **106** 132
[5] Hastie T and Tibshirani R 1993 Varying–coefficient models *Journal of the Royal Statistical Society Series B (Methodological)* **55** 757
[6] Hoover D R, Rice J A, Wu C O and Yang L-P 1998 Nonparametric smoothing estimates of time-varying coefficient models with longitudinal data *Biometrika* **85** 809
[7] Eubank R, Huang C, Maldonado Y M, Wang N, Wang S and Buchanan R 2004 Smoothing spline estimation in varying-coefficient models *Journal of the Royal Statistical Society: Series B (Statistical Methodology)* **66** 653
[8] Wu C O and Chiang C T 2000 Kernel smoothing on varying coefficient models with longitudinal dependent variable Statistica Sinica **10** 433
[9] Wu C O, Chiang C T, and Hoover D R 1998 Asymptotic confidence regions for kernel smoothing of a varying-coefficient model with longitudinal data *Journal of the American statistical Association* **93** 1388
[10] Fan J and Zhang W 1999 Statistical estimation in varying coefficient models *The Annals of Statistics* **27** 1491
[11] Chiang C T, Rice J A and Wu C O 2001 Smoothing spline estimation for varying coefficient models with repeatedly measured dependent variables *Journal of the American Statistical Association* **96** 605
[12] He X, Zhu Z Y, and Fung W-K 2002 Estimation in a semiparametric model for longitudinal data with unspecified dependence structure *Biometrika* **89** 579
[13] Zhou S, Shen X and Wolfe D 1998 Local asymptotics for regression splines and confidence regions *Ann Statist* **26** 1760
[14] Chambers J M and Hastie T J 1992 *Statistical methods in S Wadsworth and Brooks/Cole* 196-246
[15] De Boor C 1978 *A practical guide to splines, Mathematics of Computation*
[16] Fama E F and French K R 1996 Multifactor explanations of asset pricing anomalies *The Journal of finance* **51** 55
[17] Jegadeesh N and Titman S 2011 *Momentum Annu Rev Financ Econ* **3** 493
[18] Carhart M 1997 On persistence in mutual fund performance *The Journal of finance* **52** 57

IOP Conf. Series: Journal of Physics: Conf. Series **943** (2017) 012046 doi:10.1088/1742-6596/943/1/012046

ICT media design for higher grade of elementary school mathematics learning using CS6 program

M Zainil[1], R C I Prahmana[2], Y Helsa[1] and S Hendri[1]

[1]Universitas Negeri Padang, Jl. Prof. Dr. Hamka Air Tawar, Padang, Indonesia
[2]Universitas Ahmad Dahlan, Jl. Pramuka Kav. 5, Pandeyan, Yogyakarta, Indonesia

E-mail: melva_zainil@yahoo.com

Abstract. Technological innovation contributes to the emerging of new possibilities to change the learning process. The development of technology could bring the higher quality of education through the integration of technology in the learning. The purpose of this research is to create an interactive multimedia using CS6 program for mathematics learning in higher grade of elementary school. It was a development research using ADDIE model which consists of analysis, design, and evaluation stages. It has successfully developed interactive multimedia in a form of learning CD used in the material of plane figures and solid figures. The prototype has been validated and then tested for the 4th grade of elementary schools. Two schools were involved and the students taught by utilizing the prototype, and then, in the end of learning, they are examined to determine the learning result. There were 72% of the students passed the examination as they classified at good and excellent categories. Finally, the use of CS6 program is promising to help the students learning plane and solid figure in mathematics learning.

1. Introduction

Learning process requires a smooth communication between teacher and students. In order to avoid or reduce the possibility of miscommunication between them, a tool such as media is needed to help the communication process. The teacher can improve the quality of the lesson when supported by various forms of media [1]. As a means to communicate in teaching and learning activities, the presence of media is inevitable. The use of appropriate media in the learning process will help students to receive the material and increase the interest of students to the lesson [2].

Geometry is one of the mathematics topics introduced to elementary school students. The students will be introduced first to planes, then the students will get an explanation about solids. Since we talk about geometry in elementary schools, then the representation would be cloise to the children surrounding objects[3]. The solid material plays an important role in the development of science of technology. Various kinds of solid learned in elementary school such as: cube, cuboid, prism, cylinder, pyramid and cone [4].

Some solid concepts are difficult to understand by elementary school students. This happens because elementary students are still at a concrete operational stage according to Piaget [5–7]. However, most of mathematics teachers in elementary school teach solid concepts in an abstract way and less multimedia used. The use of concrete objects and virtual manipulative are promising to help students understanding mathematics concepts [8,9], including solid. The virtual manipulative is the multimedia that can be used, utilizing information technology (IT). By leveraging various IT advantages, it can improve students' ability to understand mathematical concepts, procedures, and algorithms. The presence and advancement of IT in today's global communications era has provided an opportunity and expansion of learning resources that can occur anytime and anywhere without being limited by space and time [10]. With the help of IT, the material presentation process can be more interesting and fun. On the other hand, the presence of IT challenges teachers in a professional way to be able to manage information so they can choose the media and utilize IT effectively and efficiently in the learning process. This is in accordance with the demands of the 2013 Indonesian Curriculum, where this research area focus on, to encourage taechers use IT as a tool or media in helping students to understand the material.

The development of IT has enriched sources and learning media in various forms such as OHP Transparency, Power Point Slide, Image, Animation, Film/Video, Hypertext, Webpage, Computer aided learning program, and Software supporting learning applications such as Macromedia Flash, Camtasia, Maple and CS6. Professional teachers have be able to choose, to develop, and to utilize various types of learning media in order to ease the delivery of teaching materials and to help students understand the materials [11]. The advancement of IT has also enabled the utilization of various types of media simultaneously in the form of multimedia learning. The use of interactive multimedia that includes audio-visual components (voice and display) for the delivery of learning materials can attract students to learn. Interactive multimedia can also provide opportunities for students to conduct pseudo-exploratory experiments so as to provide a learning experience rather than simply listening to the teacher's description and explanation.

Based on our observations towards several elementary schools in Bukittinggi, such as SD Sabbihisma and SD Negeri 20 Kurao Pagang, we found that the utilization of multimedia in the schools was not massively used. Both of the schools have background of parents who classified as upper middle economic level and there are many students are accustomed to using technology tools such as I-pad. Furthermore, we found that in SD Negeri 20 Kurao Pagang, each classroom has been equipped by 20 sets of computer. The school has 18 teachers, they were not skilled in using any gadgets.

We also conducted observation at SD Negeri 05 and 09 Bukittinggi. We found that similar phenomena happened in the two schools. The schools have been equipped with the multimedia supported classroom, but the teachers werenot skilled in using it. Such phenomena became challenges for us to develop the use of IT in mathematics learning at the schools. In order to help the elementary school teachers to master using IT tools in the solids learning process, in order to not just teach but make the students learn and to be motivated to learn, it can be done with the use of media-based learning media IT such as CS6 program.

The learning activities by using IT media-based learning is one of the teacher concern so that the students get ease to learn the solids material and the students are motivated to learn well. While the reason of using Macromedia Flash because it is easy to be made as a media, and Macromedia Flash media can also be animated. By using CS6 program, the students are expected to be free to learn planes and solids material anytime and anywhere without a certain time limit. The students will be able to learn by using technology tools they have such as I-pad and so on and the teachers are increasingly professional in using technology tools in addition to their profession as educators.

2. Method

This type of research was a developmental research using qualitative and quantitative descriptive data analysis techniques. The development of IT-based multimedia is an activity to develop the media for mathematics learning especially in the material of solid which includes content and procedure, and the presentation in a form of interactive CD using CS6 program as the software.

It was a development research using ADDIE model which consists of analysis, design, and evaluation stages [12]. It has successfully developed interactive multimedia in a form of learning CD used in the material of plane figures and solid figures. The prototype has been validated and then tested for the 4th grade of elementary schools. Two schools were involved and the students taught by utilizing the prototype, and then, in the end of learning, they are examined to determine the learning result.

IOP Conf. Series: Journal of Physics: Conf. Series **943** (2017) 012046 doi:10.1088/1742-6596/943/1/012046

3. Results and Discussion

The research began with an informal observation in SD Sabbihisma, SD N 20 Kurao Pagang, SD N 5 Bukittinggi and SD N 9 Bukittinggi. They are elementary schools in West Sumatera. We observed computer laboratory and the students and teacher's computer skill. Further, there were 3 stages performed in this research: analysis, design, and evaluation.

3.1. Analysis Stage

Analysis stage was conducted in a form of media program outline analysis. The analysis included two steps, they are: (1) analysis on the curriculum material and (2) learning objectives. The analysis on curriculum material was conducted to select and to adjust the material of plane figure and solid figure in order to be appropriately relevant with the Indonesian 2013 curriculum. Further, identification and selection of material which is relevant to the interactive multimedia was conducted such that the basic competence of the curriculum can be achieved. These analysis result on the formulation of learning objectives. It was to understand the forms of plane figures and solid figures. In this stage, limitation of material on the interactive multimedia was determined. This stage also determined what materials can be geometrized either in the form of animation or simulation.

3.2. Design Stage

In the design stage, we create flowchart to determine the path of research. Flowchart is a program flow made from start, content to exit / exit program; scenario was clearly illustrated in flowchart. Creating flowchart involved the following steps.

3.2.1. Making storyboard. Before creating a storyboard, the researchers did a paper based stage. This stage contained a design that started from the sketch of the image on the paper. After designing in paper-based form, it was proceed to make storyboard. The storyboard was a description that contained the visual and audio explanations of each flow in the flowchart. One column in the storyboard represented a single display on the screen. This stage aimed to get a description of the form and what was displayed on the macromedia flash. The design consisted of main menu and support menu. The main menu consisted of purpose, material, sample questions, simulation and evaluation, while the support menu consists of help feature. Here is the example of the menu provided in the media as shown in Figure 1.

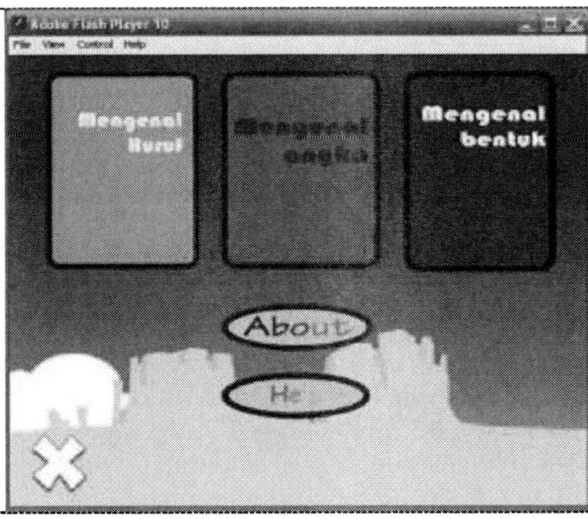

Figure 1. Some menu buttons in the interactive multimedia

3.2.2. Computer designing. This stage was included the creating ideas from the storyboard to the computer. It was divided into two stages: (1) Collecting the materials needed to complete the interactive multimedia presentation. Materials that need to be prepared included video, sound, animation, and images, (2) assembling all the existing material.

An interactive multimedia was one of the media that was able to present learning materials presented in a systematic and concrete way, making it easier for the students to understand. Prototype 1 of the initial design of interactive multimedia is shown in Figure 2.

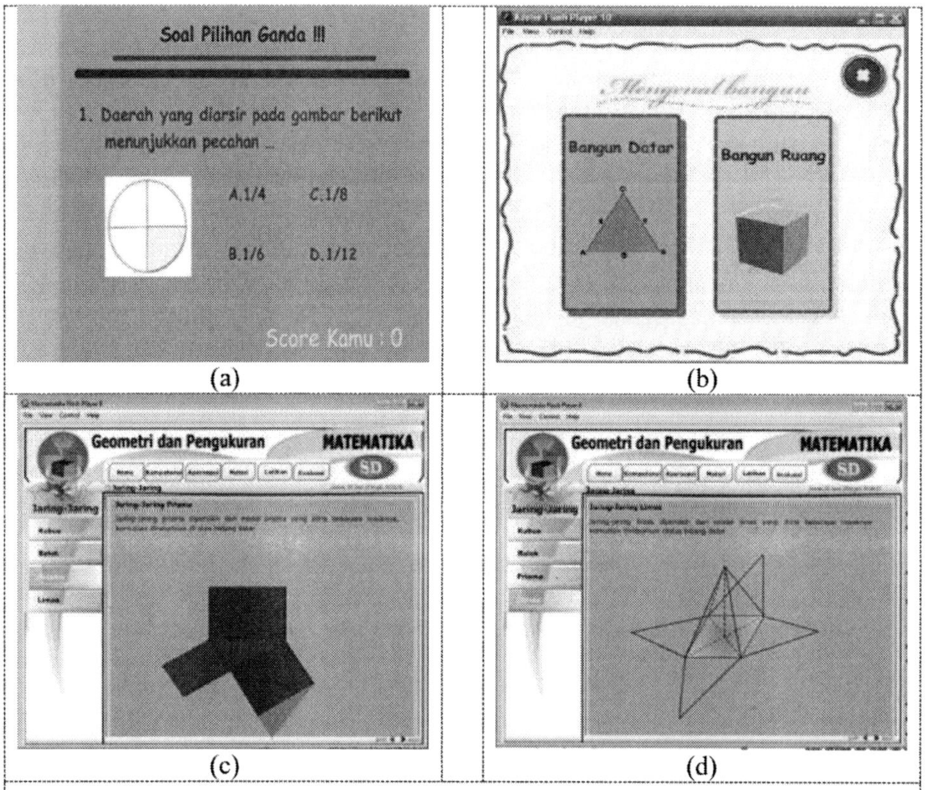

Figure 2 (a, b, c, d). Prototype 1 of the initial design of interactive multimedia

Figure 2 shows that prototype 1 has already focused on three main characteristics (content, support and lay out). The content consisted of chapters, sub-sections, paragraphs, and others about matter recognizing planes and solids form, structure was reasonable and flowing and explained from chapters and sections above. Lay out has already contained visual aspects such as images, graphics, colors and interactive and there is already granularity or icons connected with free and structured hyperlinks and used to fragment text in chapters and sections.

Van Hiele stated that there were five stages of the student learning in learning geometry, namely: introduction, analysis, informal deduction, deduction, and accuracy stage described as follows: (1) Introduction (visualization). In this stage, the students began to learn about a form of geometry as a whole, but had not been able to know the properties of the geometry that they saw; (2) Analysis. In this stage, the students started to recognize the properties possessed geometry objects that observed. The students were able to mention the regularity contained in the object of the geometry; (3) Informal Deduction. In this stage, the students begun to be able to carry out the conclusion that was known as deductive thinking. But this ability was not fully developed yet. One thing to note here that was the students at this stage started to do informal deduction; (4) Deduction. In this stage, the students was able to get conclusion deductively, namely the conclusion of the things that were general to the things that were specific; (5) Accuracy. In this stage, the students begun to realize how important the precision of the basic principles underlying a proof. For example, the students knew the importance of axioms or postulates of geometry Euclid. The accuracy stage was a high stage of thinking, complicated, and complex. It was therefore not surprising that some university students, even though they were already in high school before, yet they still have not reached this stage of thinking [13].

Each stage in Van Hiele's theory showed the characteristics of students' thinking processes in learning geometry and their understanding in the context of geometry. The quality of student knowledge was not determined by the accumulation of knowledge, but it was more determined by the geometric thinking process used. The stages of Van Hiele's thinking would be followed and passed by the students in sequence. Thus, the student must pass through a stage well before going to the next stage. The speed of moving from one stage to the next stage was depending on the number of the content and method of learning used, rather than on age and maturity. Thus, the teacher must provide a learning experience that matched the student's thinking stage. On the other hands, some research already documented that explain about mathematics learning experience to support this research result [14-17].

3.3. Evaluation

In this stage, these pre-designed products were evaluated. In this evaluation phase, the product was tested by expert, one-to-one and small group, and tested on the actual research subject. However, this product was new to the initial design, not to the stage of expert evaluation, one-to-one and small group which is the stage to see the validity and practicality of interactive multimedia developed, while field trials were tested on the actual research subject where the results of the valid and practical prototype was tested in order to see the potential effects on student learning outcomes.

After obtaining valid and practical prototype, then test field test to see the potential effect on learning result was done. This trial was conducted towards the students of 4th grade of SD Sabbihisma and SD N 20 Kurao Padang. The learning result of the 4th grade students of SDN 20 Kurao Padang is described in Table 1.

Table 1. Field test result

Score Interval (x)	Percentage	Predicate
$80 \leq x \leq 100$	30%	Excellent
$66 \leq x < 80$	42%	Good
$56 \leq x < 66$	16%	Fair
$40 \leq x < 56$	12%	Poor
$0 \leq x < 40$	0%	Failed

In the test of final prototype, there was no longer a student whose learning results fall into failed category. The percentage of students who can pass the examination was 72% of the students in the material of solid figure which the learning activities used the developed interactive multimedia in good and excellent category.

4. Conclusion

The prototype of IT-based learning media that developed has potential effects such as: the students could answer similar questions to the sample problems in IT-based learning media, and the test results also shows that the prototype of IT-based learning media that developed could bring the students learning results in good and excellent category. Finally, the use of CS6 program is promising to help the students learning plane and solid figure in mathematics learning. Based on the results of the research and the conclusion above, it can be suggested as follows: (1) Teacher needs to use IT-based learning media not only in conventional learning but also in cooperative learning. It is necessary to develop IT-based learning media not only on solid figure material but also on the other topics. (2) Students could use IT-based learning media in learning both in classroom and at home, so that their skill improved.

IOP Conf. Series: Journal of Physics: Conf. Series **943** (2017) 012046 doi:10.1088/1742-6596/943/1/012046

References

[1] Leow F T and Neo M 2014 Interactive multimedia learning: Innovating classroom education in a Malaysian university *Turkish Online J. Educ. Technol.* **13** 99–110
[2] Sholikhah M, H D A and Yusuf M 2017 Interactive Multimedia to Enhance Early Reading Skill on Children with Light Mental Retardation at 5th Grader SLB Autis Mitra Ananda Karanganyar Indonesia in Academic Year 2016/2017 *Eur. J. Spec. Educ. Res.* **2** 1–13
[3] Maričić S M and Stamatović J D 2017 The Effect of Preschool Mathematics Education in Development of Geometry Concepts in Children *EURASIA J. Math. Sci. Technol. Educ.* **13** 6175–87
[4] Özdemir B G 2017 Mathematical Practices in a Learning Environment Designed By Realistic Mathematics Education : Teaching Experiment About Cone and Pyramid *Eur. J. Educ. Stud.* **3** 405–31
[5] Uttal D H, Scudder K V. and DeLoache J S 1997 Manipulatives as symbols: A new perspective on the use of concrete objects to teach mathematics *J. Appl. Dev. Psychol.* **18** 37–54
[6] Hidayah I, Pujiastuti E and Chrisna J E 2017 Teacher's Stimulus Helps Students Achieve Mathematics Reasoning and Problem Solving Competences *J. Phys. Conf. Ser* **824**
[7] Ojose B 2008 Applying Piaget's Theory of Cognitive Development to Mathematics Instruction *Math. Educ.* **18** 26–30
[8] Collett P, Elizabeth P, Steyn C and Elizabeth P 2017 Using African Designs in Virtual Manipulatives for Geometrical Concept Development *Bridges 2017 Conference Proceedings* pp 455–8
[9] Durmus S and Karakirik E 2006 Virtual Manipulatives in Mathematics Education - A Theoritical Framework *Turkish Online J. Educ. Technol.* **5** 117–24
[10] Ekici D I 2017 The Use of Edmodo in Creating an Online Learning Community of Practice for Learning to Teach Science *Malaysian Online J. Educ. Sci.* **5** 91–106
[11] Chinapah V and Odero J O 2016 Towards Inclusive , Quality ICT-Based Learning for Rural Transformation on Education for Rural Transformation (ERT) 2016 *J. Educ. Res.* **5** 107–25
[12] Aldoobie N 2015 ADDIE Model *Am. Int. J. Contemp. Res.* **5**
[13] Chan H, Tsai P and Huang T-Y 2006 Web-based Learning in a Geometry Course *Educ. Technol. Soc.* **9** 133–40
[14] Prahmana R C I and Kusumah Y S 2016 The hypothetical learning trajectory on research in mathematics education using research-based learning *Pedagogika* **123** 42
[15] Sundayana R, Herman T, Dahlan J A, and Prahmana R C I 2017 Using ASSURE learning design to develop students' mathematical communication ability *World Transactions on Engineering and Technology Education* **15** 245
[16] Tanujaya B, Prahmana R C I, and Mumu J 2017 Mathematics instruction, problems, challenges, and opportunities: A case study in Manokwari regency, Indonesia *World Transactions on Engineering and Technology Education* **15** 287
[17] Prahmana R C I, Kusumah Y S, and Darhim 2017 Didactic trajectory of research in mathematics education using research-based learning *J. Phys.: Conf. Ser.* **893** 012001

Interactive multimedia-based teaching material for 3-dimensional geometry

A Prabowo, R P Anggoro, D Astuti and S Fahmi

Universitas Ahmad Dahlan, Jalan Prof. Dr. Soepomo SH, Warungboto, Umbulharjo, Yogyakarta 55164, Indonesia

E-mail: anggit.prabowo@pmat.uad.ac.id

Abstract. This study aims to develop the interactive multimedia-based teaching material for 3-dimensional geometry in junior high school. The product was produced through the stages of define, design, develop, and disseminate. Two media experts and two teaching experts had validated it. They judged that the product developed was valid. It had been revised based on their advice. It has been disseminated to 15 mathematics teachers and tried to 30 students of junior high school. Teachers stated that this product gives a new form of teaching material in 3-dimensional geometry. According to the student, the product is interesting. It can motivate them to study mathematics, help them to master the material and increase their interest in mathematics.

1. Introduction

Mathematics is a subject learned by students from elementary school level to higher education level. It is always related to human activities as it helps human life to be easier. Mathematics plays a main role in many branches of science such as physics, engineering, and statistics [1]. Thus, it has a main role in developing of science and technology [2]. Considering the importance of mathematics in the human life, students have to be skilled in the mathematics subject. However, the real condition of students in Indonesia does not meet the expectation. The Indonesian national examination in 2016 showed the low score of the students in mathematics. Compared to the other subjects (Bahasa Indonesia, English, and Science), the students' score in mathematics was the lowest (50.24) [3] as shown in Figure 1.

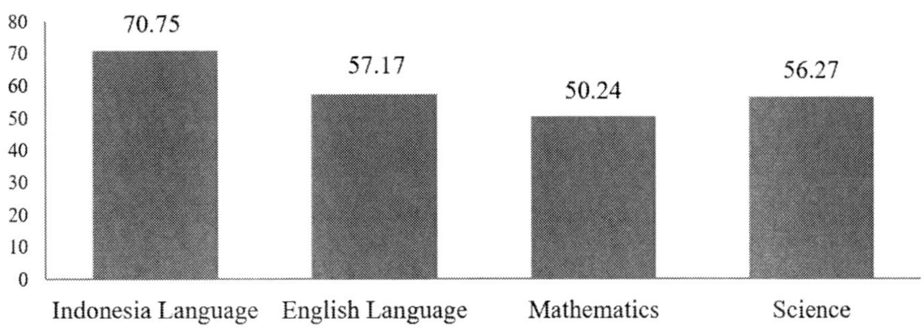

Figure 1. Score of the Indonesian national examination in 2016

This problem needs a notice from teacher, parents, government, and all part related why mathematics being the hardest matter for students. Students' competence in mathematics which measured at national examination consist 5 competencies. The percentage of competence achievement is shown in Table 1.

Table 1. Percentage of competence achievement in mathematics at national examination

Number	Competence Assessed	Percentage
1	Using arithmetic concept and number property, comparison, exponential number, root number, social arithmetic, sequence number, and the use in problem solving	60.64
2	Understand the algebra arithmetic, linear equation and inequality concept, line equation, set, relation, function, Linear equation system, and their use in problem solving	57.28
3	Understand the congruent concept, property of flat shape, and concept of relation of angles and/or line, and use them in problem solving	52.44
4	Understand the property of 3-dimensional, and use it in problem solving.	51.37
5	Understand the property of statistics and use them in problem solving	62.28
6	Understand the concept of probability of event and use it in problem solving	56.25

Based on Table 1, the lowest percentage of achievement is understanding the property of 3-dimensional shapes, and use it in problem solving. The 3-dimensional is a geometrical object with length and width (base area) and height such as cone, pyramid, sphere, cylinder, prism, cube, and cuboid [4].

Mastering competency is affected by students activity in learning. The low percentage of mastering in understanding the principle and element of 3-dimensional geometry could because of some factors. One of them is the instructional material. Instructional materials have the main role in teaching and learning of school subjects to promote teachers' efficiency and improve students' performance [5]. Instructional resource assists the teacher and allows students' interaction and make students achieve better and higher in lesson [6]. Instructional material used by the teacher was one of learning facilities which affect the students learning activity.

The most popular instructional material used by teachers is visual teaching material like handout, book, module, and students worksheet. These kinds of instructional material only present the seeing media like pictures, texts, or graphics. They are unable to give the sounds, 3-dimensional animations, video effect and interactive activity that needed in study geometry. Instructional material needs them because competency of understanding the principle and element of 3-dimensional geometry, and using it in problem solving needs good spatial ability. Students' difficulty in geometry was caused by non-availability of instructional materials and teachers' method of instruction [7]. Students generally encountered difficulties in geometry and performed poorly in senior secondary school mathematics lesson [8].

The spatial ability could be improved when a teacher teaches using the multimedia-based instructional material. It can provide media effects like audio, video, simulation, animation, and video related to 3-dimensional geometry. Therefore, it needs to develop the interactive multimedia-based for 3-dimensional geometry subject matter in senior high school.

2. Method

This was a research and development (R&D) using 4D (Define, Design, Development, and Dissemination) procedure [9] shown in Figure 2.

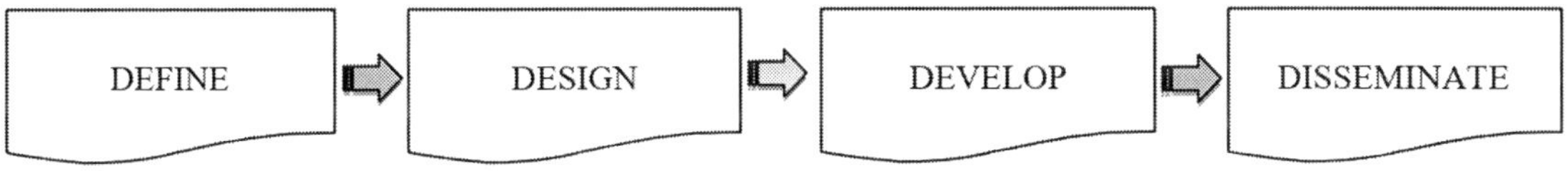

Figure 2. Research and development procedure

Step 1 was define. The product developed was an interactive multimedia-based instructional material for 3-dimensional geometry. Step 2 was design. The instructional material was designed by constructing the storyboard and preparing the tools and material. It was developed using Macromedia Flash Professional program. Step 3 was development. The product developed has been reviewed by four experts (2 multimedia experts and 2 teaching experts) and tried to 30 junior high school students. It also was disseminated to 15 junior high school mathematics teachers.

3. Results and discussion

The steps of this research started from defining the product developed. The product is an interactive multimedia-based for 3-dimensional geometry. It contains all aspects of media i.e. text, graphic, audio, video, and animation. The product is interactive. Users or students have an access to operate it and get a response base on what they want. It was developed using Macromedia Flash Professional program.

This product was designed by constructing the storyboard first. Storyboard is a visualization of the idea of the application to be built, so that gives an overview of the application to be generated [10]. The storyboard shows the stages of a story or an order of events [11]. Storyboards also defined as "…a visual reference depicting particular scene or action, usually taken from a screenplay or some other form of the written storyline." [12]. It provides visual script as an outline of the project which will be used to be guided to develop the product. It is shown in Figure 3.

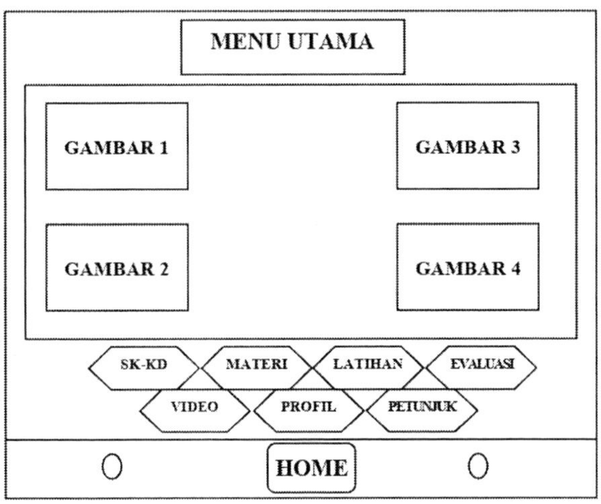

Figure 3. The main display of interactive multimedia for 3-dimensional geometry

The main display contains 7 menus (*SK-KD, MATERI, LATIHAN, EVALUASI, VIDEO, PROFIL,* and *PETUNJUK*). When the *SK-KD* menu is clicked, the page will show the competence standard and basic competence of teaching material. *MATERI* menu shows users the types flat sided spaces (cubes, beams, prisms, pyramids) and curved side spaces (cones, tubes, balls). *LATIHAN* menu shows users 10 items exercise followed by discussion. *EVALUASI* menu gives users an 10 item test to evaluate their comprehension of the 3-dimensional material. *VIDEO* menu contains videos related to a 3-dimensional geometry. *PROFIL* menu will display the profile of the instructional material and the developer. *PETUNJUK* menu shows the instructions to operate the product. After the storyboard was prepared, it needs to specify the tools and materials used to develop the product as shown in Table 2.

Table 2. Tools required

Tools	Function
Personal Computer	Used to develop interactive multimedia-based teaching materials
Printer	Used to print storyboard
Scanner	Used to scan images to be used in teaching materials
CD	Used to save data
Flash drive	Used to save data
Speaker	Used to check the functionality of audio media on instructional materials

After the tools required were prepared, the designing instructional materials were complete and ready for product development as shown in Table 3.

Table 3. Materials required

Material	Function
Macromedia Flash Professional Software	Used to create programs
3-dimensional material	Used as the content of developed teaching materials
3- dimensional shape	Used as the content of developed teaching materials
Video	Used as the content of developed teaching materials

The product development was done by implementing the plan that has been designed at the early stage. Based on the storyboard that has been compiled, the product is successfully developed as shown in Figure 4 and Figure 5.

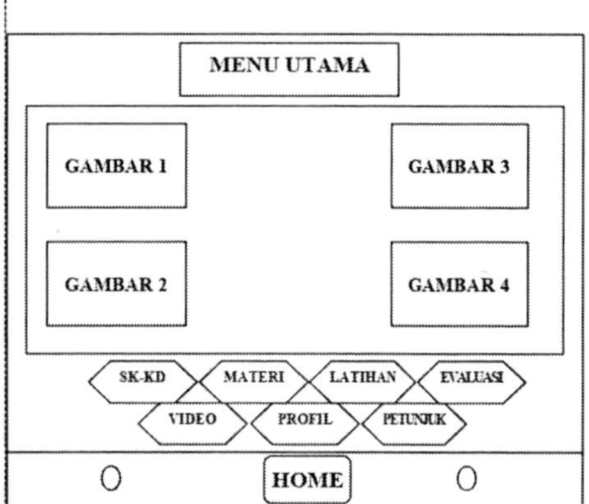

Figure 4. The storyboard of the main display of the multimedia

Figure 5. The product of the main display of the multimedia

The main menu of the product consist some 3-dimensional shape in daily life like ball, bottle, wardrobe, cone hat, tent, Rubik, and roof of the house. It also contains 7 menus (SK/KD, Materi, Latihan, Evaluasi, Video, Profil, and Petunjuk). When each menu is clicked, they will go to the page related.

After the product developed, it was assessed by 2 media learning experts and 2 mathematics material experts. The result of media learning expert assessment is shown in Table 4.

Table 4. Assessment of media experts

No	Assessment
1	The arrangement of days and dates overlaps, especially between months and years
2	The instructions for using are clear, so it is easy to operate the products.
3	The language used is simple so it is easy to understand
4	The cover of the developed product CD already represents the contents
5	The font type used is correct so it is easy to read
6	The letters size are proportional
7	Alloy colors are presented in harmony
8	The quality of audio, video, image is good
9	The presentation of the menus is presented consistently

The result of teaching expert assessment is shown in Table 5.

Table 5. Assessment of teaching experts

No	Assessment
1	The use of square and square mathematical symbols is not displayed properly
2	The material presented is suitable with the competence to be achieved
3	The material is presented in a coherent manner
4	Training questions have been completed with the discussion so as to facilitate the students to know the truth of the answer
5	The discussion of exercise questions is presented appropriately

The experts gave scores in 84 maximum scale for the product as presented in Table 6.

Table 6. Score of product based on experts judgment

Validator	Score	Category
Media expert 1	77	Very good
Media expert 2	72	Very good
Teaching expert 1	83	Very good
Teaching expert 2	70	Very good

Based on the expert assessment above and after the revision based on expert judgment, it can be concluded that the developed teaching material is very good and can be said valid. The last step, valid product disseminated to 15 teachers of mathematics in junior high school. They implemented the product as a teaching material for 3-dimensional geometry. Figure 6 and 7 show it.

Figure 6. Teacher was implementing the product in class

Figure 7. Students were using the product in learning

Nowadays, instructional was not only held in the classroom by talk and chalk. Traditionally, instructional methods have been described "presentation forms" such as lectures and discussion [13]. Traditional learning is an activity occurring in the classroom where teachers deliver learning the material to students [14]. Traditional instructional is started to be abandoned. Instructional using the more various method and instructional material had been implemented in the class. Instructional using technology is not strange. Along with these developments, there exists a prodigious increase in the amount of educational software provided for use in a class [15, 16]. Many applications are provided to develop teaching material. Every new technological development opens new opportunities for lesson planning for teachers and learning tools for student [17].

This research had produced an interactive multimedia-based instructional material of 3-dimensional geometry for junior high school. It is the kind of Computer Aided Learning (CAL). CAL means the application of computers integrated with learning system for learning and teaching process [18]. It was developed to help students in mastering 3-dimensional geometry. In the developing process of computer-based learning program, the concern is not on "what technological tools are to be used during the development process of e-learning program", but it is on "how to design and plan an e-learning program that ensures the achievement of the learning objectives" [19].

The product developed can be used the mathematics teacher of the junior high school to teach 3-dimensional geometry. It can be implemented with various models of learning in the classroom available computer facilities. It confirms that the use of teaching material should be integrated with the teaching method such that it can be implemented optimally [20].

Based on the implementation of the product in class, it helps students to master 3-dimensional geometry material. The score of student achievement in 3-dimensional geometry was 85.67 from score maximum 100.00. The minimum criteria for mastery learning is 75.00. Students got more information (material, simulation, exercise, discussion, video tutorial, and evaluation) from the product. Students used it not only in the classroom but also in their house using computer or notebook. It gives a capability for students to study at home. An interactive multimedia instructional material, the information more meaningfully and repeatedly can be obtained by students [21]. Many kinds of media are provided by interactive multimedia more systematically using rich instructional strategies [22].

Beside affect in cognitive, the product developed also affect to affective aspect of students. There are 80.00% students that state that studying using interactive multimedia more motivate them to study mathematics. The interactive multimedia instructional material also made them more interest in mathematics (76.67%). Technology-backed learning environment motivates students to participate and to interact with others in the instructional process [23]. Their motivation made them confident in studying mathematics. Interactive multimedia helps the students to strengthen self-esteem in a learning community [24-27]. From this study, 86.67% students state it.

IOP Conf. Series: Journal of Physics: Conf. Series **943** (2017) 012047 doi:10.1088/1742-6596/943/1/012047

4. Conclusion

This research has produced a valid interactive multimedia-based teaching material of 3-dimensional geometry for Junior high school. It had been disseminated to mathematics teachers and implemented in the class of 3-dimensional geometry instructional. This product help students understand the concept of 3-dimensional geometry, motivate to study mathematics, and increase their interest and self-esteem in mathematics.

References

[1] Muijs D and Reynolds D 2005 *Effective Teaching Evidence and Practice* (New Jersey: Pearson-Merrill Prentice Hall)

[2] Fabiyi T R 2017 Geometry concepts in mathematics perceived difficult to learn by senior secondary school students in Ekiti State Nigeria *IOSR Journal of Research & Method in Education (IOS-JRME)* **7** 83

[3] BSNP 2016 *Laporan Hasil Ujian Nasional Tahun Pelajaran 2015/2016* (Jakarta: Badan Standar Nasional Pendidikan)

[4] Salman M F 2009 Active learning techniques (ALT) in mathematics workshop: Nigerian primary school teacher's assessment *International Electronic Journal of Mathematics Education* **4** 23

[5] Olayinka A R B 2016 Effect of instructional materials on secondary schools students' academic achievement in social studies in Ekiti State, Nigeria *Worlds Journal of Education* **6** 32

[6] Omebe C A and Akani O 2015 Effect of instructional resource on students' achievement in physics and chemistry in secondary schools in Ebonyi state Nigeria *European Journal of Training and Development Studies* **2** 56

[7] Adolphus T 2011 Problems of teaching and learning of geometry in secondary schools in River State Nigeria *International Journal of Emerging Science* **1** 143

[8] Adegun I K and Adegun B O 2013 Students and teachers' views of difficult areas in mathematics syllabus: basic requirement for science and engineering education *Journal of Education and Practice* **4** 235

[9] Thiagarajan S, Semmel D S and Semmel M I 1974 *Instructional Development for Training Teachers of Exceptional Children: A sourcebook* (Minnesota: ERIC)

[10] Waryanto N H 2005 *Storyboard dalam Media Pembelajaran Interaktif* (Yogyakarta: Yogyakarta State University)

[11] Birchman J and Sadowski M 2006 Idea development and communication through storyboards *American Society for Engineering Education* **1723** 11.711.1

[12] Fraioli J O 2000 *Storyboarding 101: A crash course in professional storyboarding* (California: Michael Wiese Productions)

[13] Heinich R, Molenda M, Russel J D and Smaldino S E 1996 *Instructional media and technologies for learning* (New Jersey: Prentice-Hall Inc.)

[14] Astra I M, Nasbey H and Nugraha A 2015 Development of an android application in the form of a simulation lab as learning media for senior high school students *Eurasia Journal of Mathematics, Science & Technology Education* **11** 1081

[15] Bayram S and Nous A 2004 Evolution of educational software evaluation: instructional software assessment *TOJET* **3** 21

[16] Le Q and Le T 2007 Evaluation of educational software: theory into practice *Technology and Teaching* ed J. Sigafoos and V Green (New York: Nova Science Publisher) chapter 11 pp 10-11

[17] Rosado D E N V, Argüero M E and Rojas R 2014 An adaptive interactive multimedia system for intelligent environments *International Journal of Information and Education Technology* **4** 54

[18] Rahmawati E and Hartati S J 2013 The application of computer-aided learning to learn basic concepts of branching and looping on logic algorithm *The International Journal of Multimedia & Its Applications (IJMA)* **5** 15

[19] Tuparov G, Tuparova D D and Peneva J 2004 *Proceedings of the 5th International Conference on Computer Systems and Technologies* (Ruse: University of Ruse) pp. IV.14.1-6

[20] Fadillah S and Jamillah 2016 Pengembangan bahan ajar struktur aljabar untuk meningkatkan kemampuan pembuktian matematis mahasiswa *Jurnal Cakrawala Pendidikan* **35** 106
[21] Dembo M H and Seli H 2012 *Motivation and Learning Strategies for College Success: A Focus on Self Regulated Learning* (New York: Erlbaum)
[22] Qureshi E 2004 *Instructional Design Models* (Ontario: University of Windsor)
[23] Leow F T and Neo M 2014 Interactive multimedia learning: innovating classroom education in Malaysian university *The Turkish Online Journal of Educational Technology* **13** 99
[24] Suh H 2011 Collaborative learning model and support teaching technologies in the future classroom *International Journal for Educational Media and Technology* **5** 50
[25] Young J R 2017 Technology integration in mathematics education: Examining the quality of meta-analytic research *International Journal on Emerging Mathematics Education* **1** 71
[26] Fathurrohman M, Porter A L and Worthy A L 2017 Teachers' real and perceived of ICT supported-situation for mathematics teaching and learning *International Journal on Emerging Mathematics Education* **1** 11
[27] Tanujaya B, Prahmana R C I, and Mumu J 2017 Mathematics instruction, problems, challenges, and opportunities: A case study in Manokwari regency, Indonesia *World Transactions on Engineering and Technology Education* **15** 287

IOP Conf. Series: Journal of Physics: Conf. Series **943** (2017) 012048 doi:10.1088/1742-6596/943/1/012048

Almost stochastic dominance for poverty level in Central Java Province

Isnandar Slamet[1], Aryanto Agus Wibowo[2], and Mania Roswitha[3]
[1]Department of Statistics, Sebelas Maret University, Surakarta, Indonesia 57126
[2,3]Department of Mathematics, Sebelas Maret University, Surakarta, Indonesia 57126

E-mail: isnandar06@yahoo.com

Abstract. The criteria for the domination of the distribution function has been used in the investment issues, momentum, agricultural production, and so on. One criteria of domination is stochastic dominance (SD). When this criteria is applied to the dominating area that has smaller value than the dominated area, then almost stochastic dominance (ASD) can be used. It this research, we apply the ASD criteria on data of expenditure per capita based on districts/cities in Central Java. Furthermore, we determine which year the expenditure per capita in the period 2009-2013 is the most dominating to know the level of poverty in Central Java. From the discussion, it can be concluded that the expenditure per capita in Central Java in 2013 dominates expenditure per capita in Central Java in 2009-2012. In other words, the level of poverty in Central Java in 2013 is lower than in 2009-2012.

1. Introduction

Poverty is one of the major problems faced by developing countries, including Indonesia. The changes of the level of poverty in Indonesia is one of the big concern that is observed every year. According to the Indonesia Central Statistics Agency [1], the poor are the people who have an average per capita expenditure below the poverty line. Therefore, the poverty line is used to determine whether a person is poor or not.

Generally, there are three indexes that are used to measure poverty levels. These are head count index, poverty depth index, and poverty severity index. The third index is a family of poverty indices F-G-T *(Foster-Greer-Thorbecke)* which is often used to determine the changes in poverty over time and regions. According to Madden and Smith [6], the weaknesses of the three indexes in knowing the changes in poverty is sensitive to the selection of the poverty line as well as the size of the poverty level. To overcome this problem, it is required a more robust approach to the poverty line as well as the size of the poverty level. Madden and Smith [6] explained that the stochastic dominance (*SD*) method is a powerful method in the selection of the poverty line. As noted by Ravallion [7], the poverty line used in the stochastic dominance interval is the region interval which produces same conclusion when the specific poverty line is used.

The *SD* method is the method which can be used to compare two distribution functions at which certain distribution function is more dominant than other (Heyer [3]). In discussion of *SD*, there are three criteria which we call first degree stochastic dominance (*FSD*), second degree stochastic dominance (*SSD*), and third degree stochastic dominance (*TSD*). Based on the definition of *SD*, a random variable Y dominates a random variable X when the condition of the expectation of utility price Y is higher or equal than the expectation of the utility price X. However, if the criteria *SD* is not satisfied, then the alternative decision-making almost stochastic dominance (*ASD*) can be used (Leshno and Levy [4]). In *ASD*, there are almost first stochastic dominance (*AFSD*) and almost second stochastic dominance (*ASSD*). The values of *AFSD* can be calculated from the areas that violate the criteria of stochastic dominance divided by the total absolute value of the area between the cumulative distributions of X and the cumulative distribution of Y (Leshno and Levy [4]). The value of *ASSD* can be obtained from the quotient between the areas that violate the criteria stochastic dominance divided by the average area dominated.

In this paper, we apply the method *ASD* (Levy [5]) for analysing of poverty level based on the expenditure per capita per month in Central Java for period 2009 to 2013.

2. Literature Review

Method *SD* is a method to determine the criteria for decision-making when at least one criterion *SD* is negative (Levy [5]). Heyer [3] stated that *SD* is a term that refers to a relationship between two investment distribution functions F and G, where one investment dominates the other investment. If there are two investment options F and G, then G investment will be favored over F if and only if the expectation of the utility G is higher or equal to the expectation of the utility F. Systematically, it can be written as

$$E_F[u(x)] \leq E_G[u(x)], \tag{1}$$

where $E_F[u(x)]$ is the expectation of the utility F and $E_G[u(x)]$ is the expectation of the utility G. According to Bain and Engelhardt [2]), if X is a continuous random variable, equation (1) can be written as

$$\int_{-\infty}^{\infty} u(x)f(x)dx \leq \int_{-\infty}^{\infty} u(x)g(x)dx, \tag{2}$$

with $f(x)$ and $g(x)$ each of which is the probability distribution function of the random variable X and $u(x)$ is defined as the utility value of the random variable X.

Equation (2) is a formula *SD* generally with at least there is an inequality $u(x) \leq 0$. From equation (2) with an investment of G dominate investment of F, criteria of *FSD* is

$$F(t) \geq G(t). \tag{3}$$

When equation (3) is not fulfilled, then the criteria *FSD* can be used in the decision-making.

According to Levy [5], S_1 is defined as the total value when the rule criteria *FSD* is not fulfilled $S_1(F, G) = t : F(t) < G(t)$. Areas that are not fulfilled the criteria *FSD* with continuous random variables X and Y can be written follow.

$$\int_{S_1} [F(t) - G(t)]dt. \tag{4}$$

When X and Y are discrete random variables, equation (4) can be written as

$$\frac{1}{n} \sum_{i:y_i > x_i}^{n} (y_i - x_i) \tag{5}$$

where y_i is the data that is assumed to dominate and x_i is the data that is assumed to be dominated. Then it can be written as

$$\int_{S_1} [F(t) - G(t)]dt = A_1.$$

For continuous random variables X and Y, the total area of FSD between these two variables can be written as

$$\int_S [F(t) - G(t)]dt. \tag{6}$$

For discrete random variables, equation (6) can be written as

$$\frac{1}{n}\sum_{i:1}^{n} |y_i - x_i| \tag{7}$$

or

$$\int_S |F(t) - G(t)|dt = A_1 + A_2.$$

The results for the areas that do not meet the criteria FSD and the total area between random variables X and Y is continuous or symbolized by ε and formulated with

$$\varepsilon_{AFSD} = \frac{\int_{S_1}[F(t) - G(t)]dt}{\int_S [F(t) - G(t)]dt}. \tag{8}$$

If the random variables X and Y are discrete, the equation (8) can be solved using the equation (5) dan (7), and we obtain

$$\varepsilon_{AFSD} = \frac{\sum_{i:y_i > x_i}^{n}(y_i - x_i)}{\sum_{i:1}^{n}|y_i - x_i|}. \tag{9}$$

Equation (9) can be written as

$$\varepsilon_{AFSD} = \frac{A_1}{A_1 + A_2}.$$

If $0 < \varepsilon < 0.5$, equation (9) meets the criteria $AFSD$ with G dominating F with ε_{AFSD}. Thus, Y is said to dominate the X with $AFSD$ (for ε specific, or ε_{AFSD}) and is denoted by $G \geq \varepsilon_{AFSD} F$ if and only if:

$$\int_{S_1} [F(t) - G(t)]dt \leq \varepsilon \int_S |F(t) - G(t)|. \tag{10}$$

Using a discrete equation (10) can be written

$$\sum_{i:y_i > x_i}^{n} (y_i - x_i) \leq \varepsilon \sum_{i:1}^{n} |y_i - x_i|$$

provided $0 < \varepsilon < 0.5$.

According to Levy [5], G dominate the F in SSD if and only if

$$\int_{-\infty}^{t} F(X)dx \geq \int_{-\infty}^{t} G(X)dx \tag{11}$$

for all $t \in S$. It is assumed that this inequality is true for most of the range of S. Suppose that S_2 defined as the value of the area ranges SSD violated and formulated with

$$S_2 = \{t : F(t) < G(t); \int_{-\infty}^{t} G(X) \geq \int_{-\infty}^{t} F(X)dx\}$$

and complement the area of S_2 is denoted by $\overline{S_2}$ and $S_2 \cup \overline{S_2} = S$. Areas that do not meet the criteria *SSD* with random variables X and Y is continuous may be written by the equation

$$\int_{S_2} (F(t) - G(t))dt. \tag{12}$$

When the random variable X and Y discrete equation (12) can be written

$$\frac{1}{n} \sum_{i:y_i>x_i}^{n} (y_i - x_i) \tag{13}$$

with x_i is the data that is assumed to dominate and y_i is the data that is assumed to be dominated. While the area range *SSD* between random variables X and Y is continuous may be written

$$\int_S (F(t) - G(t))dt + 2\int_{S_2} (F(t) - G(t))dt. \tag{14}$$

If the total area of *SSD* between random variables X and Y is discrete, then the equation (14) can be written

$$\frac{1}{n}\sum_{i:1}^{n}(x_i) - \sum_{i:1}^{n}(y_i) + 2[\frac{1}{n} \sum_{i:y_i>x_i}^{n} (y_i - x_i)]. \tag{15}$$

The results for the areas that do not meet the *SSD* and the total area of the random variable X and Y is continuous, or can be formulated with ε_{ASSD} defined by

$$\varepsilon_{ASSD} = \frac{\int_{S_2}(F(t) - G(t))dt}{\int_S(F(t) - G(t)) + 2\int_{S_2}(F(t) - G(t))dt}. \tag{16}$$

From equation (16) can be described as follows

$$\begin{aligned}
\varepsilon_{ASSD} &= \frac{\int_{S_2}(F(t) - G(t))dt}{\int_S(F(t) - G(t)) + 2\int_{S_2}(F(t) - G(t))dt} \\
&= \frac{\int_{S_2}(F(t) - G(t))dt}{\int_{\overline{S_2}}(F(t) - G(t)) + \int_{S_2}(F(t) - G(t))dt + \int_{S_2}(F(t) - G(t))dt + \int_{S_2}(F(t) - G(t))dt} \\
&= \frac{\int_{S_2}(F(t) - G(t))dt}{\int_S(F(t) - G(t)) + 2\int_{S_2}(F(t) - G(t))dt}
\end{aligned}$$

where

$$\int_{\overline{S_2}}(F(t) - G(t)) + \int_{S_2} (F(t) - G(t))dt = \int_S (F(t) - G(t))dt.$$

Because $S_2 \cup \overline{S_2} = S$, we get

$$-\int_{S_2} (G(t) - F(t))dt = \int_S (F(t) - G(t))dt.$$

According to Levy [5]

$$\int_S (F(t) - G(t))dt = E_F - E_G \tag{17}$$

with E_F is an average value of F and E_G is an average value of G. Then the area was violated SSD range denoted more simply by VA, where

$$VA = \int_{S_2} (F(t) - G(t))dt. \tag{18}$$

Equation (17) and equation (18) are substituted in equation (16) in order to obtain

$$\varepsilon_{ASSD} = \frac{VA}{E_F - E_G + 2VA}.$$

If $\varepsilon_{ASSD} < 0.5$ G dominate the F is $ASSD$. Levy [5] stated that the lower the value ε obtained, the higher the degree of dominance. A random variable Y is said to dominate the X with $ASSD$ (for ε_{ASSD}) and is denoted by $G \geq \varepsilon_{ASSD}F$ if and only if:

$$\int_{S_2} (F(t) - G(t))dt \leq \varepsilon[\int_S (F(t) - G(t))dt + 2\int_{S_2} (F(t) - G(t))dt]. \quad (19)$$

For discrete case, equation (19) can be written as

$$\frac{1}{n}\sum_{i:y_i>x_i}^{n} (y_i - x_i) \leq \varepsilon[\frac{1}{n}\sum_{i:1}^{n}(x_i) - \sum_{i:1}^{n}(y_i) + 2[\frac{1}{n}\sum_{i:y_i>x_i}^{n} (y_i - x_i)]]$$

provided $0 < \varepsilon < 0,5$.

3. Research Methodology

In this research, we applied ASD on poverty levels in Central Java based on data of expenditure per capita per month in all districts/cities in Central Java for period 2009 until 2013. Suppose there are two investments i.e F and G. Then, we determine the cumulative distribution functions of F and G which we write as $F(t)$ and $G(t)$, respectively. Then, we calculate violation area based on SD criteria. Next, we determine the value of ε using ASD from the violation area. Finally, we apply the ASD method for data of average expenditure per capita per month to get the conclusion.

4. Results

ASD method has been applied to data of average expenditure per capita per month for all districts/cities in Central Java in 2009 until 2013 that were taken from the Central Statistics Agency (BPS) [1]. In this case, the *ASD* method is used to analyze whether the data of the average expenditure per capita per month for all districts / cities in Central Java in 2013 dominate (higher) than the average data per capita per month spending in 2009, 2010, 2011, and 2012.

4.1. Randomness Test

According to Wackerly [8], randomness test is used to determine whether the data used is random. If the data is random then data is mutually independent and identical. The steps of test is given below.

(i) H_0 : data of average expenditure per capita per month for all districts/cities in Central Java is random.
(ii) H_1 : data of average expenditure per capita per month for all districts/cities in Central Java is not random.
(iii) Level of significance $\alpha = 0.05$.
(iv) Critical area : H_0 is rejected if $Z_{Calc} \leq -Z_{0.025} = -1.96$ or $Z_{Calc} \geq Z_{0.025} = 1.96$.
(v) Test of statistic :

$$Z = \frac{R - E(R)}{\sqrt{V(R)}}$$

IOP Conf. Series: Journal of Physics: Conf. Series **943** (2017) 012048 doi:10.1088/1742-6596/943/1/012048

(vi) Conclusion.

Using a significance level $\alpha = 0.05$, we conclude that the data on average expenditure per capita per month for all districts/cities in Central Java for 2009 to 2013 are random.

4.2. Stochastic Dominance

Value development expenditure per capita per month is calculated from the assumed outcome highest expenditure reduced by expenditure will be compared. Value development expenditure per capita per month is calculated from the reduction of expenditure per capita in Central Java in 2013 to the expenditures in 2009, 2010, 2011, and 2012. As an example of the value of the development expenditure in Central Java in 2009 against 2013 in Sragen earned $-1,035$ rupiah. The percentage of development expenditure in Sragen obtained -0.004. There are areas that do not meet the criteria FSD by equation (3) and does not meet the criteria SSD by equation (11). Then, we need to calculate the value of ε through FSD and SSD.

4.3. Almost Stochastic Dominance

Criteria FSD and SSD has negative value. Therefore, it can not be used. Furthermore, the criteria $AFSD$ and $ASSD$ are used to find out if the data expenditure per capita per month in Central Java in 2013 dominate the data in 2009 2010, 2011, and 2012 based on a percentage of its development. The calculation on the values of ε_{AFSD} and ε_{ASSD} is presented in Table below.

Table 1. Values ε

Year	ε_{AFSD}	ε_{ASSD}
2009	$0,00085$	$1,078 \times 10^{-6}$
2010	$0,014$	$1,749 \times 10^{-6}$
2011	$0,079$	$1,146 \times 10^{-5}$
2012	$0,081$	$1,188 \times 10^{-5}$

Because the value of ε_{AFSD} and ε_{ASSD} lies between 0 to 0.5, it means that the criteria $AFSD$ and $ASSD$ can be used. Expenditure per capita per month by district/city in Central Java in 2013 dominates the spending in Central Java in 2009-2012 based on the percentage of development expenditure through the criteria of $AFSD$ and $ASSD$. It explains that in 2013 the lowest poverty rates compared to the 2009-2012 seen from the percentage of development expenditure per capita. Based on the results of ε obtained, it can be concluded that the value of ε_{ASSD} is better than the value of ε_{AFSD} data spending per capita per month in Central Java in 2009 until 2013. This is because the value of ε_{ASSD} is nearly zero so that the higher degree of dominance.

5. Conclusion

Based on the discussion, it can be concluded that expenditure per capita per month for all districts/cities in Central Java in 2013 dominates expenditure per capita per month in Central Java in 2009-2012 based on the percentage of development expenditure using the criteria of *almost stochastic dominance* first and second order degree. It explains that in 2013 the lowest poverty rates compared to 2009-2012.

References

[1] Badan Pusat Statistik Jawa Tengah, *Politik dan Keamanan Provinsi Jawa Tengah*, Badan Pusat Statistik Jawa Tengah, 2011.

[2] Bain,L.J., and Engelhardt, M. *Introduction to Probability and Mathematical Statistics*, Duxbury Press An Imprint of Wadsworth Publishing Company, Belmont California, 1992.

IOP Conf. Series: Journal of Physics: Conf. Series **943** (2017) 012048 doi:10.1088/1742-6596/943/1/012048

[3] Heyer D.D., Stochastic Dominance: *A tool for Evaluating Reinsurance Alternative*, Casuality Actuarial Society (CAS) Form, 2001.
[4] Leshno, M. and H. Levy, *Preferred by All and and Preferred by Most Decision Makers: Almost Stochastic Dominance*, Journal of Management Science **48** (2002), 1074-1085.
[5] Levy, M., *Almost Stochastic Dominance and Efficient Investment Sets*, American Journal of Operations Research **2** (2012).
[6] Madden, D. and F. Smith, *Poverty in Ireland, 1987-1994: A Stochastis Dominance Approach*, The Economic and Social Review 31 (2000), no. 3, 187-214.
[7] Ravallion, M., *Poverty Comparisons. Aguide to Concepts and Methods*, LSMS Working Paper Number 88, The World Bank (1992).
[8] Wackerly, D. D., W. Mandenhall III and R. L. Sceaffer. *Mathematical Statistics with Applications*, **7ed.**, Thonson Brooks/Cole, Dukbury, 2008.

IOP Conf. Series: Journal of Physics: Conf. Series **943** (2017) 012049 doi:10.1088/1742-6596/943/1/012049

Mathematical modeling in realistic mathematics education

B Riyanto, Zulkardi, R I I Putri and Darmawijoyo

University of Sriwijaya, Jl. Srijaya Negara, Bukit Lama, Palembang, Indonesia.

Email: bambangriyantomath@gmail.com

Abstract. The purpose of this paper is to produce Mathematical modelling in Realistics Mathematics Education of Junior High School. This study used development research consisting of 3 stages, namely analysis, design and evaluation. The success criteria of this study were obtained in the form of local instruction theory for school mathematical modelling learning which was valid and practical for students. The data were analyzed using descriptive analysis method as follows: (1) walk through, analysis based on the expert comments in the expert review to get Hypothetical Learning Trajectory for valid mathematical modelling learning; (2) analyzing the results of the review in one to one and small group to gain practicality. Based on the expert validation and students' opinion and answers, the obtained mathematical modeling problem in Realistics Mathematics Education was valid and practical.

1. Introduction

By modeling, mathematics education for student becomes more meaningful and justification of modeling is the essential goal of teaching mathematics [1, 2]. In fact, mathematical modeling learning in schools is very few. Everyday, in many countries, teaching mathematics seldom apply the modeling because it is difficult for the teachers to do [1, 2]. In addition, modeling is actually rather difficult for students). Then Blum [1] states that students find it difficult to mathematical modeling due to the demands of modeling tasks. Modeling is inseparable from other mathematical competencies, i.e. reading and communicating, designing and implementing problem-solving strategies or working mathematically such as reasoning, computing, etc. [3]. It is closely related to PISA, Stacey [4] state that PISA has moved away from the dubious traject of concluding students'ability to solve problems in modeling from a measure of students'ability to perform mathematical heuristics in the abstract (e.g. solving equations, performing calculations).

The 2013 Curriculum states that Mathematics in schools is expected to support the achievement of the competence of elementary and secondary education graduates through learning experiences in order to be able to understand concepts and apply mathematical procedures in everyday life, make generalizations based on patterns, facts, phenomena, or existing data, conduct mathematical operations for simplification, and analysis of existing components, do mathematical reasoning that includes making allegations and verifying them, solve problems by communicating ideas through symbols, tables, diagrams, or other media to clarify circumstances or problems, and grow positive attitude such as logical, critical, meticulous, and not easily give up in solving problems.

To achieve these competencies requires an innovation, one of which is by modeling. According to Bliss [5] that 21st century skills are creativity and innovation, critical thinking and problem-solving, and communication and collaboration which are all achievable through modeling. The importance of preparing students for mathematical literacy is by equipping them with 21st century competencies [6]. Mathematical modeling is one way to achieve this goal [7]. Thus, to achieve the competency goals of elementary and secondary school graduates and 21st century mathematics and literacy skills are by learning mathematical modeling in schools.

Freudhental [8] argues that the traditional problem-solving practice in mathematics learned in schools does not quite fit into the idea of mathematical modeling since the mathematical modeling is the structuralization of reality by the use of mathematics, whereas the type of problem solving does not contain mathematical heuristic strategies. Furthermore, according to Mousoulides [9] modeling activities not only can help students to use essential mathematical ideas in problem-solving, and but also can help teachers to develop students' sense of thinking. Traditionally, students do not learn mathematical modeling until they reach secondary school [10]. Whereas mathematical modeling can and should be started early when the students already have basic competence then modeling can be developed [11, 12]. It is also supported by [5] that mathematical modeling should be taught in every level of mathematics education for the students. This means modeling learning in primary and secondary schools has to be taught. In fact in Indonesian, mathematical modeling is not formally introduced at any school level [13].

This means that junior high schools have to include modeling in learning. Based on the previous description, the authors were interested in conducting a study entitled "Mathematical modeling in Realistic Mathematics Education".The problems of the study were formulated in the following questions as follows: 1. How is the Mathematical modeling Problem in Realistic Mathematics Education made valid? 2. How is the Mathematical modeling Problem in Realistic Mathematics Education made practical? This study aimed to produce valid and practical Mathematical modeling Problem in Realistic Mathematics Education for junior high schools.The expected benefits of the results of this study were for teachers, students, policy makers, and researchers.

2. Method

This study used development research methods developed by Akker, et al. [14] consisted of 3 stages, namely analysis, design and evaluation. In the analysis step, student analysis, curriculum and mathematical modeling are carried out. The second step designed and produced (Mathematical modeling). The final step used a formative evaluation design (Figure 1) consisting of self-evaluation, one-to-one, expert review, small group, and field tests [15, 16]. This study was only conducted until the small group phase.

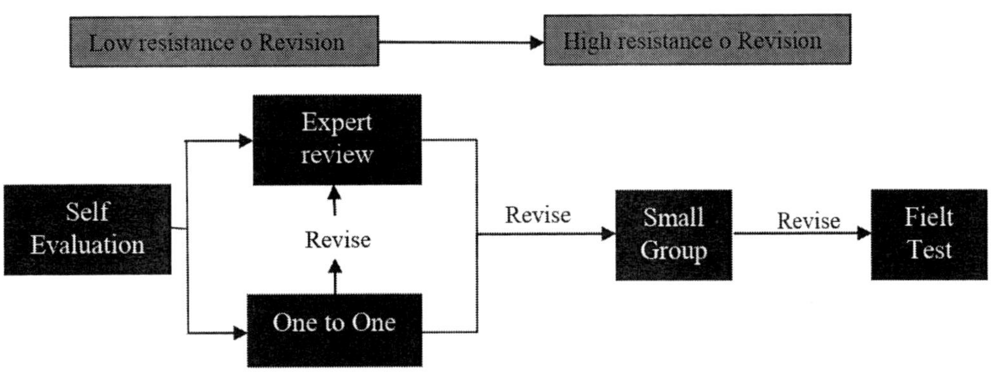

Figure 1. Design of Formative Evaluation [15, 16]

The success criteria of this study used the form of Mathematical modeling for learning modeling that was valid and practical for students. The validity was obtained from the validation of experts and the practicality was obtained from the students' opinions and observations of the small group and one-to-one. Practicality means easy to use, interpretable, and unambiguous.

The subjects of this study were students of SMP Negeri 6 Kayuagung, Ogan Komering Ilir District, South Sumatra. The techniques of collecting the data were (1) walk through based on the expert review to get a valid mathematical modeling problem in content, construct and language aspects. (2) interview derived from one to one and small group to find out the practicality of the problem. The collected data were analyzed using descriptive analysis method: (1) walk through sheet analysis based on the expect comments in expert review to get valid mathematical modeling problem; (2) analyze the results of the review in one to one and small group to get practicality.

3. Results and Discussion

3.1 Results

Validity of the mathematical modeling problem in Indonesian Realistic Mathematics Education was done by Zulkardi [16]. Zulkardi [16] suggested using Uber context while my problem is using bus hire context and book pen spending in Mathematics modeling problem in Indonesian Realistic Mathematics Education on Material of Two Variable Linear Equation System. Therefore, the design of the mathematical modeling in IRME used Uber context. Darmawijoyo suggested using real photos in contextual use and immediately raising questions from the photo. Therefore, the images were used in designing modeling problems. The validation of these two experts obtained a valid mathematical modeling problem in the Indonesian realistic mathematics education using Uber context. The mathematical modeling problem is as follows:

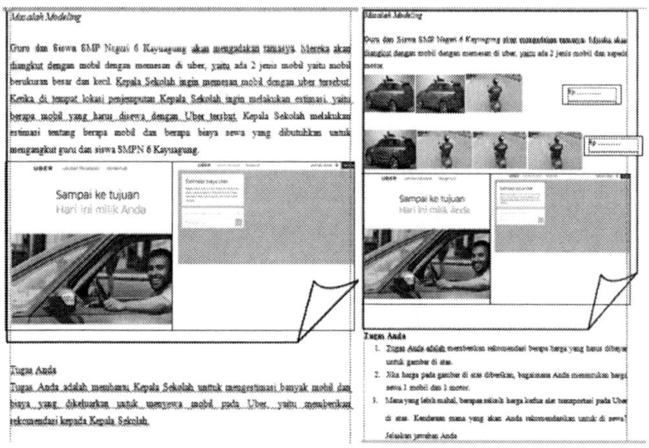

Figure 2. Mathematical Modeling Problem in Realistic Mathematics Education

After design being validated by the experts, then one-to-one was conducted to find out the practicality of Mathematical modeling Problem in Realistic Mathematics Education. One-to-One was conducted at SMP Negeri 6 Kayuagung on Tuesday, October 3, 2016. There were three students of grade VIII.2 selected, namely Kaysha Frida Pracillia, M. Rahman Aria Pratama and Orin Allysia. The following are one-to-one photographs.

IOP Conf. Series: Journal of Physics: Conf. Series **943** (2017) 012049 doi:10.1088/1742-6596/943/1/012049

M. Rahman Aria Pratama

Orin Allysia

Kaysha Frida Pracillia

Figure 3. One-to-One photos

The one-to-one result based on the student opinion showed that the modeling problem was good and could make the students think reasonable. However, there were some terms that had to be explained first because the students were not familiar with the terms such as assumptions, parameters, and essential variables. In addition, they did not yet understand the language of identification and specify the problem to be resolved. Thus, it should be given an explanation first in order to make them understand the language. The following are the student's responses to the modeling problem.

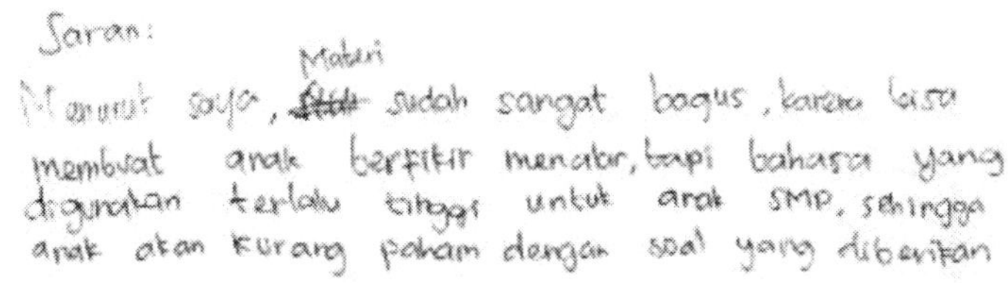
Saran:
Menurut saya, Materi sudah sangat bagus, karena bisa membuat anak berfikir menalar, tapi bahasa yang digunakan terlalu tinggi untuk anak SMP, sehingga anak akan kurang paham dengan soal yang diberikan

Figure 4. The student's comment

Judging from the student's answer, it shows that the students were able to make identification and problem specifications, make assumptions, do math to get the solution and provide recommendations although they cannot be generalized. Based on the validation and one-to-one results, we obtained valid and practical mathematical modeling problem in realistic mathematics education for learning mathematical modeling for material equations and system of linear equations of two variables. The student answers seen in Figure 5.

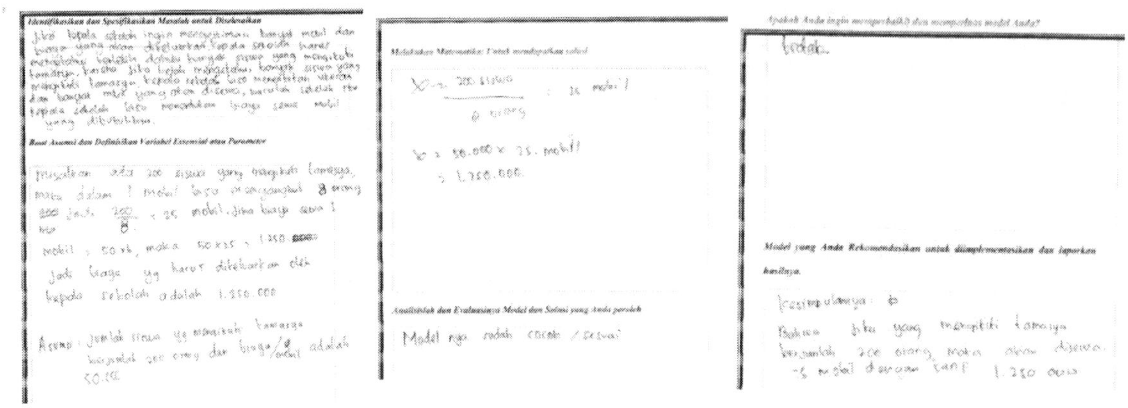

Figure 5. The student's solution

3.2 Discussion

The Traditional Approach to teaching mathematics is not sufficient for strengthening student problem-solving skills and thus can not help students to develop competence in math and application [17, 18, 19]. In the traditional approach of learning there were more emphases on the mechanistic and memorize solutions and operations of mathematics. For example, according to Bahmaei [20] in elementary schools in general, early arithmetic teaching focused on computational skills. Further, it is also supported by the assertion that the results of traditional approaches in mathematics are mechanistic or reminding of solutions to story problems/word problem [21, 22]. Furthermore, according to Zulkardi [23] in traditional programs, objectives are classified as low-level goals based on the skills of formulas, simple algorithms, and definitions. Bustang [24] state that learning process not only takes place within individuals themselves, but it also involves the social interactions among them. Furthermore, in the traditional approach, students solve without understanding the problem [25], as well as in traditional dominant learning strategies the dominantly solving problems look for keyword interpretation rather than thinking deeply about the problem [17, 26]. Based on this it is necessary to have a new learning approach. Based on the study conducted by Mouselides [27] it recommends that the development of more specific problem solving strategies (modeling process) is needed. Furthermore Zulkardi [28] states that there are 4 steps of doing mathematics or the use of mathematics in everyday life are identifying where math can be applied, changing the practical problem into a problem or mathematical model, solving math problems, and interpreting and evaluating the obtained results.

Furthermore Zulkadi [28] states that during the focus of mathematics education is in step 3 it can all be done by a computer. Thus, the focus of mathematics education is important to move on to the other three steps. It also means that there are needs for innovation in mathematics learning in schools in order to achieve competence or skills in the digital community. This is in accordance with Zulkardi's [28] statement that mathematical competence is known in mathematics education as the competence required for the digital community: application or modeling, understanding and checking. In other words it is very important to apply the modeling competence in mathematics education in schools.

Furthermore Bliss [5] states that modeling in pre-kindergarten up to grade 8 is a way to float and maintain the mathematical disposition of students to mathematics. This is in line with Gravemeijer [29] assertion that there is a tendency to shift mathematical learning approaches in mathematics education in primary schools from "knowledge transfer" to "construction of knowledge" and also this certainly happens in junior high schools. This fits perfectly with learning by modeling. This is also supported by Bliss [5] that mathematical modeling should be taught in every level of student math education. Mathematical modeling (the process of translating between the real world and mathematics in both directions) is a topic in mathematics education that is most discussed and developed over the last few decades [1].

Everyday the teaching of Mathematics in many countries is very few with modeling, and this is due to the fact that the modeling is difficult for teachers [1]. Also Sumarto [30] states that It is not easy for a teacher to facilitate all the students' needs by giving an individual coach for every student in the class. Then, traditionally, students were not introduced Mathematical modeling through junior high schools [10]. Furthermore, Mathematical modeling is not formally introduced at the school level in Indonesia[13]. In fact, Mathematical modeling should be and could be started early, when students already have basic competencies then modeling can be developed [11, 12]. It is also supported by Bliss [5] that Mathematical modeling should be taught in every level of mathematics education. On the Competency Standards of the 2013 Curriculum students must have 21st Century Skills [31]. The 21st century skills are creativity and innovation, critical thinking and problem solving, communication and collaboration which are all accessible via modeling [5, 32-34].

This study conducted the designing process of mathematical modelling problem in Realistic Mathematics Education which was valid and practical, i.e. on linear equation material and system of linear equation of two variables which were recommended to be done in small group and field test further. With small groups and field tests it can be obtained mathematical modelling problem that has a potential effect [23, 35]. In this case researchers conducted the small groups and field tests further to obtain valid and practical mathematical modelling problems, had potential effects, and could prepare students having 21st century skills and improve their high-order thinking skill (HOTS) according to the hope of revised edition of 2013 curriculum. This study as a continuation of the study Mouselides [27] that study about to develop a smaller and more focused theoretical model and modeling as problem -solving activities for teaching, learning and development in Elementary and junior high school. This study about how to design mathematical modeling in realistic mathematics education in junior high school. Morever, this study also relate to Zulkardi [23].

4. Conclusion

The validation and one-to-one results produced valid and practical mathematical modeling problem in realistic mathematics education for learning mathematical modeling of material equations and system of linear equations of two variables.

References

[1] Blum W 2009 Mathematical modeling: Can It Be Taught And Learnt? *Journal of Mathematical modeling and Application* **1** 45

[2] Ayla A 2015 Mathematical modeling Approach in Mathematics Education *Universal Journal of Educational Research* **3** 973

[3] Niss M 2003 Mathematical Competencies and the Learning of Mathematics: The Danish KOM *3rd Mediterranean conference on mathematical education*

[4] Stacey K 2015 The Real World and the Mathematical World *In Stacey. Assessing Mathematical Literacy: the PISA Experience* Springer 57-80

[5] Bliss K et. al 2016 *GAIMME: Guidelines for Assessment & Instruction in Mathematical modeling Education* (United State America: COMAP & SIAM)

[6] Ministry of Education [MOE] 2014 Information sheet on 21st century competencies Retrieved Oktober, 2017, from http://www.moe.gov.sg/media/ press/2014/04/information-sheet-on-21st-century.php

[7] Dawn N K E and Lee N H 2015 Introduction: Mathematical modeling Outreach in Singapore. *In Hoe, Lee Ngan Mathematical modeling: From Theory to Practice*1-13

[8] Freudenthal H 1991 *Revisiting Mathematics Education* (Dordrecht: Kluwer Academic Publishers)

[9] Mousoulides N G 2009 *Mathematical modeling for Elementary dan Secondary School Teachers* (University of Cyprus)

[10] Stillman G 1998 *The emperor's new clothes? Teaching and assessment of mathematical applications at the senior secondary level.* In P. L. Galbraith et al. (Eds). Mathematical modeling: Teaching and assessment in a technology-rich world (West Sussex: Horwood Publishing Ltd) 243-254

[11] Diezmann C J J, Watters, and English L D 2002 Teacher behaviours that influence young children's reasoning *Proc of the 26th International PME Conference* (Norwich: University of EastAnglia) 289-296

[12] Lehrer R and L Schauble 2003 Origins and evolutions of model-based reasoning in mathematics and science. *In R. A. Lesh and H. Doerr (Eds.), Beyond constructivism:Models and modeling perspectives on mathematics problem solving, learning, andteaching* (Mahwah, NJ: Lawrence

Erlbaum) 59-70

[13] Widjaya W 2013 Building Awareness of Mathematical modeling in Teacher Education: A Case Study in Indonesia. In Stillman, A.G., et al. *Teaching Mathematical modeling: Connecting Research and Practice* 583-592

[14] Akker, et al 2006 *Educational Design Research* (Enschede: SLO)

[15] Tessmer M 1993 *Planning and Conducting Formative Evaluation* (Philadelphia: Kogan Page)

[16] Zulkardi 2006 *Formative Evaluation: What, Why, When, and How* Retrieved Nopember 2016, from http://reocities.com/zulkardi/books.html

[17] Schoenfeld A 1987 Confessions of an accidental theorist. *For the Learning of Mathematics* **7** 30-38.

[18] Schoenfeld A 1991 On mathematics as sense-making: An informal attack on the unfortunate divorce of formal and informal mathematics *In J Voss, D Perkin and J Segal (Eds.) Informal reasoning and education* (Hillsdale N J: Lawrence Erlbaum Associates) 311-343

[19] Lesh R and Zawojewski J S 2007 Problemsolving and modeling *In F. Lester (Ed.), Second Handbook of research on mathematics teaching and learning* (Greenwich, CT: Information Age Publishing) p 763

[20] Bahmaei F 2011 Mathematical Modeling in Primary school, advantages and challenges *Journal of Mathematical and Application* **1** 3-13

[21] Verschaffel L, De Corte E and Lasure S 1994 Realistic considerations in mathematical modeling of school arithmetic word problems *Learning and Instruction* **4** 273-294

[22] Greer B 1997 Modeling Reality in Mathematics Education: The case of word problem *Learning and Instruction* **7** 293

[23] Zulkardi 2002 Developing a Learning Environment on Realistic Mathematics Education for Indonesian Student Teacher *Dissertation* (Enschede: University of Twenty)

[24] Bustang, et. al 2013 Developing a Local Instruction Theory for Learning the Concept of Angle through Visual Field Activities and Spatial Representations *International Education Studies* **6** 58

[25] Reusser K and R Stebler 1997 Every word problem has a solution - The social rationality of mathematical modeling in schools *Learning and Instruction* **7** 309

[26] English L D 2003 Reconciling theory, research, and practice: A models and modeling perspective Educational Stidies inMathematics **54** 225.

[27] Mouselides N G 2007 The Modeling Perspective in the Teaching and Learning of Mathematical Problem Solving *Dissertation* (Cyprus: University of Cyprus)

[28] Zulkardi, et. All 2017 Pengembangan Perangkat Pembelajaran yang Inovatif untuk Meningkatkan Kemampuan modeling dan Problem Solving Mahasiswa Calon Guru Matematika Abad 21 *Hibah Profesi* (Palembang: Universitas Sriwijaya)

[29] Gravemeijer K 1993 The Empty Number-line as an Alternative Means of Rerpresenting Addition and Substruction *Innovation in Math Education by modeling and Application* (New York London)

[30] Sumarto S N, et.al 2013 Proportional Reasoning: How do the 4th Graders Use Their Intuitive Understanding? *International Educational Studies* **7** 69

[31] Kemendikbud 2016 *Permendikbud Nomor 20 Tahun 2016 tentang Standar Kompetensi Lulusan Pendidikan Dasar dan Menengah* (Jakarta: Kemendikbud)

[32] Sundayana R, Herman T, Dahlan J A and Prahmana R C I 2017 Using ASSURE learning design to develop students' mathematical communication ability *World Transactions on Engineering and Technology Education* **15** 245

[33] Maharani H R, Sukestiyarno, and Waluya B 2017 Creative thinking process based on Wallas model in solving mathematics problem *International Journal on Emerging Mathematics Education* **1** 177

[34] Runisah, Herman T and Dahlan J A 2017 Using the 5E learning cycle with metacognitive technique to enhance students'mathematical critical thinking skills *International Journal on Emerging Mathematics Education* **1** 87

[35] Tanujaya B, Prahmana R C I, and Mumu J 2017 Mathematics instruction, problems, challenges, and opportunities: A case study in Manokwari regency, Indonesia *World Transactions on Engineering and Technology Education* **15** 287

Interactive basic mathematics web using Wordpress

Tika Septia, Husna and Anna Cesaria

STKIP PGRI Sumatera Barat, Jl. Gunung Pangilun, Padang 25172, Indonesia

E-mail: tikaseptia2589@gmail.com

Abstract. Wordpress is a popular open source tool that can be used for developing learning media. Basic Mathematics is the difficult subject for a physics student. The students need an interactive learning to improve their knowledge. The aims of this study were to develop the interactive media using Wordpress and to know the effectiveness of web as a learning media to improve the ICT Literacy students. This study used ADDIE models. The effectiveness of interactive web can be described as the students' equipness of ICT literacy. The population is physics students. The findings show that the interactive web is valid for the content, presentation, linguistic, and graphic aspects. The results concluded that basic mathematic interactive web is effective to equip the learners ICT literacy of categories of high, medium, and low with the observations and questionnaires are in very good criteria.

1. Introduction

Basic Mathematics is the difficult subject for a physics student. The results of interviews with students who had been taken basic mathematics regarding the difficulties in understanding the material [1-3]. Students often only find numbers, formulas, charts, and die pictures, thus all things make the students feel less interested and think the materials are boring. Lack of teaching materials specifically for this subject is the reasons faced by the students [4]. This condition makes the activity and results of student learning to be low. Students commonly use the textbook as an aid in studying for exams, they do not read all of the books but only partially read according to their needs [5]. Therefore, the attractiveness of the students can increase by the media so that it can improve students' motivation and understanding can improve. The use of media in the learning process can clarify the message that is not too verbal, overcoming the limitations of space and time, and excite student's learning [6]. The students need an interactive learning to improve their knowledge. The aims of this study were to develop the interactive media using Wordpress and to know the effectiveness of web as a learning media to improve the ICT Literacy students. ICT (Information and Communications Technology) stands in interesting relation to literacy, being as it is capable both of supporting and promoting the basic skills of reading and writing - the dominant classroom definition of literacy [7]. The idea for becoming successful in XXI century required ICT literacy skills [8]. ICT literacy is defined as the ability to effectively evaluate, navigation, and build the information by using various digital technologies. ICT literacy becomes an important precondition for the socialization and professional career [9]. Therefore, ICT literacy plays an important role in social development in education. ICT literacy differentiates among three major dimensions: knowledge of technology, skills relevant to using the technology, and attitudes of using technology [10].

IOP Conf. Series: Journal of Physics: Conf. Series **943** (2017) 012050 doi:10.1088/1742-6596/943/1/012050

Based on preliminary studies with the students, it is known that there are no web-based media yet to learn basic mathematics subject. The development of Internet-based information technology can make it easier for students to better understand the learning materials well. With advances in technology such as increased bandwidth, wider internet coverage and increasing number of stand-alone and web-based education-related software, teachers are expected to be able to use the technology available in schools to improve teaching and engage students in learning. Based on above conclusion, we need to innovate interactive media so that it can help the students to understand the concept, one of them using Wordpress. Wordpress is an online, open source website creation tool written in PHP [11]. But in non-geek speak, it's probably the easiest and most powerful blogging and website content management system (or CMS) in existence today [11]. Wordpress has thousands of *plugin* pieces of software and it's customizable so you can use your site for just about anything. With appropriate choice of media or learning resources, it can affect the ability of students to engage in learning through interaction with the media [12], [13].

The web that is going to be developed is the web which can involve the students in its use Orin other words, the interactive web. The interactive web can help students to achieve a better understanding of the topic because they can interact with the teacher. The interactive web can also facilitate long-distance learning because it encourages interaction between teacher-student, within students and student-content. The interactive web makes learning remains student-centered, so they can understand the content well. Accordingly, we need to develop interactive web for basic mathematics subject.

2. Method

The development procedures in this study use a model ADDIE (Analyze, Design, Develop, Implement and Evaluation). The steps of design ADDIE model of the analysis are analyze, design, develop, implement, and evaluate [14]. The study was conducted in Physics Education Study Program. Analyze stage was carried out by analysis syllabus, interviews with an associate friend, determine the students' need for students' need questionnaire, and analysis textbook. Design stage was carried out by design the interactive media. After having a self-evaluation completely, validation activities carried out in the form of validation fill out a web-based learning media by the validated-lectures to obtain a valid web. Validation of web-based learning media was carried out by four lecturers. On sheet validation, they also took a note on some of the links found on the web.

The effectiveness of interactive web can be described with the students' equipness of ICT literacy. The indicator of ICT Literacy with using web can be seen in Table 1.

Table 1. The indicator of ICT literacy with using web

Variable	Dimension	Indicator
ICT Literacy	Knowledge of web	a. Have knowledge of web b. Be able to identify content management system (or CMS) that able to create web. c. Be able to provide an actual assessment of using web d. Know the basic feature of web e. Able to distinguish the virtual world and the real world.
	Relevant skills to using the web	a. Able to use web features b. Able to search and access web c. Able to take advantage of basic services on the internet, such as creating accounts, using e-mail, attaching and downloading data, participating in discussion forums and social networking sites d. Able to use computer & internet to support critical thinking, creativity, and innovation in work.
	attitudes of using technology	a. Have the ability to use web for individual or group. b. Utilize web responsibly/wisely. c. Understand the consequences of using technology. For example, the habits of communicating with the lecturer, able to understand the effect of using technology.

All of the population in this study are physics students. The data collection is done by the observation and questionnaires of the students' equipness in ICT literacy with effective criteria only if it is in good or excellent criteria.

3. Result and discussion

Validation of web-based learning media was carried out by four lecturers. On sheet validation, they also took a note on some of the links found on the web. On sheet validation, they also took a note on some of the links found on the web.

Untuk website nya, sudah sangat bagus sekali. Pewarnaan, grafik, dan animasinya sudah lebih dari cukup. Saran dari saya adalah *link* yang diberikan sebaiknya langsung ke web aslinya, tidak usah pakai *localhost* lagi, sehingga tidak terjadi kegagalan akses pada beberapa menu penting, seperti:

1. http://localhost/matdastes/pages/team/;
2. http://localhost/matdastes/about/
3. http://localhost/matdastes/evaluasi/
4. http://localhost/matdastes/materi/

Figure 1. Advice from a validated-lecturer on a link on the web

The next suggestion is the evaluation tests.

Saya sudah mencoba tes evaluasinya. Saran dari saya, diantaranya sebagai berikut:

1. Evaluasi 1, Soal No. 8, gak ada jawabannya atau salah kunci jawaban. Secara keseluruhan, soalnya terlalu mudah (saya bisa ngerjain dengan mencongak). Kunci jawabannya gak ada pengecoh. tuliskan tanda "=" disetiap akhir soal.
2. Evaluasi 2, Soal No. 4 dan 5, gak ada jawabannya atau salah kunci jawaban. Soal No. 8, sebaiknya kata "atau" atau tanda "," nya dihapus. pakai salah satu saja. Soal No. 9 merupakan pertidaksamaan bukan persamaan. Selain itu, jawabannya gak ada karena harus dalam bentuk interval. Secara keseluruhan, soalnya terlalu mudah (sebahagian besar soal, saya bisa ngerjain dengan mencongak). Kunci jawabannya gak ada pengecoh. Coba dipahami pembuat nol dalam penyebut soal.
3. Evaluasi 3, gak bisa masuk. Apakah password nya salah atau emang belum bisa di akses.
4. Evaluasi 4, Soal No. 2 dan 4, gak ada jawabannya atau salah kunci jawaban.
5. Evaluasi 5, Soal No. 2, 4, dan 5, gak ada jawabannya atau salah kunci jawaban. sebahagian besar soal membingungkan alias memiliki banyak arti (bias makna). Soal yang bisa saya kerjakan menggunakan asumsi saya sendiri dalam

Figure 2. Advice from a validated-lecturer on the question of evaluation

The lecturers who validated it also provide advice to increase the loading time when accessing the web. It is intended that the learning process becomes efficient without losing a lot of time in accessing each page of the web. Based on the advice of them, the researchers conducted a revision to make small the size of each of the constituent components in order to improve the speed of loading web. After doing the revision, the speed of the web was tested by using the service GTmetrix (www.gtmetrix.com). Here is a picture of the web speed test results GTmetrix.

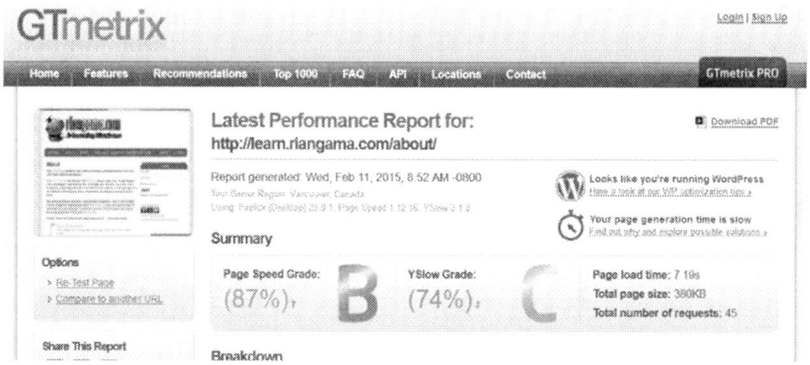

Figure 3. Screenshot Loading Speed

In Figure 3, it shows that the web speed test results of grade 87%, load time is 7:19 seconds and the total size of pages is 380 KB. Results Validation by the lecturers for all aspects can be seen Table 2.

Table 2. Results Validation All Aspects

ASPECTS	TOTAL	SCORE
Material/content	86	89.58%
Presentation	94	87.04%
Linguistics	67	79.76%
Graphic	115	87.12%
Total	**362**	**86.19%**

The results of validation by the four lectures, it indicated that the validity of web-based learning media included in the criteria developed is very valid. The results of validation from the lectures as a whole is 86.19%.The view of interactive basic Mathematics web can be seen in Figure 4.

Figure 4. The Interactive Basic Mathematics Web

The equipness of ICT literacy using observation sheet and questionnaire ICT literacy. Validity test shows there are 52 valid statements of 55 statements with the level of reliability is very high (0.942). The respondents' knowledge of web is in the pretty good category (value 86%). The respondents' technical skills relevant to using the web are in low category (value 49%). The last, the respondents' attitudes accruing from the critical reflection of using web skills are in pretty good category (value 80%).

4. Conclusion

Related to the findings and analysis of data, it can be concluded that web-based learning media for an introduction to mathematics subject in physics education department included in the criteria very valid. The results and discussion of this study concluded that basic mathematics interactive web is effective to equip the learners ICT literacy of categories of high, medium, and low with the observations and questionnaires are in very good criteria. Learning integrated with the basic mathematics interactive web can be applied as early as possible to equip ICT literacy because it is one of the XXI century competencies that must be owned by the current learner.

5. Acknowledgments

Thanks for Mr. I Made Arnawa (Lecturer at Andalas University), Mr. Rully Charitas Indra Prahmana (Lecturer at Ahmad Dahlan University), Mr. Syafdi Maizora (Lecturer at Bengkulu State University), and Mrs. Nurmi (Lecturer at STKIP PGRI Sumatera Barat) who have been willing to be validator of interactive web for Basic Mathematics Subject, and thanks for DP2M Dikti which have been funding the grant.

References

[1] Burchfield C M, and Sappington J 2000 Compliance with required reading assignments. *Teaching of Psychology* **27** 58
[2] Clump M A, Bauer H, and Bradley C 2004 The extent to which psychology students read textbooks: A multiple class analysis of reading across the psychology curriculum *Journal of Instructional Psychology* **31** 227
[3] Sappington J, Kinsey K, and Munsayac K 2002 Two studies of reading compliance among college students *Teaching of Psychology,* **29** 272
[4] Lestiana H, Rejeki S, and Setyawan F 2016 Identifying Students' Errors on Fractions *Journal of Research and Advances in Mathematics Education* **1** 131
[5] Sikorski J F, Rich K, Saville B K, Buskist W, Drogan O and Davis S F 2002 Student use of introductory texts: Comparative survey findings from two universities *Teaching of Psychology* **29** 312
[6] Daryanto 2010 *Media Pembelajaran.* (Yogyakarta: Gava Media) p 5
[7] Papert S 1993 *The Children's Machine: Rethinking School in the Age of the Computer,* New York: Basic Books
[8] Wilson M, Scaliseb K, and Gochyyev P 2015 Rethinking ICT literacy: From computer skills to socialnetwork settings. *Thinking Skills and Creativity* **18** 65
[9] Ivankovića A Š 2013 ICT literacy among the students of the Faculty of Philosophy, University of Mostar *Procedia - Social and Behavioral Sciences* **93** 684
[10] Pernia E E 2008 Strategy Framework for Promoting ICT Literacy in the Asia-Pacic. UNESCO Bangkok: Thailand
[11] Jones K M and Alida-Farrington P 2011 Getting started with wordpress *Library Technology Reports* **47** 8
[12] Kizito R 2003 A personal experience of learning with print and learning with electronic media in open and distance education *Progressio* **25**, 29
[13] Fathurrohman M, Porter A L, and Worthy A L 2017 Teachers' real and perceived of ICT supported-situation for mathematics teaching and learning *International Journal on Emerging Mathematics Education* **1** 11
[14] Maribe Branch, Robert 2009 *Instructional Design: The Addie Approach* USA: Springer

Modeling Tetanus Neonatorum case using the regression of negative binomial and zero-inflated negative binomial

Luthfatul Amaliana, Umu Sa'adah and Ni Wayan Surya Wardhani

Brawijaya University, Jl. Veteran, Malang, 65145, Indonesia

E-mail: luthfatul@ub.ac.id

Abstract. Tetanus Neonatorum is an infectious disease that can be prevented by immunization. The number of Tetanus Neonatorum cases in East Java Province is the highest in Indonesia until 2015. Tetanus Neonatorum data contain over dispersion and big enough proportion of zero-inflation. Negative Binomial (NB) regression is an alternative method when over dispersion happens in Poisson regression. However, the data containing over dispersion and zero-inflation are more appropriately analyzed by using Zero-Inflated Negative Binomial (ZINB) regression. The purpose of this study are: (1) to model Tetanus Neonatorum cases in East Java Province with 71.05 percent proportion of zero-inflation by using NB and ZINB regression, (2) to obtain the best model. The result of this study indicates that ZINB is better than NB regression with smaller AIC.

1. Introduction

Tetanus Neonatorum (TN) is one of the main caused of infant death in Indonesia. The high infant mortality is mostly caused by unclean birth and cord handling. Tetanus is characterized by a painful muscle stiff. It was an infection that caused by Neurotoxin bacteria produced by Clostridium Tetani on a closed wound. The systematic effort of the government to eliminate TN begins with the TT immunization of pregnant and prospective bride. It followed by School Immunization Month program with DT vaccine and TT vaccine for elementary school children as a long-term control of TN.

The World Health Organization (WHO) together with UNICEF invites developing countries in the world to implement the Elimination of Tetanus Maternal and Neonatal (ETMN) program. Previously, Indonesia was one of the countries with the highest cases in Asia. Based on data from the Ministry of Health in the Health Profile of Indonesia 2015, East Java is the province with the highest TN case in Indonesia, which is 22 cases. Provinces with the second and the third rank of TN cases were Banten and West Kalimantan, with twelve and five cases, respectively. Overall, East Java was also the first province in Indonesia with 663 TN cases [1]. TN data in East Java province showed the proportion of zero-inflation was large (71.05 percent). Therefore, modeling the number of TN cases requires a method that can accommodate zero-inflation problems in the data.

Count data commonly modeled by Poisson regression. However, when over dispersion problem happens in Poisson regression, Negative Binomial regression often becomes an alternative solution. Over dispersion is a condition when the sample mean is smaller than the sample variance. NB distribution is a mixture of Poisson-Gamma distribution [2]. It has a dispersion parameter that capable to overcome the over dispersion problem [3]. However, data on TN cases containing over dispersion and a large proportion of zero-inflation. Thus it needs a method to analyze a large zero-inflation and over dispersion data. Astuti and Mulyanto [4] using Zero Inflated Negative Binomial regression to model TN case in East Java province in 2012. Another study of ZINB regression is the comparison of two zero-inflation models, which is ZIP and ZINB regression, to obtain the best model with over dispersion problem and 73.67 percent proportion of zero-inflation on TN data in 2013 [5]. The purpose of this study is to model the number of TN cases in East Java province with 71.05 percent proportion of zero-inflation by using NB and ZINB regression. Then, find the best model between NB and ZINB regression.

This paper is structured as follows. Section 2 discusses the literature review about over dispersion, NB and ZINB regression that will be used to model later. Then Section 3 discusses the result of modeling TN case by using NB and ZINB regression.

2. NB and ZINB regression

In this section, both of models that will be used in this study are explained. This study uses two models that have a dispersion parameter to overcome over dispersion problem. However, over dispersion problem will be discussed first.

2.1. *Over dispersion*

A condition while the sample variance is greater than the sample mean called as over dispersion. This condition is a violation of the equality of mean and variance assumption in Poisson regression. The occurrence of over dispersion can be identified by finding the deviance model and dividing by degrees of freedom model. If the result is greater than 1, then there is over dispersion. Otherwise, if the result is less than 1, then it is said to be under dispersion. The deviance could be written as [6]:

$$D = 2\sum_{i=1}^{n} y_i \ln\left(\frac{y_i}{\hat{\mu}_i}\right) \tag{1}$$

n is the number of observation, y_i is dependent variable for $i = 1,2,\dots,n$, and $\hat{\mu}_i$ is mean of dependent variable which is influenced by independent variables.

2.2. *NB regression*

A generalization of Poisson regression with no restrictive assumption known as Negative Binomial regression. The restrictive assumption of Poisson model is the equality of mean and variance. The traditional NB regression model commonly known as NB-2 is written as [7]:

$$\ln \hat{\mu}_i = \hat{\beta}_0 + \sum_{j=1}^{p} \hat{\beta}_j x_{ij} \tag{2}$$

$i = 1,2,\dots,n$, $j = 1,2,\dots,p$, p is the number of independent variables, β is unknown parameter, and a set of x (independent variables). The parameters of NB regression are estimated using Maximum Likelihood Estimation (MLE) method with Fisher Scoring method to maximize the ln-likelihood function.

The likelihood function of NB model could be written as:

$$\mathrm{P}(\mathrm{Y_i} = y_i) = \frac{\Gamma\left(y_i+\frac{1}{\kappa}\right)}{\Gamma\left(\frac{1}{\kappa}\right)y_i!}\left(\frac{1}{1+\kappa\mu_i}\right)^{\frac{1}{\kappa}}\left(\frac{\kappa\mu_i}{1+\kappa\mu_i}\right)^{y_i} \tag{3}$$

$\Gamma(.)$ is gamma function, κ is dispersion parameter, and $Y_i \sim NB(\mu_i, \kappa)$.

2.3. *ZINB regression*

Zero-inflated negative binomial regression is a model formed from a Poisson-Gamma mixture distribution. The ZINB model can modeling a count data contains zero-inflation in the dependent variable and over dispersion [8]. The observation y_1, y_2,..., y_n from dependent variable occured in this two condition:

$$y_i = \begin{cases} 0 & \text{, with probability } \pi_i \\ NB(\mu_i, \kappa), & \text{with probability } (1-\pi_i) \end{cases} \quad (4)$$

The first condition is called as zero-inflation state. It happened when the dependent variable is zero, while the probability is π_i. The other condition called as NB state and the probability is $(1-\pi_i)$. The likelihood function of ZINB model could be expressed as:

$$\mathrm{P}(\mathrm{Y_i} = y_i) = \begin{cases} \pi_i + (1-\pi_i)\left(\frac{1}{1+\kappa\mu_i}\right)^{\frac{1}{\kappa}}, & \text{for } y_i = 0 \\ (1-\pi_i)\frac{\Gamma\left(y_i+\frac{1}{\kappa}\right)}{\Gamma\left(\frac{1}{\kappa}\right)y_i!}\left(\frac{1}{1+\kappa\mu_i}\right)^{\frac{1}{\kappa}}\left(\frac{\kappa\mu_i}{1+\kappa\mu_i}\right)^{y_i}, & \text{for } y_i > 0 \end{cases} \quad (5)$$

$0 \le \pi_i \le 1$, $\mu_i \ge 0$, κ is dispersion parameter, while $\Gamma(.)$ is gamma function. Random variable $Y_i \sim NB(\mu_i, \kappa)$ for $\pi_i = 0$. μ_i and π_i are assumed depend on independent variable x_i that is defined by:

$$\mu_i = \exp\left(\mathbf{x_i^T}\boldsymbol{\beta}\right) \quad (6)$$

$$\pi_i = \frac{\exp(\mathbf{x_i^T}\boldsymbol{\gamma})}{1+\exp(\mathbf{x_i^T}\boldsymbol{\gamma})} \text{ and } (1-\pi_i) = \frac{1}{1+\exp(\mathbf{x_i^T}\boldsymbol{\gamma})} \quad (7)$$

The ZINB model for Negative Binomial state $\hat{\mu}_i$ and zero-inflation state $\hat{\pi}_i$ for $i = 1,2,\dots,n$ and $j = 1,2,\dots,p$:

$$\ln\hat{\mu}_i = \hat{\beta}_0 + \sum_{j=1}^{p}\hat{\beta}_j x_{ij} \quad (8)$$

$$\text{logit}\,\hat{\pi}_i = \hat{\gamma}_0 + \sum_{j=1}^{p}\hat{\gamma}_j x_{ij} \quad (9)$$

β and γ are parameters of ZINB model. Parameters of ZINB model is estimated by MLE completed with EM (Expectation-Maximization) algorithm. The maximum ln-likelihood function is:

$$\ln L(\kappa,\beta,\gamma) = \begin{cases} \sum_{i=1}^{n}\ln\left(\frac{e^{\left(\mathbf{x_i^T}\boldsymbol{\gamma}\right)}}{1+e^{\left(\mathbf{x_i^T}\boldsymbol{\gamma}\right)}} + \frac{1}{1+e^{\left(\mathbf{x_i^T}\boldsymbol{\gamma}\right)}}\left(\frac{1}{1+\kappa e^{\left(\mathbf{x_i^T}\boldsymbol{\beta}\right)}}\right)^{1/\kappa}, \text{for } y_i = 0\right) \\ \sum_{i=1}^{n}\ln\left(\frac{1}{1+e^{\left(\mathbf{x_i^T}\boldsymbol{\gamma}\right)}}\frac{\Gamma\left(y_i+\frac{1}{\kappa}\right)}{\Gamma\left(\frac{1}{\kappa}\right)\Gamma(y_i+1)}\left(\frac{1}{1+\kappa e^{\left(\mathbf{x_i^T}\boldsymbol{\beta}\right)}}\right)^{\frac{1}{\kappa}}\left(\frac{\kappa e^{\left(\mathbf{x_i^T}\boldsymbol{\beta}\right)}}{1+\kappa e^{\left(\mathbf{x_i^T}\boldsymbol{\beta}\right)}}\right)^{y_i}, \text{for } y_i > 0\right) \end{cases} \quad (10)$$

3. Method

This study using secondary data based on East Java Province Health Profile 2015 which is published by East Java Province Health Department in 2016 [9]. The data is a cross-sectional data including 38 regency/ city in East Java consist of 29 regencies and 9 cities. The dependent variable is the number of Tetanus Neonatorum (TN) case in every regency/city. The steps of analysis in this study can be described as follows:

1. Do the descriptive analysis of variables
2. Identify proportion of zero-inflation
3. Identify over dispersion
4. Identify multicollinearity

5. Modeling TN case using NB regression
6. Modeling TN case using ZINB regression
7. Testing the significance parameter of NB and ZINB model simultaneously and partially
8. Decide the best model

4. Result and discussion

Data on TN case in every regency/city is a non-negative integer ranging from zero to eight in the sample. Numbers of TN case in East Java is 22 cases from 11 regency/ city.

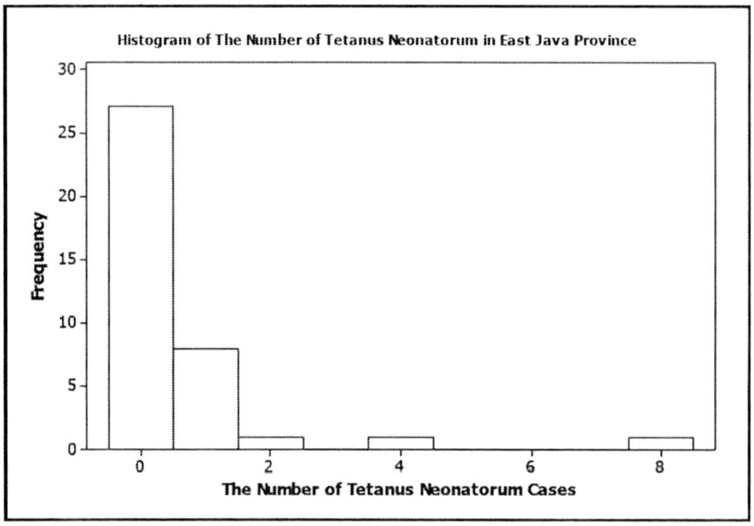

Figure 1. Histogram of Numbers of TN cases

Figure 1 shows frequency distribution of numbers of TN cases. The graph indicates a highly frequency of zeros and long right tailed. The graph also gives the evidence of over dispersion in data.

The independent variables in this study are percent of TT2+ immunization for pregnant women (X_1), maternity mothers assisted by health personnels in health care facility (X_2), K1 visit of pregnant women (X_3), TT1 immunization of pregnant women (X_4), and TT immunization of fertile age woman (X_5).

Table 1. Descriptive analysis of variables

Variable	Mean	Standard Deviation	Min	Max
Y	0.579	1.464	0	8
X_1	47.720	52.980	0	281.90
X_2	95.048	3.849	86.150	100.65
X_3	99.724	4.004	91.900	109.40
X_4	2.910	6.280	0	32.33
X_5	26.310	46.740	0	257.71

Table 1 summarize the descriptive statistics of variables consist of the dependent and independent variables. The highest TN cases occur in Bangkalan regency with eight cases, while TN cases not occur in 27 regencies/cities.

IOP Conf. Series: Journal of Physics: Conf. Series **943** (2017) 012051 doi:10.1088/1742-6596/943/1/012051

Table 2. Zero-inflation checking

The number of TN case	Frequency	Percent	Cumulative Percentage
0	27	71.05%	71.05%
1	8	21.05%	92.11%
2	1	2.63%	94.74%
4	1	2.63%	97.37%
8	1	2.63%	100.00%

The result of zero-inflation checking on numbers of TN case in East Java is summarized in Table 2. The sample data has a great proportion of zero-inflation, about 71.05 percent.

Table 3. Over dispersion checking

Deviance	Degrees of Freedom	Deviance / Degrees of Freedom
38.13502	32	1.192

Table 3 summarizes the result of over dispersion checking. The result of deviance divided by degrees of freedom (df) which is greater than 1 indicates the evidence of over dispersion in the data. It is also indicated by the sample variance of dependent variable that greater than the sample mean in Table 1.

It will be identified in advance whether there is multicollinearity problem among independent variables, before conduct to further analysis. There are several ways to identify multicollinearity and one of them is by looking at the value of Variance Inflation Factor (VIF).

Table 4. Multicollinierity identification

Variable	X_1	X_2	X_3	X_4	X_5
VIF	1.597	6.723	7.367	1.789	1.341

Table 4 shows that all the predictor variables have a VIF value < 10 [10], so it could be said that there are no multicollinearity among five independent variables. Therefore, all the independent variables can be used to modeling numbers of TN cases by using NB and ZINB regression.

Table 5. Parameter estimation of NB regression

Parameter	NB regression			
	Estimate	S.E.	Z-value	p-value
$\hat{\beta}_0$	-13.628	9.477	-1.438	0.15043
$\hat{\beta}_1$	-0.008	0.009	-0.896	0.37038
$\hat{\beta}_2$	0.210	0.171	1.234	0.21736
$\hat{\beta}_3$	-0.075	0.176	-0.425	0.67058
$\hat{\beta}_4$	0.128	0.046	2.784	< 0.010*
$\hat{\beta}_5$	0.0003	0.007	0.047	0.96222
	Kappa = 1.248			
	Log-likelihood = -26.190 ; df = 32			
	AIC = 75.863			

*significant at α = 0.05

Table 6. Parameter estimation of ZINB regression

Parameter	ZINB regression			
	Estimate	S.E.	Z-value	p-value
$\hat{\beta}_0$	3.051	9.881	0.309	0.7575
$\hat{\beta}_1$	-0.028	0.016	-1.738	0.0821**
$\hat{\beta}_2$	0.265	0.095	2.800	< 0.010*
$\hat{\beta}_3$	-0.275	0.100	-2.753	< 0.010*
$\hat{\beta}_4$	0.117	0.033	3.564	< 0.001*
$\hat{\beta}_5$	0.009	0.008	1.124	0.2608
$\hat{\gamma}_0$	31.620	26.527	1.192	0.2333
$\hat{\gamma}_1$	-0.072	0.047	-1.530	0.1261
$\hat{\gamma}_2$	1.139	0.824	1.381	0.1672
$\hat{\gamma}_3$	-1.374	0.822	-1.672	0.0945**
$\hat{\gamma}_4$	-0.645	0.413	-1.560	0.1187
$\hat{\gamma}_5$	0.045	0.028	1.622	0.1048
	Log-Kappa = 15.5308			
	Log-likelihood = -24.00 ; df = 13			
	AIC = 74. 00883			

*significant at α = 0.05
**significant at α = 0.10

Table 5 and Table 6 summarize the results of NB and ZINB regression using software R 3.3.0. According to Table 5, the number of TN cases was found to be associated only with TT1 immunization of pregnant women (X_4). While it was associated with maternity mothers assisted by health personnels in health care facility (X_2), K1 visit of pregnant women (X_3), and TT1 immunization of pregnant women (X_4) according to ZINB regression on Table 6 (at significant level $\alpha = 0.05$) for NB state model. Even TT2+ immunization of pregnant women (X_1) for NB state model and K1 visit of pregnant women (X_3) for zero-inflation state model also significant at level $\alpha = 0.10$.

According to Table 5, the significance test of NB parameters estimate simultaneously ($\alpha = 0.05$) using G test statistics, G = -2 (Log-likelihood) = 61.863 and χ^2-table with df = 32 is 46.19426. Thus, it could be concluded that simultaneously all the independent variables in NB regression have a significant influence on numbers of TN cases in East Java Province in 2015. While based on Table 6 the significance test of ZINB parameters estimate simultaneously ($\alpha = 0.05$) using G test statistics, G = -2 (Log-likelihood) = 48.00 and χ^2-table with df = 13 is 22.36. Therefore, simultaneously all the independent variables in ZINB regression also have a significant influence on the number of TN case in East Java Province in 2015.

This study use AIC (Akaike Information Criterion) value to decide the best between NB and ZINB model. The smaller AIC value of a model shows the better model. The AIC value of NB and ZINB model in Table 5 and Table 6, show that the AIC value of ZINB model is smaller than the AIC value of NB model. This suggests that in this study, the ZINB model is better used to modeling numbers of TN case in East Java Province in 2015.

The Zero-Inflated Negative Binomial model can be expressed as:

- NB state model

$$\hat{\mu} = \exp\,(3.051 - 0.028X_1 + 0.265X_2 - 0.275X_3 + 0.117X_4 + 0.009X_5) \tag{11}$$

- Zero-inflation state model

$$\hat{\pi} = \frac{exp\,(31.620 - 0.072X_1 + 1.139X_2 - 1.374X_3 - 0.645X_4 + 0.045X_5)}{1 + exp\,(31.620 - 0.072X_1 + 1.139X_2 - 1.374X_3 - 0.645X_4 + 0.045X_5)} \tag{12}$$

In zero-inflation state model (12), there is no independent variables that significant at level of significance $\alpha = 0.05$. Therefore, only NB state model is to be interpreted where maternity mothers assisted by health personnels in health care facility (X_2), K1 visit of pregnant women (X_3), and TT1 immunization of pregnant women (X_4) are significant in the model.

The average of TN cases expected will be increased by 1.303 times from the previous cases for a percent increase maternity mothers assisted by health personnels in health care facility, will be decreased by 1.317 times from the previous cases for a percent increase K1 visit of pregnant women, and will be increased by 1.124 times from the previous cases for a percent increase TT1 immunization of pregnant women when other variables are constant.

5. Conclusion

According to the result of modeling numbers of TN cases in East Java Province in 2015, ZINB regression is better than NB regression. The result also indicate that maternity mothers assisted by health personnels in health care facility, K1 visit of pregnant women, and TT1 immunization of pregnant women are associated with numbers of TN cases.

Acknowledge

This work was supported by BOPTN No. 27/UN10.F09.01/PN/2017 by Faculty of Mathematics and Natural Sciences, Brawijaya University, Malang, Indonesia.

References

[1] Kementerian Kesehatan RI 2016 *Profil Kesehatan Indonesia Tahun 2015* (Jakarta: Kementerian Kesehatan RI)
[2] Hilbe, J M 2011 *Negative Binomial Regression 2nd Edition* (New York: Cambridge University Press)
[3] Amaliana, L 2017 Perbandingan ZIP dan ZINB pada Jumlah Kasus Tetanus Neonatorum di Jawa Timur Tahun 2013 *Prosiding Seminar Nasional Matematika Indo-MS Wilayah Sulawesi* vol 1 (Makassar: Universitas Hasanuddin) 428-38
[4] Astuti, C C, Mulyanto, A D 2016 Pemodelan Regresi ZINB pada Kasus Tetanus Neonatorum di Provinsi Jawa Timur *Cauchy–Jurnal Matematika Murni dan Aplikasi* **4** 115
[5] Keswari, N M R, Sumarjaya, I W, and Suciptawati, N L P 2014 Perbandingan Regresi Binomial Negatif dan Regresi Generalisasi Poisson dalam Mengatasi Overdispersi *E-Jurnal Matematika* **3** 107
[6] Agresti, A 2002 *Categorical Data Analysis 2nd Edition* (New York: John Wiley & Sons)
[7] Cameron, A C and Trivedi, P K 2013 *Regresson Analysis of Count Data 2nd Edition* (Cambridge: Cambridge University Press)
[8] Garay, A M, Hashimoto, E M, Ortega, E M M, and Lachos, V H 2011 On Estimation and Influence Diagnostics for ZINB Regression Models *Computational Statistics and Data Analysis* **55** 1304
[9] Dinas Kesehatan Provinsi Jawa Timur 2016 *Profil Kesehatan Provinsi Jawa Timur Tahun 2015* (Surabaya: Dinas Kesehatan Jawa Timur)
[10] Gujarati, D N 2004 *Basic Econometrics, 4th Edition* (New York: Mc.Graw-Hill Companies)

IOP Conf. Series: Journal of Physics: Conf. Series **943** (2017) 012052 doi:10.1088/1742-6596/943/1/012052

Visualization on triangle concept using Adobe Flash Professional SC6

Laela Sagita and Adha Ratih Kusumarini

Universitas PGRI Yogyakarta, Jl. PGRI No 117, Yogyakarta 55182, Indonesia

E-mail: laelasagita@upy.ac.id

Abstract. The purpose of this paper is to develop teaching aids using Adobe Flash Professional CS6 emphasize on Triangle concept. A new alternative way to deliver a basic concept in geometry with visualization is software Adobe Flash Professional CS 6. Research method is research and development with 5 phase of Ploom's model, namely (1) preliminary, (2) design, (3) realization/ construction, (4) test, evaluation and revision, and 5) implementation. The results showed that teaching aids was valid, practice, and effective. Validity: expert judgement for material score is 3.95 and media expert judgement produce an average score of 3,2, both in the category are valid. Practically: the average of questionnaire response is 4,04 (good). Effectiveness: n-gain test value is 0,36 (medium). It concluded that developed of teaching aids using Adobe Flash CS6 on triangle can improve student achievement.

1. Introduction

The one of development in science and technology today is computer which has been widely used in all sectors of human life. The widespread popularity of entertaining computer is in various sectors, such as not limited on health, education, and military [1]. Computers have become a living part of today's society, not only adults, but also children. The study conducted by [2] about the challenge to education to exploit computer for doing, learning, and teaching mathematics. To use the computer as an interactive learning suggestion needs hardware as a support facility [2]. One of transformation role in education through IT, one way to enhance these changes are by curriculum design may be reconsidered in such a context [3]. Using computer as a teaching aid at school needs some requirements, because teaching aid has the ability to increase student learning motivation, should be able stimulate students to remember what has been learnt, and can provide new learning stimuli for students [2]. Thus, good media should have the ability to enable students to provide response, feedback, and encourage students to practice properly. These will be easily to enable teachers and students to bring image from the outside into the mathematic classroom.

Basic skill in mathematics by National Council of Teachers Mathematics is Geometry [4]. It is taught in Indonesian primary and secondary school curriculum. Spatial ability is a main role in teaching mathematics and geometry [5]. Many various sub dimension in geometry ability, such as spatial visualization, spatial perception, spatial perception, spatial perception, spatial relation, spatial orientation, and mental rotation [6]. Abilities on visualization are the ability to represent, transform, generalize, communicate, document, and reflect on visual information [7]. It focused on their figural properties, i.e., their properties as figures perceived through the senses and interpreted by mental reflection, with spatial (sensorial) representation. By [8] clearly that visualization plays a main role in the understanding of geometry. Also [9] emphasize the essential part of geometric thought is visualization.

Research by [10] conducted the difficulties teachers in the area of geometry, the difficulties are a weak background and dislike teaching geometry. TRIAD (Technology-enhanced Research-based, Instruction, Assessment, and professional Development) model is used to change that promotes depth and quality of children's thinking, placing research-based learning trajectories at the core of the teacher/child/curriculum [10]. The difficulties that occur as a result of their spontaneous processes of visual perception in cases in which they contradict the geometric concepts/knowledge aimed at by the teacher and the tasks [10]. Students fail to accomplish a dimensional deconstruction of the figures in order to infer mathematical properties in axiomatic geometry [8]. The study by [11] conducted 3D model and the multimedia courseware aims to support the students geometric thinking by Van Hille [12] on level 1 (visualization) by providing a learning aid for this topic. The student ability to master the first level is very important in order for them to progress to the higher level of geometric understanding.

Visualization focused on their figural properties, i.e., their properties as figures perceived through the senses and interpreted by mental reflection, with spatial (sensorial) representation. The preced conducted research about geometry visualization generally used available apps such as Geogebra, Cabri 3D, and Google SketchUp, by examining the use of google Sketchup to develop spatial visualization. On the other study, it has found the effect of teaching Geometry using Cabri 3D [13]. It is determined that activities on Cabri 3D contributed significantly to the spatial ability of the teacher.

A new alternative to deliver a basic concept in geometry with visualization of 3D is software Adobe professional CS 6. Adobe Flash Professional CS6 is used to create and deliver interactive content [14]. Adobe Flash Professional CS6 is the authoring environment for creating rich, interactive content and advertisement for digital, web delivery. There are key feature areas in flash CS6 Professional (1) drawing environment. Flash features a complete set of drawing tools to handle intricate illustration and typography, (2) Animation. Flash light weight animation that incorporates images, sound, and video, and can be quickly downloaded through the Web [14]. Based on the features presented by Adobe Flash Professional CS6, the development of instructional media using Adobe Flash Professional CS6 software can produce visualization for the learning of geometry on Triangle material in 3-D.

2. Method

The method used in this research is research and development. Research and development method is a method used to produce a certain product, and test the effectiveness of the product. According to Nieveen [15], a product is said to be good/proper if it meets aspects of quality, among others validity, practicality, and effectiveness.

The development model being used refers to Ploom's development model in [9] this model consists of five phases, namely preliminary investigation phase, design phase, construction phase, evaluation and revision, and implementation phase. In the preliminary investigation phase is (1) curriculum analysis, (2) analysis of student characteristics, and (3) analysis of computer utilization. The next design phase is making the product design to be developed, media flowchart, and storyboard which the general design are covering the design of the template, the location of the menu, the navigation key, and the materials to be presented. Furthermore, the construction phase is realized in the design to become a product, so that the teaching aids obtained by the first prototype. After that, the result of this construction is reviewed by the researcher for the improvement. The next phase of the test, evaluation and revision are conducted in two main activities (1) the teaching aids validation activities, (2) limited experimental activity to use teaching aids. Furthermore, the implementation phase is evaluated and the product obtained, so the product can be implemented. In limited trials, products that have been revised and valid for use are then tested on field (classroom) learning. Flowchart' development teaching aids described on Figure 1.

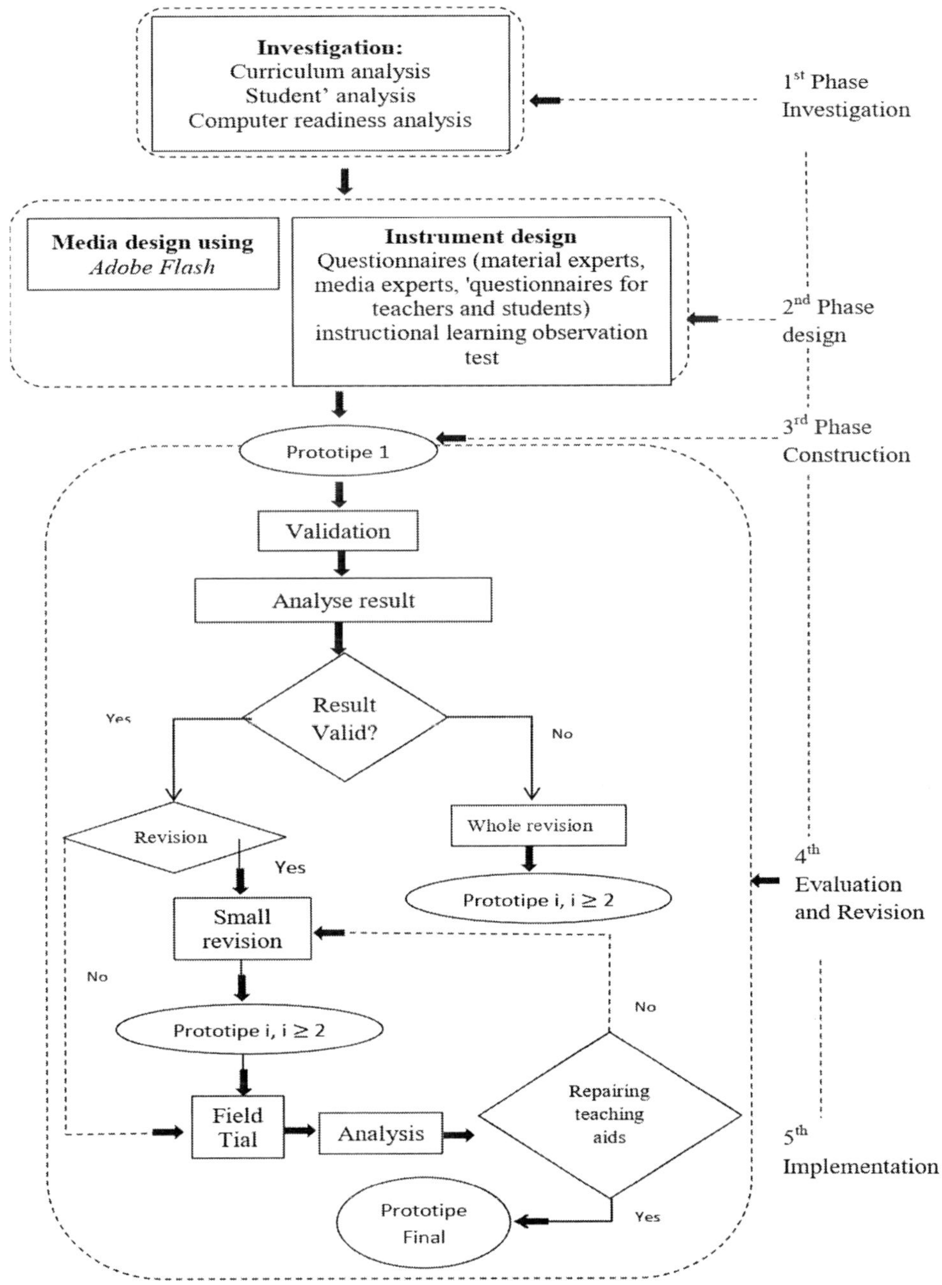

Figure 1. Flowchart of Development Teaching Aids

Data collection techniques in this development research are interviewing, observing of learning, giving questionnaire, testing and documenting. Instruments being used include questionnaires for material experts, questionnaires for media experts, respondents' questionnaires for teachers, and respondents' questionnaires for students, instructional learning observation sheets, and test. Then, the data obtained are analyzed descriptively.

The purpose of the questionnaires material experts and questionnaires media experts are to review the media design, consideration to media revisions, and media validation. The questionnaire for the material expert is used to measure the validity of teaching aids. Aspects of assessment used for the questionnaires material expert include content quality and objectives, instructional quality, and instructional design [16]. The questionnaire media expert is used to measure the validity of teaching aids that has been developed. In order to measure the validity of the media are visual communication, software engineering, content quality and objectives and instructional quality.

Respondents' questionnaires are the assessment instrument used to measure the practicality of learning media that has been developed. The purpose of this assessment is to review the accuracy of teaching aids content and visualization on teaching aids. Respondents' questionnaire is given to teacher of mathematics and students at SMPN 3 Banguntapan. Aspects used to construct the questionnaire are instructional design, content quality and purpose, visual communication, and software engineering Instructional learning observation sheet is an instrument used to measure the practicality of teaching aids in terms of easy to use. It is filled by observer teacher and student activity in class. Teaching aids validation analysis was obtained using Likert Scale 1-5, and scoring using mean ideal formula. N-gain of the test is an instrument used to measure the effectiveness of developed teaching aids where is consists of preliminary and final tests Data obtained from the process of developing teaching aids on the subject matter of triangle for students of SMP class VII in accordance with predetermined will be analyzed descriptively.

3. Result and discussion

The development of teaching aids using Adobe Flash CS 6 was conducted on mathematic subjects of triangle subject in SMP Negeri 3 Banguntapan. The process of development is conducted in stages and to produce a decent and quality teaching aids is done material expert validation, media expert validation, limited testing, field trials, learning implementation observation, and test. All processes are intended to obtain media quality data viewed from the validity, practicality, and effectiveness of the next revision or improvement in order to achieve a decent, quality and useful teaching aids for its users.

After the teaching aids process is completed, it was reviewed by an expert who will assess the validity of the teaching aids (i.e. material experts and media experts). Each expert fills out a validation questionnaire that has been compiled based on predetermined aspects. Data from media questionnaire by material experts are presented in Table 1.

Table 1. The result of material experts

Aspect	Average Value	Criteria
Quality of content and purpose	3,87	Good
Instructional	4	Good
Learning design	4	Good
Average	3,95	Good
Validity	3,95	Valid

From the Table 1, it can be seen the average value of each aspect, then from all these aspects average is 3.95. Media validity can be revealed by converting score to scale 5. According to expert judgment result of conversion of average score, it can be concluded that the media stated valid. Expert judgment stated that instructional media deserve to be tested with little revision. Furthermore, data from media questionnaire results by media experts presented in Table 2.

Table 2. The result of media experts

Aspect	Average value	Criteria
Visual communication	3,5	Very Good
Software engineering	3	Good
Quality of content and purpose	3	Good
Instructional	3,3	Good
Average	3,2	Good
Validity	3,2	Valid

From Table 2, it can be seen the average value of each aspect, then from all these aspects average is 3.2. Teaching aids validity can be determined by converting scores to scale 4. According from material expert judgment result of conversion of average score, it can be concluded that the teaching aids stated valid. Material expert judgment stated that teaching aids is worthy to be tested without revision.

Trial for teaching aids conducted after the teaching aids is said to be valid and feasible to be tested by a lecturer of material experts and lecturers media. It is intended to determine the practicality of the media developed. Trial conducted in SMP Negeri 3 Banguntapan on 10 students of 7th Grade class G. Table 3 is the result of the questionnaire of the student's response to a limited trial.

Table 3. A limited student response questionnaire

Aspect	Average value	Criteria
Learning design	3,9	Good
Quality of content and purpose	4,5	Very Good
Instructional quality	3,9	Good
Visual communication	4	Good
Software engineering	3,9	Good
Average	4,04	Good

The result from Table 3 can be seen the average value of each aspect, then from all aspects in the average is 4.04. The quality of the media can be determined by converting that average score with the conversion guide to scale 5. From the results of the average conversion of the score can be concluded that according to a limited scale trial the media has a good. After the product has been revised and valid for use, then it is tested on the learning in the class. The trial was conducted in class 7th Grade class F with the number of students that is 26 students and some teachers. This trial class is conducted to see the effectiveness of learning kits developed by using questionnaire responses of students, teacher response questionnaires and student achievement test. Questionnaire response student field trials aims to determine whether the teaching aids developed can be used properly. Table 4 is the results of the field questionnaire response questionnaire.

Table 4. A field response questionnaire

Aspect	Average value	Criteria
Learning design	4,1	Good
Quality of content and purpose	4,4	Very Good
Instructional quality	4,2	Good
Visual communication	4,2	Good
Software engineering	4,1	Good
Average	4,2	Good

The results of the average conversion of the score in Table 4 can be concluded that according to the results of questionnaire responses of field trials, teaching aids has a good. Questionnaire response teacher field trials aims to determine whether the teaching aids developed can be used properly. Table 5 is the results of the field teacher response questionnaire.

Table 5. Master field response questionnaire

Aspect	Average value	Criteria
Learning design	4,3	Very Good
Quality of content and purpose	4	Good
Visual communication	4	Good
Software engineering	4,5	Very Good
Average	4,2	Good
Validity	4,2	Good

From the quality of the teaching aids can be known by converting the average score with the conversion guidelines to scale 5. From the results of the average conversion of the score can be concluded that according to the results of questionnaire responses of field trials students, teaching aids have good and declared valid without revision.

Test consists of two kinds, namely the initial test and the final test. The test is conducted before and after the use of teaching aids. The test was given to 24 students which aimed at knowing the improvement of student learning achievement and assess the effectiveness of teaching aids developed. Table 6 is the result of preliminary and final test.

Table 6. Preliminary and final test results

	Pre-test	**Post-test**	**N-Gain test**	**Criteria**
Mean	54,8	71,25	0,36	Medium

The results of the initial test score analysis obtained an average value of 54.8 and the final test obtained an average score of 71.28. The n-gain test scores obtained an average of 0.36 with moderate (medium). This indicates that the initial test to the final test has increased.

Teaching aids is valid which states that teaching aids can be used with little revision or no revision. The results showed that the expert analysis of the material obtained an average score of 3.95 so that the validity of the media stated valid. Media experts obtain an average score of 3.2 so that the validity of the media is declared valid. From the analysis of material experts and media experts can be concluded that the teaching aids is valid and can be used. In the aspect of practicality, the validator states that the media can be used in learning with a little revision. Questionnaire response of the limited trial students obtained an average score of 4.04 and expressed by of good. It shows that the developed teaching aids can be used well. In the aspect of effectiveness, student test results have increased. Judging from the results of the initial test score analysis obtained an average value of 54.8 and the final test obtained an average score of 71.28. The N-gain test scores obtained an average of 0.36 with moderate (medium). Questionnaire of field trial student responses earned an average of 4.2 otherwise with good. Furthermore, the questionnaire responses of subject teachers earned an average of 4.2 otherwise with good. It can be concluded that the developed teaching aids can be used well and declared effective.

Based on the criteria of validity, practicality, and effectiveness that have been explained before, this research has generated a new product teaching aids for teaching geometry by using software Adobe Flash CS6. It is different with research conducted by [6, 10, 13, 17] that was used available teaching aids such as Geogebra, Cabri 3D, Google Sketchup. Teaching aids generated by this research has many strengths such as: (1) teaching aids has packed on Interactive CD that can be used on a computer without installing, (2) the content consists of materials and also questions based on indicator, (3) The material are presented interestingly with animation, narration, and background music to decrease the students' boredom, (4) the student can operate it everywhere. It is in accordance with statement regarding revolution that should be conducted by teachers in combination learning method with information technology to be more interactive, efficient and pleasant. It is stated that software used in development the animation in teaching aids is Macromedia Flash, a software that can create graphic media, this facilitating their transmission to any user connected to the network [18-20].

4. Conclusion

To develop teaching aids in Triangle subject with Adobe Media Flash CS 6 can use Ploom's model, which consists of five phases, namely preliminary investigation phase, design phase, construction phase, evaluation and revision, and implementation phase. Teaching aids to be good is have minimum three aspects of quality (1) validity; both of expert judgment said that teaching aids valid, (2) practicality; student response and field response said that teaching aids on Criteria Good, it should be practice, and (3) effectiveness: final test result on medium Criteria, it should be teaching aids effective for student's ability. This develop teaching aids can be supporting teachers in class. But it could not be primary aid to deliver triangle concept to student.

References

[1] Kebritchi M, Atsusi H and Haiyan B 2010 The effects of modern mathematics computer games on mathematics achievement and class motivation *Computers & education* **55** 427

[2] Ahmed A, Clark-Jeavons A and Oldknow A 2004 How to teaching aids improve the quality of mathematics education *Journal of Educational Studies In Mathematics* **56** 313

[3] Fluck A 2010 *Researching IT in Education Theory, Practice and Future Directions* (London and New York: Routledge)

[4] National Council of Teachers of Mathematics 2000 *Curriculum and Evaluation Standars for Schools Mathematics*

[5] Kurtulus A and Candas U 2010 The effects of Google Sketchup based geometry activities and projects on spatial visualization ability of student mathematics teachers *Procedia-Social and Behavioral Sciences* 9 pp 384-389

[6] Contero M, Ferran N, Pedro C, Jose L S and Julian C 2005 Improving visualization skills in engineering education *IEEE Computer Graphics and Applications* **25** 24

[7] Hershkowitz R, Ben Haim D, Holes C, Lappan G, Mitchelmore M, and Vinner S 1990 *International Group for the Psychology of Mathematics Education* p 70

[8] Gal H and Linchevski L 2010 To see or not see: analyzing difficulties in geometry from the perspective of visual perception *Journal Educ Stud Math* **74** 163

[9] Boakes N J 2009 Origami instruction in the middle school mathematics classroom: its impact on spatial visualization and geometry knowledge of students *RMLE Online* **32** 1

[10] Clements D H and Sarama J 2011 Early childhood teacher education: the case of geometry *Journa Math Teacher Educ* **14** 133

[11] Noordin S, Wan Fatimah W A and Hooi Y K 2011 Study of effectivenes and usability of multimedia coursware integrated with 3-dimensional model as a teaching aid *International Journal of Computer and Application Computer&Educatio* **16** 0975

[12] Erdogan H 2006 Sex-related differences in the acquisition of the van hiele levels and motivation in learning geometry *J. Asia Pacific Education Review* **7** 173

[13] Kosa T and Karakus F 2010 Using dynamic geometry software Cabri 3D for teaching analytic geometry *Procidia Procedia Social and Behavioral Science* *2* 1385

[14] Gerantabee F and Team A C 2012 *Adobe Flash Professional CS6 Digital Classroom* (Indianapolis: John Willey&Sons)

[15] Nieveen, N 2013 *Improving curriculum developers' formative evaluation through an electronic performance support system* (the Netherlands: SLO)

[16] Prahmana R C I, Kusumah Y S, and Darhim 2017 Didactic trajectory of research in mathematics education using research-based learning *J. Phys.: Conf. Ser.* **893** 012001

[17] Ismail Z and Rahman S N A 2017 Learning 2-dimensional and 3-dimensional geometry with Geogebra: Which would students do better? *International Journal on Emerging Mathematics Education* **1** 121

[18] Garcia R R, Quiroz J S, Santos R G, Gonzalez S M and Fernanz S M 2007 Interactive multimedia animation with Macromedia Flash in descriptive geometry teaching *Journal Computer and Education* **49** 615

[19] Fathurrohman M, Porter A L and Worthy A L 2017 Teachers' real and perceived of ICT supported-situation for mathematics teaching and learning *International Journal on Emerging Mathematics Education* **1** 11

[20] Young J R 2017 Technology integration in mathematics education: Examining the quality of meta-analytic research *International Journal on Emerging Mathematics Education* **1** 71

IOP Conf. Series: Journal of Physics: Conf. Series **943** (2017) 012053 doi:10.1088/1742-6596/943/1/012053

The instruments of higher order thinking skills

S Ahmad[1], R C I Prahmana[2], A K Kenedi[1], Y Helsa[1], Y Arianil[1], M Zainil[1]

[1]Universitas Negeri Padang, Jl. Prof. Dr. Hamka Air Tawar, Padang, Indonesia
[2]Universitas Ahmad Dahlan, Jl. Pramuka Kav. 5, Pandeyan, Yogyakarta, Indonesia

E-mail: syafriahmad95@yahoo.co.id

Abstract. This research developed the standard of instrument for measuring the High Order Thinking Skill (HOTS) ability of PGSD students. The research method used is development research with eight steps namely theoretical studies, operational definition, designation construct, dimensions and indicators, the preparation of the lattice, the preparation of grain, an analysis of legibility and Social desirability, field trials, and data analysis. In accordance with the type of data to be obtained in this study, the research instrument using validation sheet, implementation observation, and questionnaire. The results show that the instruments are valid and feasible to be used by expert and have been tested on PGSD students with 60% of PGSD students with low categorization.

1. Introduction

Mathematics learning is one of the main subjects that need special attention in Indonesia [1]. It is because mathematics is a basic knowledge needed by learners to support the success of their learning process in pursuing higher education [2-3]. Mathematics is one of the tools or tools in scientific thinking necessary to develop critical, logical and, and systematic thinking skills in learners [4]. It aims to learners can analyze a problem, take the right conclusions and use systematic thinking.

NCTM (National Council of Teachers of Mathematics) recommended four principles of mathematics, namely mathematics as problem solving, mathematics as reasoning, mathematics as communication, and mathematics as a relationship [5]. It can be concluded that the learning of mathematics should be given to the learners so that they can solve a problem, think logically, analytically, systematically, critically, creatively, able to communicate correctly (communicative) and can cooperate as well.

The improvement of mathematics learning achievement needs to get serious attention in every education level. Mathematics is a subject taught in every level and type of education, according to the level of need of each level and type of education [6-8]. However, at this time there are still many found mathematics achievement in schools at the levels of elementary, middle, high school / vocational school and even university level is still low, not least in for pre-service primary school teacher (PGSD).

Mastery of mathematics by PGSD students becomes an obligation that must be implemented in the management of reason and decision making in the era of increasingly fierce and competitive competition. But in fact, the math still considered subject is unpopular or even most hated by most students [9]. It becomes the duty of the lecturer as a teacher to correct the assumption to change into a positive assumption.

Negative side on mathematics learning also occurs in the course of high grade Mathematics Learning is also experienced by PGSD students. Mathematics is considered a difficult and scary course. The assumption has an impact on the results of the lectures are always less satisfactory, even though they are prospective teachers in primary schools who will teach the concept of mathematics in elementary. This fact if left unabated then PGSD graduates will be difficult to compete with graduates from universities or even other countries. It is more worrying that one day our graduates become spectators in their own country because of the inability of our graduates to be competitors in their own country.

Based on the observations data conducted by researchers in the department of PGSD, the assessment of learning achievement is conducted by lecturers only based on the test scores. Assessment of student achievement that is only based on the value of the test causes the lecturer lacks precisely understand what has been achieved and what is needed by students in learning mathematics. It is because the mastery of the mathematics material measured in the test generally does not measure the thinking skills of the students.

Several researchers already documented their research about students' mathematics thinking ability [10-14]. Based on their studies, we can conclude that the thinking ability of high-level PGSD students is still low. Human thinking skills can be classified into two, namely the low level thinking skills / lower order thinking skills (LOTS) and high-level thinking skills or higher order thinking skills (HOTS). Additionally Higher order thinking skills include critical, logical, reflective, metacognitive, and creative thinking. They are activated when individuals encounter unfamiliar problems, uncertainties, questions, or dilemmas [15].

Higher order thinking skills is an important aspect in teaching and learning especially at higher education institutions. Thinking skills practices are part of the generic skills that should be infused in all technical subjects. Students with higher order thinking skills are able to learn improve their performance and reduce their weaknesses [16]. Therefore student's HOTS ability who studies mathematics needs to know, so it can be arranged qualified mathematics learning. Newman and Wehlage [17] states that HOT requires students to manipulate information and ideas that transform their meaning and implications, such as when students combine facts and ideas in order to synthesize, generalize, explain, hypnotize, or arrive at some conclusion or interpretation.

HOTS important role in learning Mathematics is also shown by research conducted by Rooney [18] which uses inquiry learning model-based learning to improve student`s HOTS ability. Furthermore it is said that in the action research done, the researchers measure the students' HOTS ability based on the mathematics assigned by the students. HOTS that aims to make students solve problems is a matter that is needed by students, so that when students graduate from a level of education then students are ready in facing greater challenges [19]. HOTS also stressed that teaching students how to think is a very critical effort to be taken seriously in preparing the students to become a Better future workers and problem solver [20].

HOTS consist of two components namely critical and creative thinking skills [16]. When students apply both skills it means students apply HOTS. The same was stated by Rosnawati [21] who stated that the ability of critical thinking and creative thinking are two important components of HOTS. Furthermore, in the planning of University North Carolina State in order to develop students' HOTS ability, it also uses only two variables, namely critical thinking skills and creative thinking skills. Others argue that HOTS is sufficiently measured using critical thinking skills [22].

Based on several studies that have been done, it appears that there is no standard instrument developed specifically to measure of HOTS ability of PGSD students on learning Mathematics Education. The instruments are intended specifically for measuring of HOTS ability of PGSD students. The results of these measurements can be used to conduct a comprehensive evaluation of PGSD students' thinking skills in high school mathematics learning.

As a variable, HOTS is a latent variable that cannot be measured directly as a physical variable. In order to measure the characteristics of latent variables, it takes a manifest variable commensurate with the latent variable requiring a standard instrument [23]. The main problem is how lecturers can have raw instances about HOTS and able to perform measurements properly and correctly.

On the other hand, the limited knowledge and time of the lecturers of mathematics in developing valid and reliable instruments for measuring HOTS become the obstacles less attention to the achievement of the basic goals of mathematics lessons. Therefore it is necessary to develop HOTS instrument in High class Mathematics learning in the Department of PGSD.

2. Method

This research was carried out in the Department of PGSD UPP I FIP UNP, i.e., even semester students amounted to 30 people enrolled in the learning year 2017/2018. This research is development research using eight steps namely development of theoretical studies, development of operational definition, designation construct, dimensions and indicators, the preparation of the lattice, the preparation of grain, an analysis of legibility and Social desirability, field trials, as well as data analysis. In accordance with the type of data to be obtained in this study, the research instrument using validation sheet, implementation observation, and questionnaire.

3. Result and discussion

The results of the higher order thinking skills instrument development in learning high class mathematics in the PGSD majors are described in the following steps:

1. Theoretical Review.

Theoretical study in this study aims to develop the conceptual definition of HOTS instrument compiler variables. This theoretical assessment get the sense that HOTS is a broader thinking skills, not just to remember, understand, and the ability to apply a concept but also the ability to think for analyzing a concept, evaluate and even to create a concept.

2. Operational Definition Development.

The next step is to develop operational definitions of HOTS instrument compiler variables. The development and determination of operational definitions of HOTS is based on the conceptual definition that has been obtained that is the ability to analyze a concept, evaluate and even create a concept. After the researchers collide from various opinions of experts then for the operational definition of HOTS is composed of critical thinking skills and creative thinking skills which will then be developed into indicators.

3. Determination of Constructs, Dimensions, and Indicators.

The determination of collisions, dimensions and indicators is based on the withdrawal of operational definitions. The operational definition of HOTS is contracted into critical thinking ability and creative thinking ability.

The dimensions of critical thinking skills are all things that can make a person think critically namely the use of the concept, the use of the principle, the prediction of the impact, problem solving, and decision making. Therefore, after the researchers analyzed the dimensions of critical thinking, the indicator of critical thinking is to have the basic skills, provide an explanation, concluded the assumptions required, questioned the concept, analyze concepts, to synthesize the relationships between concepts, determine the result of the assumption, using the concept is good and balanced , find the source of the problem, guess the cause of the problem, gather the information in solving the problem, find some alternative solutions, choose the best alternative / solution, and evaluate the decision that has been taken.

The dimension of the ability to think creatively is anything that can make people think creatively. After the analysis of the researchers as for the dimensions of creative thinking is to work on the limits of competence, try new things, divergent mindset (spread) and imaginative mindset. for indicators of creative thinking is to reject the standard technique, optimizing knowledge, high motivation and contains, broad interests, orientation fore/optimistic, like the challenge/new ideas, to think freely, not rigid, develop concepts, modifying concepts, approaches trial and error, have original ideas, and have new ideas.

4. Grid Arrangement

Preparation of the lattice is held after the establishment of the constructs, dimensions and indicators. The HOTS instrument grille can be seen in the following Table 1.

Table 1. The HOTS instrument grille

Dimensions	Indicator	Number
Concept Utilization	1. Have basic skills	1,2
	2. Provide an explanation	3.4
	3. Summing up the necessary assumptions	5
Principle Utilization	1. Questioning the concept	6,7
	2. Analyze the concept	8,9
	3. Synthesizing relationships between concepts	10,11

Predicted Impact	1. Determine the consequences of the assumption	12,13
	2. Assumptions	14,15
	3. Using the concept well and balanced	16,17
Solution to problem	1. Find the source of the problem	18, 19
	2. Suspect the cause of the problem	20, 21
	3. Gathering information in solving problems	22, 23
Decision-making	1. Find some alternative solutions	24, 25
	2. Choosing the best alternative / solution	26, 27
	3. Evaluate the decisions that have been taken	28, 29
Working limited competence	1. Refuse standard techniques	30, 31
	2. Optimizing knowledge	32, 33
	3. High motivation and contents	34, 35
Try new things	1. Broad interest	36, 37
	2. Future orientation / optimism	38,
	3. Love new challenges / ideas	39, 40
Divergent mindset (spread)	1. Think freely, not rigidly	41,
	2. Develop a concept	42, 43
	3. Modify the concept	44,
Imaginative Mindset	1. Approaching trial and error	45, 46
	2. Have original ideas	47, 48
	3. Have new ideas	49, 50
	AMOUNT	

5. Preparation of Items

The preparation of the instrument item is based on the planned grid. The examples of HOTS instruments that have been made are:

a. Examples of critical Thinking with the dimensions of concept utilization with indicators summarizing assumptions required.
   "If the circumference of the equilateral triangle wakes up, circle, and square are equal. Determine the largest flat wake. Explain your reasons!"
   The reason for this represent the critical thinking with the dimension of concept utilization with the indicator conclude the assumption needed is in this matter the students are asked to think critically by utilizing the concept of geometry wake flat then from the concept student can conclude assumption of a fact.
b. Examples of creative thinking with the dimension of trying new things with indicators like challenges with new ideas. Andi gives a problem to Lusi. As for the case as follows:
   1 + 4 = 4
   2 + 5 = 12
   3 + 6 = 21
   8 + 11 =?
   Help the search for answers above using mathematics.
   The reason for this represents the problem of creative thinking with the dimension of trying new things with indicators like the challenge with new ideas is that students are required to be able to think creatively about new things with the problem students will feel energized to find new ideas and think creatively.

6. An analysis of legibility And Social desirability.

Analysis of legibility is done by a pre-trial in the form of qualitative evaluation conducted by expert judgment. Expert assessment is conducted to test whether the items that have been compiled meet the rules of the Indonesian language is good and true, and whether the items already reflect the measured indicators or dimensions or not. It also carried out an examination of whether the items are arranged still have a Social desirability. The implementation of the validation is done several times by the expert. In the execution there are several HOTS instruments commented explain in Table 2.

Table 2. Advice from experts for prototyping and decision 1st revision

Comment	Revision Decision
1. The language used is good and correct some letters are missing	1. Inspect and repair instruments is ketinggaln letters.
2. There are some questions that do not reflect HOTS such as numbers: 1,2,3,4,6,7,12,13,14,15,16,17,18,21,22,23, 26,29,30,34,35,36,38,41,42,44,45,46,48,49,50.	2. Fixed issues that did not comply with HOTS
3. The question is not biased from the indicator set.	

Examples of problems that do not conform to the HOTS Indicators are as follows:

a. "A rectangular photo placed on a cardboard measuring 45 cm x 30 cm. To the right, top and bottom of the photo there are 7 cm of cardboard remaining. If the photo and carton are real, specify the length of the remaining cardboard left."
The above problem represents the ability to think critically, the dimension of the use of concepts, indicators have basic skills said not HOTS by expert experts because the problem can be solved by using only the basic formula, not the basic concept that uses basic skills.
Therefore the problem is revised to be:
"Find all nonnegative integer pairs (a, b, x, y) that satisfy the system of equations

$$\begin{cases} a + b = xy \\ x + y = ab \end{cases}$$

About the Issuer representing the ability to think critically, the dimensions of the use of concepts, indicators have basic skills HOTS by expert said the matter because it can solve by utilizing the concepts and basic skills but students must be able to think critically to find a relationship between mathematical systems."

b. A quadratic number when divided by 3, then the likelihood of the rest is ...
About the Issuer represents creative thinking ability, competence limited to dimensional works with high motivation indicators and contains said not HOTS because of their working process is not reflected by the high motivation.
Therefore the problem is revised to be:
"Andy wants to make a right triangle. The number of right triangles having vertical edges p, q and the oblique side q + 1, where p, q is an integer and b <101, is ...."
About the Issuer represents creative thinking ability, competence limited to dimensional works with high motivation indicators and contains said HOTS by expert because of their working process students are required to be able to think creatively and students should require high motivation because in this matter required a high willingness.

The instrument would be ready as an instrument to measure HOTS after three revisions by expert and all instrument declared valid and worthy to be used.

7. Field Trials

Field trials conducted as much as 2 times in the class 15 AT 1 Department PGSD FIP UNP. Data collections by giving HOTS instruments that have been validated by expert experts. The first meeting lasted for 90 minutes with a total of 9 questions. And the second meeting in the next week with time for 90 minutes with the number of questions 9 pieces. The student answers the question on the answer sheet provided and the lecturer declares the time to begin the process of completion of the answer and is collected after the time stated otherwise. As for example student's answer at first meeting can be seen in Figure 1.

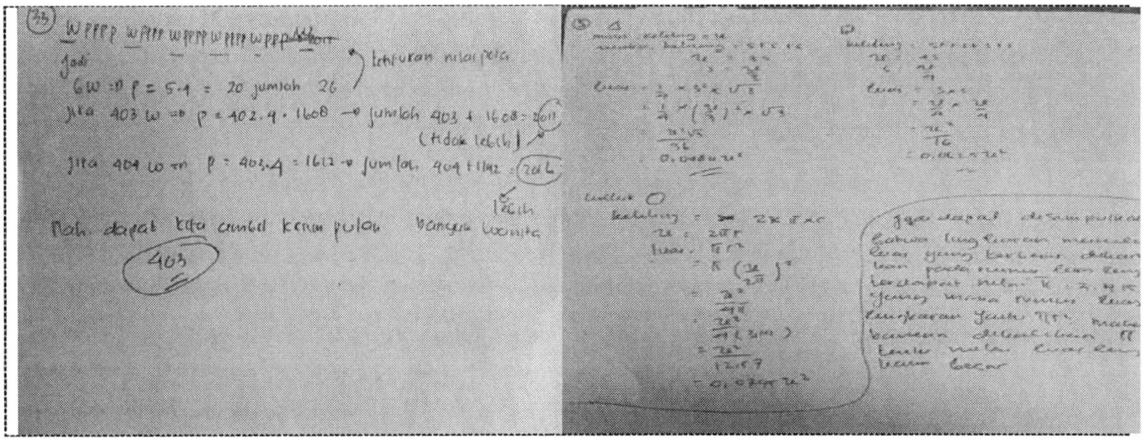

Figure 1. Sample student answers on the matter of critical thinking 1st meeting

In addition, sample student answers at the second meeting can be explained in Figure 2.

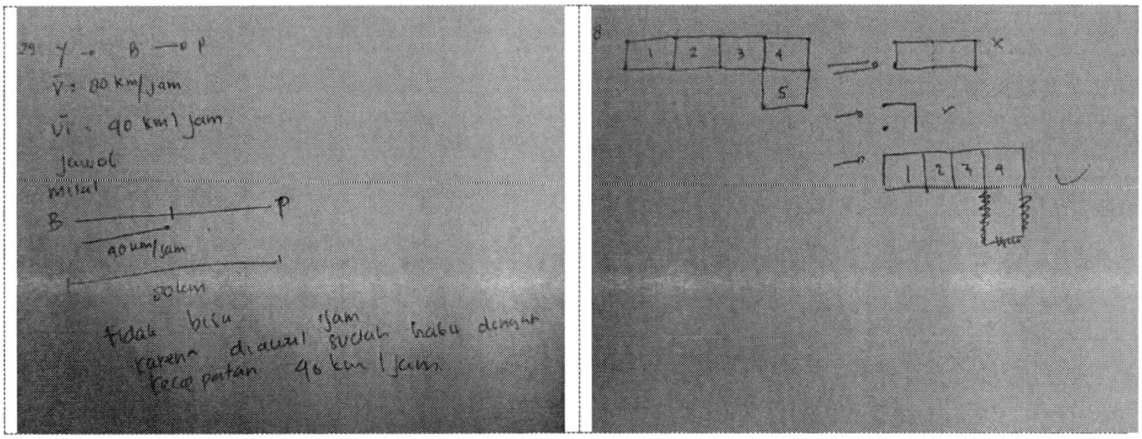

Figure 2. Sample student answers on the matter of critical thinking 2nd meeting

8. Data analysis
Data analysis was performed twice in accordance with the trial. The first analysis of the results of answers from students is there are some problems that students have difficulty that is a matter of number 6. The item is, the operation * for the set of numbers S = {0, 1, 2, 3, 4, 5, 6} is defined according to the Table 3.

Table 3. The operation of *

*	0	1	2	3	4	5	6
0	0	0	0	0	0	0	0
1	0	1	2	3	4	5	6
2	0	2	4	6	1	3	5
3	0	3	6	2	5	1	4
4	0	4	1	5	2	6	3
5	0	5	3	1	6	4	2
6	0	6	5	4	3	2	1

If for every integer n which is greater than 1 defined x n = x n -1 * x, then 5 2015 = ...

Of the 30 students, none of the students were able to answer this question. This proves that the problem requires higher thinking for its accomplishment. In the solve process, students are required to think critically, dementia work indefinitely with high motivation indicators and contains. This fact also proves that PGSD students are still not maximized in work and lack of high motivation.

HOTS measurement using three sections, namely, understanding the question, answer procedure, and the truth of the answers. Each type has a value of 0-4 per item. For the matter of Assessment of Work Results Time value received by students are grouped into five categories of HOTS can be seen in Table 4.

Table 4. HOTS Criteria

No	HOTS value	Category
1	≥ 76	Very high
2	≤ 61 HOTS <76	High
3	$46 \leq$ HOTS <61	Medium
4	HOTS ≤ 31 <46	Low
5	<31	Very low

Students are said to be successful in mathematics if 50% of the value obtained HOTS students included in the category of very high or at least reach the high category [1]. The results of the first meeting is only 1 person (3.33%) who is moderate category, 11 people (36.67%) are low and 18 people (60%) are very low. This proves that PGSD UNP students still have low HOTS level. Furthermore, for the second meeting PGSD students have been able to solve about HOTS. The scores obtained at the second meeting were 5 people (16.67%), 18 people (60%) low category, and 7 people (23.34%) were very low category.

The researcher's analysis of this result is the low level of HOTS students of PGSD FIP UNP. This is caused by PGSD students are students who come from high school majoring in science, social studies, and vocational school. This fact cannot be denied because the department of PGSD FIP UNP does not restrict prospective students who enroll in PGSD. So the basic mathematics of each individual is different. In addition, the development of HOTS instruments can be utilized and meaningful for PGSD majors to measure HOTS level of PGSD students.

4. Conclusion

The developed instrument is categorized as valid and unbiased by expert. This validity is illustrated by expert test results by expert experts in terms of construction, content, language and practically seen during field trials. In addition, the instrument can prove that the ability of HOTS PGSD UNP students is still low category based on the development process.

Acknowledgment

Thanks to PGSD UNP department which has provided an opportunity for researchers to conduct research. Thanks to the researchers. Thanks to the students of 15 AT 01 and all those who have assisted this research.

Reference

[1] Tanujaya B, Prahmana R C I, and Mumu J 2017 Mathematics instruction, problems, challenges, and opportunities: A case study in Manokwari regency, Indonesia *World Transactions on Engineering and Technology Education* **15** 287

[2] Maharani H R, Sukestiyarno and Waluya B 2017 Creative thinking process based on Wallas model in solving mathematics problem *International Journal on Emerging Mathematics Education* **1** 177

[3] Prahmana R C I, Kusumah Y S and Darhim 2017 Didactic trajectory of research in mathematics education using research-based learning *J. Phys.: Conf. Ser.* **893** 012001
[4] Nurhasanah F, Kusumah Y S and Sabandar J 2017 Concept of triangle: Examples of mathematical abstraction in two different contexts *International Journal on Emerging Mathematics Education* **1** 53
[5] National Council of Teachers Mathematics 2000 *Principles and standards for school mathematics* (Reston: National Council of Teachers of Mathematics) p 304
[6] Kamarullah 2007 Pendidikan matematika di sekolah kita *Al Khawarizmi: Jurnal Pendidikan dan Pembelajaran Matematika* **1** 21
[7] Runisah, Herman T and Dahlan J A 2017 Using the 5E learning cycle with metacognitive technique to enhance students'mathematical critical thinking skills *International Journal on Emerging Mathematics Education* **1** 87
[8] Wahyu K, Amin S M and Lukito A 2017 Motivation cards to support students' understanding on fraction divisions *International Journal on Emerging Mathematics Education* **1** 99
[9] Ruseffendi E T 1990 *Pengantar kepada membantu guru mengembangkan kompetensinya dalam pengajaran matematika untuk meningkatkan CBSA* (Bandung: Tarsito)
[10] Syamsuri, Purwanto, Subanji, and Irawati S 2017 Using APOS theory framework: Why did students unable to construct a formal proof? *International Journal on Emerging Mathematics Education* **1** 135
[11] Putra H D, Herman T and Sumarmo U 2017 Development of student worksheets to improve the ability of mathematical problem posing *International Journal on Emerging Mathematics Education* **1** 1
[12] Widodo S A, Purnami A S, and Prahmana R C I 2017 Team accelerated instruction, initials and problem-solves ability in junior high school *International Journal on Emerging Mathematics Education* **1** 193
[13] Kusdinar U, Sukestiyarno, Isnarto, and Istiandaru A 2017 Krulik and Rudnik model heuristic strategy in mathematics problem solving *International Journal on Emerging Mathematics Education* **1** 205
[14] Sundayana R, Herman T, Dahlan J A and Prahmana R C I 2017 Using ASSURE learning design to develop students' mathematical communication ability *World Transactions on Engineering and Technology Education* **15** 245
[15] King F J, Goodson L, Rohani F 1997 Higher order thinking skills: Definition, teaching strategies, and assessment (Tallahasee: Center for Advancement of Learning and Assessment Florida State University)
[16] Heong Y M 2011 The level of marzano higher order thinking skills among technical education students *International Journal of Social Science and Humanity* **1** 121
[17] Newman F M and Wehlage G G 1993 Cinco estándares para una auténtica enseñanza *Guía educacional. Recuperado el* **16** 2015
[18] Rooney C 2012 How am i using inqury-based learning to improve my recites and to encourage higher order thinking among my students of mathematics? *Educational Journal of Living Theories* **5** 99
[19] Lailly N R and Asih W W 2015 Analisis soal tipe Higher Order Thinking Skill (HOTS) dalam soal UN Kimia SMA rayon B Tahun 2012/2013 *Jurnal Kaunia* **11** 38
[20] Kerka S 1992 *Higher Order Thinking Skills in Vocational Education* (ERIC Clearinghouse)
[21] Rosnawati R 2009 Enam tahapan aktivitas dalam pembelajaran matematika untuk mendayagunakan berpikir tingkat tinggi siswa In *Seminar Nasional Penelitian, Pendidikan, dan Penerapan MIPA 2009* (Yogyakarta: Universitas Negeri Yogyakarta)
[22] North Carolina State University 2014 *Higher Order Skill in critical and creative thinking, quality enhancement plan* (Carolina: NCSU)
[23] Naga D S 2012 *Teori skor pada pengukuran mental* (Jakarta: Nagarani Citrayasa)

IOP Conf. Series: Journal of Physics: Conf. Series **943** (2017) 012054 doi:10.1088/1742-6596/943/1/012054

The models for financial crisis detection in Indonesia based on import, export, and foreign exchange reserves

Sugiyanto[1], Supriyadi Wibowo[2] and Vivi Rizky Aristina Suwardi[2]

[1]Departement of Statistics, Sebelas Maret University, Solo, Indonesia
[2]Departement of Mathematics, Sebelas Maret University, Solo, Indonesia

E-mail: sugiy50@yahoo.co.id

Abstract. The severity of the financial crisis that occurred in Indonesia required an early warning system of financial crisis. The financial crisis in Indonesia can be detected based on imports, exports, and foreign exchange reserves. The purpose of the research is to determine an appropriate model to detect the financial crisis in Indonesia based on imports, exports, and foreign exchange reserves. Markov switching is an alternative framework for the approach often used in financial crisis detection. Combined volatility and Markov switching model with three states assumptions can be established if an AR and volatility models have been obtained. Imports, exports, and foreign exchange reserves data from January 1990 to December 2016 have the heteroscedasticity effect so that an ARCH model is used as a volatility model. Research shows that SWARCH(3.1) model is an appropriate model for detecting financial crisis in Indonesia based on imports, exports, and foreign exchange reserves.

1. Introduction

Economic crisis ever hit the countries in Asia, one of Indonesia in 1997-1998. The fall of Thai baht exchange rate in the middle of July 1997 was the beginning of the financial crisis in several countries in Asia, one of them is Indonesia. The IMF stated that a device for early detection systems is needed to detect crisis as early as possible. The early detection system is an order of delivery of predicted information to a threat community before the occurrence of an event that can cause a recipe so as to warn the recipient of information can act immediately according to the conditions and situations.

SWARCH model or Markov switching ARCH model is combined volatility and Markov switching model with ARCH model is a volatility model. Kaminsky et al. [4] states that Markov switching is an alternative framework for the approach often used in early detection systems for economic crisis. There are 15 indicators of crisis including foreign exchange reserves, exports, and imports. According to Pradini and Raharjo [6] exports and imports have an important role in the economy of the country because it can affect the amount of foreign exchange. The detection of the financial crisis in Indonesia has been developed by Hermawan [2] based on the indicator of the amount of import value. From the result of these research the appropriate model to detect the financial crisis in Indonesia based on the import value indicator of SWARCH (2.1) and SWARCH (3.1). This study determines the appropriate model for detecting financial crises in Indonesia based on indicators of import, export, and foreign exchange reserve.

2. Experimental details

2.1. Autoregressive Model
Autoregressive model or AR(p) can be written as

$$r_t = \phi_0 + \phi_1 r_{t-1} + \phi_2 r_{t-2} + \cdots + \phi_p r_{t-p} + a_t, \quad (1)$$

with r_t is a transformation data, ϕ_0 is a constant, and ϕ_p are parameter of the autoregressive model (Tsay [7]).

2.2 Volatility Model
Residue of AR model containing heteroscedasticity effect can be modelled using a volatility model.

2.2.1 ARCH Model. Model ARCH(m) can be written as

$$\sigma_t^2 = \alpha_0 + \sum_{i=1}^{m} \alpha_i a_{t-i}^2, \quad (2)$$

where σ_t^2 is the residual variance to the period-t, α_0 is a constant, α_i are parameter of the ARCH model, and m is an order of ARCH (Tsay [7]).
2.2.2 GARCH Model. High order of ARCH model can be overcome using a GARCH(m,s) which can be written as

$$\sigma_t^2 = \alpha_0 + \sum_{i=1}^{m} \alpha_i a_{t-i}^2 + \sum_{j=1}^{s} \beta_j \sigma_{t-j}^2 \quad (3)$$

where β_j are a parameters of GARCH model (Tsay [7]).
2.2.3 EGARCH Model. Leverage problem on GARCH effect can be overcome by model EGARCH(m,s) which can be written as

$$\ln \sigma_t^2 = \alpha_0 + \sum_{i=1}^{m} \alpha_i \left(\left| \frac{a_{t-i}}{\sqrt{\sigma_{t-i}^2}} \right| - \sqrt{\frac{2}{\pi}} \right) + \sum_{i=1}^{s} \beta_i \frac{a_{t-i}}{\sqrt{\sigma_{t-i}^2}} + \sum_{j=1}^{m} \gamma_j \ln \sigma_{t-j}^2, \quad (4)$$

where γ_j are parameter Leverage effect (Tsay [7]).

2.3 SWARCH Model. According to Hamilton and Susmel [1], Markov switching model for the average conditional can be written as

$$r_t = \mu_{s_t} + \tilde{r}_t, \quad (5)$$

where r_t is the log return at time t, $\tilde{r}_t$ is following the process of ARMA (p, q) with an average of zero, and μ_{s_t} is the average of state at time t. While SWARCH model can be written as

$$r_t = \mu_{st} + a_t, a_t = \sigma_{t,st} \epsilon_t \quad (7)$$

$$\sigma_{t,s_t}^2 = \alpha_{0,s_t} + \sum_{i=1}^{m} \alpha_{i,s_t} a_{t-i}^2, \quad (8)$$

where μ_{st} is the conditional averagee on a state and σ_{t,s_t}^2 is a variance of residue in a state at time t.

2.4 Smoothed Probability
According to Kuan [5], smoothed probability value is defined as

$$P(s_t = i|Z^T;\theta) = \sum_{s_t=1}^{3} P(s_{t+1} = i|Z^T;\theta) P(s_t = i|s_{t+1} = i, Z^T;\theta). \quad (9)$$

Based on Hermosillo and Hesse [3] while the probability of a low-volatility regime has decrease under 0.4 means that the indicator is stabil, the probability of a medium-volatility regime (though declining) still remains at around 0.4-0.6 means that the indicators is prone condition, and if the probability of a high-volatility regime has increased to over 0.6 means that the indicators on a crisis condition.

3. Method

The research used data of import and export from January 1990 to December 2016 were obtained from the International Monetary Fund. While the data of foreign exchange reserve from January 1990 to December 2016 were obtained from Bank Indonesia. These data are used to build the model. The steps of the research are follows. 1) Create a data plot and then test the stationary of data. If the data are not stationary then transform the data using log return. 2) Make a plot of partial autocorrelation function (PACF) of log return data, then make the AR model. 3) Test the heteroscedasticity effects on the residue of AR model using Lagrange multiplier test. 4) If there are heteroscedasticity effects on the residue of AR model then estimate the parameter of ARCH model. 5) Establish the combination of volatility and Markov switching models with three states assumption. 6) Determine the conditions of crisis based on the smoothed probability.

4. Result and discussion

4.1 Data

Plot data of import, export, and foreign exchange reserve can be seen in Figure 1.

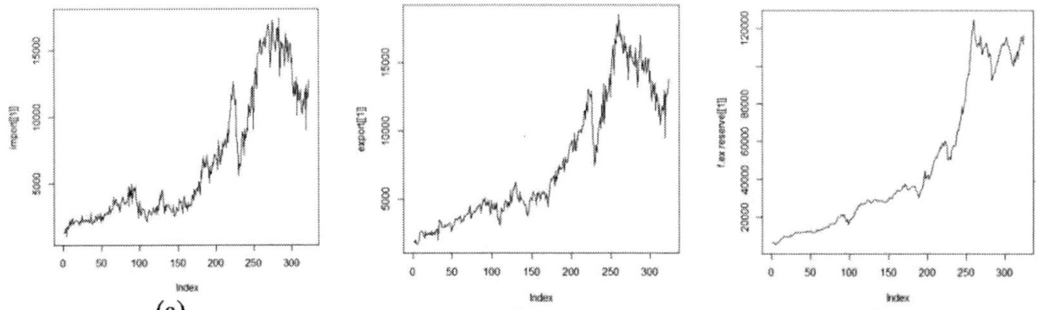

(a) (b) (c)

Figure 1. Plot of (a) import, (b) export, and (c) foreign exchange reserve

Figure 1 indicated that the data are not stationary. Based on the ADF test, the probability values for import, export, and foreign exchange reserve are 0.4446, 0.4622, and 0.5251 respectively. Because the probabilities more than $\alpha = 0.05$, so it can be concluded that the data are not stationary. Then, the data were transformed using log return. Based on ADF test for log return data, the probability values are 0.01 where the probability values are less than $\alpha = 0.05$ so it can be concluded that the data stationary. The next step is to form an AR model.

4.2 Establishment of AR Model

AR Model can be identified from PACF plot of log return data. Based on import, it was obtained an AR(2) model is $r_t = -0.52619r_{t-1} - 0.20209r_{t-2} + a_t$. Based on export, it was obtained an AR(2) model is $r_t = -0.37879r_{t-1} - 0.11493r_{t-2} + a_t$. For the foreign exchange reserve, it obtained an AR(1) model ie $r_t = 0.16803r_{t-1} + a_t$. Based on Lagrange multiplier test on the residue of model, it were obtained the each probability of import, export, and foreign exchange reserve as 0.00338, 1.113×10^{-5}, and 7.688×10^{-6}. This probability values are less than $\alpha = 0.05$, so it can be concluded that there are heteroscedasticity effects on the residue of AR model for import, export, and foreign exchange reserve, so the next step is estimate the parameter of volatility model.

4.3 Establishment of Volatility Model

For import, export, and foreign exchange reserve, the best model is ARCH(1) which can be written as $\sigma_t{}^2 = 0.00944 + 0.28387a_{t-1}{}^2$, $\sigma_t{}^2 = 0.00551 + 0.24750a_{t-1}{}^2$, and $\sigma_t{}^2 = 8.739 \times 10^{-4} + 0.31240a_1{}^2$ respectively. Based on Kolmogorov-Smirnov test, the probability values of import, export, and foreign exchange reserve are 0.8313, 0.8625, and 0.729 that the probability values are more than $\alpha = 0.05$, so it can be concluded that the residue of ARCH model for import, export, and foreign exchange reserve are normally distributed. Based on L-Jung Box test, the probability values for import, export, and foreign exchange reserve are 0.8253, 0.7659, and 0.4944. The probability values are more than $\alpha = 0.05$, so it can be concluded that there are no autocorrelation on the residue of ARCH model for import, export, and foreign exchange reserve. Based on Lagrange multiplier test, the probability values for import, export, and foreign exchange reserve are 0.9011, 0.2469, and 0.6676 respectively. The probability values are more than $\alpha = 0.05$, so it can be concluded that there are no heteroscedasticity effects on the residue of ARCH model. Based on these tests, it can be concluded that ARCH(1) is the appropriate models for import, export, and foreign exchange reserve.

4.4 Establishement of SWARCH Model

The volatility model combined with Markov switching model with three states assumption for resolve the change condition of crisis, not, crisis, and prone of crisis. The change of condition in Markov switching is an unobserved random variable namely state, where state used id low, high, and medium. States can be formed by transition probability. The transition probability can be formed as a matrix notation and usually called as matrix of transition probability. For import, transition probability matrix can be written as the following

$$P_1 = \begin{bmatrix} 0.31172 & 0.89372 & 0.31492 \\ 0.65723 & 0.00510 & 0.40596 \\ 0.03106 & 0.10118 & 0.27912 \end{bmatrix}$$

Based on P_1 obtained that the probability to survive on low volatility state is 0.31172. The transition probability from low to medium volatility state is 0.65723. The transition probability from low to high volatility state is 0.03106. The transition probability from medium to low volatility state is 0.89372. Probability survive in medium volatility state is 0.00510. As well as the transition probability from medium to high volatility state is 0.10118. The transition probability from high to low volatility state is 0.31492. Probability transition state from high to medium volatility state is 0.40596. Probability survive in high volatility state is 0.27912.The matrix of transition probability for export and foreign exchange reserve stated in P_2 and P_3 as follows

$$P_2 = \begin{bmatrix} 0.70125 & 0.35182 & 0.24608 \\ 0.21744 & 0.47754 & 0.28097 \\ 0.08131 & 0.17064 & 0.47295 \end{bmatrix}, P_3 = \begin{bmatrix} 0.80906 & 0.16466 & 0.00099 \\ 0.19060 & 0.80095 & 0.11560 \\ 0.00033 & 0.03439 & 0.88340 \end{bmatrix}.$$

Estimation of parameter SWARCH(3.1) model can be written as

$$\mu_{1,t} = \begin{cases} 0.01623, \text{ state 1} \\ 0.03540, \text{ state 2} \\ -0.20903, \text{ state 3} \end{cases} \quad \sigma_{1,t}^2 = \begin{cases} 3.194 \times 10^{-5} + 0.03142a_{t-1}^2, \text{ state 1} \\ 0.00015 + 0.78234a_{t-1}^2, \text{ state 2} \\ 0.04051 + 0.419276a_{t-1}^2, \text{ state 3} \end{cases}$$

$$\mu_{2,t} = \begin{cases} 0.00217, \text{ state 1} \\ 0.01219, \text{ state 2} \\ 0.00651, \text{ state 3} \end{cases} \quad \sigma_{2,t}^2 = \begin{cases} 0.00115 + 0.19954a_{t-1}^2, \text{ state 1} \\ 1.637 \times 10^{-5} + 0.03272a_{t-1}^2, \text{ state 2} \\ 0.00489 + 0.47281a_{t-1}^2, \text{ state 3} \end{cases}$$

$$\mu_{3,t} = \begin{cases} 0.00705, \text{ state 1} \\ 0.01061, \text{ state 2} \\ 0.00955, \text{ state 3} \end{cases} \quad \sigma_{3,t}^2 = \begin{cases} 0.00244 + 0.03119a_{t-1}^2, \text{ state 1} \\ 2.911 \times 10^{-6} + 0.02593a_{t-1}^2, \text{ state 2} \\ 1.085 \times 10^{-6} + 0.00010a_{t-1}^2, \text{ state 3} \end{cases}$$

where $\mu_{1,t}$ and $\sigma_{1,t}^2$ is the conditional mean and variance models SWARCH(3.1) for import, $\mu_{2,t}$ and $\sigma_{2,t}^2$ is the conditional mean and variance models SWARCH(3.1) for export, and $\mu_{3,t}$ and $\sigma_{3,t}^2$ is the conditional mean and variance models SWARCH(3.1) for foreign exchange reserve.

4.5 Detection of Crisis

Crisis can be detected with smoothed probability value. Plot of smoothed probability for import, export, and foreign exchange reserve can be seen as Figure 2.

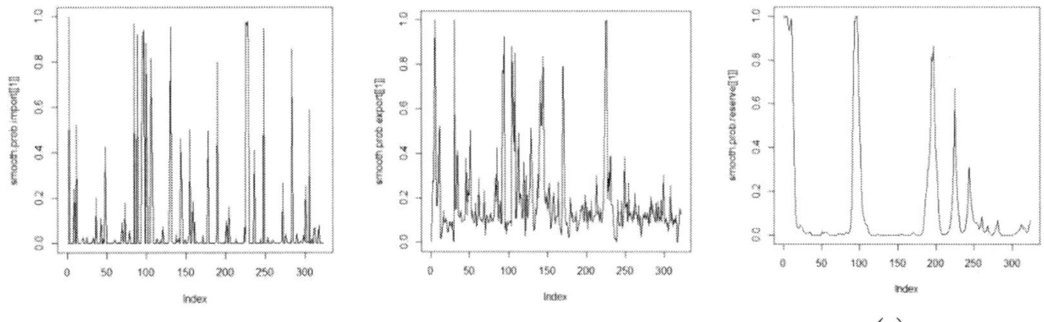

(a) (b) (c)

Figure 2. Plot smoothed probability of (a) import, (b) export, and (c) foreign exchange reserve

Crisis condition signed with value of smoothed probability more than 0.6 for import, export, and foreign exchange reserve as shown in Figure 2. Table 1 showed the crisis period that has been detected on 1997, 1998, and 2008 based on the smoothed probability greater than 0.6 by import, export, and foreign exchange reserve.

Table 1. The crisis detection based on smoothed probability

Year	Import	Export	Foreign Exchange Reserve
1997	February, June, December	November, December	September, October, November, December
1998	January, February, November	January, November, December	January, February, March, April
2008	November, December	November, December	October

Table 1. showed that smoothed probability for import, export, and foreign exchange reserve can detect the financial crisis that hit Indonesia in 1997, 1998, and 2008.

5. Conclusion

The result of research showed that SWARCH(3.1) model is an appropriate model for import, export, and foreign exchange reserve that can be used to detect the financial crisis that hit Indonesia in 1997, 1998, and 2008 based on the indicators.

References

[1] Hamilton J D and R Susmel 1994 Autoregressive Conditional Heteroscedasticity and Changes in Regime, *Journal of Econometrics* **64** 307

[2] Hermawan D W 2015 *Pendeteksian Krisis Keuangan di Indonesia Menggunakan Gabungan Model Volatilitas dan Markov Switching Berdasarkan Indikator Jumlah Nilai Impor* (Solo: Sebelas Maret University)

[3] Hermosillo B G and H Hesse 2009 *Global Market Conditions and Systemic Risk*, IMF Working Paper **230**

[4] Kaminsky G, S Lizondo, and C M Reinhart 1998 *Leading Indicators of Currency Crisis*, IMF Working Paper **45**

[5] Kuan C M 2002 The Markov Switching Model, *Journal of Economics Academia Sinica* **1** 1

[6] Pradini D Y and S Raharjo 2012 Peramalan Nilai Ekspor dan Impor Indonesia ke Jepang menggunakan Model Varima, *Journal Online Matematika Universitas Negeri Malang* **1** 1

[7] Tsay R S 2005 *Analysis of Financial Time Series* (Canada: John Wiley and Sons)

IOP Conf. Series: Journal of Physics: Conf. Series **943** (2017) 012055 doi:10.1088/1742-6596/943/1/012055

Student's rigorous mathematical thinking based on cognitive style

H Fitriyani and U Khasanah

Mathematics Education Department of Universitas Ahmad Dahlan, Jalan Prof. Dr. Soepomo, SH. Warungboto, Yogyakarta 55164, Indonesia

E-mail: harina.fitriyani@pmat.uad.ac.id

Abstract. The purpose of this research was to determine the rigorous mathematical thinking (RMT) of mathematics education students in solving math problems in terms of reflective and impulsive cognitive styles. The research used descriptive qualitative approach. Subjects in this research were 4 students of the reflective and impulsive cognitive style which was each consisting male and female subjects. Data collection techniques used problem-solving test and interview. Analysis of research data used Miles and Huberman model that was reduction of data, presentation of data, and conclusion. The results showed that impulsive male subjects used three levels of the cognitive function required for RMT that were qualitative thinking, quantitative thinking with precision, and relational thinking completely while the other three subjects were only able to use cognitive function at qualitative thinking level of RMT. Therefore the subject of impulsive male has a better RMT ability than the other three research subjects.

1. Introduction

Thinking is a routine activity that people do every day. Almost every issue of human life involves thinking activity to solve it. Thinking means mental activities which happen in the human mind involve remembering, considering, giving reason, reflecting, and deciding something tend to solving a problem. In the face of 21st century challenges, every individual should have a thinking skill to solve every problem. The thinking skills that have to be mastered at the moment are called high order thinking skill or HOTS. Higher order thinking (HOT) involves a variety of thinking processes which applied to complex situations and having multiple variables [1]. When an individual thinks of solving a mathematical problem, it does not rule out that he is doing a mathematical thinking. Katagiri [2] states that mathematical thinking is the ability of scholastic to think and make judgments independently. Furthermore, Sumarmo [3] defined mathematical thinking as a way of thinking with regard to the process of math (doing math) or in solving mathematical tasks both simple and complex.

In mathematical thinking, a person translates incoming external information into symbols for subsequent symbols to be processed according to the rules in mathematics already compiled. In learning and solving mathematical problems, need for precision, while the prerequisite to be precise and logical is rigor. Kinard [4] reveals that mathematical thinking integrates and harnesses cognitive processes that increase higher abstraction levels, and therefore they must be rigorous in nature. With regard to the necessity of rigor in synthesizing and utilizing cognitive processes to increase the level of abstraction function it is necessary to have rigorous mathematical thinking (RMT). Furthermore, RMT is characterized by three levels of cognitive function such as cognitive function for qualitative thinking, cognitive function for quantitative thinking with accuracy and cognitive function for abstract relational thinking [4, 5]. Therefore RMT is a mathematical thinking for high level like HOT.

The first level of cognitive function for RMT consists of the general cognitive functions necessary for qualitative thinking. Before students engage in rigor conceptual reasoning, the cognitive process takes place at the concrete level and is dominated by existing natural psychological functions. The cognitive functions in this first level are labeling-visualizing, comparing, searching systematically to collect information clearly and completely, making use more than one source of information, and encoding-decoding. The second level consists of the cognitive functions necessary for quantitative thinking and accuracy. These functions are more structured than general cognitive functions. The cognitive functions in this second level are conserving constancy, quantifying space and spatial relationships, quantifying time and temporal relationships, analyzing-integrating, generalizing, and being precise. The third level of cognitive function integrates processes related to quantity and precision into unique structures and generalized abstract relational thinking. The cognitive functions in this third level consist of activating previous mathematically related knowledge, providing and articulating mathematical logical evidence, defining the problem, inferential-hypothetical thinking, projecting and restructuring relationships, Forming proportional quantitative relationships, forming a functional relationship, forming a unit functional relationship, mathematical inductive- deductive thinking, mathematical analogical thinking, mathematical syllogistic thinking, mathematical transitive relational thinking, and elaborating mathematical activity through cognitive categories [4, 5]. These three levels of cognitive function together define the mental processes of general cognitive skills to higher level specific mathematical cognitions [6].

Each individual is born with a unique characteristic attached to himself that is not owned by another individual. This uniqueness is included in terms of cognitive style. The cognitive style according to Warli [7] defined as individual characteristics in terms of feeling, remembering, organizing, processing, and solving problems, in an attempt to differentiate, understand, store, embody, and utilize information. Furthermore, Goldstein and Blackman in Amstrong [8] suggest that cognitive styles relate to the characters and habits in which individuals process and evaluate information, solve the problem and make decisions. Witkin [9] classify cognitive style in several types, one of which is reflective-impulsive cognitive style. The classify was in term of conceptual tempo. Conceptual tempo can be interpreted as the speed and accurate in thinking.

Warli [7] states that there are two important aspects to be considered in measuring reflective-impulsive cognitive style: 1) the level of the subject in describing the accuracy of the problem-solving allegation or the time to make a decision in solving the problem; 2) contains uncertain answers. When the time aspect is divided into two, namely fast and slow, then the uncertainty aspect is divided into two that is accurate and inaccurate then the students can be grouped into four categories: fast-accurate, slow and accurate (reflexive), fast and inaccurate (impulsive), slow-inaccurate. In this research, only focus on reflective-impulsive categories because both of categories most dominant [7, 10-13] The characteristics of reflective cognitive-style students are slow in answering questions but the answers inclined to precise. whereas impulsive cognitive style students are fast in answering questions but the results tend to be inaccurate. The test to map reflective-impulsive cognitive style used MFFT (matching familiar figure test) wich was develop by Warli [7]

2. Method

This research method used descriptive qualitative research. Subjects in this study were 4 first year students in the mathematics education study program of FKIP UAD, each consisting of 2 students of reflexive and impulsive cognitive style in different gender. The subjects were MR (male reflective), FR (female reflective), MI (male impulsive), and FI (female impulsive). Data collection techniques in this study using problem-solving test and interview. To see the consistency of research data used time triangulation. Analysis of research data used Miles and Huberman [14] model that was reduction of data, presentation of data, and conclusion.

3. Result and discussion

The RMT process requires the use of cognitive function from low-level to high-level cognitive functions. The cognitive function is categorized in three levels that are general cognitive functions for qualitative thinking, cognitive functions for quantitative thinking in the manner of precision, and cognitive functions for generalized, logical abstract relational thinking in the mathematics culture. The following is a summary of the use of cognitive function of the subjects.

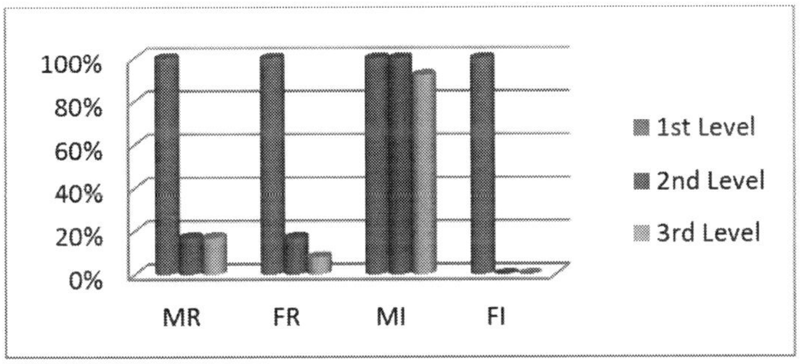

Figure 1. Percentage of the using of cognitive function of subject for RMT

3.1 Characteristics of cognitive function of reflective cognitive style subject for RMT

Reflective subject consist of male and female reflective cognitive style which labeled by MR and FR respectively. The use of cognitive function for RMT at first level is indicated by the ability of the both subjects in giving something a name (labeling) when given the picture by taking the critical attributes of the object but MR looks still unsure of the explanation. Subjects make constructed of cone, sphere, equilateral triangles, cubes, cuboid when names given but the picture made only sketches without using ruler. The subjects compared the cube and cuboid in terms of the diagonal, the shape of the sides and their dimensions. Subjects paid attention to the picture carefully to collect and complete the information, even the subjects added streaks on the image to gather information. Furthermore, subjects used more than one source of information by mentally working with several concepts at the same time when solving a problem, but have not been able to discover the right answer.

Subjects able to symbolize radius with r; the measure of cuboid edge with p, l, t; right angle with "□"; same length with "/"; angle symbol but the symbol of angle size was less precise. MR paid less attention to the symbol of the vertex with capital letters. Subjects were able interpreting the symbols "/", ">>", and measure of segment accurately. The subject defined the symbol "/" as the same long side symbol; "□" as a right angle symbol; ">>" on some segments as symbols of two parallel lines; MR interpreted the XY symbol on the picture as the length of the X segment plus the length of the Y line segment. However, he was able to understand the symbol of measurement of segment of the constructed picture itself. This shown the subject's lack of understanding about the symbol of the segment and its size. Overall, reflective subject able to use all of cognitive function at the level qualitative thinking for RMT.

At second level cognitive function for RMT, both subject only able to use one cognitive function appropriately. FR used cognitive function of integrating by building rectangles by combining 3 congruent square, trapezoid with respect to two parallel edge while MR used cognitive function of generalizing by stated the orthocenter of the equilateral triangle divided the altitude into 2 parts by 2: 1 ratio and cuts the right side of it in the middle. Generally, MR comprehended that three special lines of equilateral triangule were coincide. When given the problem of explaining the relationship between the cube and the cuboid, both subjects were unable to name the cuboid when the size of the edge was the same length. FR identified that the edge of the cuboid and cubes were congruent to differentiate were their dimensions and constituents (the cube was hollow and in 3 dimension, the cuboid was solid and in 2 dimension). Subjects were less able to use internal references as a guide to analyze the relationship between cone, cone meridians and sphere in cone, and the relationship between cubes and cuboid. Subjects were not able to analyze the critical attributes of equilateral triangles, right triangles, cubes and cuboid appropriately. Subjects can not distinguish the terms face and edge on the context of cubes and cuboid. FR was less able to generalize the features of cubes and cuboid without referring to the picture. Both subjects were less focused and meticulous when working on the problem. Therefore, reflective subject still not fulfilled second level of cognitive function for RMT yet.

As the previous level of cognitive function for RMT, reflective subject not able to use all cognitive function at the third level of cognitive functions for generalized, logical abstract relational thinking in the mathematics culture. While solve on the problem, MR has activated prior mathematically related knowledge but the solution was not quite right. In addition, MR has also been able to use cognitive function of mathematical deductive thinking by applying formulas in problem solving though not yet precise. While FR was only capable of doing mathematical inductive thinking by determining the height of the cone using the height of the triangle as its meridian slices using a ruler. In addition, mathematical inductive thinking was done by assuming that the angles of the right triangle were $90^0,45^0,45^0$ or 90^0, $30^0,60^0$. It was constructed by FR from her experience that often saw a right triangle with such an angle size. Others cognitive function for RMT still not appeared in the both of subject. Therefore, reflective subject still not fulfilled third level of cognitive function for RMT. Although both of the reflective cognitive style subjects have different gender, but their RMT process almost the same.

3.2 Characteristics of cognitive function of impulsive cognitive style subject for RMT

Impulsive subject consist of male and female impulsive cognitive style which labeled by MI and FI respectively. Both of subjects were able to give name the object based on their critical attributes such as symbols of equal edge, size of angle, and parallel. Subjects construct cone, sphere, equilateral triangles, cubes, cuboid, spheres in cones and meridian cones in equilateral triangles when the name was given. MI was able to determine the similarities of cubes and cuboid appropriately while FI determined the similarities of cubes and cuboid on the same edge, whereas the difference lied in its dimensions and constructions. MI has a well concept understanding so that he be able to compare both objects geometry properly. When solving the problem, beside to paying attention to the picture carefully to collect and complete the information, both of subjects also mentally worked with some concepts at the same time but there were concepts used FI was still not quite right. In the use of cognitive functuion of encoding, both of subjects were able to interpret the object by using a particular symbol, such as the area of the triangle with L, the length of the triangle with a, half of the circumference of the triangle with S, the surface area of the sphere with Lp, the same length of the segment by "/ ", the radius by "r", the right angle with" □ ", congruent, the name of the object by encoding the vertex with capital letters, angle notation but not appropriate in the use of notation angle size. While for the cognitive decoding function, the subjects were able to interpret the mathematical symbol correctly, but FI failed to interpret the symbol of the size of the segment. This was because she was used to denoting the size of a segment with segment notation. Even though the notation was totally different.

There was the difference of using cognitive function at the second level of cognitive function for RMT. FI failed using cognitive function at this level. Such as impulsive subject characteristics, FI tend to fast in solve the given problem but the answer tend to inaccurate. FI did not able to identify and describe what stays the same in terms of an attribute and concept of cube and cuboid while some other things were changing. FI did not being precise in solve the problem. FI cannot quantify space and spatial relationships between cube and cuboid, radius of sphere in cone and high of cone, cannot analyze the cone and sphere in cone, cube and cuboid, cannot integrate the square and quadrilateral, and cannot describe the nature of cube and cuboid without referring to specific details. While MI can using all of cognitive functions at second level of cognitive function for RMT. MI was able to identify cuboid even though its similar edges, use internal referential as an internal of reference as a guide to analyze parts relationships based on whole-to-parts relationships, do decomposing a quantity into its critical attributes and constructing a whole by merging its parts. MI was able to generalize that altitude of equilateral triangle will divide the edge in front of it into two equal lengths, without regard to critical attributes of equilateral triangle. Beside it, MI also can generalize that cube was cuboid with equal edges. During solve the problem, MI strived to be focused and exact.

On the third level of cognitive function for RMT, only MI was capable of using all of cognitive functions needed for RMT. MI able to do mobilizing previously acquired mathematical knowledge by searching through past experiences to solve the problem, explain the logical supporting details to prove the truth of his preposition. MI can define the problem by analyzing and seeing relationships to figure out precisely what has to be done mathematically. MI also can form a mathematical proposition that a cube was a cuboid and search for mathematical logical evidence to support the proposition. He also able to develop valid generalizations and proofs based on a number of mathematical events. MI has formed connections between congruent triangles and reconstructing existing connections between triangles to find the measurement of its edges. In solve the problem, MI also took aspects from various mathematical details that seem to form a pattern, categorizing them into general relationships of attributes, and organizing the results to form a general mathematical rule; applying some formulas ie Pythagorean theorem, area of triangle theorem, the length of radius of sphere in triangle formula, area and volume of the sphere, the rule of congruent triangle to a specific situation that connect only with the rule in terms of belonging to categories of attributes by the rule. MI used cognitive function mathematical relational thinking by considering relationship among some objects in geometry ie relationship between parallelepipedum, cube and cuboid. In the whole of solving problem, MI able to elaborate mathematical activity through cognitive categories by reflecting on and analyzing mathematical activity and discovering, labeling, and articulating, orally and in writing, underlying mathematical principles and concepts using the language of mathematics and cognitive functions. Therefore, both of the impulsive subjects have different ability in using cognitive function needed for RMT. Male impulsive subject have better ability than female impulsive subject.

Based on the above exposure, from the fourth research subjects obtained that only male impulsive subject was capable to use all of cognitive function needed for RMT, while the other three subjects were only capable to use cognitive function at first level (qualitative thinking). Although if noted the characteristics of reflective-impulsive cognitive style, reflective subject need more time to answer and tend to accurate. Therefore, it should be more careful and thorough in solve the problem with the result that the answer was more accurate. But in this research the opposite was true, impulsive subject was more accurate and thorough in solve the problem so that cognitive function needed to solve the problem can used properly. Therefore the finding of this research was in accordance with the findings of Witkin[9], Taram [15], Fitriyani [11], Agoestanto [16], and Warli [7] that the difference of cognitive style influences how the student solve the problem. The problem in this research was geometry problem. The research result shown that students have difficulties to solve the given geometry problem. Students still have difficultys in construct concepts of geometry, make relation between the concepts, etc [17-19]. This finding was suitable with the finding of Hardianti [20].

4. Conclusions

Based on the result and discussion, impulsive male subjects used three levels of the cognitive function required for RMT that were qualitative thinking, quantitative thinking with precision, and relational thinking completely while the other three subjects were only able to use cognitive function at the level of qualitative thinking for RMT. Therefore the subject of impulsive male has a better RMT ability than the other three research subjects.

5. Acknowledgments
This study was supported by *Penelitian Dosen Pemula* (PDP) grant from The Directorate of Research and Community Service. Ministry of Research,Technology and Higher Education.

References
[1] King F J, Goodson L and Rohani F 1998 Higher order thinking skills: Definition, strategies, and assesment *A publication of the Educational Services Program*, now known as the Center for Advancement of Learning and Assessment
[2] Katagiri S 2004 *Mathematical Thinking and How to Teach it* (Tokyo: Meijitosyo Publishers)
[3] Sumarmo U 2010 *Berpikir dan Disposisi Matematika: Apa, Mengapa, dan Bagaimana Dikembangkan pada Peserta Didik.* (Bandung: FMIPA UPI)
[4] Kinard J T 2007 Method and Apparatus for Creating Rigorous Mathematical Thinking *United States patent application US 11/584,367*
[5] Kinard J T and Kozulin A 2008 *Rigorous Mathematical Thinking: Conceptual Formation in the Mathematics Classroom* (New York: Cambridge University Press)
[6] Fitriyani H 2013 Profil Berpikir Matematis Rigor Siswa SMP dalam Memecahkan Masalah Matematika Ditinjau dari Perbedaan Kemampuan Matematika *Jurnal AdMathEdu* **3** 37
[7] Warli 2013 Kreativitas Siswa SMP yang Bergaya Kognitif Reflektif atau Impulsif dalam Memecahkan Masalah Geometri *Jurnal Pendidikan dan Pembelajaran* **20** 190
[8] Amstrong S J 2000 The Influence of Individual Cognitive Style on Performance in Management Education *Journal of Educational Psychology* **20** 323
[9] Witkin H A 1973 *The Role of Cognitive Style in Academic Performance and in Teacher-Student Relations* (New Jersey: Educational Testing Service)
[10] Rozencwajg P and Corroyer D 2005 Cognitive Processes in the Reflective–Impulsive Cognitive Style *The Journal of Genetic Psychology* **166** 451
[11] Fitriyani H and Khasanah U 2016 Analisis Kesalahan Newman (NEA) Pada Pemecahan Masalah Geometri Mahasiswa Ditinjau Dari Gaya Kognitif *Proc Seminar Nasional Pendidikan Matematika 2016* (Surabaya: Universitas Negeri Surabaya) pp 129-134
[12] Prihastanto A R and Fitriyani H 2017 Profil Kemampuan Koneksi matematis Siwa SMP yang bergaya Kognitif reflektif-Impulsif dalam Menyelesaikan Soal Geometri *Jurnal Didaktika* **23** 89
[13] Apriyanti S and Fitriyani H 2017 Teori Van Hiele: Tingkat Berpikir Siswa Smp Bergaya Kognitif Refleksif Dan Impulsif Pada Materi Segiempat *Proc Seminar Nasional Pendidikan, Sains dan Teknologi* (Semarang: Universitas Muhammadiyah Semarang) pp 364-370
[14] Milles B M , Huberman A M and Saldana J 2014 *Qualitative Data Analysis : An Expanded Sourcebook* (Los Angeles: SAGE Publications Inc.p)
[15] Taram A 2017 Probabilistic Thinking Ability of Students Viewed from Their Field Independent and Field Dependent Cognitive Style *J. Phys.: Conf. Ser.* **824** 012050
[16] Agoestanto A, Sukestiyarno Y L, and Rochmad 2017 Analysis of Mathematics Critical Thinking Students in Junior High School Based on Cognitive Style *J. Phys.: Conf. Ser.* **824** 012052
[17] Tanujaya B, Prahmana R C I, and Mumu J 2017 Mathematics instruction, problems, challenges, and opportunities: A case study in Manokwari regency, Indonesia *World Transactions on Engineering and Technology Education* 15 287
[18] Sundayana R, Herman T, Dahlan J A and Prahmana R C I 2017 Using ASSURE learning design to develop students' mathematical communication ability *World Transactions on Engineering and Technology Education* **15** 245
[19] Nurhasanah F, Kusumah Y S and Sabandar J 2017 Concept of triangle: Examples of mathematical abstraction in two different contexts *International Journal on Emerging Mathematics Education* **1** 53
[20] Hardianti D, Priatna N, and Priatna B A 2017 Analysis of Geometric Thinking Students' and Process-Guided Inquiry Learning Model *J. Phys.: Conf. Ser.* 895 012088

IOP Conf. Series: Journal of Physics: Conf. Series **943** (2017) 012056 doi:10.1088/1742-6596/943/1/012056

Edmodo social learning network for elementary school mathematics learning

Y Ariani[1], Y Helsa[1], S Ahmad[1] and RCI Prahmana[2]

[1]Universitas Negeri Padang, Jl. Prof. Dr. Hamka Air Tawar, Padang, Indonesia
[2]Universitas Ahmad Dahlan, Jl. Pramuka Kav. 5, Pandeyan, Yogyakarta, Indonesia

Email: arianiyetti@gmail.com

Abstract. A developed instructional media can be as printed media, visual media, audio media, and multimedia. The development of instructional media can also take advantage of technological development by utilizing Edmodo social network. This research aims to develop a digital classroom learning model using Edmodo social learning network for elementary school mathematics learning which is practical, valid and effective in order to improve the quality of learning activities. The result of this research showed that the prototype of mathematics learning device for elementary school students using Edmodo was in good category. There were 72% of students passed the assessment as a result of Edmodo learning. Edmodo has become a promising way to engage students in a collaborative learning process.

1. Introduction

The ways on how learning is implemented are remarkably developed along with the development of information technology [1]. Learning, which is defined as an activity such that students can achieve educational purposes, has adopted these technologies in a form of e-learning [1] and integration with mobile technology [2]. The education purpose itself cannot be separated from intellectual, behavioral, spiritual, attitude, moral, and social changing for the students to be independent in society life [3]. In order to achieve the purpose, learning concept and environment need to be designed such that the interaction between students and the learning environment could benefit their learning result [4, 5].

Teaching in elementary school need a well preparation [6] including the objectives, materials, methods, models, media, and assessment of learning [7]. All them need to be designed carefully in order to successfully reach the learning objectives. In the mathematics learning process for elementary school level, instructional media (e.g. manipulatives, virtual web-based manipulative) is one of the necessary components in supporting students' understanding of mathematical thinking [8-10]. Instructional media is a tool designed, planned, and regulated by the educators as an introduction in delivering materials to improve the quality of the learning process and learning outcomes of students [11]. The educators must be able to use the learning media that has been available in the school according to the needs of the learning process. In addition, the educators must also be skilled and innovative in creating and developing instructional media so that the learning process is more meaningful for the students.

A professional educator must be able to deal with various different types of students [12]. Various types of students can be found in our learning in terms of their behavioral engagement, which means how the students participate in the classroom. Behavioral engagement is the students' participation in the classroom discussion, academic task, and listening to teacher instruction. Some students may be active during the learning, but the others may appear passive and inattentive [13].

Based on the above opinion, we can see that a classroom can be regarded as a community. Classroom is a means of training the students to become good classroom members. The mathematics teacher has to be creative in managing the classroom, because not all students like to learn mathematics. Teacher needs to utilize the instructional media such that the students engage effectively [14]. The development of instructional media can also take advantage of technological developments by utilizing social networks such as Facebook, Edmodo, Schoology, and so on. The utilization aims to support the learning process and ease the students to access the material information quickly.

Our team tried to observe and to conduct an interview with the mathematics teacher of SD N 4 Birugo. It is a public elementary school in Bukittinggi, Indonesia, which has adequate facilities to support e-learning. The school has utilized social networking such as Edmodo as a means for educators and students in the learning process as seen in Figure 1.

(a) (b) (c) (d)

Figure 1 (a, b, c, d). Learning process by utilizing Edmodo

In the Figure 1, we can see that the school has been supported by e-learning facilities. One teacher at the school had an opportunity to join short course and comparative studies in Australia, through the cooperation between the Education Department of Bukittinggi and the Australian government, studying how the digital class was developed there. Then the Australian government provided guide and help to model schools that used digital class. It was not only using equipment such as gadgets, laptops, and projector but also the existing digital class comes with applications and media needed to support learning. Though the learning has been supported with information technology, the use of social network such as Edmodo was still less optimal since the material was limited. The adequate information and materials provided by teacher will greatly assist the students in understanding the material provided.

Mathematics is regarded as the basic foundation of the development of other science lessons and it is compulsory in Indonesian elementary school curriculum. Basically, learning mathematics is challenging for the students. Therefore, mathematics learning at the elementary school level should be fun and exciting for students. This demand requires an innovation in the learning process, one of them is the utilization of Edmodo. Thus, development of sophisticated digital class by utilizing Edmodo in elementary school mathematics learning process is needed to increase the interest of the students and the quality of learning process.

2. Method

The research was part of a developmental research with the aim of producing and developing new products in learning. The research and development stage used refers to the ADDIE development model which consists of 5 stages: analyses, design, develop, implement, and evaluate [15]. ADDIE is a generating process because it applies concepts and theories to specific contexts. ADDIE is used in an educational setting to facilitate the formation of knowledge and skills during the learning coaching stage. The basic principle of ADDIE is all planning activities on coaching or building knowledge in some learning spaces [15]. This paper focuses on the result of product testing as an impact towards the students achievement which is in the evaluation stage.

In this research, after the implementation of Edmodo learning, the implementation stage was conducted to test the success of the development. We provide a percentage of the students who passed the test after the Edmodo learning.

3. Results and Discussion

Learning by using Edmodo platform is a learning that can be done anywhere, anytime and can be done from any device that supports. This provides considerable benefits for the students, the educators, the school, and the parents of the students themselves. The result of the fourth grade students' learning achievement of SD N 4 Birugo can be seen in table 1.

Table 1. Students' achievement in the field test

Score (x)	Frequency
$80 \leq x \leq 100$	9
$66 \leq x < 80$	14
$56 \leq x < 66$	4
$40 \leq x < 56$	5
$0 \leq x < 40$	0
Average Score	71

Table 1 summarizes the student learning outcomes using Edmodo with the category of Very Good 30%, Good category 42%, Fair 16% and Poor 12%. In the test prototype 3, there were no students whose learning results fall into Fail category. If the percentage of students viewed from the completeness of the learning outcomes set by math teachers was 71, then 72% of the students passes in understanding the material of cubes and cuboids which the learning activities were using Edmodo. Based on the results of the average score of learning outcomes, it can be concluded that prototype 3 was being developed categorized in good category.

Edmodo is a social media platform that can be used as a tool in the learning process. Edmodo is an app that appeals to the teachers and students with social elements that resemble Facebook, but there is actually a greater value in this social-based educational app. Through this platform, it will be easier to monitor the interaction of students in the Edmodo learning environment [14]. Edmodo is an interactive platform that allows users to upload and share links to websites and an array of digital files not permitted in some platforms. Further, it also allows students to form groups to work on projects collaboratively. Edmodo is a micro blog that is similar in appearance to Facebook. By using Edmodo, the students perceive it as a clearly academic platform, which is not the case with Facebook, which is mostly used for non-educational purposes. Edmodo allows for bi-directional teacher-student interaction. The teacher can communicate to the class or to any individual student. Likewise, a student can communicate to the class or send a private message to the teacher [16].

Edmodo is a learning tool in the form of an online platform to encourage teacher in learning activities, or it can be a more creative way to engage students in collaborative learning and distributed cognition. This platform provides student pathways to interact with fellow students and teachers in an academic setting. Furthermore the use of this platform can teach the students how to behave online and be responsible for organizing their learning activities with their secure system. Edmodo provides an environment with teaching and learning atmosphere that can generate students' joy, the students become more independent, without forgetting the standard of measuring the success of students. Edmodo is a secure learning platform for the teachers, students, and schools based on social media. Edmodo provides a safe and easy way for the class to connect and collaborate, share content and work access, grades and school notifications. The goal of Edmodo is to help the educators harness the power of social media to tailor the class to each student. Edmodo can help the teachers build a virtual classroom based on classroom real-life at school, where there are assignments, quizzes, and grades at the end of each lesson. On the other hands, some research already documented that explain about mathematics learning process supported the Edmodo instruction learning research result [17-20].

4. Conclusion

Edmodo is a learning tool in the form of an online platform to encourage teacher in teaching learning process, or it can be a more creative way to engage students in collaborative learning and distributed cognition. If the percentage of students viewed from the completeness of the learning outcomes set by the math teachers were 71, then 72% of students passed in understanding the material of cubes and cuboids which used Edmodo in the learning activities. Based on the results of the average score of learning outcomes, it can be concluded that prototype 3 was being developed categorized in good category.

References

[1] Paulin D and Haythornthwaite C 2016 Crowdsourcing the Curriculum: Redefining E- Learning Practices Through Peer-Generated Approaches *Sch. Inf. Stud. Fac. Scholarsh* **173** 1–37

[2] Sung Y T, Chang K E and Liu T C 2016 The effects of integrating mobile devices with teaching and learning on students' learning performance: A meta-analysis and research synthesis *Comput. Educ.* **94** 252–75

[3] Turmudi I, Hamzah R, Ismail S, Afzamiman M, Hasrina N and Sulaiman M 2017 Purpose of Education and Spiritual Potential; What is Your Definition *J. Sains Humanika* **9** 23–31

[4] Darmawanti Y, Siahaan P and Widodo A 2017 The Effect of Generate Argument' Instruction Model to Increase Reasoning Ability of Seventh Grade Students on Interactions of Living Thing with their Environment *J. Phys. Conf. Ser.* **812** 012042

[5] Casagrand J and Semsar K 2017 Redesigning a course to help students achieve higher-order cognitive thinking skills: from goals and mechanics to student outcomes *Adv. Physiol. Educ.* **41** 194–202

[6] Gaikhorst L, Beishuizen J, Roosenboom B and Volman M 2017 The challenges of beginning teachers in urban primary schools *Eur. J. Teach. Educ.* **40** 46–61

[7] Bold T *et al.* 2017 What Do Teachers Know and Do in Primary Schools in Sub-Saharan Africa? in *World Bank's World Development Report 2015: Realizing the Promise of Education for Development* (Washington DC: Education Global Practice Group, World Bank)

[8] Hidayah I, Pujiastuti E and Chrisna J E 2017 Teacher's Stimulus Helps Students Achieve Mathematics Reasoning and Problem Solving Competences *J. Phys. Conf. Ser* **824** 012042

[9] Jitendra A K, Nelson G, Pulles S M, Kiss A J and Houseworth J 2016 Is Mathematical Representation of Problems an Evidence-Based Strategy for Students With Mathematics

Difficulties? *Except. Child.* **83** 8–25

[10] Tucker S I, Lommatsch C W, Moyer-Packenham P S, Anderson-Pence K L and Symanzik J 2017 Kindergarten Children's Interactions with Touchscreen Mathematics Virtual Manipulatives: An Innovative Mixed Methods Analysis *Int. J. Res. Educ. Sci.* **3** 646

[11] Sholikhah M, H D A and Yusuf M 2017 Interactive Multimedia to Enhance Early Reading Skill on Children with Light Mental Retardation at 5th Grader SLB Autis Mitra Ananda Karanganyar Indonesia in Academic Year 2016/2017 *Eur. J. Spec. Educ. Res.* **2** 1–13.

[12] Hill M and Robertson J 2004 Great Expectations: Working in Partnership to Enhance Learning and Strengthen Teaching in Diverse Primary Schools *Teach. Curric.* **7** 9–14

[13] Gregory A and Allen J P 2017 Effect of A Professional Development Program on Behavioral Engagement of Students in Middle and High School *Psychol. Scholarsh.* **51** 143–63

[14] Ekici D I 2017 The Use of Edmodo in Creating an Online Learning Community of Practice for Learning to Teach Science *Malaysian Online J. Educ. Sci.* **5** 91–106

[15] Branch R M 2009 *Instructional Design* (New York: Springer)

[16] Holland C and Muilenburg L 2011 Supporting Student Collaboration: Edmodo in the Classroom *Society for Information Technology & Teacher Education Int Conf* (Nashville: Association for the Advancement of Computing in Education)

[17] Sundayana R, Herman T, Dahlan J A, and Prahmana R C I 2017 Using ASSURE learning design to develop students' mathematical communication ability *World Transactions on Engineering and Technology Education* **15** 245

[18] Prahmana R C I, Kusumah Y S, and Darhim 2017 Didactic trajectory of research in mathematics education using research-based learning *J. Phys.: Conf. Ser.* **893** 012001

[19] Tanujaya B, Prahmana R C I, and Mumu J 2017 Mathematics instruction, problems, challenges, and opportunities: A case study in Manokwari regency, Indonesia *World Transactions on Engineering and Technology Education* **15** 287

[20] Prahmana R C I and Kusumah Y S 2016 The hypothetical learning trajectory on research in mathematics education using research-based learning *Pedagogika* **123** 42

IOP Conf. Series: Journal of Physics: Conf. Series **943** (2017) 012057 doi:10.1088/1742-6596/943/1/012057

On factoring RSA modulus using random-restart hill-climbing algorithm and Pollard's rho algorithm

M A Budiman and D Rachmawati

Departemen Ilmu Komputer, Fakultas Ilmu Komputer dan Teknologi Informasi, Universitas Sumatera Utara, Jl. Universitas No. 9-A, Kampus USU, Medan 20155, Indonesia

E-mail: mandrib@usu.ac.id

Abstract. The security of the widely-used RSA public key cryptography algorithm depends on the difficulty of factoring a big integer into two large prime numbers. For many years, the integer factorization problem has been intensively and extensively studied in the field of number theory. As a result, a lot of deterministic algorithms such as Euler's algorithm, Kraitchik's, and variants of Pollard's algorithms have been researched comprehensively. Our study takes a rather uncommon approach: rather than making use of intensive number theories, we attempt to factorize RSA modulus n by using random-restart hill-climbing algorithm, which belongs the class of metaheuristic algorithms. The factorization time of RSA moduli with different lengths is recorded and compared with the factorization time of Pollard's rho algorithm, which is a deterministic algorithm. Our experimental results indicates that while random-restart hill-climbing algorithm is an acceptable candidate to factorize smaller RSA moduli, the factorization speed is much slower than that of Pollard's rho algorithm.

1. Introduction

RSA is one of the first algorithms that employs Diffie-Hellman's concept of asymmetrical cryptography [1] and has, until now, been one of the most widely used public key crypto-system since its publication in 1978 [2]. There are a considerable number of variants of RSA [3] such as Batch RSA [4], Multi-power RSA [5], Rebalanced RSA [6], RSA-CRT [7], and R-Prime RSA [8].

The original RSA and all of its variants have at least one thing in common: their security depends on the hardness of factoring a large integer into its prime factors. Although Boneh and Venkatesan [9] have pointed out that breaking RSA may be easier than factoring, they have also highlighted that this does not lead to any weakness of the RSA crypto-system. In fact, [10] has stated that there are no obliterating attacks on RSA crypto-system have been found. It should be noted, nevertheless, that RSA crypto-system may be vulnerable if the nowadays' hypothetical quantum computers can eventually become real in the future [11].

A fair number of deterministic algorithms to factorize integers into their prime factors have been researched in the field of number theory, such as Fermat's, Euler's, Kraitchik's, and some variants of Pollard's algorithms [12]. Meanwhile, there are currently lack of researches regarding the use of metaheuristic algorithms to factorize integers. As argued by Siarry [13], metaheuristic algorithms can be applied to all kinds of discrete problems and can be adapted to continuous problems. Some examples of metaheuristic algorithms are tabu search, hill-climbing, simulated annealing, and ant colony optimization algorithms.

The purpose of this work is to give some suggestions on how efficient metaheuristic algorithms can be used to factorize n, *i.e.*, the RSA public key. Here we test random-restart hill-climbing algorithm, which belongs to the class of metaheuristic algorithms, to factor small to large values of n. Random-restart hill-climbing algorithm is an improvement of hill-climbing algorithm, in which hill-climbings are repeated with random initial candidate solutions that are distant from each other in order to escape from local optimum solution. As a benchmark, we also work with a deterministic integer factorization algorithm called Pollard's rho algorithm [14] in order to determine which algorithm is faster to factorize n.

2. Methods

In this section we explain the RSA computation, the hill-climbing and the random-restart hill-climbing techniques, and the Pollard's rho algorithm. As an addition, we also provide our Python code for the Pollard's rho algorithm.

2.1 RSA

The RSA public key cryptography algorithm has three stages: key generation, encryption, and decryption. Suppose we have two parties that want to communicate, which are a sender called "Alice" and a recipient called "Bob". The computation involved in each stage, which can be found in [2] and [15], are explained below.

In the key generation stage, Bob will do the following:

1. Randomly generate two very large prime numbers, p and q, such that $p \neq q$. The values of p and q are Bob's private keys and, therefore, should be kept secret.
2. Compute $n = pq$. The value of n is Bob's public key and should be published so anyone who wants to send a message to Bob knows it.
3. Compute $\Phi(n) = (p - 1)(q - 1)$. The value of $\Phi(n)$ should also be kept secret.
4. Randomly select a value of e, such that $1 < e < \Phi(n)$ and $gcd(\Phi(n), e) = 1$. The value of e will be used in the encryption stage by Alice. Therefore, e should be published.
5. Compute d, such that $de \equiv 1 \pmod{\Phi(n)}$. The value of d should be kept private. It will be used in the decryption stage by Bob.

In the encryption stage, assuming that Alice has a message m, in this stage, she will do the following:

1. Get Bob's public keys, n and e.
2. Compute the cipher-text, $c = m^e \bmod n$.
3. Send c to Bob.

In the decryption stage, Bob will do the following:

1. Receive c from Alice.
2. Compute $m = c^d \bmod n$. Thus, Bob can now read the message from Alice.

2.2 Hill-climbing algorithm and random-restart hill-climbing algorithm

Hill-climbing is a simple metaheuristic optimization algorithm that works as follows [16].

1. Take a candidate solution and test the quality.
2. Generate a new random and slightly different candidate solution and test the quality.
3. If the a quality of the new candidate solution is better than the quality of the previous candidate solution, then this new candidate solution becomes the candidate solution.
4. Repeat step 2 and 3 until time is up or the ideal solution is found.

The second step (generating a new candidate solution that is only slightly different from the candidate solution is called 'tweak' or 'walk'). Since hill-climbing only takes a slightly different new candidate solution in each trial, it often only leads to a local optimum solution.

IOP Conf. Series: Journal of Physics: Conf. Series **943** (2017) 012057 doi:10.1088/1742-6596/943/1/012057

To minimize the probability of getting only a local optimum solution, hill-climbing is repeated for some random amounts of time. When the time is up, a new random solution that is more distant from the previous best solution will be generated, and hill-climbing is done again for some random amounts of time, and so on until the ideal solution is found. This whole technique is known as random-restart hill-climbing [16].

The step of generating a new random solution that is more distant from the previous best solution is called 'explore'. Doing the 'exploration' more than once is expected to maximize the chance of getting the global optimum solution, since there are many 'hills' to be explored to see whether or not the currently explored hill is better than the previously explored hills.

2.3 Pollard's rho algorithm

Pollard's rho algorithm is a method to factorize an odd integer. It works as follows [12].

1. Input n (the odd integer to be factored).
2. Select a polynomial with degree >= 2 and with all coefficients modulo n. In general, one may simply use $f(x) \equiv x^2 + 1 \pmod n$.
3. Select a random value x_0. In general, simply use $x_0 = 2$.
4. Set $i = 0$, $x_i = x_0$, $y_i = x_0$, and $d = 1$.
5. Compute $x_i = f(x_{i-1})$ and $y_i = f(f(y_{i-1}))$.
6. Compute $d = gcd(abs(x_i - y_i))$.
7. Set $i = i + 1$.
8. If $d = 1$, go to step 5. If not, go to step 9.
9. If $d = n$, then output failure or go back to step 2 to choose a different polynomial. If not, output d as one factor of n.

If n is RSA modulus and if in the last step $d \neq n$, then the value of d is either the value of p or q. Our Python code of Pollard's rho algorithm is as follows.

```
#title: Factoring RSA Modulus with Pollard's Rho Algorithm
#author: Mohammad Andri Budiman
#version: 1.01
#date: October 1st 2017
#time: 16:20

import time

def gcd(m, n):
    if n == 0:
            return m
    return gcd(n, m % n)

def f(x, n):
    return (x * x + 1) % n

def PollardRho(n):
    iterations = 0
    x = 2
    y = 2
    d = 1
    start = time.time()
    while d == 1:
            x = f(x, n)
            y = f(f(y, n), n)
            d = gcd(abs(x - y), n)
            iterations += 1
    stop = time.time()

    if d == n:
```

```
        print "failure"
else:
        print n, "=", d, "*", n // d
print "iterations =", iterations
print "time =", stop - start, "secs"
```

3. Results and Discussions

In the case of factoring the RSA modulus, we set the random-restart hill-climbing to work as follows. Two candidate solutions are generated in each trial. These two candidate solutions are random prime numbers p and q that are generated using Fermat's little theorem [15]. When testing the quality of these new candidate solutions, we actually multiply the p and the q, and check whether or not the result of this multiplication is close enough to the value of n. If it is close enough, then the 'walk' will be done for 10 times, but if it is not close enough, then in the next exploration, the new random values of p and q that are distant from the previous best p and q values will be generated.

Since RSA modulus is an exact number, then the only ideal solution that can be accepted is when $pq = n$. If the time limit is reached and pq is still not equal to n, then we let the algorithm end without a result.

The results of factoring RSA moduli with random-restart hill-climbing are shown in Figure 1-4.

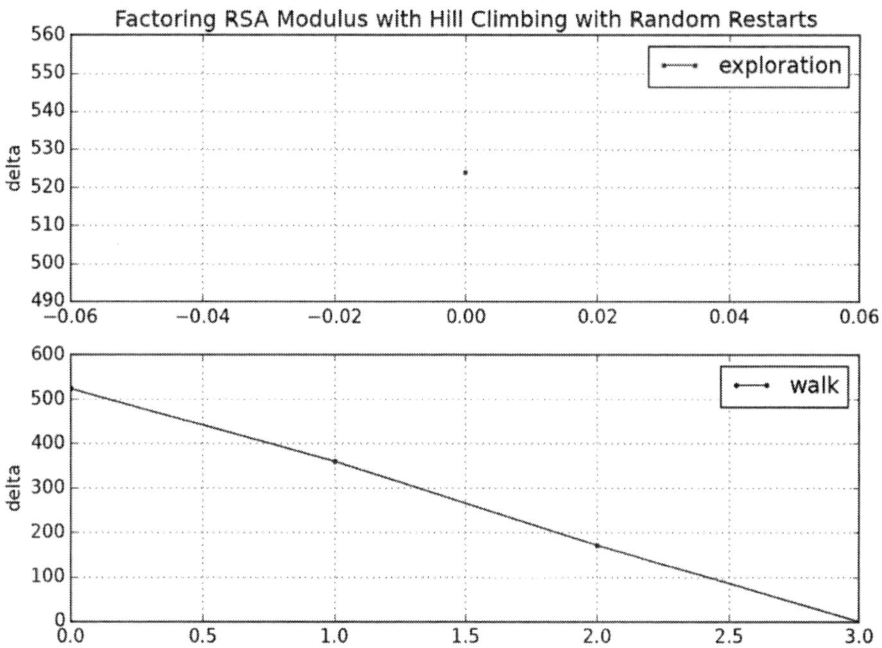

Figure 1. Factoring n = 989 with random-restart hill-climbing algorithm

It can be seen from Figure 1, that only one red dot (which means one exploration) and four blue dots (which mean four walks) are needed to factor n = 989 = 23 * 43. The running time needed is 0.00448393821716 secs.

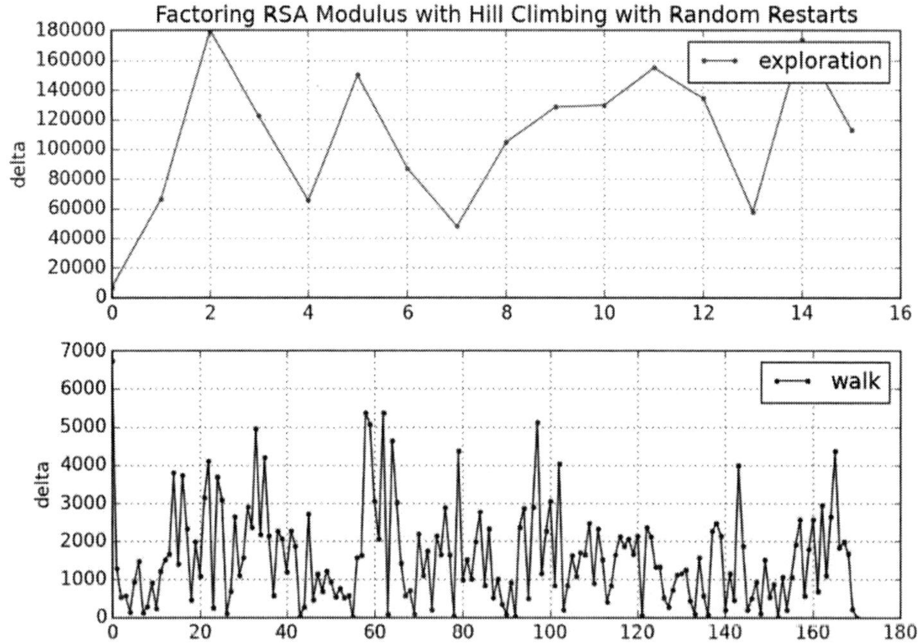

Figure 2. Factoring n = 13289 with random-restart hill-climbing algorithm

It can be seen from Figure 2 that 16 explorations and more than 160 walks are needed to factor $n = 13289 = 137 * 97$. The running time needed is 0.00215601921082 second. Note that the time needed to factor $n = 13289$ is lower than the time needed to factor $n = 989$. Therefore, the correlation between n and running time may not be directly proportional.

IOP Conf. Series: Journal of Physics: Conf. Series **943** (2017) 012057 doi:10.1088/1742-6596/943/1/012057

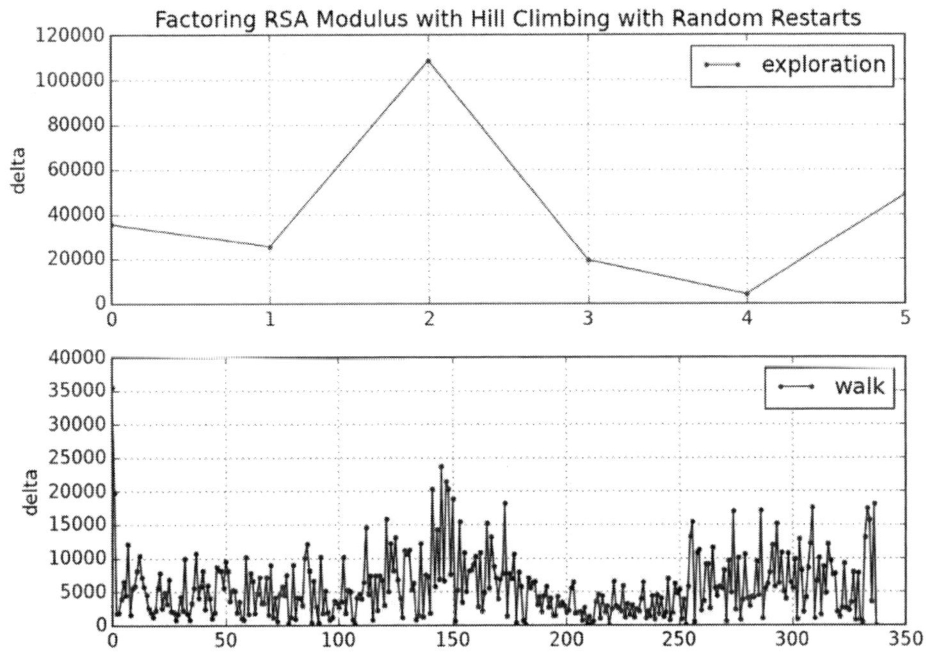

Figure 3. Factoring n = 70531 with random-restart hill-climbing algorithm

It can be seen from Figure 3, that 6 explorations and more than 300 walks are needed to factor $n = 70531 = 251 * 281$. The running time needed is 0.0644018650055 second. Note that the number of explorations to factor $n = 70531$ is lower than the number of explorations needed to factor $n = 13289$. Therefore, the correlation between n and the number of explorations may not be directly proportional.

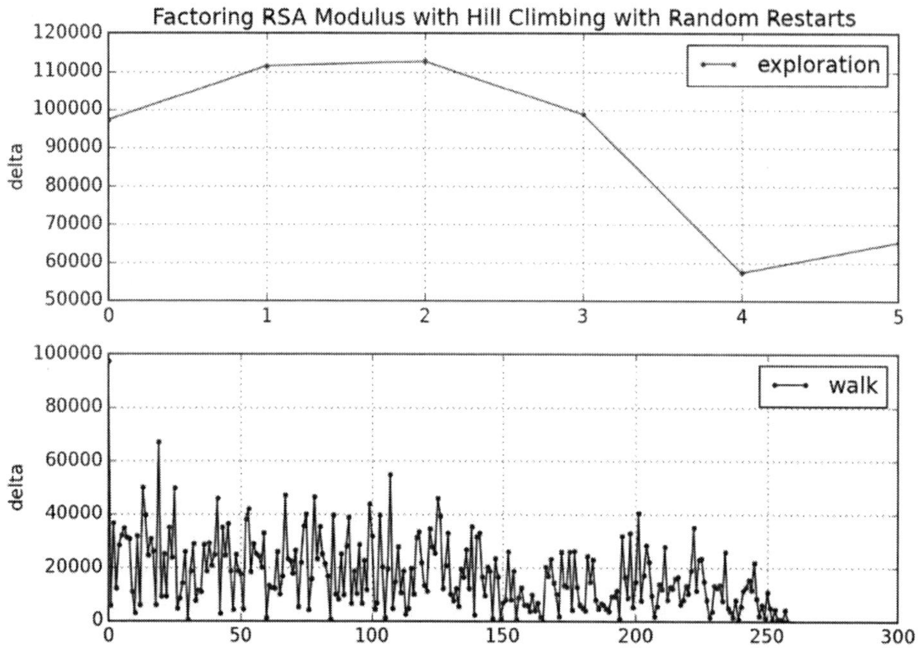

Figure 4. Factoring n = 194111 with random-restart hill-climbing algorithm

It can be seen from Figure 4 that 6 explorations and more than 250 walks are needed to factor $n = 194111 = 389 * 499$. The running time needed is 0.000250101089478 second. Note that the number of walks needed to factor $n = 194111$ is lower than the number of explorations needed to factor $n = 70531$. More surprisingly, the time needed to factor $n = 194111$ is lower than the time needed to factor $n = 989$. Therefore, the correlation between n and the number of walks may not be directly proportional, and the correlation between n and running time may also not be directly proportional.

We also try to use random-restart hill-climbing algorithm to factor a very large value of RSA modulus ($n = 1356992669$), but the algorithm fails to find the corresponding p and q within 20 seconds, so the program automatically ends.

In Table 1, we show the results of factoring RSA moduli with Pollard's rho algorithm.

Table 1. Factoring RSA moduli with Pollards's rho algorithm

n	p * q	iterations	time (seconds)
989	23 * 43	4	0.00002.19345092773
13289	97 * 137	3	0.0000181198120117
70531	281 * 251	8	0.0000488758087158
194111	389 * 499	13	0.0000498294830322
1356992669	24691 * 54959	215	0.000957012176514
11352765949520300619931	19401210197 * 58515761822	122130	1.52793002129

From Table 1, it can be seen that the larger the value n, the number of iterations and the running time needed to factor n tend to increase.

Table 2. Time comparison between random-restart hill-climbing and Pollard's rho

n	Random-Restart Hill-Climbing (secs)	Pollard's rho (secs)
989	0.00448393821716	0.00002.19345092773
13289	0.00215601921082	0.0000181198120117
70531	0.0644018650055	0.0000488758087158

194111	0.000250101089478	0.0000498294830322
1356992669	failed	0.000957012176514
1135276594952030061993	failed	1.52793002129

From Table 2, it can be seen that in RSA modulus factorization, the Pollard's rho algorithm is much faster than random-restart hill-climbing algorithm.

4. Conclusions

The conclusions of this work are as follows. Firstly, in random-restart hill-climbing algorithm, the correlation between RSA modulus n and the number of explorations, the number of walks, or running time is not directly proportional. Secondly, in Pollard's rho algorithm, the correlation between RSA modulus n and the number of iterations or running time tends to be directly proportional. Thirdly, random-restart hill-climbing algorithm is only acceptable to factor smaller values of n, while Pollard's rho algorithm is suitable to factor larger values of n. Fourthly, Pollard's rho algorithm is much faster and more reliable than random-restart hill-climbing to factor n.

Acknowledgments

We gratefully acknowledge that this research is funded by Lembaga Penelitian Universitas Sumatera Utara. The support is under the research grant TALENTA USU of Year 2017 Contract Number : 5338/UN5.1.R/PPM/2017.

References

[1] Diffie W and Hellman M 1976 New directions in cryptography *IEEE Transactions on Information Theory* **22** 644-54
[2] Rivest R L, Shamir A and Adleman L 1978 A method for obtaining digital signatures and public-key cryptosystems," *Communications of the ACM* **21** 120-26
[3] Boneh D and Shacham H 2002 Fast variants of RSA *CryptoBytes* **5** 1-9
[4] Fiat A 1997 Batch RSA *Journal of Cryptology* **10** 75–88
[5] Takagi T 1998 Fast RSA-type cryptosystem modulo p^k*q *Advances in Cryptology — CRYPTO 98* Lecture Notes in Computer Science 318–26
[6] Wiener M J 1990 Cryptanalysis of short RSA secret exponents *IEEE Transactions on Information Theory* **36** 553-8
[7] Quisquater J J and Couvreur C 1982 Fast decipherment algorithm for RSA public-key cryptosystem *Electronics Letters* **18** 905-07
[8] Paixao C A M and Filho D L G 2003 An efficient variant of the RSA cryptosystem *IACR Cryptology ePrint Archive*
[9] Boneh D and Venkatesan R 1998 Breaking RSA may not be equivalent to factoring *Advances in Cryptology - EUROCRYPT ' 98* **1233** 59–71
[10] Boneh D 1999 Twenty years of attacks on the RSA cryptosystem *Notices of the AMS* **46** 203-13
[11] Kiviharju M 2017 On the fog of RSA key lengths: Verifying public key cryptography strength recommendations *International Conference on Military Communications and Information Systems (ICMCIS) IEEE*
[12] Batten L M 2013 *Public key cryptography applications and attacks* (Hoboken, N.J: Wiley-Blackwell)
[13] Siarry P 2016 *Metaheuristics* (Cham: Springer)
[14] Sarnaik S, Bhakkad R and Desai C 2015 Comparative study on integer factorization algorithm-Pollard's rho and Pollard's P-1 *2nd International Conference on Computing for Sustainable Global Development (INDIACom) IEEE* 677-79
[15] Mollin R A 2007 *An introduction to cryptography* (Boca Raton, FL: Chapman & Hall/CRC)
[16] Luke S 2013 *Essentials of Metaheuristics* 2nd ed (Lulu) available at http://cs. gmu. edu/~ sean/book/metaheuristics/

IOP Conf. Series: Journal of Physics: Conf. Series **943** (2017) 012058 doi:10.1088/1742-6596/943/1/012058

Helping students mathematical construction on square and rectangle's area by using Sarong motive chess

Eka Zuliana[1,2], Fariz Setyawan[3] and Arsaythamby Veloo[4]

[1]Universitas Muria Kudus, Gondangmanis, Bae, Kudus 59327, Indonesia
[2]Universitas Negeri Yogyakarta, Jl. Colombo, Yogyakarta 55281, Indonesia
[3]Universitas Ahmad Dahlan, Jl. Prof. Dr. Soepomo, S.H. Yogyakarta 55164, Indonesia
[4]Universiti Utara Malaysia, Sintok 06010, Malaysia

E-mail: eka.zuliana@umk.ac.id

Abstract. The aim of this study is developing the learning trajectory to construct students' understanding of the concept of the area of square and rectangle by using Sarong Motive Chess. This research is a design research which is consists of three stages. The stages are preparing for the experiment, designing experiment, and making a retrospective analysis. The activities started by the activity of using sarong motive chess as the manipulative measurement unit. The Sarong motive chess helps students to understand the concept of area of square and rectangle. In the formal stage of cognitive level, students estimate the area of square and rectangle by determining the square unit at the surface area of sarong through many ways. The result of this study concludes that Sarong motive chess can be used for mathematics learning process. It helps the students to construct the concept of a square and rectangle's area. This study produces learning trajectory to construct the concept of a square and rectangle's area by using Sarong motive chess, especially for elementary school students.

1. Introduction

Elementary school teachers should understand the cognitive development level of their students because they need to apply the appropriate strategies of learning for the students. Piaget stated that the logical stages consist of pre-operational, operational concrete and operational formal stages [1]. Elementary school students in Indonesia are children whose ages are 7-11 years old. Based on Piaget's logical stages, elementary school students are classified to the operational concrete stage of Piaget's cognitive theory [2]. However, teachers have a key role to manage the classroom instruction [3]. A teacher has a major impact on student's learning activities. *Kemendikbud* [4] stated that many teachers in Indonesia have not involving their students in teaching and learning process. Students, as the subject of education, have not actively participated in the learning process. Thus, teacher, especially in elementary school, should aware to their students that they are children who have still in operational concrete stages. In fact, teaching and learning process in mathematics at the elementary school is still delivered by the teacher using abstract concept [5–7]. The elementary school teacher was just only giving the concept or formula, giving examples and doing exercises. This condition makes students only memorize the given formulas, substitute what is known and unknown variables into the formulas and then calculate it [4]. Consequently, the process of constructing concepts of mathematics, the competence of students' reasoning and problem-solving were not well-established. Besides, the literacy of Indonesian students in mathematics subject is still in the lower grade. it can be seen on PISA's and TIMSS Study derived by Stacey [8] that in 2015, the average score on mathematics, Indonesian students only reach 386. This result ranks the competence of Indonesian student in literacy is in 65th of 72 countries. Moreover, Indonesian score on TIMSS 2015 only reaches score 397. This condition ranks Indonesia in 45th of 50 countries.

In the elementary school, subject matter such as mathematics, science, civic, social science, and *Bahasa Indonesia* subject have been integrated into thematic learning process. The teacher will be able to teach and decides the best learning strategies according to the cognitive levels of their students, especially in the operational concrete stage of Piaget's cognitive theory [9]. Students, whose still in the operational concrete stage, are more interested in manipulating objects [10]. The teacher has a responsibility to prepare the learning strategies which is appropriate with student's cognitive level. Learning strategy by using manipulative is a way for a teacher to teach mathematics materials by considering the cognitive level of students whom still in the operational concrete stage. Many research [10–12] had shown that manipulative were used effectively, students are able to construct their own mathematical concept through a valuable learning experience. In the topic of the area of the plane figure, the square unit is a unit area that most widely used [13]. In that topic, the teacher can mix the concept of the square unit using manipulative. Sarong is Indonesian famous traditional cloth. it has many motives. One of them is chess motive like used by Balinese. The chess motif of Sarong can represent the concept of the square unit. It can be used to help students in constructing the concept of square and rectangle's area. The aim of this study is to develop a learning trajectory of a square and rectangle's area by using sarong motive chess.

2. Method

The type of this research is a design research study. The process of design research, as well as in the educational design process, consists of analysis, design, evaluation, and revision [14-16]. This process is a cyclical process that ends until theory and practice has the same result [14, 16]. Design research is a methodology which has some characteristics. They are interventionist nature, process-oriented, a reflective component, cyclic character, and theory-oriented [17]. There are two important components of a design research, Hypothetical Learning Trajectory (HLT) and Local Instruction Theory (LIT). Both of them will exist on learning activities that may be taken by students in their learning process [18]. The learning trajectories are underlying the curriculum [19]. Because of that, learning trajectories must be prepared in the learning process through the level of thinking [20]. There are some components at Hypothetical learning trajectory: the goal of learning, the learning activities, and the thinking in which students might hypotheses about the process of students' learning [17].

The stages of design research are described as follows. Firstly preparing for the experiment, in this stage, the hypothetical learning trajectory (HLT) is created as a learning activity. While formulating HLT, a preliminary design is required which implements the initial idea derived from the literature review before designing the learning activities, discussions with experienced teachers, researchers and experts in the related fields, secondly is design experiment to collect data which is used to answer the research question, and lastly is retrospective analysis (analysis of data obtained from previous stages), in this stage the researchers analyzed the data which is obtained from the experimental design. Then, the data is used based on the results of the analysis to develop the next design.

3. Result and discussion

The result of this study indicates that learning design of square and rectangle's area, in mathematics learning process; by using sarong motive chess have an important role to construct the concept the area of a plane. The square unit, as a motive of Sarong, can help the students to determine the area of square and rectangle.

IOP Conf. Series: Journal of Physics: Conf. Series **943** (2017) 012058 doi:10.1088/1742-6596/943/1/012058

3.1. Preliminary design

The first stage of design research is preliminary design. The aim of this stage is constructing the concept as a result of conjectured Local Instruction Theory (LIT) which is consisting of three components: (a) learning objectives for the students, (b) planning of the instructional activities and the tools or the manipulative, and (c) a hypothetical learning trajectory (HLT), so that the teacher is able to anticipate students' thought and to understand which instructional activity that can be used in the classroom. In the first phase, the process has begun by formulating the objectives of learning, combined with the experimental anticipatory thinking activities. The teacher as a learning designer imagines how the process of teaching and learning activities is conducted in the classroom [21]. Preliminary design consists of the following steps. Firstly, researcher determine the objective of learning. The objective of learning is helping students' construction process to build the concept of square and rectangle's area by using Sarong motive chess. Secondly, researcher determine some instructional activities. In this instructional activities researcher measuring the area of square and rectangle using Sarong motive chess and determine the HLT.

The area of a plane is a measure of the amount of space a two-dimensional shape. That is the space that is enclosed by its boundary. Constructing the concept of area of a plane, students activities is derived as follow. First, students must understand the characteristics of the area of a plane before they were measuring it. Second, they cut the plane into two parts then rearrange it into many various shapes. It showed up that the shapes have the same area, before and after they cut it. In measuring area, the students use a unit area. Using a square unit as the unit area is a common way to help students to build their understanding of a concept of a topic in mathematics [22]. Some people can use colored square tiles, the other can use square pieces of cardboard. In measuring the area, the students used sarong motive chess. The square and rectangle's area construction showed at Table 1 and Table 2.

Table 1. Square's area construction

No	Square	Side	Square unit	The area
1	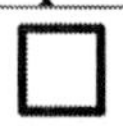	1	1	1 = 1 x 1
2		2	4	4 = 2 x 2
3		3	9	9 = 3 x 3
4	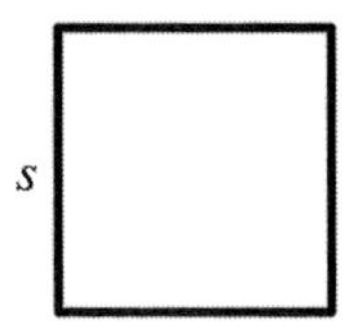	s		Area = s x s

Table 2. Rectangle's area construction

No	Square	length	width	Square unit	The area
1		2	1	2	2 = 2 x 1
2		4	2	8	8 = 4 x 2
3		5	3	15	15 = 5 x 3
4	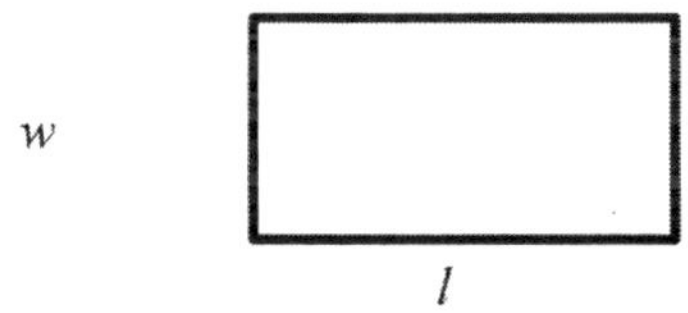	*l*	*w*		Area = *l x w*

In the instructional activity, researcher derived the Hypothetical learning trajectory (HLT) of the square and rectangle's area for elementary school students (Table 3). The HLT was formulated as the teaching and learning activities (Figure 1).

Table 3. HLT of square and rectangle's area by using sarong motive chess

No	Learning objectives	Learning activities	Hypothesis	Concept
1	Students are able to recall the square and rectangular shape	By observation, students observed a few plane and then matched sarong with suitable plane	- Students are able to mention the shape of the plane while they find the right frame of square and rectangle - There are inverted concept between square and rectangle	Accuracy of type and shape of plane: square and rectangle
2	Students are able to find the concept of plane's area	By experiments, students find the concept of the plane's area.	- Students are able to find the concept of area and differentiate with perimeter concept - There are inverted concept between area and perimeter	Area concept
3	Students are able to find the concept of square and rectangle's area	By experiments using sarong motive chess students find the concept of square and rectangle's area	- By experiment, students can determine the square and rectangle's area by using sarong motive chess - There is an inaccuracy of the concept of square and rectangle's area	The concept of square and rectangle's area

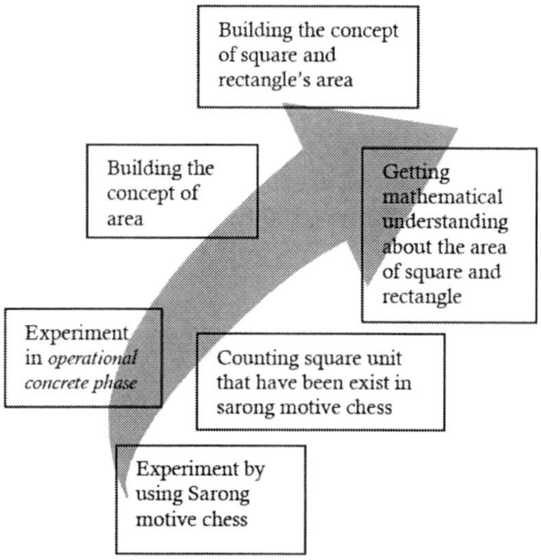

Figure 1. HLT square and rectangle's area by using sarong motive chess

3.2. Teaching Experiment

The learning activities were formulated in the previous phase. Teaching experiment was conducted for the 3rd-grade students of 1 Pati Kidul Elementary School. Researchers are assisted by the team, teacher of the class and undergraduate students of Universitas Muria Kudus. The teacher, as the partner in the research, observes the management of class as well as contributes giving questionnaire response. The research team observes students' learning activities, and some undergraduate students make photograph documentation of learning activities. The learning stages are described as follows. Firstly, researcher uses sarong motive chess as realistic problem context to discover the concept of square and rectangle's area. Secondly, researcher use of models for progressive mathematization. In this stage reasearcher is assisted by discovery board concept of the square and rectangle's area. Thirdly, researcher uses student construction result, in this condition students use the given context to obtain the concept of square and rectangle's area. The student's paperwork and construction result were used as the basic knowledge of the development of mathematical concepts. The next step, researcher develop interactivity and character. At the learning process occurs social interaction between students to students and students to the teacher. Social interaction is one example of character education for students to respect the opinions of others and be democratic. The demands of presenting the idea of the discovery rectangular area concept in the discussion developed into a form of awareness and responsibility in communicating ideas to the environment. Lastly is intertwinning concept. In this stage, while students were discovering the concept, several interrelated mathematical concepts such as geometry concept, square area, and multiplication are used.

3.3. Retrospective Analysis

Student's activities, in constructing the concept of the square and rectangle's area by using Sarong, have different steps with the common activities which are implemented in class before. In this study, Sarong motive chess is used as a context to build the concept of the square and rectangle's area. Both researchers and teacher conduct a retrospective analysis which is used to reflect and analyze the learning process. Both of them were also comparing the HLT and the instructional design with the reality that occurs in teaching and learning process. Overall, the learning process is well-organized. Some findings which were related to student activity in constructing the concept of square and rectangle's area is described as follows.

IOP Conf. Series: Journal of Physics: Conf. Series **943** (2017) 012058 doi:10.1088/1742-6596/943/1/012058

3.3.1. Observation Activity of Sarong Motive Chess to find the concept of square and rectangle
Rani, a student who sat in front of the class, was asked to mention two shapes in the Sarong. She was able to answer the shape in Sarong, which is a square and a rectangle, but Rani confused to point out which one is square and rectangle. Fortunately, Soleh helps Rani by pointing out that the rectangle has long edges (called the length) and short edges (called the width) while the square has the same edges of the length.

3.3.2 The concept of Area
Based on the observation conducted by the researcher, it is difficult for students to derive the concept of Sarong's area. After the teacher gives a hint, Ifana, one of the students in the class, was able to answer that the area of Sarong is the surface itself. Then, the teacher asks the students what the surface area it was? The students answer that the surface area of Sarong is its fabricants. By using these guided questions, students are able to construct the concept of area of a plane. Students explained that the surface area of Sarong is the surface area of its fabricants.

3.3.3 The concept of Square and Rectangle's Area
In this activity, each group tried to find the concept of square and rectangle's area by using Sarong and student's discovery board (Figure 2 and 3). Teachers have a role as mentors and facilitators. The student's paperwork presented the work belonging to several groups of students (Figure 4).

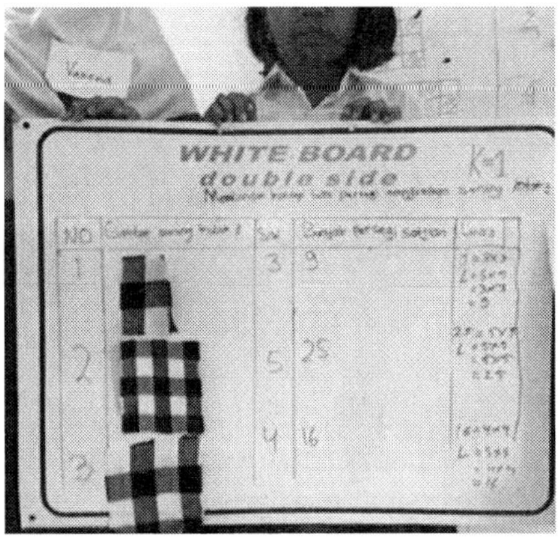

Figure 2. First group' cardboard

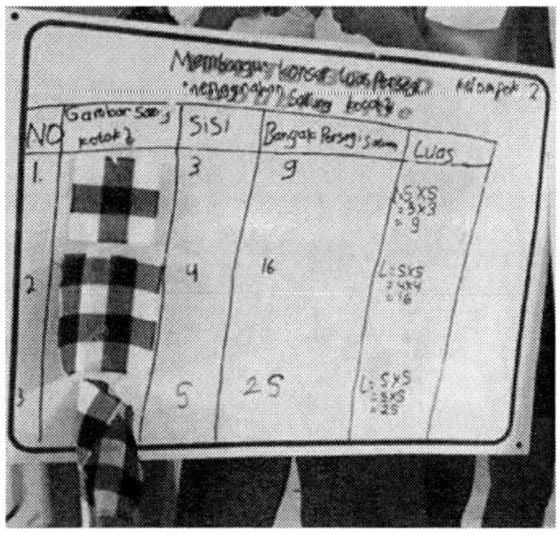

Figure 3. Second group's cardboard

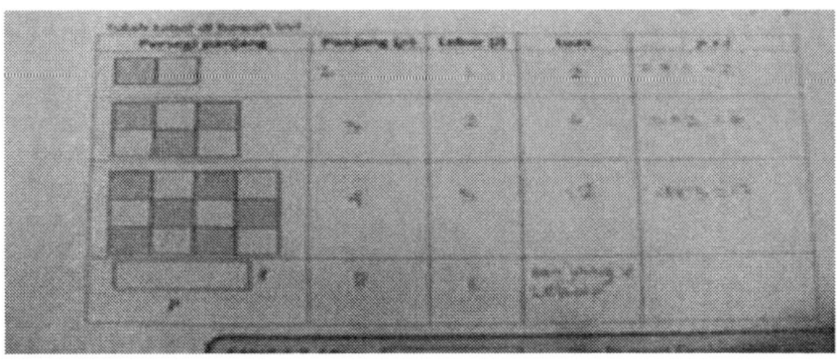

Figure 4. Students' paperwork

There are three activities to construct the concept of square and rectangle's area by using Sarong motive Chess. There are determining the number of the square unit as a motive at Sarong one by one, determining the number of the square unit as a motive at Sarong based on the number of rows, determining the number of the square unit as a motive at Sarong based on the number of columns. By doing this activity, students are not afraid to give new ideas, share with each other about their ideas, and produce products [23, 24]. In addition, based on the student's paperwork, it shows that the student was able to show how many square units indicate the side and length, and how many square units indicate the width and the square and rectangle's area which is indicated by the number of square units in it.

In comparison with the previous studies, students used Caping to construct the concept of the area of a circle [11]. It found that the implementation of manipulative helps students to construct the concept of area of a plane as well as Sarong. Since Sarong can represent the square unit of a plane. It should be possible for researchers to implement it in the circle for the further study. The process of mathematization constructed by the students through the use of a model is illustrated in the iceberg of the invention square and rectangle's area in Figure 5.

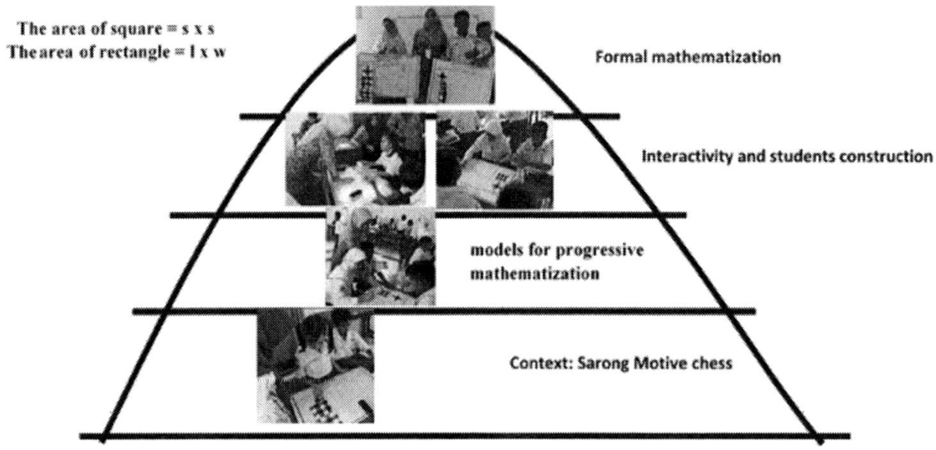

Figure 5. Iceberg of Students Construction on Square And Rectangle's Area By Using Sarong Motive Chess

4. Conclusions

Based on the data can be concluded that sarong motive chess can be used for mathematics learning process. It helps the students to construct the concept of a square and rectangle's area. This study produced learning trajectory to construct the concept of a square and rectangle's area by using sarong motive chess, especially for elementary school students.

5. Acknowledgment

The researchers would like to thank and give appreciation to The Directorate for Research and Community Service Indonesian, Directorate General of Higher Education, for the financial support in this study.

References

[1] Demetriou A, Shayer M and Efklides A 2017 *Neo-Piagetian theories of cognitive development : implications and applications for education* (Routledge)
[2] Piaget J 1973 *Main trends in psychology* (Allen and Unwin)
[3] Setyawan F 2015 Conceptual Understanding Profile of LEOV Junior High School Students Based on Kolb's Learning Style *International Conference on Mathematics, Science, and Education (ICSME)* (Semarang: Universitas Negeri Semarang) pp 61–3
[4] Kemendikbud RI 2015 *Rencana Strategis Kementrian Pendidikan dan Kebudayaan 2015-2019*
[5] Pukjiwati 2017 Upaya Meningkatkan Aktivitas Hasil Belajar Siswa pada Materi Pecahan dengan Pendekatan Contextual Teaching and Learning Class IV SDN SUMUR 03 *J. Refleks. Edukatika* **7** 85

[6] Fitriani C, Murniati A R and Usman N 2017 Kompetensi Profesional Guru dalam Pengelolaan Pembelajarn di MTs Muhammadiyah Banda Aceh *J. Magister Adm. Pendidik.* **5** 16
[7] Anam R S, Widodo A and Sopandi W 2017 Representation of Elementary School Teachers on Concept of Heat Transfer *Journal of Physics: Conf. Series* p 895
[8] Stacey K 2011 The PISA view of mathematical literacy in Indonesia *Indones. Math. Soc. J. Math. Educ.* **2** 95
[9] Jehan S and Butt M N 2014 Attainment of Conservation Ability Among Primary School Children in the Light of Piaget ' S Cognitive Theory *VFAST Trans. Educ. Soc. Sci.* **5** 26
[10] Ariyadi Wijaya 2008 Design Research in Mathematics Education Indonesian Traditional Games as Preliminaries in Learning Measurement of Length *Konferensi Nasional Matematika* ed Zulkardi (Palembang: IndoMS) pp 24–7
[11] Zuliana E 2014 E – 12 Caping Kalo As Kudus Cultural Heritage To Construct Circle Concept Of Primary School Students *International Seminar on Innovation in Mathematics and Mathematics Education* (Yogyakarta: Yogyakarta State University) pp 83–8
[12] Prahmana R C I, Zulkardi Z and Hartono Y 2012 Learning Multiplication Using Indonesian Traditional game in Third Grade *J. on Math. Educ.* **3** 115
[13] Van de Walle J A, Karp K S and Bay-Williams J M 2012 *Elementary and middle school mathematics : teaching developmentally* (Allyn & Bacon)
[14] Prahmana R C I, Kusumah Y S and Darhim 2017 Didactic trajectory of research in mathematics education using research-based learning *J. Phys.: Conf. Ser.* **893** 012001
[15] Prahmana R C I 2017 *Design Research (Teori dan Implementasinya: Suatu Pengantar)* (Jakarta: Rajawali Pers)
[16] Prahmana R C I and Kusumah Y S 2016 The hypothetical learning trajectory on research in mathematics education using research-based learning *Pedagogika* **123** 42
[17] Akker J van den, Brenda B, Anthony E K, Nienke N and Tjeerd P 2013 Educational Design Research *Educational Design Research, Part A: An introduction* ed N Nieveen and T Plomp (Enschede: SLO, Netherlands institute for curriculum development) p 72
[18] Prahmana R C I 2013 Designing Division Operation Learning in the Mathematics of Gasing *1st SEA-DR* (IndoMS) pp 391–8
[19] Confrey J, Maloney A P and Corley A K 2014 Learning trajectories: a framework for connecting standards with curriculum *ZDM - Int. J. Math. Educ.* **46** 719
[20] Sarama J and Clements D H 2011 *Research-based Education* (New York: Information Age Publishing)
[21] Pramudiani P, Zulkardi, Hartono Y and van Amerom B 2011 A Concrete Situation for Learning Decimals *Indones. Math. Soc. J. Math. Educ.* **2** 215
[22] Wahyu K, Amin S M and Lukito A 2017 Motivation Cards to Support Students ' Understanding on Fraction Division *International. Journal on Emerging Mathematics Education* **1** 99
[23] Collins A, Joseph D and Bielaczyc K 2009 Design Research: Theoretical and Methodological Issues *J. Learn. Sci.* **13** 15
[24] Tanujaya B, Prahmana R C I and Mumu J 2017 Mathematics instruction, problems, challenges, and opportunities: A case study in Manokwari regency, Indonesia *World Transactions on Engineering and Technology Education* **15** 287

IOP Conf. Series: Journal of Physics: Conf. Series **943** (2017) 012059 doi:10.1088/1742-6596/943/1/012059

Teachers' ability in using math learning media

Masniladevi[1], R C I Prahmana[2], Y Helsa[1] and M Dalais[1]

[1]Universitas Negeri Padang, Jl. Prof. Dr. Hamka Air Tawar, Padang, Indonesia
[2]Universitas Ahmad Dahlan, Jl. Pramuka Kav. 5, Pandeyan, Yogyakarta, Indonesia

E-mail: masniladevi@yahoo.com

Abstract. The studies aim to enhance teachers' knowledge and skill in making math instructional media, develop math instructional media, train and assist the use of instructional media in learning math in the classroom. The method used in the activities adopted the pattern of preventive implementation, planning stage, program implementation, observation and evaluation and reflection. The research results show that the evaluation of teachers' ability is still in average category. The result required more intensive training.

1. Introduction

The development thinking of primary school students is still in a concrete operational stage [1, 2]. It is because the logical thinking process is even based on the physical manipulation of the objects [3]. Students can learn math based on clear and visible facts when using teaching aids [4, 5]. The subject that students learn will be easier to learn and understand when the learning process seems real to them [6, 7]. Therefore, students need learning media in the form of actual objects (teaching aids) during the learning process.

Cope [8] says that the review of the research on the effectiveness of using math manipulative supported by the use of concrete, pictorial, and virtual manipulative. Furthermore, primary school teachers who use manipulative to help teach math can positively affect student learning manipulative objects in learning mathematics have a positive impact on the students' learning process [2, 7, 9]. So, the teacher who can design the learning of mathematics that is abstract to change learning to be concrete with teaching aids would be able to improve student learning outcomes.

Several studies have documented the results of their research regarding the use of learning media that is manipulative [2, 10, 11]. Indonesian traditional games could be used in teaching mathematics like learning multiplication [2] and measuring time [10]. On the other hands, Haris and Putri [11] teach measuring area using one of Indonesian context namely *anyaman*. Furthermore, several contexts can be used as starting point in learning mathematics.

In reality, the government has made efforts to assist Math Learning Media (MLM) tools in primary schools in Indonesia especially in the Pariaman city for the mathematics learning process becomes more meaningful for students. Observations made by the researcher shows that teachers rarely use

MLM that has been given by the government. Teachers do not use the visible tool in MLM during the learning process. Though the teaching materials taught can be taught by using MLM. The condition of MLM is quite a lot and well maintained, but only display in the warehouse storage. It makes the big question for the researchers why it can happen.

Based on the interviews, the researchers found that teachers strongly agreed in the use of MLM tools. The tool of MLM is essential for the learning process, but they are sometimes confused in matching learning materials with MLM used. Also, the problems faced by teachers are the teacher's ignorance in using the available tools so that teachers hesitate in the utilization of the MLM. Even some teachers feel bothered when it comes to explaining learning materials using MLM.

Yanti et al. [12] stated that the utilization of MLM has been implemented by teachers in schools but not yet optimal. Teachers' preparation is limited to checking the availability of MLM alone not considering the completeness of MLM or testing. More implementation of the Standards Competence wakes up, and some wake up only while other competencies do not take advantage of existing MLM. The ability of teachers in utilizing the MLM still needs to be improved. MLM storage requires attention to be more organized and well maintained. Support from school principals and supervisors should be done to motivate teachers to utilize MLM at school.

It proves that the use of MLM is still not optimal. Continuous use of aids in mathematics learning in primary schools can provide satisfactory results on student learning outcomes. It is following the results of a study conducted by Nurhanif [13] which states that the use of MLM could support learning in free materials and build a flat area in class IIIA Primary School 16 Banda Aceh. In line with research conducted by Manurung, et al. [14] state that the use of MLM can improve student learning outcomes. Based on the explanation above, researchers interested to see the ability of teachers in using the MLM. Furthermore, researchers provide treatment to teachers to maximize any existing learning media.

2. Method

Learning implementation phase is divided into three major stages. First, the researcher explained to the teacher basic theoretical training and working principle of various kinds MLM. Furthermore, the researcher gives an opportunity to the teachers to try using MLM. Finally, the researcher motivates the trainees to develop or modify the MLM so that it can add a collection of teaching aids in their school.

The form of learning evaluation conducted consists of three major stages. First, the researchers did an initial assessment that was used to measure the first ability of the prospective teachers. Furthermore, the researcher evaluates the process used to measure the knowledge of research subjects at each stage of learning using MLM. Process evaluation is used to refine the next activity. Finally, the researcher conducts a final useful assessment to measure learning achievement with success indicator is a component of the learning outcome.

3. Results and discussion

The implementation of mathematics learning using MLM divided into two major sections. Each section is held in the form of a workshop meeting, described as follows.

The first session was held on August 19, 2017. Workshop and training activities on developing the ability to make and using media for these teachers focused on doing media contextual and IT-based media training. It aims to improve the quality of mathematics learning of primary school students through teachers who have mastered MLM. This activity was held in one classroom at SD Negeri 11

Koto Marapak. The material that has been delivered during the learning process is the definition of instructional media, the function of instructional media, the study of the making and utilization of mathematics learning media in primary school, the making of contextual media, training of ICT-based media development, and learning simulation using media. Teachers must be able to select and use media in accordance with the material presented. Teachers are expected to be ready to try to use the media accessible to handle students.

Furthermore, teachers need to take advantage of learning media because it can improve student responsiveness and improve the quality of education [15]. Learning process using only lecture method will cause boredom for students so that the laziness of students in receiving learning. The quality and student learning outcomes would be reduced if we could no using MLM.

The second session was held on September 9, 2017. The teacher's ability in using MLM was done by one teacher from SD 11 Koto. She presented the results of KPK and FPB teaching aids which can be used for grade 4 of primary school as shown in Figure 1. Media made of cardboard and used plastic bottles are used to find Lowest Common Multiple (LCM) and Greatest Common Divisor (GCD) from a number. The number used is represented by using candy. Next, how to use the teaching aids by placing a candy that represents the multiples of each number that the KPK and FPB want to find out. Each bag on the media is filled with some sweets by the multiples of numbers. Then look at the most candy bags.

Figure 1. Teacher present the teaching aids of LCM and GCD

Based on the evaluation form provided, the learning activities are implemented successfully, although there are still some obstacles. The success of the learning process, seen from most of the targets can be achieved, that teachers can make one or several mathematical learning tools from the cheap tools and materials provided, but the tools and functions are quite functional. Next, teachers can understand various accurate elements or concepts in primary schools. Lastly, the teacher can design the learning activities for the implementation of the MLM that is produced as outlined in the form of student worksheets. The response of the research subjects to the learning process is positive, indicating that although many participants have enough experience related to MLM, they still need the empowerment to broaden their insights and experiences in developing the media of mathematics model.

Furthermore, the researcher explained that the innovation of learning the mathematics of primary school oriented to the learning objectives in the curriculum [16]. Then the characteristics of learning mathematics must also be adjusted with the level of thinking the development of elementary students, as well as the most critical point, is the process of student-centered learning, so that students are expected to be active and creative.

During the presentation, the teachers were involved in providing other ideas and suggestions on the performance of their colleagues. The assessment process is using Macromedia Flash 8 software. Teachers are guided in installing Macromedia Flash according to the guide seen in Figure 2.

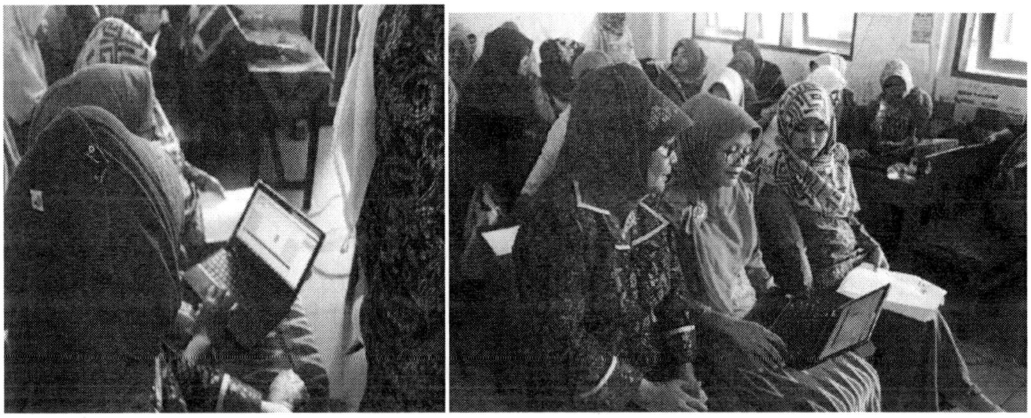

Figure 2. All participants install Macromedia Flash 8 according to the guide

The presenters explained that to understand the creation of flash animation media, student-assisted participants first install the software to each participant's laptop (Figure 3).

Figure 3. Installing process of Macromedia Flash 8

Finally, the evaluation of the implementation of learning activities divided into four things. First, attendance and participation rate of participants through presence attendance is 90% because some participants cannot attend. Furthermore, based on the implementation schedule, as well as the timeliness of the implementation, this devotion activity has been done well, as planned. Third, the participants gain new insights about the media of mathematics learning at the end of the event. Mathematical tools in the form of games and real concept make media and positive chip teaching aids in understanding the concept of integers, as well as interactive learning media which is the development of IT-based learning media [15-17]. Lastly, the speaker explained that to follow the creation of flash animation media, student-assisted participants first install the software to the laptop.

4. Conclusion

Teachers can create and use MLM, especially various kinds of visuals such as volume and flat builds and numbers and so on. Besides, teachers have been able to maximize the use of learning media that can be used in the development of the concept of learning mathematics. Furthermore, the teacher has been able to choose teaching media in mathematics learning that can be used in the cultivation of the idea of learning math. Finally, teachers can develop and create mathematics learning media associated with abstract concepts using IT-based learning media, so that learning becomes interactive.

References

[1] Tall D 2002 The psychology of advanced mathematical thinking In *Advanced mathematical thinking* (Netherlands: Springer) pp 3-21

[2] Prahmana R C I, Zulkardi and Hartono Y 2012 Learning multiplication using Indonesian traditional game in third grade *Journal on Mathematics Education* **3** 115

[3] Reiner M and Gilbert J 2000 Epistemological resources for thought experimentation in science learning *International Journal of Science Education* **22** 489

[4] Biggs J B 2011 *Teaching for quality learning at university: What the student does* (London: McGraw-Hill Education)

[5] Loewenberg Ball D, Thames M H and Phelps G 2008 Content knowledge for teaching: What makes it special? *Journal of Teacher Education* **59** 389

[6] Entwistle N and Ramsden P 2015 *Understanding Student Learning* (Abingdon: Routledge)

[7] Tanujaya B, Prahmana R C I, and Mumu J 2017 Mathematics instruction, problems, challenges, and opportunities: A case study in Manokwari regency, Indonesia *World Transactions on Engineering and Technology Education* **15** 287

[8] Cope L 2015 Math manipulative: Making the abstract tangible *Delta Journal of Education* **5** 10

[9] Boggan M, Harper S and Whitmire A 2010 Using manipulatives to teach elementary mathematics *Journal of Instructional Pedagogies* **3** 1

[10] Jaelani A, Putri R I I and Hartono Y 2013 Students' strategies of measuring time using traditional Gasing game in third grade of primary school *Journal on Mathematics Education* **4** 29

[11] Haris D and Putri R I I 2011 The role of context in third graders' learning of area measurement *Journal on Mathematics Education* **2** 55

[12] Yanti S, Rif'at M and Suratman D 2015 Pemanfaatan media kit dalam proses pembelajaran matematika di SMP Negeri Kota Pontianak *Jurnal Pendidikan dan Pembelajaran* **4** 5

[13] Nurhanif 2015 Pembelajaran keliling dan luas bangun datar menggunakan media KIT matematika di Kelas III A Sekolah Dasar Negeri 16 Banda Aceh *Undergraduate Thesis* (Aceh: Universitas Syiah Kuala)

[14] Manurung E, Sugiatno and Yani A 2013 Pengembangan aktivitas belajar matematika berbantuan alat peraga kartu KIT di Kelas IV SD *Jurnal Pendidikan dan Pembelajaran* **2** 24

[15] Young J R 2017 Technology integration in mathematics education: Examining the quality of meta-analytic research *International Journal on Emerging Mathematics Education* **1** 71

[16] Fathurrohman M, Porter A L and Worthy A L 2017 Teachers' real and perceived of ICT supported-situation for mathematics teaching and learning *International Journal on Emerging Mathematics Education* **1** 11

[17] Khomsiatun S and Retnawati H 2015 Pengembangan perangkat pembelajaran dengan penemuan terbimbing untuk meningkatkan kemampuan pemecahan masalah *Jurnal Riset Pendidikan Matematika* **2** 92

Super (a^*, d^*)-$\mathcal{H}$-antimagic total covering of second order of shackle graphs

Ika Hesti Agustin[1,3], **Dafik**[1,2], **Rosanita Nisviasari**[1,3], **R.M. Prihandini**[1,4]

[1] CGANT University of Jember, Indonesia
[2] Mathematics Edu. Depart. University of Jember, Indonesia
[3] Mathematics Depart. University of Jember, Indonesia
[4] Elementary School Edu. Depart. University of Jember, Indonesia

E-mail: ikahesti.fmipa@unej.ac.id, d.dafik@unej.ac.id

Abstract. Let H be a simple and connected graph. A *shackle* of graph H, denoted by $G = shack(H, v, n)$, is a graph G constructed by non-trivial graphs $H_1, H_2, \ldots, H_n$ such that, for every $1 \leq s, t \leq n$, H_s and H_t have no a common vertex with $|s - t| \geq 2$ and for every $1 \leq i \leq n - 1$, H_i and H_{i+1} share exactly one common vertex v, called *connecting vertex*, and those $k - 1$ connecting vertices are all distinct. The graph G is said to be an (a^*, d^*)-H-antimagic total graph of second order if there exist a bijective function $f : V(G) \cup E(G) \rightarrow \{1, 2, \ldots, |V(G)| + |E(G)|\}$ such that for all subgraphs isomorphic to H, the total H-weights $W(H) = \sum_{v \in V(H)} f(v) + \sum_{e \in E(H)} f(e)$ form an arithmetic sequence of second order of $\{a^*, a^* + d^*, a^* + 3d^*, a^* + 6d^*, \ldots, a^* + (\frac{n^2 - n}{2})d^*\}$, where a^* and d^* are positive integers and n is the number of all subgraphs isomorphic to H. An (a^*, d^*)-H-antimagic total labeling of second order f is called super if the smallest labels appear in the vertices. In this paper, we study a super (a^*, d^*)-H antimagic total labeling of second order of $G = shack(H, v, n)$ by using a partition technique of second order.

1. Introduction

All graphs in this study are simple, connected an undirected. A graph G is said to be an (a^*, d^*)-H-antimagic total graph of second order if there exist a bijective function $f : V(G) \cup E(G) \rightarrow \{1, 2, \ldots, |V(G)| + |E(G)|\}$ such that for all subgraphs of G isomorphic to H, the total H-weights $w(H) = \sum_{v \in V(H)} f(v) + \sum_{e \in E(H)} f(e)$ form an arithmetic sequence of second order $\{a^*, a^* + d^*, a^* + 3d^*, a^* + 6d^*, \ldots, a^* + (\frac{n^2 - n}{2})d^*\}$, where a and d are positive integers and n is the number of all subgraphs of G isomorphic to H. If such a function exist then f is called an (a^*, d^*)-H-antimagic total labeling of second order of G. An (a^*, d^*)-H-antimagic total labeling of second order f is called super if $f : V(G) \rightarrow \{1, 2, \ldots, |V(G)|\}$. By this notion, the super $(a, d) - H$ antimagic total labeling is classified as the super $(a, d) - H$ antimagic total labeling of first order.

We initiate to study this concept, thus we have not found any relevant results yet. But for the super $(a, d) - H$ antimagic total labeling, we can find many published results, some of them can be cited in [2, 3, 8, 9] and [10, 11, 12, 13, 15]. Inayah *et al.* in [8] proved that, for H is a non-trivial connected graph and $k \geq 2$ is an integer, $shack(H, v, k)$ which contains exactly k subgraphs isomorphic to H is H-super antimagic. All this papers only dealt with d derived from the sequence of order one. Our paper attempt to solve a super (a^*, d^*)-H antimagic total labeling of order two of $G = shack(H, v, n)$.

To show those existence, we will use a special technique, namely *an integer set partition technique*. We consider the partition $\mathcal{P}^n_{m,d}(i,j)$ of the set $\{1, 2, \ldots, mn\}$ into n columns with $n \geq 2$, m-rows such that the difference between the sum of the numbers in the $(j+1)$th m-rows and the sum of the numbers in the jth m-rows is always equal to the constant d, where $j = 1, 2, \ldots, n-1$. We need to establish some lemmas related to the partition $\mathcal{P}^n_{m,d}(i,j)$. Furthermore, the partition will be developed into a second order partition. These lemmas are useful to develop the super (a^*, d^*)-H antimagic total labeling of second order of $G = shack(H, v, n)$.

Let G be a *shackle* of graph denoted by $G = shack(H, v, n)$. Let G and H be a connected graph with $|V(G)| = p_G$, $|E(G)| = q_G$, $|V(H)| = p_H$, and $|E(H)| = q_H$. The vertex set and edge set of the graph $G = shack(H, v, n)$ can be split into following sets: $V = \{x_j; 1 \leq j \leq n+1\} \cup \{x_{ij}; 1 \leq i \leq p_H - 2, 1 \leq j \leq n\}$ and $E = \{e_{ij}; 1 \leq i \leq q_H, 1 \leq j \leq n\}$. Let n, i, j be positive integers with $n \geq 2$. Thus $p_G = |V(G)| = n + 1 + (p_H - 2)n = 1 + np_H + n - 2n = 1 + n(p_H - 1)$ and $q_G = |E(G)| = nq_H$.

We recall a partition $\mathcal{P}^n_{m,d}(i,j)$ introduced in [4]. We will use the partition for a linear combination in developing a bijection of vertex and edge label of the main theorem.

Lemma 1.1. [4] *Let n and m be positive integers. The sum of $\mathcal{P}^n_{m,d_1}(i,j) = \{(i-1)n + j,\ 1 \leq i \leq m\}$ and $\mathcal{P}^n_{m,d_2}(i,j) = \{(j-1)m + i;\ 1 \leq i \leq m\}$ form an aritmatic sequence of difference $d_1 = m$ and $d_2 = m^2$, respectively.*

2. Main Results

Lemma 2.1. *Let G be a simple graph of order p and size q. If G is super (a^*, d^*)-H-antimagic total labeling of second order then $d \leq \frac{(p_G - p_H)p_H + (q_G - q_H)q_H}{(\frac{n^2-n}{2})}$, for $p_G = |V(G)|$, $q_G = |E(G)|$, $p_H = |V(H)|$, $q_H = |E(H)|$, and $n = |H|$.*

Proof. Given the function $f(V) = 1, 2, 3, \ldots, p_G$ and $f(E) = p_G + 1, p_G + 2, p_G + 3, \ldots, p_G + q_G$. Let (p_G, q_G) admit a super $(a^*, d^*) - \mathcal{H}$ antimagic total labeling with the total second order function, $f(total) = 1, 2, 3, \ldots, p_G + q_G$ then the set of edge weight of a graph is $\{a^*, a^* + d^*, a^* + 3d^*, a^* + 6d^*, \ldots, a + (\frac{n^2-n}{2})d\}$ with a^* is the smallest weight thus:

$$\begin{aligned}
1 + 2 + \cdots + p_H + (p_G + 1) + (p_G + 2) + \cdots + (p_G + q_H) &\leqslant a^* \\
\frac{p_H}{2}(1 + p_H) + q_H p_G + \frac{q_H}{2}(1 + q_H) &\leqslant a^* \\
\frac{p_H}{2} + \frac{p_H^2}{2} + q_H p_G + \frac{q_H}{2} + \frac{q_H^2}{2} &\leqslant a^*
\end{aligned}$$

$$
\begin{aligned}
a^* + (\frac{n^2-n}{2})d^* &\leqslant p_G + (p_G - 1) + (p_G - 2) + \dots + (p_G - (p_H - 1)) \\
&\quad +(p_G + q_G) + (p_G + q_G - 1) + (p_G + q_G - 2) + \dots \\
&\quad +(p_G + q_G - (q_H - 1)) \\
&= p_H p_G - \frac{p_H - 1}{2}(1 + (p_H - 1)) + q_H p_G + q_H q_G \\
&\quad -\frac{q_H - 1}{2}(q + q_H - 1)) \\
&= p_H p_G - \frac{p_H - 1}{2}(p_H) + q_H p_G + q_H q_G - \frac{q_H - 1}{2}(q_H) \\
(\frac{n^2-n}{2})d^* &\leqslant p_H p_G - \frac{p_H - 1}{2}(p_H) + q_H p_G + q_H q_G - \frac{q_H - 1}{2}(q_H) - a \\
&\leqslant p_H p_G - \frac{p_H - 1}{2}(p_H) + q_H p_G + q_H q_G - \frac{q_H - 1}{2}(q_H) - \\
&\quad (\frac{p_H}{2} + \frac{p_H^2}{2} + q_H p_G + \frac{q_H}{2} + \frac{q_H^2}{2}) \\
&= p_H p_G - \frac{p_H^2}{2} + \frac{p_H}{2} + q_H p_G - \frac{q_H^2}{2} + \frac{q_H}{2} - \\
&\quad (\frac{p_H}{2} + \frac{p_H^2}{2} + \frac{q_H}{2} + \frac{q_H^2}{2}) \\
&= p_H p_G + q_H q_G - p_H^2 - q_H^2 \\
&= p_H p_G - p_H^2 + q_H q_G - q_H^2 \\
&= (p_G - p_H)p_H + (q_G - q_H)q_H \\
d^* &\leqslant \frac{(p_G - p_H)p_H + (q_G - q_H)q_H}{(\frac{n^2-n}{2})}
\end{aligned}
$$

Based on the inequality above, we get the upper bound of feasible difference $d^* \leq \frac{(p_G-p_H)p_H+(q_G-q_H)q_H}{(\frac{n^2-n}{2})}$ for the graph G to be a super $(a^*, d^*) - \mathcal{H} -$ *antimagic* total *covering* of second order. □

Corollary 2.2. *If the graph* $G = shack(H, v, n)$ *admits super* $(a^*, d^*) - H$*-antimagic total labeling of second order for integer* $n \geq 2$*, then* $d \leq \frac{2(p_H^2 - p_H + q_H^2)}{n}$

The following lemmas are useful for showing the existence of super (a^*, d^*)-H antimagic total labeling $G = shack(H, v, n)$.

Lemma 2.3. *Let* n *and* m *be positive integers. For* $1 \leq j \leq n$*, the sum of* $\mathcal{P}^n_{m,d_3}(i,j) = \{1 + ni - j; 1 \leq i \leq m\}$ *and* $\mathcal{P}^n_{m,d_4}(i,j) = \{mn + i - mj; 1 \leq i \leq m\}$ *form an arithmetic sequence of differences* $d_3 = -m$ *and* $d_4 = -m^2$.

Proof. By simple calculation, for $j = 1, 2, ..., n$, it gives $\sum_{i=1}^m \mathcal{P}^n_{m,d_1}(i,j) = \mathcal{P}^n_{m,d_3}(j) \longleftrightarrow \mathcal{P}^n_{m,d_3}(j) = \{\frac{n}{2}(m^2 + m) + m - mj\} \longleftrightarrow \mathcal{P}^n_{m,d_1}(j) = \{\frac{n}{2}(m^2 + m), \frac{n}{2}(m^2 + m) - m, \frac{n}{2}(m^2 + m) - 2m, \dots, \frac{n}{2}(m^2 + m)m - mn\}$ and $\sum_{i=1}^m \mathcal{P}^n_{m,d_4}(i,j) = \mathcal{P}^n_{m,d_2}(j) \longleftrightarrow \mathcal{P}^n_{m,d_4}(j) = \{\frac{m}{2}(2mn + m + 1) - m^2 j\} \longleftrightarrow \mathcal{P}^n_{m,d_4}(j) = \{\frac{m}{2}(2mn + m + 1) - m^2, \frac{m}{2}(2mn + m + 1) - 2m^2, \dots, \frac{m}{2}(2mn + m + 1) - m^2 n\}$. It is easy to see that the differences of those sequences are $d_3 = -m, d_4 = -m^2$. It concludes the proof. □

Lemma 2.4. *Let* n*,* m *be positive integers and* $n = m$*. The sum of*

$$
\mathcal{P}^{*n}_{m,d_5}(i,j) = \begin{cases} \frac{i(2m+2j+1) - j(2m-1) - i^2 - j^2}{2}; & for\ i - j \geq m - 1 \\ \frac{m^2 - m + j(2m+1-j) + i(-2m+2j+1-i)}{2}; & for\ m - 1 \geq i - j \end{cases}
$$

form an arithmetic sequence of second order with common difference $d_5 = m$.

Proof. By simple calculation, for $j = 1, 2, ..., n$, it gives $\sum_{i=1}^{m} \mathcal{P}^n_{m,d_1}(i,j) = \mathcal{P}^n_{m,d_3}(j) \longleftrightarrow \mathcal{P}^n_{m,d_3}(j) = \{\frac{n}{2}(m^2+m)+m-mj\} \longleftrightarrow \mathcal{P}^n_{m,d_1}(j) = \{\frac{n}{2}(m^2+m), \frac{n}{2}(m^2+m)-m, \frac{n}{2}(m^2+m)-2m, \dots, \frac{n}{2}(m^2+m)m-mn\}$ and $\sum_{i=1}^{m} \mathcal{P}^n_{m,d_4}(i,j) = \mathcal{P}^n_{m,d_2}(j) \longleftrightarrow \mathcal{P}^n_{m,d_4}(j) = \{\frac{m}{2}(2mn+m+1)-m^2 j\} \longleftrightarrow \mathcal{P}^n_{m,d_4}(j) = \{\frac{m}{2}(2mn+m+1)-m^2, \frac{m}{2}(2mn+m+1)-2m^2, \dots, \frac{m}{2}(2mn+m+1)-m^2 n\}$. It is easy to see that the differences of those sequences are $d_3 = -m, d_4 = -m^2$. It concludes the proof. □

Lemma 2.5. *Let n, m be positive integers and $n = m$. The sum of*

$$\mathcal{P}^{*n}_{m,d_6}(i,j) = \begin{cases} \frac{2m^2-2mi+i^2-i+2j}{2}; 1 \le i \le m, i+j \le m+1 \\ \frac{4m^2+4m-4mi-4mj-3i-j+2ij+i^2+j^2+2}{2}; 1 \le i \le m, m+1 < i+j \end{cases}$$

form an arithmetic sequence of second order with common difference $d_6 = -m$.

PROOF. By simple calculation, for $j = 1, 2, ..., n$, it gives $\sum_{i=1}^{m} \mathcal{P}^n_{m,d_6^*}(i,j) = \mathcal{P}^n_{m,d_6^*}(j) \longleftrightarrow \mathcal{P}^n_{m,d_6^*}(j) = \{\frac{4m^3-3m^2-(3j^2-9j+1)m}{6}\} \longleftrightarrow \mathcal{P}^n_{m,d_6^*}(j) = \{\frac{2m^3}{3} - \frac{m^2}{2} + \frac{5m}{6}, \frac{2m^3}{3} - \frac{m^2}{2} + \frac{5m}{6}, \frac{2m^3}{3} - \frac{m^2}{2} - \frac{m}{6}, \dots, \frac{2m^3}{3} - \frac{m^2}{2} - \frac{(3n^2-15n+13)m}{6}, \frac{2m^3}{3} - \frac{m^2}{2} - \frac{(3n^2-9n+1)m}{6}\}$. It is easy to see that the differences of those sequences are $d_6^* = -m$. It concludes the proof. □

Lemma 2.6. *Let d^* be the common difference of arithmetic sequence of second order and d be the common difference of arithmetic sequence of first order, the sum of corresponding terms will form an arithmetic sequence of second order with common difference d^*.*

Proof. An arithmetic sequence of first order is a sequence of the form:

$$a, a+d, a+2d, a+3d, \dots, a+(n-1)d$$

where a is the first term and d is common difference of the sequence. Whilst an arithmetic sequence of second order is of the form $a^*, a^*+d^*, a^*+3d^*, a^*+6d^*, \dots, a^*+(\frac{(n^2-n}{2})d^*$, where a^* and d^* are the first term and common difference of the sequence, respectively. Now, add the corresponding terms of these two expression:

Sequence	a	$a+d$	$a+2d$	...	$a+(n-1)d$
Sequence	a^*	a^*+d^*	a^*+3d^*	...	$a^*+(\frac{n^2-n}{2})d^*+$
Sequence	$a+a^*$ $+d+d^*$	$a+a^*$ $+2d+3d^*$	$a+a^*$ $+3d+6d^*$	...	$a+a^*+(n-1)d$ $+(\frac{(n^2-n)}{2})d^*$
First order difference	$d+d^*$	$d+2d^*$	$d+3d^*$	...	$(n-1)d$ $+(\frac{(n^2-n)}{2})d^*$
Second order difference	d^*	d^*	d^*	...	d^*

It conclude the proof. □

Theorem 2.7. *Let H be a connected graph, then the shackle of the connected graph $G = shack(H, v, n)$ admits super $(a^*, d^*) - H$ antimagic total labeling.*

Proof. Let G be a *shackle* of graph denoted by $G = shack(H, v, n)$. Let G and H be a connected graph with $|V(G)| = p_G$, $|E(G)| = q_G$, $|V(H)| = p_H$, and $|E(H)| = q_H$. The vertex set and edge set of the graph $G = shack(H, v, n)$ can be split into following sets: $V = \{x_j; 1 \leq j \leq n+1\} \cup \{x_{ij}; 1 \leq i \leq p_H - 2, 1 \leq j \leq n\}$ and $E = \{e_{ij}; 1 \leq i \leq q_H, 1 \leq j \leq n\}$. Let n, i, j be positive integers with $n \geq 2$. Thus $p_G = |V(G)| = n + 1 + (p_H - 2)n = 1 + np_H + n - 2n = 1 + n(p_H - 1)$ and $q_G = |E(G)| = nq_H$. Construct a total labeling f,$f : V(shack(H, v, n)) \cup E(shack(H, v, n)) \to \{1, 2, \ldots, 1 + n(p_H + q_H - 1)\}$ constitute the following set:

$$\begin{aligned}
f_1(V_{p_H}) &= \{j; 1 \leq j \leq n+1\} \\
f_2(V_{p_H}) &= \{{\mathcal{P}^*}^n_{m_1,d_v^*}(i,j) \oplus n+1\} \cup \{\mathcal{P}^n_{m_2,d_v}(i,j) \oplus n(m_1+1)+1\} \\
f(E_{q_H}) &= \{{\mathcal{P}^*}^n_{c_1,d_e^*}(i,j) \oplus (p_H-1)n+1\} \cup \{\mathcal{P}^n_{c_2,d_e}(i,j) \oplus (p_H+r_1-1)n+1\}
\end{aligned}$$

where $m_1 + m_2 = p_H - 2$, $c_1 + c_2 = q_H$, d_v and d_e depends on $p_H - 2$ and q_H, respectively. Furthermore the weight of the subgraphs H_i, $i = 1, 2, \ldots, p_L$ in the following way:

$$\begin{aligned}
W &= \sum_{v \in V(H_i)} f(v) + \sum_{e \in E(H_i)} f(e) \\
&= (2j+1) + (\sum_{i=1}^{m_1} ({\mathcal{P}^*}^n_{m_1,d_v^*}(j) \oplus n+1) + (\sum_{i=1}^{m_2} (\mathcal{P}^n_{m_2,d_v}(j) \oplus n(m_1+1)+1) \\
&\quad + (\sum_{i=1}^{c_1} ({\mathcal{P}^*}^n_{c_1,d_e^*}(j) \oplus (p_H-1)n+1)(\sum_{i=1}^{c_2} (\mathcal{P}^n_{c_2,d_e}(j) \oplus (p_H+r_1-1)n+1) \\
&= [2j+1] + [{C^*_{m_1,d_v^*}}^n + d_v^*(\frac{j^2-j+2}{2}) + m_1(n+1)] + [C^n_{m_2,d_v} + d_v j + \\
&\quad m_2(n(m_1+1)+1)] + [{C^*_{c_1,d_e^*}}^n + d_e^*(\frac{j^2-j+2}{2}) + c_1((p_H-1)n+1)] \\
&\quad + [C^n_{c_2,d_e} + d_e j + c_2(n(p_H+r_1-1)+1)]
\end{aligned}$$

based on Lemma 2.6 we obtained:

$$\begin{aligned}
&= 1 + {C^*_{m_1,d_v^*}}^n + {C_{m_2,d_v}}^n + {C^*_{c_1,d_e^*}}^n + {C_{c_2,d_e}}^n + m_1(n+1) + m_2(n(m_1+1)+1) \\
&\quad + c_1((p_H-1)n + 1 + c_2(n(p_H+r_1-1)+1) + [d_v^* + d_e^* + 2]j
\end{aligned}$$

3. Special Families

Theorem 3.1. *Suppose $G = shack(C_m, v, n)$, with $s \geq 3$ dan $2s \geq n+1$, graph G admits super (a^*, d^*)-$\mathcal{H}$-antimagic total covering of second order with* $a = 3 + [\frac{2{m_1}^3+4m_1+(3m_1-3m_1)}{6} + m_1(n+1)] + [\frac{{m_2}^2-m_2n}{2} + m_2 + m_2(n(m_1+1)+1)] + [\frac{m_3-{m_3}^2}{2} + {m_3}^2 + m_3(n(m_1+m_2+1)+1)] + [\frac{n}{2}(m_4^2 + m_4) + m_4 - m_4 + m_4(n(\sum_{t=1}^3 m_t + 1)+1)] + [\frac{m_5}{2}(2m_5n + m_5 + 1) - m_5^2 + m_5(n(\sum_{t=1}^4 m_t + 1) + 1)][\frac{2{c_1}^3+4c_1+(3c_1-3c_1)}{6} + c_1(n(m+1)+1)] + [\frac{{c_2}^2-c_2n}{2} + c_2(n(m+c_1+1)+1)] + [\frac{c_3-{c_3}^2}{2} + c_3(n(m+c_1+c_2+1)+1)] + [\frac{n}{2}(c_4^2+c_4) + c_4 + c_4(n(m+\sum_{t=1}^3 c_t+1)+1)] + [\frac{c_5}{2}(2c_5n+c_5+1) + c_5(n(m+\sum_{t=1}^4 c_t+1)+1)]$ *and* $d = m_1^* + c_1^*$.

Proof. The graph $G = shack(C_m, v, n)$ have vertex set $V = \{x_j; 1 \leq j \leq n+1\} \cup \{x_{ij}; 1 \leq i \leq m-2, 1 \leq j \leq n\}$ and edge set $E = \{e_{ij}; 1 \leq i \leq m, 1 \leq j \leq n\}$. Thus $p_G = |V(G)| = mn - n + 1$ and $q_G = |E(G)| = mn$ where $p_H = m - 2$ and $q_H = m$ respectively are the cardinality of the vertex and edge on one cover H. We can define the vertex labeling

$f_1 : V(G) \cup E(G) \to \{1, 2, \ldots, p_G + q_G\}$ by using the linear combination of ${\mathcal{P}^*}^n_{m,m^*}$, ${\mathcal{P}^*}^n_{m,-m^*}$, $\mathcal{P}^n_{m,m}$, $\mathcal{P}^n_{m,-m}$, $\mathcal{P}^n_{m,m^2}$ and $\mathcal{P}^n_{m,-m^2}$. By Lemma 1.1,2.3,2.4 and 2.5 we use m_1 and c_1 for the partition ${\mathcal{P}^*}^n_{m,m^*}(i,j)$, m_2 and c_2 for the partition $\mathcal{P}^n_{m,m}(i,j)$, m_3 and c_3 for the partition $\mathcal{P}^n_{m,m^2}(i,j)$, m_4 and c_4 for the partition ${\mathcal{P}^*}^n_{m,-m^*}(i,j)$, m_5 and c_5 for the partition $\mathcal{P}^n_{m,-m}(i,j)$ and we use m_6 and c_6 for the partition $\mathcal{P}^n_{m,-m^2}(i,j)$. For $i = 1, 2, \ldots, m$, $l = 1, 2, \ldots, c$ and $j = 1, 2, \ldots, n$, the total labels can be expressed as follows

$$\begin{aligned}
f_1(V_{p_H}) &= \{j; 1 \le j \le n+1\} \\
f_2(V_{p_H}) &= \{{\mathcal{P}^*}^n_{m_1,d_v^*}(i,j) \oplus n+1\} \cup \{\mathcal{P}^n_{m_2,d_v}(i,j) \oplus n(m_1+1)+1\} \\
f(E_{q_H}) &= \{{\mathcal{P}^*}^n_{c_1,d_e^*}(i,j) \oplus (p_H-1)n+1\} \cup \{\mathcal{P}^n_{c_2,d_e}(i,j) \oplus (p_H+r_1-1)n+1\}
\end{aligned}$$

The total vertex and edge-weights of $G = shack(C_m, v, n)$ under the labeling f_1, for $1 \le j \le n$, constitute the following sets:

$$\begin{aligned}
W &= \sum_{v \in V(H_i)} f(v) + \sum_{e \in E(H_i)} f(e) \\
&= (2j+1) + (\sum_{i=1}^{m_1} ({\mathcal{P}^*}^n_{m_1,d_v^*}(j) \oplus n+1) + (\sum_{i=1}^{m_2} (\mathcal{P}^n_{m_2,d_v}(j) \oplus n(m_1+1)+1) \\
&\quad + (\sum_{i=1}^{c_1} ({\mathcal{P}^*}^n_{c_1,d_e^*}(j) \oplus (p_H-1)n+1)(\sum_{i=1}^{c_2} (\mathcal{P}^n_{c_2,d_e}(j) \oplus (p_H+r_1-1)n+1) \\
&= [2j+1] + [{C^*_{m_1,d_v^*}}^n + d_v^*(\frac{j^2-j+2}{2}) + m_1(n+1)] + [C^n_{m_2,d_v} + d_v j + \\
&\quad m_2(n(m_1+1)+1)] + [{C^*_{c_1,d_e^*}}^n + d_e^*(\frac{j^2-j+2}{2}) + c_1((p_H-1)n+1)] \\
&\quad + [C^n_{c_2,d_e} + d_e j + c_2(n(p_H+r_1-1)+1)]
\end{aligned}$$

The total weights of $G = shack(C_m, v, n)$ constitute the following sets:

$$\begin{aligned}
W_{f_1} &= w^1_{f_1} + w^2_{f_1} + w^3_{f_1} \\
&= [2j+1] + w^2_{f_1} + w^3_{f_1} \\
&= C^* + C + 1 + [2 + \frac{3m_1 j - 3m_1}{2} + m_2 + m_3^2 - m_4 - m_5^2 + \frac{3c_1 j - 3c_1}{2} \\
&\quad + c_2 + c_3^2 - c_4 - c_5^2]j; 1 \le j \le n
\end{aligned}$$

By simple calculation, for $j = 1, 2, \ldots, n$, it gives $W_{f_1} = C^* + C + 1 + [2 + \frac{3m_1 j - 3m_1}{2} + m_2 + m_3^2 - m_4 - m_5^2 + \frac{3c_1 j - 3c_1}{2} + c_2 + c_3^2 - c_4 - c_5^2]j \longleftrightarrow \{C^* + C + 1 + m_2 + m_3^2 - m_4 - m_5^2 + c_2 + c_3^2 - c_4 - c_5^2, C^* + C + 1 + m_1 + c_1 + m_2 + m_3^2 - m_4 - m_5^2 + c_2 + c_3^2 - c_4 - c_5^2, C^* + C + 1 + 3(m_1 + c_1) + m_2 + m_3^2 - m_4 - m_5^2 + c_2 + c_3^2 - c_4 - c_5^2, \ldots, C^* + C + 1 + 3\frac{3(m_1 n^2 + 3c_1 n^2) - m_1 n - c_1 n}{6} + m_2 + m_3^2 - m_4 - m_5^2 + c_2 + c_3^2 - c_4 - c_5^2$. with $C^* = [\frac{2{m_1}^3 + 4m_1}{6} + m_1(n+1)] + [\frac{2{c_1}^3 + 4c_1}{6} + c_1(n(m+1)+1)]$ $C = [\frac{{m_2}^2 - m_2 n}{2} + m_2(n(m_1+1)+1)] + [\frac{m_3 - {m_3}^2}{2} + m_3(n(m_1+m_2+1)+1)] + [\frac{n}{2}(m_4^2 + m_4) + m_4 + m_4(n(\sum_{t=1}^{3} m_t + 1)+1)] + [\frac{m_5}{2}(2m_5 n + m_5 + 1) + m_5(n(\sum_{t=1}^{4} m_t + 1)+1)] + [\frac{{c_2}^2 - c_2 n}{2} + c_2(n(m+c_1+1)+1)] + [\frac{c_3 - {c_3}^2}{2} + c_3(n(m+c_1+c_2+1)+1)] + [\frac{n}{2}(c_4^2 + c_4) + c_4 + c_4(n(m + \sum_{t=1}^{3} c_t + 1)+1)] + [\frac{c_5}{2}(2c_5 n + c_5 + 1) + c_5(n(m + \sum_{t=1}^{4} c_t + 1)+1)]$

It is easy that the set of total edge-weights W_{f_1} consists of an arithmetic sequence of second order with the smallest value $a = 3 + [\frac{2{m_1}^3 + 4m_1 + (3m_1 - 3m_1)}{6} + m_1(n+1)] + [\frac{{m_2}^2 - m_2 n}{2} + m_2 +$

IOP Conf. Series: Journal of Physics: Conf. Series **943** (2017) 012060 doi:10.1088/1742-6596/943/1/012060

$m_2(n(m_1+1)+1)]+[\frac{m_3-m_3{}^2}{2}+m_3{}^2+m_3(n(m_1+m_2+1)+1)]+[\frac{n}{2}(m_4^2+m_4)+m_4-m_4+m_4(n(\sum_{t=1}^{3}m_t+1)+1)]+[\frac{m_5}{2}(2m_5n+m_5+1)-m_5^2+m_5(n(\sum_{t=1}^{4}m_t+1)+1)][[\frac{2c_1{}^3+4c_1+(3c_1-3c_1)}{6}+c_1(n(m+1)+1)]+[\frac{c_2{}^2-c_2n}{2}+c_2(n(m+c_1+1)+1)]+[\frac{c_3-c_3{}^2}{2}+c_3(n(m+c_1+c_2+1)+1)]+[\frac{n}{2}(c_4^2+c_4)+c_4+c_4(n(m+\sum_{t=1}^{3}c_t+1)+1)]+[\frac{c_5}{2}(2c_5n+c_5+1)+c_5(n(m+\sum_{t=1}^{4}c_t+1)+1)]$ when the total edge weights at $j=1$ and the difference $d=[2+m_1^*+c_1^*]$. It concludes the proof. □

4. Concluding Remarks

We have shown the existence of super antimagic labeling of second order for graph operation $G=shack(H,v,n)$. We have found that $G=shack(H,v,n)$ admits a super(a^*,d^*)-H antimagic labeling of second order for all differences $d=2+d_v^*+d_e^*$ where d_v^* and d_e^* are respectively feasible difference of second order of integer set partition. We have not found the result for another graph operations. Thus, we propose the following open problems.

Open Problem 4.1. *Analyse the existence of super (a^*,d^*)-H antimagic total labeling of second order of other graph operations.*

4.1. Acknowledgments

We gratefully acknowledge the support from CGANT University of Jember of year 2017.

References

[1] M. Bača, L. Brankovic, M. Lascsáková, O., Phanalasy, A. Semaničová-Feňovčíková 2013 On d-antimagic labelings of plane graphs, *Electr. J. Graph Theory Appli.*, **1** 28-39
[2] Dafik, A.K. Purnapraja, R Hidayat 2015 Cycle-Super Antimagicness of Connected and Disconnected Tensor Product of Graphs, *Procedia Computer Science*, **74** 9399
[3] Dafik, Slamin, Dushyant Tanna, Andrea Semaničová-Feňovčíková, Martin Bača 2017 Constructions of H-antimagic graphs using smaller edge-antimagic graphs, *Ars Combinatoria* **100** In Press
[4] Dafik, Moh. Hasan, Y.N. Azizah, Ika Hesti Agustin 2016 A Generalized Shackle of Any Graph H Admits a Super H-Antimagic Total Labeling, *Mathematics in Computer Science Journal.* Submitted
[5] Dafik, Ika Hesti Agustin, A. I. Nurvitaningrum, R. M. Prihandini 2017 On super H-antimagicness of an edge comb product of graphs with subgraph as a terminal of its amalgamation , *Journal of Physics: Conf. Series* **855**
[6] Dafik, R. M Prihandini, Ika Hesti Agustin 2016 $P_2 \rhd H$-super antimagic total labeling of comb product of graphs, *AKCE International Journal of Graphs and Combinatorics*, submitted.
[7] N. Inayah, A.N.M. Salman and R. Simanjuntak 2009 On $(a,d)-H$-antimagic coverings of graphs, *J. Combin. Math. Combin. Comput.* **71** 273281
[8] Inayah, N., R. Simanjuntak, A.N.M Salman. 2013. Super (a,d)-$\mathcal{H}$-Antimagic Total labelings For Shackles of a Connected Graph H. *The Australian Journal of Combinatorics*, **57** 127-138
[9] P. Jeyanthi, P. Selvagopal 2010 More classes of H-supermagic Graphs, *Intern. J. of Algorithms, Computing and Mathematics* **3** 93-108
[10] A. Lladó and J. Moragas 2007 Cycle-magic graphs, *Discrete Math.* **307** 2925 2933
[11] T.K. Maryati, A. N. M. Salman, E.T. Baskoro, J. Ryan, M. Miller 2010 On H- supermagic labelings for certain shackles and amalgamations of a connected graph, *Utilitas Mathematica* **83** 333-342
[12] A. A. G. Ngurah, A. N. M. Salman, L. Susilowati 2010 H-supermagic labeling of graphs, *Discrete Math.* **310** 1293-1300
[13] S.T.R. Rizvi, K. Ali, M. Hussain 2014 Cycle-supermagic labelings of the disjoint union of graphs, *Utilitas Mathematica*, in press.
[14] Rafiantika M. Prihandini, I. H Agustin, Dafik 2017 Construction of $P_2 \rhd H$-Antimagic graph using smaller edge-antimagic vertex labeling, *ICONSSED*, submitted.
[15] Roswitha, M. and Baskoro, E. T 2012 H-magic covering on some classes of graphs, *American Institute of Physics Conference Proceedings* **1450** 135-138

Institute of Physics
Dirac House, Temple Back
Bristol BS1 6BE UK

ISSN: 1742-6588
ISBN 978-1-5108-5847-3

9 781510 858473

CURRAN ASSOCIATES INC.
proceedings
.com